AF327250

Nanometer CMOS ICs

Nanometer CMOS ICs

From basics to ASICs

Harry Veendrick

mybusinessmedia

Nanometer CMOS ICs

Author:
Dr. Ir. H.J.M.Veendrick
NXP Semiconductors-Research
E-mail: h.veendrick@upcmail.nl

Cover design: Bram Veendrick
Photographs used in cover: NXP Semiconductors

Typesetting and layout: Harold Benten and Dré van den Elshout
Illustrations: Kim Veendrick and Henny Alblas

First English edition: 2008

This book is based on various previous publications. The first original 1990 publication (Delta Press b.v.) was in the Dutch language. In 1992 a revised, updated and translated English edition of that book was jointly published by VCH Verlagsgesellschaft (Weinheim, Germany) and VCH Publishers Inc. (NY, USA). The third book, entitled Deep-Submicron CMOS ICs: from Basics to ASICs, was a joint publication of Ten Hagen en Stam, Deventer, The Netherlands, and Kluwer Academic Publishers, Boston, USA) and published in two editions (1998 and 2000).

This new book covers the same subjects, but then they are completely revised and updated with the most recent state-of-the-art material. It covers all subjects, related to nanometer CMOS ICs: physics, technologies, design, testing, packaging and failure analysis. The contents have increased by almost one third, leading to a much more detailed and complete description of most of the subjects. This new book is almost full colour.

ISBN 978-1-4020-8332-7   NUR 950

Although this book and its contents were produced with great care, neither the author nor the publisher can guarantee that the information contained therein is free from errors. Readers are advised to keep in mind that statements, data, illustrations, procedural details or other items may inadvertently contain inaccuracies. This book contains many sources and references of text, photographs and illustrations. Although the author has given a lot of attention to carefully refer to the source of related material, he already apologizes for the one or few individual occasions that this has slipped his final review.

Foreword

CMOS scaling is now entering the deca-nanometer era. This enables the design of systems-on-a-chip containing more than 10 billion transistors. However, nanometer level device physics also causes a plethora of new challenges that percolate all the way up to the system level.

Therefore system-on-a-chip design is essentially teamwork requiring a close dialogue between system designers, software engineers, chip architects, intellectual property providers, and process and device engineers. This is hardly possible without a common understanding of the nanometer CMOS medium, its terminology, its future opportunities and possible pitfalls. This is what this book provides.

It is a greatly extended and revised version of the previous edition that was addressing deep-submicron CMOS systems. So besides the excellent coverage of all basic aspects of MOS devices, circuits and systems it leads the reader into the novel intricacies resulting from scaling CMOS down to the deca-nanometer level. New in this edition is the attention to the issues of increased leakage power and its mitigation, to strain induced mobility enhancement and to sub-45 nm lithographic techniques. Immersion and double patterning litho the use of high index fluids as well as of extreme UV and other alternative litho approaches for sub 32 nm are extensively discussed together with their impact on circuit layout. The design section now also extensively covers design techniques for improved robustness, yield and manufacturing in view of increased device variability, soft errors and decreased reliability when reaching atomic dimensions. In the packaging section attention is paid to rapidly emerging 3D integration techniques. Finally the author shares his thoughts on the challenges of further scaling when approaching the end of the CMOS roadmap by 2015.

This book is unique in that it covers in a very comprehensive way all aspects of the trajectory from process technology to the design and packaging of robust and testable systems in nanometer scale CMOS.

It is the reflection of the author's own research in this domain but also of almost 30 years experience in interactive teaching of CMOS design to NXP and PHILIPS system designers and process engineers alike. It provides context and perspective to both sides.

I strongly recommend this book to all engineers involved in the design and manufacturing of future systems-on-silicon as well as to engineering undergraduates who want to understand the basics that make electronics systems work.

Leuven, February 2008

Hugo De Man
Professor Emeritus K.U. Leuven
Senior Fellow IMEC
Leuven
Belgium

Preface

An integrated circuit (IC) is a piece of semiconductor material, on which
a number of electronic components are interconnected. These interconnected 'chip' components implement a specific function. The semiconductor material is usually silicon but alternatives include gallium ar-
senide.

ICs are essential in most modern electronic products. The first IC was
created by Jack Kilby in 1959. Photographs of this device and the in-
ventor are shown in figure 3. Figure 1 illustrates the subsequent progress
in IC complexity. This figure shows the numbers of components for advanced ICs and the year in which these ICs were first presented. This
doubling in complexity every two years was predicted by Moore (Intel
1964), who's law is still valid today for the number of logic transistors
on a chip. However, due to reaching the limits of scaling, the complexity
doubling of certain memories now happens at a three-year cycle. This
is shown by the complexity growth line which is slowly saturating.

Figure 2 shows the relative semiconductor revenue per IC category.
CMOS ICs take about 75% of the total semiconductor market. Today's
digital ICs may contain several hundreds of millions to more than a
billion transistors on one single $1\,cm^2$ chip. They can be subdivided into
three categories: logic, microprocessors and memories. About 10% of
the CMOS ICs are of an analog nature.

Figures 4 to 7 illustrate the evolution in IC technology. Figure 4
shows a discrete BC107 transistor. The digital filter shown in figure 5
comprises a few thousand transistors while the Digital Audio Broadcasting (DAB) chip in figure 6 contains more than six million transistors.
The Intel Pentium4 Xeon dual-core processor in figure 7.25 (section 7.6),
contains 1.3 billion transistors. Figure 7 shows an 8 Gb 63 nm multi-level
NAND-flash memory chip.

Figure 8 illustrates the sizes of various semiconductor components,
such as a silicon atom, a single transistor and an integrated circuit, in
perspective. The sizes of an individual MOS transistor are approaching
the details of a virus.

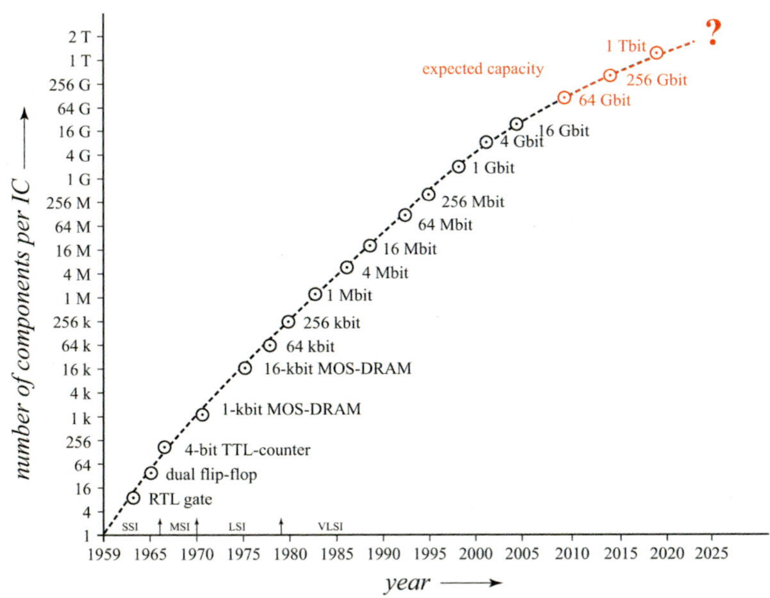

Figure 1: *Growth in the number of components per IC*

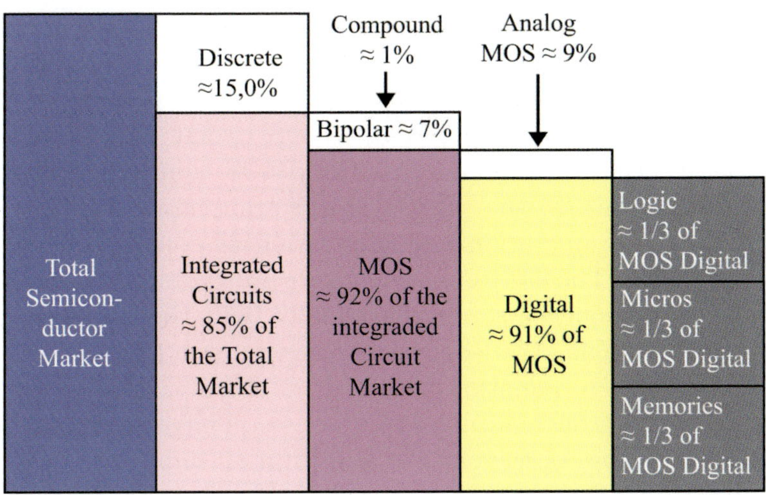

Figure 2: *Relative semiconductor revenue by IC category (Source: IC Insights)*

This book provides an insight into all aspects associated with CMOS ICs. The topics presented include relevant fundamental physics. Technology, design and implementation aspects are also explained and applications are discussed. CAD tools used for the realisation of ICs are described while current and expected developments also receive attention.

The contents of this book are based on the CMOS section of an industry-oriented course entitled 'An introduction to IC techniques'. The course has been given almost three decades, formerly in Philips, currently in NXP Semiconductors. Continuous revision and expansion of the course material ensures that this book is highly relevant to the IC industry. The level of the discussions makes this book a suitable introduction for designers, technologists, CAD developers, test engineers, failure analysis engineers, reliability engineers, technical-commercial personnel and IC applicants. The text is also suitable for both graduates and undergraduates in related engineering courses.

Considerable effort has been made to enhance the readability of this book and only essential formulae are included. The large number of diagrams and photographs should reinforce the explanations. The design and application examples are mainly digital. This reflects the fact that more than 90% of all modern CMOS ICs are digital circuits. However, the material presented will also provide the analogue designer with a basic understanding of the physics, manufacture and operation of nanometer CMOS circuits. The chapters are summarised below. For educational purposes the first four chapters each start with a discussion on nMOS physics, nMOS transistor operation, nMOS circuit behaviour, nMOS manufacturing process, etc. Because the pMOS transistor operation is fully complementary to that of the nMOS transistor, it is then easier to understand the operation and fabrication of complementary MOS (CMOS) circuits. The subjects per chapter are chosen in a very organised and logical sequence so as to gradually built the knowledge, *from Basics to ASICs*. The knowledge gathered from each chapter is required to understand the information presented in the next chapter(s). Each chapter ends with a reference list and exercises. The exercises summarise the important topics of the chapter and form an important part of the complete learning process.

Chapter 1 contains detailed discussions of the basic principles and fundamental physics of the MOS transistor. The derivation of simple current-voltage equations for MOS devices and the explanation of their characteristics illustrates the relationship between process parameters and circuit performance.

The continuous reduction of transistor dimensions leads to increased deviation between the performance predicted by the simple MOS formulae and actual transistor behaviour. The effects of temperature and the impact of the continuous scaling of the geometry on this behaviour are explained in chapter 2. In addition to their influence on transistor and circuit performance, these effects can also reduce device lifetime and reliability.

The various technologies for the manufacture of CMOS ICs are examined in chapter 3. After a summary on the available different substrates (wafers) used as starting material, an explanation of the most important associated photolithographic and processing steps is provided. This precedes a discussion of an advanced nanometer CMOS technology for the manufacture of modern VLSI circuits.

The design of CMOS circuits is treated in chapter 4. An introduction to the performance aspects of nMOS circuits provides an extremely useful background for the explanation of the CMOS design and layout procedures.

MOS technologies and their derivatives are used to realise the special devices discussed in chapter 5. Charge-coupled devices (CCDs), CMOS imagers and MOS power transistors are among the special devices. Chapter 5 concludes the presentation of the fundamental concepts behind BICMOS circuit operation.

Stand-alone memories currently represent about 25% of the total semiconductor market revenue. However, also in logic and microprocessor ICs embedded memories represent close to 80% of the total transistor count. So, of all transistors produced in the world, today, about 90% end up in either a stand-alone, or in an embedded memory. This share is expected to stay at this level or to increase. The majority of available memory types are therefore examined in chapter 6. The basic structures and the operating principles of the various types are explained. In addition, the relationships between their respective properties and application areas is made clear.

Developments in IC technology now facilitate the integration of complete systems on a chip, which contain several hundreds of millions to more than a billion of transistors. The various IC design and realisation techniques used for these VLSI ICs are presented in chapter 7. The advantages and disadvantages of the techniques and the associated CAD tools are examined. Various modern technologies are used to realise a separate class of VLSI ICs, which are specified by applicants rather than manufacturers. These application-specific ICs (ASICs) are examined in this chapter as well. Motives for their use are also discussed.

As a result of the continuous increase of power consumption, the maximum level that can be sustained by cheap plastic packages has been reached. Therefore, all CMOS designers must have a 'less-power attitude'. Chapter 8 presents a complete overview of less-power and less-leakage options for CMOS technologies, as well as for the different levels of design hierarchy.

Increased VLSI design complexities, combined with higher frequencies create a higher sensitivity to physical effects. These effects dominate the reliability and signal integrity of nanometer CMOS ICs. Chapter 9 discusses these effects and the design measures to be taken to maintain both reliability and signal integrity at a sufficiently high level.

Finally, testing, yield, packaging, debug and failure analysis are important factors that contribute to the ultimate costs of an IC. Chapter 10 presents an overview of the state-of-the-art techniques that support testing, debugging and failure analysis. It also includes a rather detailed summary on available packaging technologies and gives an insight into their future trends. Essential factors related to IC production are also examined; these factors include quality and reliability.

The continuous reduction of transistor dimensions associated with successive process generations is the subject of the final chapter (chapter 11). This scaling has various consequences for transistor behaviour and IC performance. The resulting increase of physical effects and the associated effects on reliability and signal integrity are important topics of attention. The expected consequences of and road blocks for further miniaturisation are described. This provides an insight into the challenges facing the IC industry in the race towards nanometer devices.

Not all data in this book is completely sprout from my mind. A lot of books and papers contributed to make the presented material state-of-the-art. Considerable effort has been made to make the reference list complete and correct. I apologize for possible imperfections.

Acknowledgements

I wish to express my gratitude to all those who contributed to the realisation of this book; it is impossible to include all their names. I greatly value my professional environment: Philips Research labs, of which the semiconductor research department is now part of NXP Semiconductors.

It offered me the opportunity to work with many internationally highly valued colleagues who are all real specialists in their field of semiconductor expertise. Their contributions included fruitful discussions, relevant texts and manuscript reviews. I would like to make an exception, here, for my colleagues Marcel Pelgrom and Maarten Vertregt, who greatly contributed to the discussions held on trends in MOS transistor currents and variability matters throughout this book and Roger Cuppens and Roelof Salters for the discussions on non-volatile and random-access memories, respectively.

I would especially like to thank Andries Scholten and Ronald van Langevelde for reviewing chapter 2 and for the discussions on leakage mechanisms in this chapter, Casper Juffermans and Johannes van Wingerden for their inputs to and Ewoud vreugdenhil (ASM Lithography) for his review of the lithography section in chapter 3. I would also like to sincerely thank Robert Lander for his detailed review of the section on CMOS process technologies and future trends in CMOS devices and Gerben Doornbos for the correct sizes and doping levels used in the manufacture of state-of-the-art CMOS devices. I appreciate the many circuit simulations that Octavio Santana has done to create the tapering-factor table in chapter 4. I am grateful for the review of chapter 5 on special circuits and devices based on MOS transistor operation: Albert Theuwissen (Harvest Imaging) for the section on CCD and image sensors, Johan Donkers and Erwin Hijzen for the BICMOS section and Jan Sonsky for the high voltage section. I also appreciate their willingness to supply me with great photographic material. Toby Doorn and Ewoud Vreugdenhil are thanked for their review of the memory chapter (chapter 6). I appreciate Paul Wielage's work on statistical simulations with respect to memory yield loss. I thank Ad Peeters for information on and reviewing the part on asynchronous design in the low-power chapter (chapter 8). Reliability is an important part of chapter 9, which discusses the robustness of ICs. In this respect I want to thank Andrea Scarpa for reviewing the hot-carrier and NBTI subjects, Theo Smedes for the ESD and latch-up subjects and Yuang Li for the part on electromigration. I also greatly value the work of Bram Kruseman, Henk Thoonen and Frank Zachariasse for reviewing the sections on testing, packaging and failure analysis, respectively. I also like to express them my appreciation for supplying me with a lot of figures and photographs, which support and enrich the discussions on these subjects in chapter 10. Finally, I want to thank Chris Wyland and John Janssen, for their remarks and additions on electrical and thermal aspects of IC packages, respectively

I am very grateful to all those who attended the course, because their feedback on educational aspects, their corrections and constructive criticism contributed to the quality and completeness of this book.

In addition, I want to thank Philips Research and NXP Semiconductors, in general for the co-operation I was afforded. I thank my son Bram for the layout of the cover and the layout diagrams in chapter 4, and Ron Salfrais for the correctness of a large part of the English text.

I would especially like to express my gratitude to my daughter Kim and Henny Alblas for the many hours they have spent on the creation of excellent and colourful art work, which contributes a lot to the quality and clarity of this book.

Finally, I wish to thank Harold Benten and Dré van den Elshout for their conscientious editing and type-setting work. Their efforts to ensure high quality should not go unnoticed by the reader.

However, the most important appreciation and gratitude must go to my family, again, and in particular to my wife, for her years of exceptional tolerance, patience and understanding. The year 2007 was particularly demanding. Lost hours can never be regained, but I hope that I can give her now a lot more free time in return.

Eindhoven, February 2008   Harry J.M. Veendrick

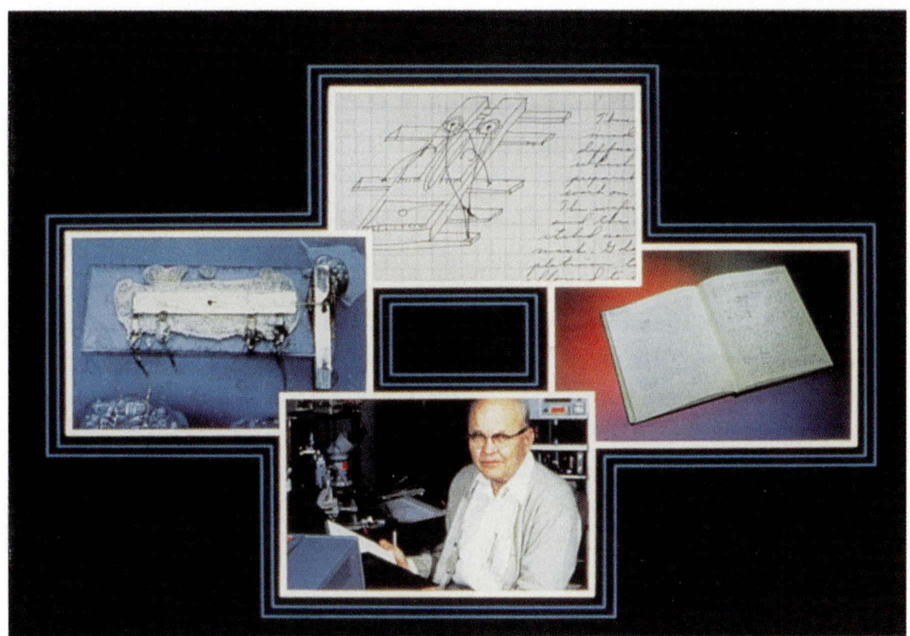

Figure 3: *The development of the first IC: in 1958 Jack Kilby demonstrated the feasibility of resistors and capacitors, in addition to transistors, based on semiconductor technology. Kilby, an employee of Texas Instruments, submitted the patent request entitled 'Miniaturized Electronic Circuits' in 1959. His request was honoured. Recognition by a number of Japanese companies in 1990 means that Texas Instruments is still benefiting from Kilby's patent (Source: Texas Instruments / Koning & Hartman).*

Figure 4: *A single BC107 bipolar transistor (Source: NXP Semicon-ductors)*

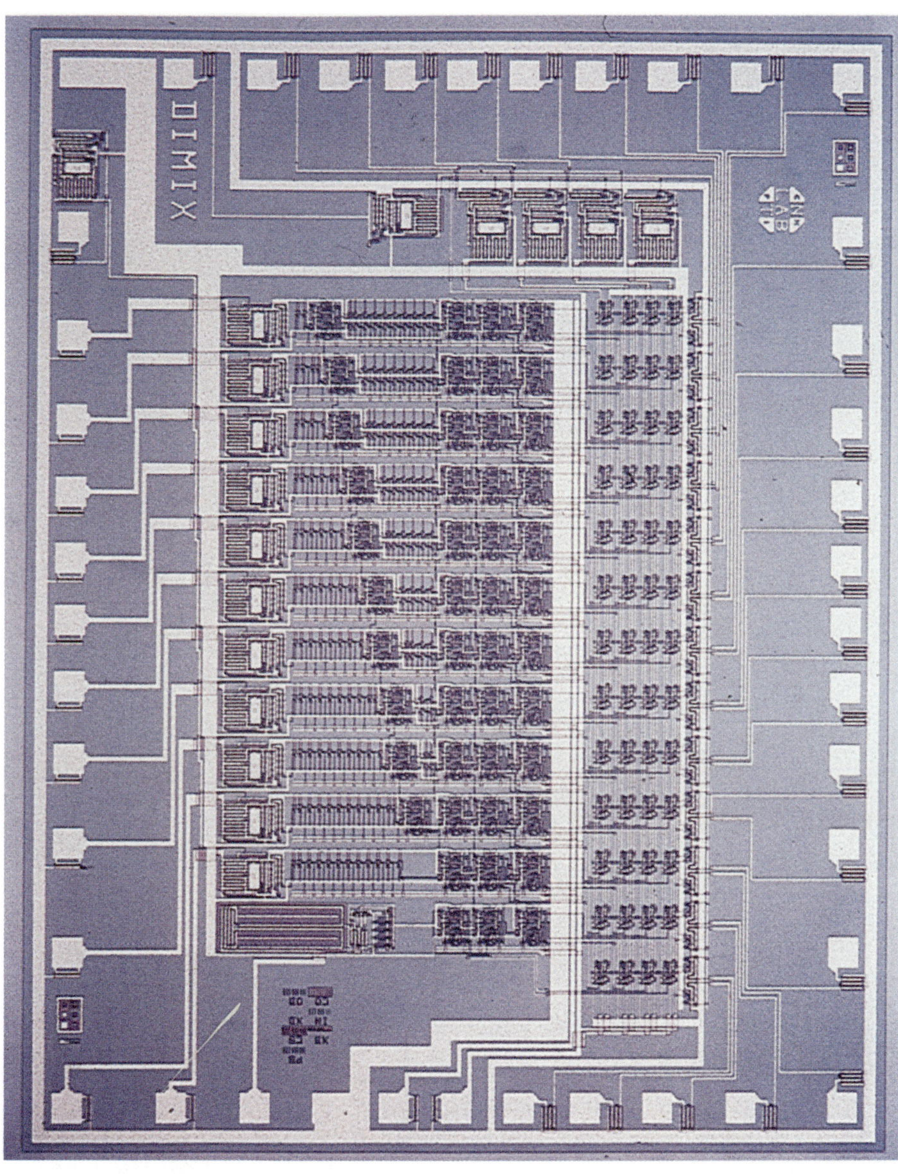

Figure 5: *A digital filter which comprises a few thousand transistors (Source: NXP Semiconductors)*

Figure 6: *A Digital Audio Broadcasting (DAB) chip, which comprises more than six million transistors (Source: NXP Semiconductors)*

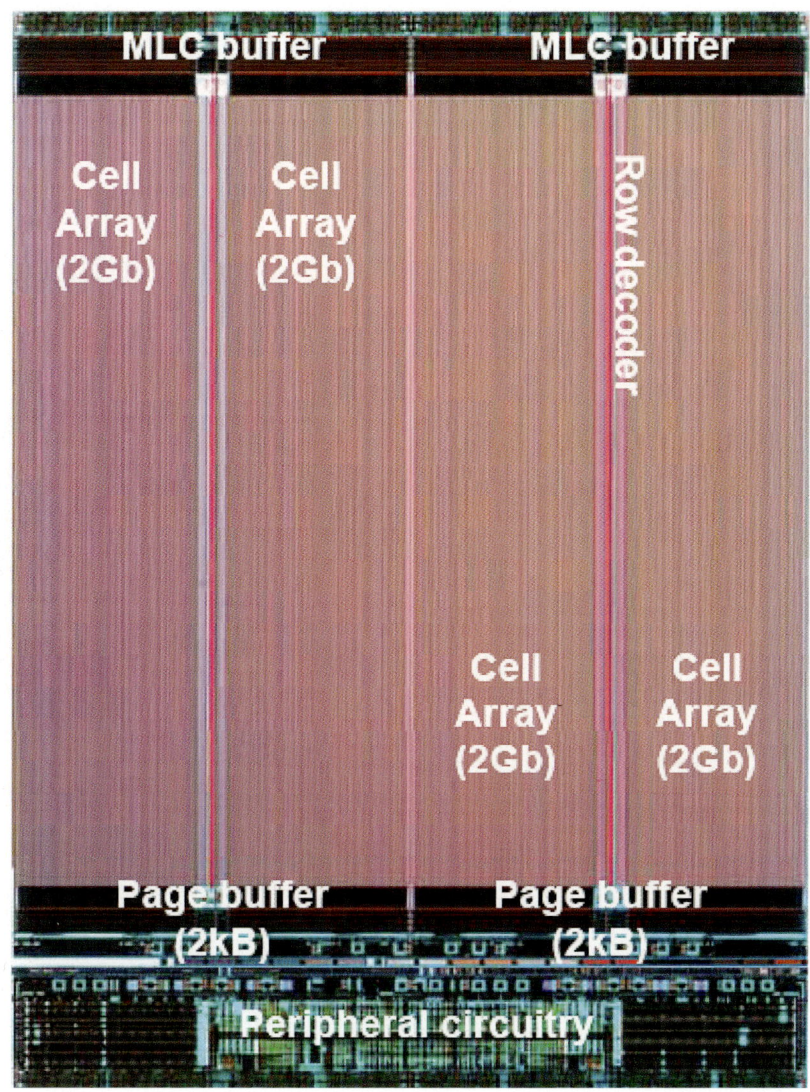

Figure 7: *An 8 Gb 63 nm MLC NAND Chip Layout (Source: Samsung)*

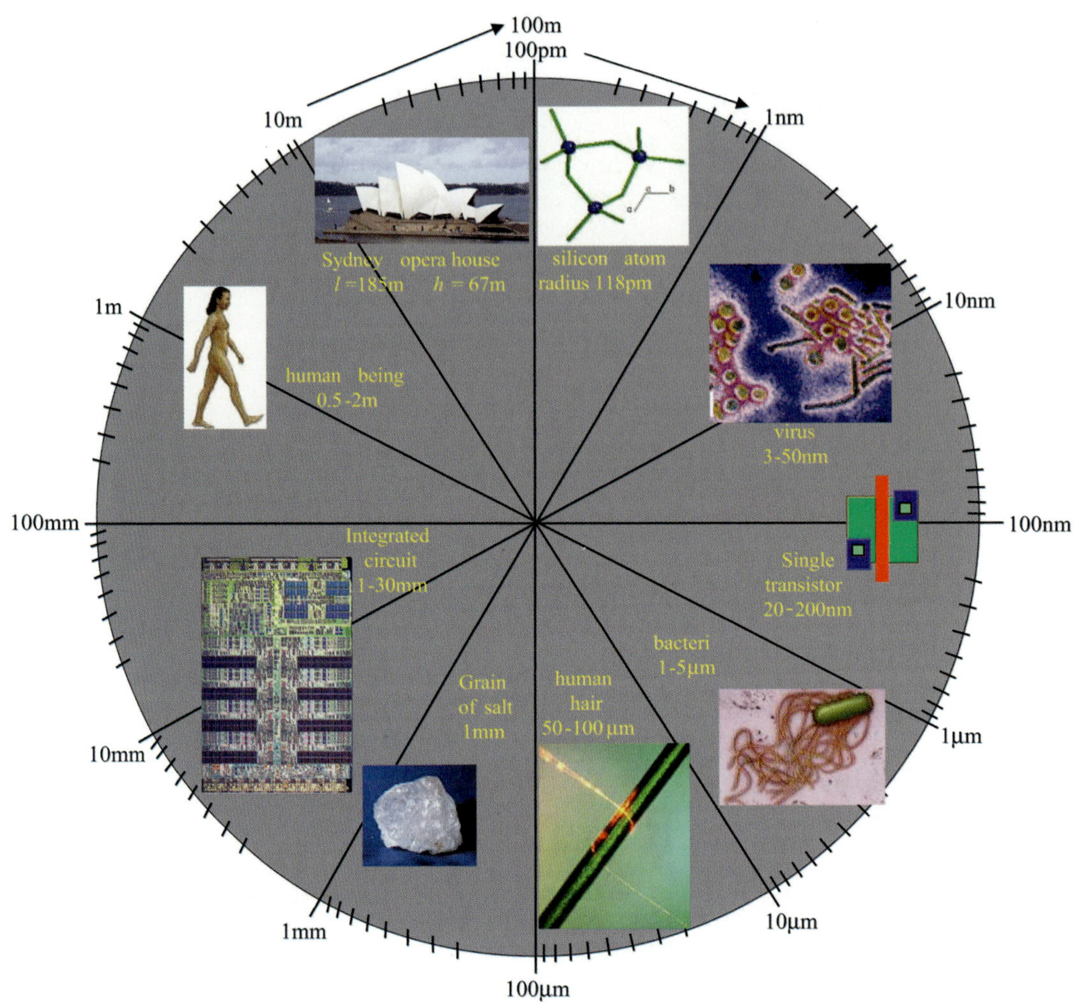

Figure 8: *Various semiconductor component sizes (e.g., atom, transistor, integrated circuit) in perspective*

Overview of symbols

α	channel-shortening factor or clustering factor
A	area
A	aspect ratio
a	activity factor
β	MOS transistor gain factor
$\beta_\square$	gain factor for MOS transistor with square channel
β_n	nMOS transistor gain factor
β_p	pMOS transistor gain factor
β_total	equivalent gain factor for a combination of transistors
BV	breakdown voltage
C	capacitance
C_b	bitline capacitance
C_d	depletion layer capacitance
C_db	drain-substrate capacitance
C_g	gate capacitance
C_gb	gate-substrate capacitance
C_gd	gate-drain capacitance
C_gs	gate-source capacitance
C_gdo	voltage-independent gate-drain capacitance
C_gso	voltage-independent gate-source capacitance
C_par	parasitic capacitance
C_min	minimum capacitance
C_s	scaled capacitance
C_ox	oxide capacitance
C_s	silicon surface-interior capacitance
C_sb	source-substrate (source-bulk) voltage
C_t	total capacitance
CD	critical dimension
ΔL	difference between drawn and effective channel length
ΔV_T	threshold voltage variation
D_0	defect density for uniformly distributed errors (dust particles)

D_l threshold-voltage channel-length dependence factor
D_w threshold-voltage channel-width dependence factor
ϵ dielectric constant
ϵ_0 absolute permittivity
ϵ_ox relative permittivity of oxide
ϵ_r relative permittivity
ϵ_si relative permittivity of silicon
E electric field strength
E_c conduction band energy level
E_f Fermi energy level
E_i intrinsic (Fermi) energy level
E_mx maximum horizontal electric field strength
E_ox electric field across an oxide layer
E_v valence band energy level
E_x horizontal electric field strength
$E_\mathrm{x_c}$ critical horizontal field strength
E_z vertical electric field strength
ϕ electric potential
ϕ_f Fermi potential
ϕ_s surface potential of silicon w.r.t. the substrate interior
ϕ_MS contact potential between gate and substrate
F feature size (= size of a half pitch used for stand-alone memories))
f clock frequency
f_max maximum clock frequency
γ factor which expresses relationship between drain-source
               voltage and threshold-voltage variation
g_m transconductance
I current
I_b substrate current
I_ds drain-source current
I_ds0 characteristic sub-threshold current for gate-substrate voltage of $0\,\mathrm{V}$
I_dsD driver transistor drain-source current
I_dsL load transistor drain-source current
I_dssat saturated transistor drain-source current
I_dssub sub-threshold drain-source current
I_max maximum current
I_on on current
I_R current through resistance
$i(t)$ time-dependent current

j	current densisty
k	Boltzman's constant
K	K-factor; expresses relationship between source-substrate voltage and threshold voltage
K	amplification factor
λ	wavelength of light
L	effective transistor channel length and inductance
L_{CLM}	channel length reduction due to channel length modulation
L_{eff}	effective channel length
L_{ref}	effective channel length of reference transistor
M	yield model parameter
μ_0	substrate carrier mobility
μ_{n}	channel electron mobility
μ_{p}	channel hole mobility
N_A	substrate doping concentration
N.A.	numeric aperture
ρ	charge density
P	power dissipation
P_{dyn}	dynamic power dissipation
P_{stat}	static power dissipation
p	voltage scaling factor
Q	charge
q	elementary charge of a single electron
Q_{d}	depletion layer charge
Q_{g}	gate charge
Q_{m}	total mobile charge in the inversion layer
Q_{n}	mobile charge per unit area in the channel
Q_{ox}	oxide charge
Q_{s}	total charge in the semiconductor
R	resistance
R_{JA}	junction-to-air thermal resistance
R_{JC}	junction-to-case thermal resistance
R_{L}	load resistance
R_{out}	output resistance or channel resistance
R_{therm}	thermal resistance of a package
r	tapering factor
s	scale factor
s_{subthr}	sub-threshold slope
τ	delay time

τ_f	fall time
τ_r	rise time
τ_R	dielectric relaxation time
T	clock period
T_min	minimum clock period
$Temp$	temperature
$Temp_\mathrm{A}$	ambient temperature
$Temp_\mathrm{C}$	case temperature
$Temp_\mathrm{J}$	junction temperature
T_lf	transistor lifetime
t	time
t_cond	conductor thickness
t_d	depletion layer thickness
$t_\mathrm{dielectric}$	dielectric thickness
t_ox	gate-oxide thickness
t_is	isolator thickness
U	computing power
v	carrier velocity
v_sat	carrier saturation velocity
V	voltage
V_B	breakdown voltage
V_r	scaled voltage
V_0	depletion layer voltage
V_bb	substrate voltage
V_dd	supply voltage
V_c	voltage at silicon surface
V_ds	drain-source voltage
V_dssat	drain-source voltage of saturated transistor
V_E	Early voltage
V_{fb}	flat-band voltage
V_g	gate voltage
V_gg	extra supply voltage
V_gs	gate-source voltage
$V_\mathrm{gs_L}$	load transistor gate-source voltage
V_H	high voltage level
V_in	input voltage
V_j	junction voltage
V_L	low voltage level
V_PT	transistor punch-through voltage

V_{sb}	source-substrate (back-bias) voltage
V_{ss}	ground voltage
V_{ws}	well-source voltage
V_{T}	threshold voltage
$V_{\text{T}_{\text{D}}}$	driver transistor threshold voltage
$V_{\text{T}_{\text{dep}}}$	depletion transistor threshold voltage
$V_{\text{T}_{\text{enh}}}$	enhancement transistor threshold voltage
$V_{\text{T}_{\text{L}}}$	load transistor threshold voltage
$V_{\text{T}_{\text{n}}}$	nMOS transistor threshold voltage
$V_{\text{T}_{\text{p}}}$	pMOS transistor threshold voltage
V_{Tpar}	parasitic transistor threshold voltage
V_{out}	output voltage
$V(x)$	potential at position x
V_{x}	process-dependent threshold voltage term
$V_{\text{X}_{\text{L}}}$	process-dependent threshold voltage term for load transistor
$V_{\text{X}_{\text{D}}}$	process-dependent threshold voltage term for driver transistor
W	transistor channel width
W_{n}	nMOS transistor channel width
W_{p}	pMOS transistor channel width
W_{ref}	reference transistor channel width
$\dfrac{W}{L}$	transistor aspect ratio
$\left(\dfrac{W}{L}\right)_{\text{n}}$	nMOS transistor aspect ratio
$\left(\dfrac{W}{L}\right)_{\text{p}}$	pMOS transistor aspect ratio
x	distance w.r.t. specific reference point
Y	yield
Z_{i}	input impedance

List of physical constants

$$\begin{aligned}
\epsilon_0 &= 8.85 \times 10^{-12}\ \text{F/m} \\
\epsilon_{\text{ox}} &= 4 \text{ for silicon dioxide} \\
\epsilon_{\text{si}} &= 11.7 \\
\phi_{\text{f}} &= 0.5\,\text{V for silicon substrate} \\
k &= 1.4 \times 10^{-23}\ \text{Joule/K} \\
q &= 1.6 \times 10^{-19}\ \text{Coulomb}
\end{aligned}$$

Contents

Chapter 1

Basic Principles

1.1 Introduction

The majority of current VLSI Very Large Scale Integration) circuits are manufactured in CMOS technologies. Familiar examples are memories (1 Gb, 4 Gb and 16 Gb), microprocessors and signal processors. A good fundamental treatment of basic MOS devices is therefore essential for an understanding of the design and manufacture of modern *VLSI* circuits. This chapter describes the operation and characteristics of MOS devices. The material requirements for their realisation are discussed and equations that predict their behaviour are derived.

The acronym MOS represents the <u>M</u>etal, <u>O</u>xide and <u>S</u>emiconductor materials used to realise early versions of the MOS transistor. The fundamental basis for the operation of MOS transistors is the field-effect principle. This principle is quite old, with related publications first appearing in the nineteen-thirties. These include a patent application filed by J.E. Lilienfeld in Canada and the USA in 1930 and one filed by O. Heil, independently of Lilienfeld, in England in 1935. At that time, however, insufficient knowledge of material properties resulted in devices which were unfit for use. The rapid development of electronic valves probably also hindered the development of the MOS transistor by largely fulfilling the transistor's envisaged role.

1.2 The field-effect principle

The field-effect principle is explained with the aid of figure 1.1. This figure shows a rectangular conductor, called a channel, with length L, width W and thickness t_{cond}. The free electrons present in the channel are the mobile charge carriers. There are n electrons per m^3 and the charge q per electron equals -1.602×10^{-19} C(coulomb). The application of a horizontal electric field of magnitude E to the channel causes the electrons to acquire an average velocity $v = -\mu_{\text{n}} \cdot E$. The electron mobility μ_{n} is positive. The direction of v therefore opposes the direction of E. The resulting current density j is the product of the average electron velocity and the mobile charge density ρ:

$$j = \rho \cdot v = -n \cdot q \cdot \mu_{\text{n}} \cdot E \tag{1.1}$$

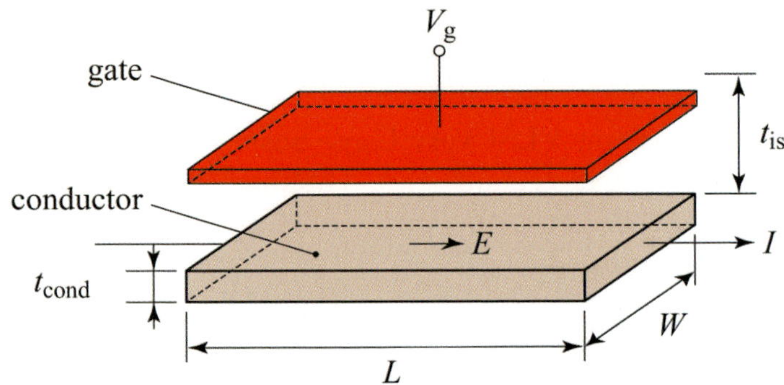

Figure 1.1: *The field-effect principle*

A gate electrode situated above the channel is separated from it by an insulator of thickness t_{is}. A change in the gate voltage V_{g} influences the charge density ρ in the channel. The current density j is therefore determined by V_{g}.

Example:
Suppose the insulator is silicon dioxide (SiO$_2$) with a thickness of 2 nm ($t_{\text{is}} = 2 \times 10^{-9}$ m). The gate capacitance will then be about 17 mF/m^2. The total gate capacitance C_{g} is therefore expressed as follows:

$$C_{\text{g}} = 17 \times 10^{-3} \cdot W \cdot L \quad [\text{F}]$$

A change in gate charge $\Delta Q_{\text{g}} = -C_{\text{g}} \cdot \Delta V_{\text{g}}$ causes the following change in channel charge:

$$+C_{\text{g}} \cdot \Delta V_{\text{g}} = 17 \times 10^{-3} \cdot W \cdot L \cdot \Delta V_{\text{g}} = W \cdot L \cdot t_{\text{cond}} \cdot \Delta\rho$$

Thus:

$$\Delta\rho = \frac{17 \times 10^{-3} \cdot \Delta V_g}{t_{cond}} \ \text{C/m}^3$$

and:

$$|\Delta n| = \left|\frac{\Delta\rho}{q}\right| = \frac{10.6 \times 10^{16} \cdot \Delta V_g}{t_{cond}} \ \text{electrons/m}^3$$

If a 0.5 V change in gate voltage is to cause a hundred times increase in current density j, then the following must apply:

$$\frac{\Delta j}{j} = \frac{\Delta\rho}{\rho} = \frac{\Delta n}{n} = \frac{10.6 \times 10^{16} \cdot 0.5}{t_{cond} \cdot n} = 100$$

$$\Rightarrow t_{cond} = \frac{5.3 \times 10^{14}}{n}$$

Examination of two materials reveals the implications of this expression for t_{cond}:

Case a The channel material is copper.

This has $n \approx 10^{28}$ electrons/m^3 and hence $t_{cond} \approx 5.3 \times 10^{-14}$ m. The required channel thickness is thus less than the size of one atom ($\approx 3 \times 10^{-10}$ m). This is impossible to realise and its excessive number of free carriers renders copper unsuitable as channel material.

Case b The channel material is 5Ωcm n-type silicon.

This has $n \approx 10^{21}$ electrons/m^3 and hence $t_{cond} \approx 530$ nm.

The *transconductance* g_m of a MOS transistor is the ratio of a change in channel (drain) current to the corresponding change in gate voltage:

$$g_m = \frac{\Delta I}{\Delta V_g}$$

$$\text{However} \quad \frac{\Delta I}{I} = \frac{\Delta j}{j}$$

$$\text{Therefore} \quad g_m = \frac{I}{\Delta V_g} \cdot \frac{\Delta j}{j}$$

If $I = j \cdot W \cdot t_{cond} = 1\,\text{mA}$, $\Delta j/j = 100$ and $\Delta V_g = 0.5\,\text{V}$ then:

$$g_m = 200 \ \text{mA/V}$$

In this case, a transconductance of $200\,\mathrm{mA/V}$ requires a channel thickness of $t_{\mathrm{cond}} = 530\,\mathrm{nm}$. Modern IC technologies allow the realisation of much thinner channels.

From the above example, it is clear that field-effect devices can only be realised with semiconductor materials. Aware of this fact, Lilienfeld used copper sulphide as a semiconductor in 1930. Germanium was used during the early fifties. Until 1960, however, usable MOS transistors could not be manufactured. Unlike the transistor channel, which comprised a manufactured thin layer, the channel in these *inversion-layer transistors* is a thin conductive layer, which is realised electrically. The breakthrough for the fast development of MOS transistors came with advances in planar silicon technology and the accompanying research into the physical phenomena in the semiconductor surface.

Generally, circuits are integrated in silicon because widely-accepted military specifications can be met with this material. These specifications require products to function correctly at a maximum operating temperature of $125\,^{\circ}\mathrm{C}$. The maximum operating temperature of germanium is only $70\,^{\circ}\mathrm{C}$, while that of silicon is $150\,^{\circ}\mathrm{C}$. A comparison of a few other germanium (Ge) and silicon (Si) material constants is presented below:

Material constant	Germanium	Silicon
Melting point $[^{\circ}\mathrm{C}]$	937	1415
Breakdown field $[\mathrm{V}/\mu\mathrm{m}]$	8	30
Relative expansion coeff. $[^{\circ}\mathrm{C}]^{-1}$	5.8×10^{-6}	2.5×10^{-6}
ϵ_r	16.8	11.7
Max. operating temp. $[^{\circ}\mathrm{C}]$	70	150

1.3   The inversion-layer MOS transistor

A schematic drawing of the inversion-layer nMOS transistor, or simply 'nMOSt', is shown in figure 1.2, which is used to explain its structure and operation. The two n^+ areas in the p-type substrate are called the *source* and *drain*. The *gate* electrode is situated above the p area between them. This electrode is either a metal plate, e.g., aluminium or molybdenum, a heavily doped and thus low-ohmic polycrystalline silicon layer, or a combination of both. Normally, the source and drain areas are also heavily doped to minimise series resistance. The resistance R

of a $10\,\mu$m long and $2\,\mu$m wide track is $\frac{10}{2}\cdot R_\square$, where $R_\square$ is the sheet resistance of the track material. The sheet resistance of the source and drain areas usually ranges from 3 to 100 $\Omega/\square$ with doping levels upto $5\cdot 10^{19}$ to $2\cdot 10^{20}$ atoms per cm^3. The dope concentration in the p-type substrate is approximately $10^{14}-10^{16}$ atoms per cm^3, while the channel dope (by threshold adjustment implantation, etc.) is between $10^{17}-10^{18}$ atoms per cm^3. A p-channel transistor differs from the above n-channel type in that it contains a p$^+$ source and drain in an n-type substrate.

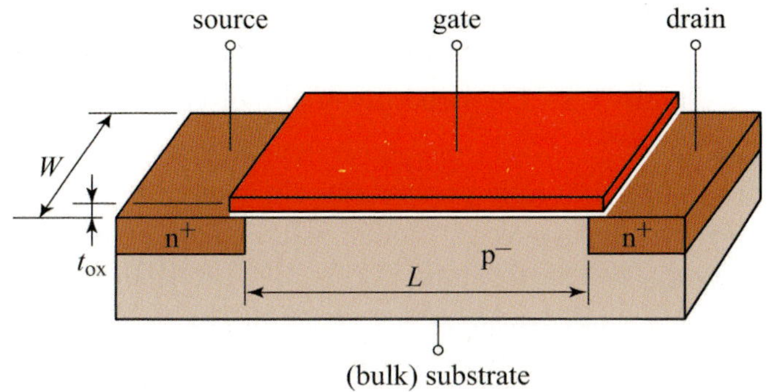

Figure 1.2: *Cross-section of an inversion-layer nMOS transistor*

Characteristic parameters of a MOS transistor are indicated in figure 1.2. These include the width W and length L of the channel and the thickness t_{ox} of the insulating oxide which separates the gate and channel. In modern CMOS VLSI circuits, the minimum values of W and L range from $40\,$nm to $120\,$nm and $t_{\mathrm{ox}} \approx 1.2\,$nm - $2.5\,$nm. Continuous development will reduce these values in the future. The depth of the source and drain junctions varies from $50\,$nm to $200\,$nm.

The energy band theory and its application to the MOS transistor are briefly summarised below. An understanding of this summary is a pre-requisite for a detailed discussion of the behaviour of the MOS transistor.

The structure of a free silicon atom is shown in figure 1.3. This atom comprises a nucleus, an inner shell and an outer shell. The nucleus contains 14 protons and 14 neutrons while the shells contain 14 electrons. Ten of the electrons are in the inner shell and four are in the outer shell. The positive charge of the protons and the negative charge of the electrons compensate each other to produce an atom with a net neutral charge.

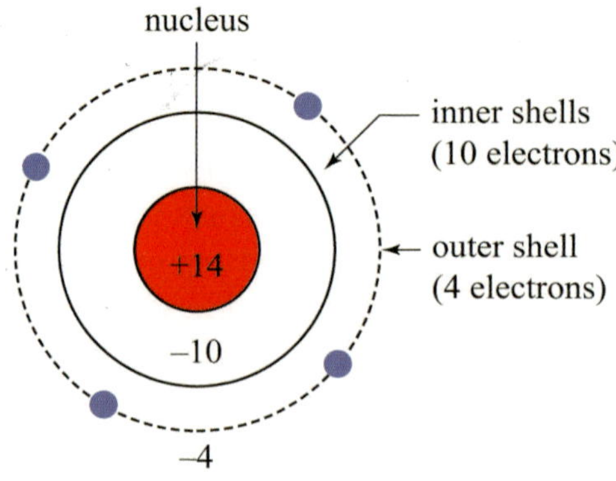

Figure 1.3: *The structure of a free silicon atom*

The electrons in an atom may possess certain energy levels. These energy levels are grouped into energy bands, which are separated by energy gaps. An energy gap represents impossible levels of electron energy. The energy bands that apply to the electrons in an atom's outer shell are *valence* and *conduction* bands. Figure 1.4 shows these bands and the energy gap for a typical solid material. The valence electrons determine the physical and chemical properties of a material.

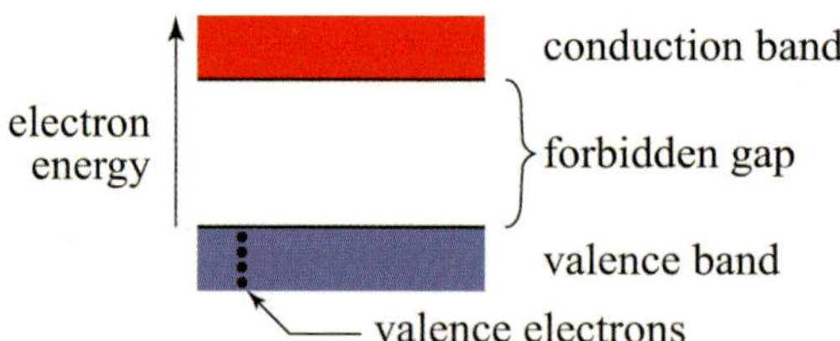

Figure 1.4: *Schematic representation of electron energy bands in a typical solid material*

The four electrons in the outer shell of a silicon atom are in the material's valence band. Figure 1.5 shows the bonds that these electrons form with neighbouring atoms to yield a silicon crystal.

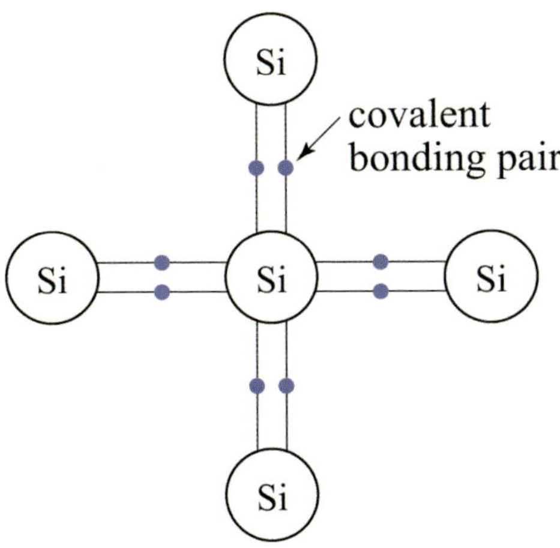

Figure 1.5: *Silicon crystal*

The electrons in a conductor can easily go from the valence band to the conduction band. Therefore, the conduction and valence bands in a conductor partly overlap, as shown in figure 1.6a. In an insulator, however, none of the valence electrons can reach the conduction band. Figure 1.6b shows the large band gap generally associated with insulators. A semiconductor lies somewhere between a conductor and an insulator. The associated small band gap is shown in figure 1.6c. Valence electrons may acquire sufficient thermal energy to reach the conduction band and therefore leave an equal number of positively-charged ions, or 'holes', in the valence band. This produces a limited conduction mechanism in semiconductors.

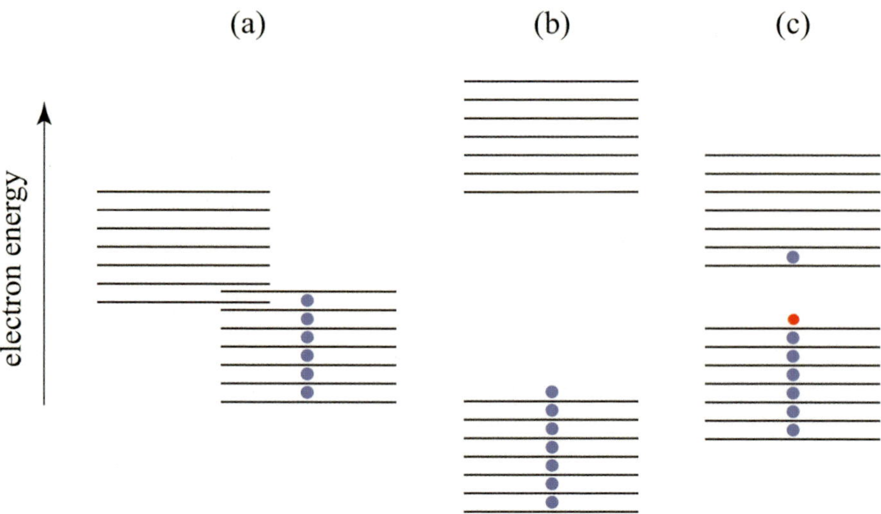

Figure 1.6: *Energy bands of a conductor, an insulator and an intrinsic semiconductor*

Semiconductor materials are located in group IV of this system. The introduction of an element from group III or V in a semiconductor crystal produces an 'acceptor' or a 'donor' atom. This *semiconductor doping process* dramatically changes the crystal properties. The following table shows the location of semiconductor materials in the periodic system of elements.

Group		
III (Acceptors)	IV	V (Donors)
Boron	Carbon	Nitrogen
Aluminium	Silicon	Phosphorous
Gallium	Germanium	Arsenic
Indium	Stannic (tin)	Stibnite

The presence of a group III atom in a silicon crystal lattice is considered first. The situation for boron (B) is illustrated in figure 1.7a. Boron has one electron less than silicon and cannot therefore provide an electron required for a bond with one of its four neighbouring silicon atoms. The hole in the resulting p-type semiconductor is a willing 'acceptor' for an electron from an alternative source. This hole can be removed relatively easily with the ionisation energy of approximately 0.045 eV shown in the energy band diagram of figure 1.7a.

Similar reasoning applies when a group V atom, such as phosphorus (P), is present in the silicon lattice. This situation is illustrated in figure 1.7c. The extra electron in the phosphorus atom cannot be accommodated in the regular bonding structure of the silicon lattice. It is therefore easy to remove this 'donor' electron in the resulting n-type semiconductor. The mere $0.037\,\text{eV}$ ionisation energy required is much lower than the $1.11\,\text{eV}$ band gap energy of silicon. Figure 1.7b shows the energy band diagram of an intrinsic silicon lattice, which contains no donor or acceptor 'impurity' atoms.

The energy level indicated by E_f in figure 1.7 is called the *Fermi level*. An electron with this energy has an equal probability of location in the valence band and the conduction band. This probability is exactly 0.5.

The Fermi level of an intrinsic semiconductor is often referred to as the intrinsic Fermi level E_i. The Fermi level E_f in a p-type semiconductor is situated near the valence band E_v, while it is close to the conduction band E_c in an n-type semiconductor. The above theory concerning the different types of semiconductors and their respective energy band diagrams will now be used to explain the behaviour of the MOS transistor. This explanation is preceded by a description of the structure and operation of the MOS capacitor.

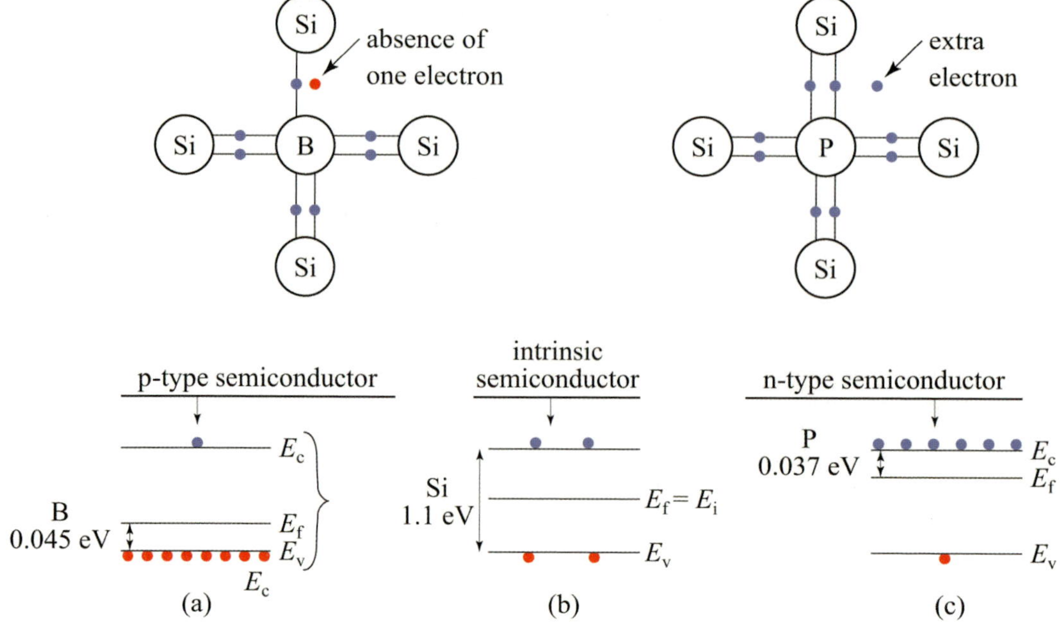

Figure 1.7: *Energy band diagrams for p-type, intrinsic, and n-type semiconductor materials*

1.3.1   The Metal-Oxide-Semiconductor (MOS) capacitor

Figure 1.8 shows a cross-section of a basic MOS capacitor. This structure is identical to a MOS transistor except that the source and drain diffusion regions are omitted.

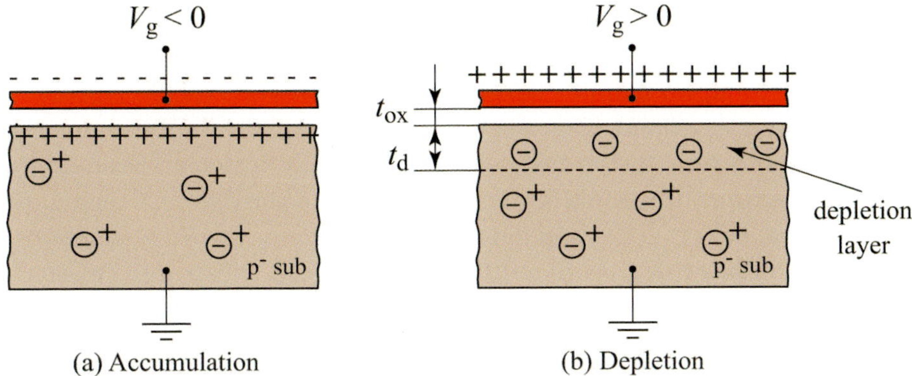

Figure 1.8: *Cross-section of a MOS structure without source and drain areas. There is a capacitance between the gate and substrate.*

The p-type substrate is made with an acceptor dope material, e.g., boron. The substrate is assumed to behave as a normal conductor and contains many free holes. The situation which occurs when the p-type substrate is grounded and a negative voltage is applied to the gate electrode is shown in figure 1.8a. The negative charge on the gate is compensated by an equal but positive charge in the substrate. This is accomplished by positively charged holes which accumulate at the Si-SiO_2 interface. These holes are the majority charge carriers in the substrate. This 'accumulation' process continues until the positive charge at the substrate surface equals the negative charge on the gate electrode. Extra holes are supplied through the ground contact to the substrate. The resulting accumulation capacitor can be viewed as an ideal parallel plate capacitor.

A different situation occurs when the potential on the gate electrode is made positive with respect to the grounded substrate. This situation is shown in the cross-section of figure 1.8b. The positive charge which is present on the gate must be counter-balanced by a negative charge at the Si-SiO$_2$ interface in the substrate. Free positively-charged holes are pushed away from the substrate surface to yield a negatively-charged depletion layer. This *'depletion'* process stops when the negative charge of the depletion layer equals the positive charge on the gate electrode. Clearly, the thickness t_d of the depletion layer in the equilibrium situation is proportional to the gate voltage. It is important to realise that a depletion layer only contains a fixed charge, i.e., ions fixed in the solid state lattice, and no mobile charge carriers.

Various energy band diagrams are used to explain the behaviour of the inversion layer MOS transistor. To provide a better understanding of these diagrams, Poisson's law is first applied to the different regions of the MOS capacitor. These regions include the gate, the SiO$_2$ insulator, the depletion layer in silicon and the p-type silicon substrate. Poisson's law is used to investigate the charge distribution $Q(z)$, the electric field $E(z)$ and the electric potential $\phi(z)$ in these regions as a function of the distance z from the Si-SiO$_2$ interface.

In its one dimensional form, Poisson's law is formulated as follows:

$$\frac{d^2\phi(z)}{dz^2} = -\frac{\rho}{\epsilon} \tag{1.2}$$

$$
\begin{aligned}
\text{where} \quad \phi(z) \ &= \ \text{electrical potential at position z;} \\
z \ &= \ \text{distance from the } Si - SiO_2 \text{ interface;} \\
\rho \ &= \ \text{space charge;} \\
\epsilon \ &= \ \text{dielectric constant.}
\end{aligned}
$$

The situation in which no space charge is present is considered first. This is almost true in the SiO$_2$ insulator, in which case $\rho = 0$. Integration of formula (1.2) once gives the electric field:

$$E(z) = C_1, \qquad C_1 = \text{integration constant.}$$

Integration of formula (1.2) twice gives the electric potential in SiO$_2$:

$$\phi(z) = C_1 \cdot z + C_2$$

The electric field in the insulator is thus constant and the electric potential is a linear function of the distance z from the Si-SiO$_2$ interface.

Next, the situation in which a constant space charge is present is considered. This is assumed to be true in the depletion layer, whose width is W_D. In this case:

$$\rho \quad = \quad -q \cdot N_A$$

$$\text{where } q \quad = \quad \text{the charge of an electron}$$

$$\text{and } N_A \quad = \quad \text{the total number of fixed ions}$$

$$\text{in the depletion layer of thickness } t_d.$$

Integrating formula (1.2) once gives the electric field:

$$E(z) = \frac{q \cdot N_A}{\epsilon} \cdot z + C_1$$

Integrating formula (1.2) twice gives the electric potential in the depletion layer:

$$\phi(z) = \frac{q \cdot N_A}{2\epsilon} \cdot z^2 + C_1 \cdot z + C_2$$

Therefore, the electric field in a depletion layer with constant space charge is a linear function of z, while the electric potential is a square function of z. The space charge in a depletion layer is only constant when the dope of the substrate has a constant value at all distances z from the Si-SiO$_2$ interface. In practice, the space-charge profile is related to the dope profile which exists in the substrate.

The gate and the substrate region outside the depletion layer are assumed to behave as ideal conductors. The electric potentials in these regions are therefore constant and their electric fields are zero.

The above results of the application of Poisson's law to the MOS capacitor are illustrated in figure 1.9. Discontinuities in the diagrams are caused by differences between the dielectric constant of silicon and silicon dioxide. The electric charge, the electric field and potential are zero in the grounded substrate outside the depletion region. The observation that the electric potential is a square function of z in the depletion layer is particularly important.

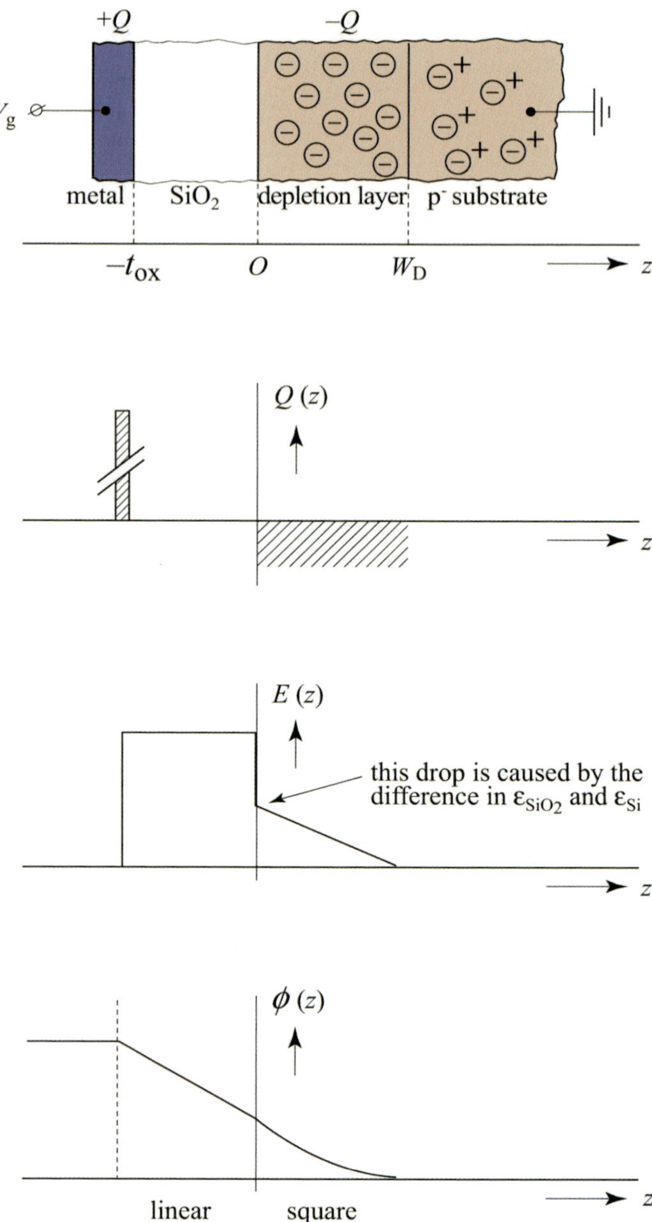

Figure 1.9: *The sections of a MOS capacitor and the associated charge distribution $Q(z)$, electric field $E(z)$ and electric potential $\phi(z)$*

14

1.3.2 The inversion-layer MOS transistor

Figure 1.10 shows a cross-section of an nMOS transistor with $0\,\mathrm{V}$ on all of its terminals. The figure also contains the associated energy band diagram.

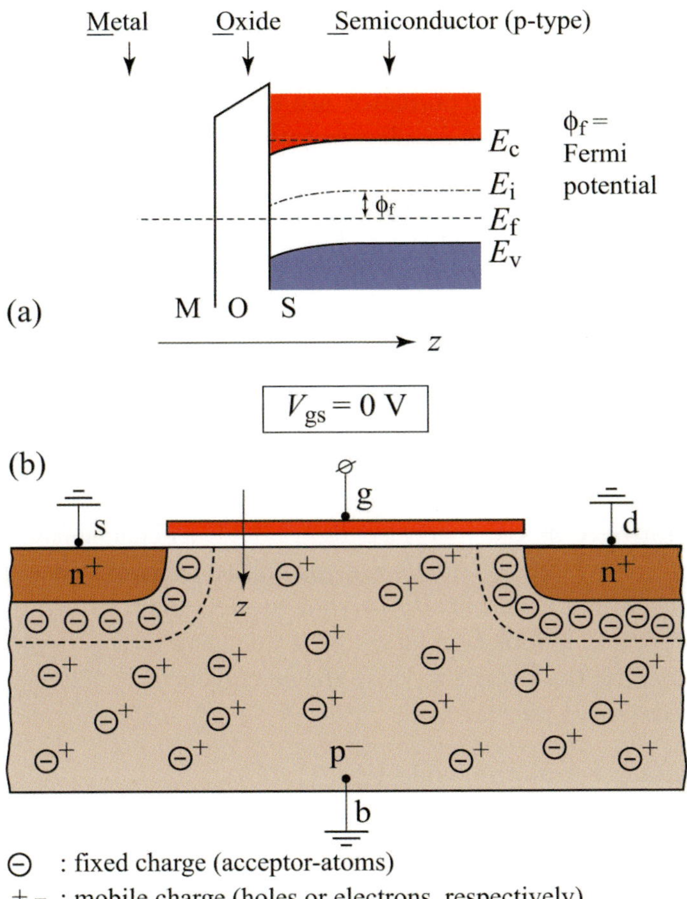

Figure 1.10: *Cross-section of a MOS transistor with $V_\mathrm{gs}{=}V_\mathrm{ds}{=}V_\mathrm{sb}{=}0\ V$ and the associated energy band diagram*

It is assumed that the presence of the gate does *not* affect the distribution of holes and electrons in the semiconductor. With the exception of the depletion areas around the n^+ areas, the entire p-substrate is assumed to be homogeneous and devoid of an electric field ($E = 0$). There is no charge on the gate and no surface charge in the silicon. Generally, the electron energies at the Fermi levels of the different materials in the structure will differ. Their work functions (i.e., the energy required to remove an electron from the Fermi level to vacuum) will also differ. When the voltage between the gate and source is zero ($V_{gs}= 0$) and the metal gate is short circuited to the semiconductor, electrons will flow from the metal to the semiconductor or vice versa until a voltage potential is built up between the two materials. This voltage potential counter-balances the difference in their work functions. The Fermi levels in the metal and the semiconductor are then aligned.

Therefore, there will be an electrostatic potential difference between the gate and substrate which will cause the energy bands to bend. The *'flat-band condition'* exists when there is no band-bending at the metal-semiconductor interface. The *'flat-band voltage'* V_{fb} is the gate voltage required to produce the flat-band condition. It is the difference between the work functions of the metal (ϕ_M) and the semiconductor (ϕ_S), i.e., $V_{fb} = \phi_{MS} = \phi_M - \phi_S$. Since equilibrium holds, the Fermi level in the semiconductor remains constant regardless of the value of the gate voltage.

A negative charge is induced in the semiconductor surface when a small positive voltage is applied to the gate, while the source, drain and substrate are at 0V, see also figure 1.11. The negative charge is caused by holes being pushed away from the insulator interface. The negatively charged acceptor atoms that are left behind form a negative space charge, i.e., a depletion layer. The thickness of this depletion layer is determined by the potential V_c at the silicon surface. The gate voltage V_{gs} now consists of two parts:

a. The voltage across the oxide $V_g - V_c$;

b. The voltage across the depletion layer V_c.

The capacitance between the gate and substrate now consists of the series connection of the oxide capacitance C_{ox} and the depletion-layer capacitance C_d.

The term V_T in figure 1.11 represents the *threshold voltage* of the transistor. This is the gate voltage at which the band-bending at the silicon surface is exactly $2\phi_f$. At this band bending, the electron concentration at the semiconductor surface becomes equal to the hole concentration in the bulk. This situation is called *(strong) inversion*, and the layer of free electrons created at the surface is called an inversion layer. For the present, V_T is assumed to be positive for an inversion-layer nMOS transistor. This assumption is confirmed later in the text.

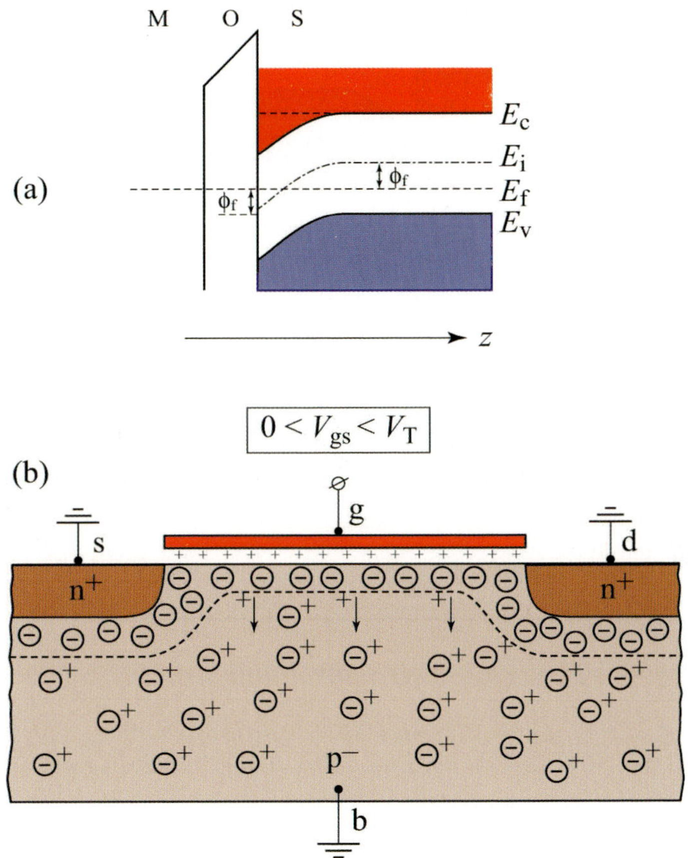

Figure 1.11: *Cross-section of a MOS transistor with $0<V_{gs}<V_T$ and $V_{ds}=V_{sb}=0\,V$ and its corresponding energy band diagram*

If the gate voltage is further increased ($V_{gs} > V_T$), then the band-bending at the silicon surface will be larger than $2\phi_f$. This situation is illustrated in figure 1.12. A comparison of figure 1.12 and figure 1.7c reveals that the energy band at the silicon surface corresponds to an n-type semiconductor.

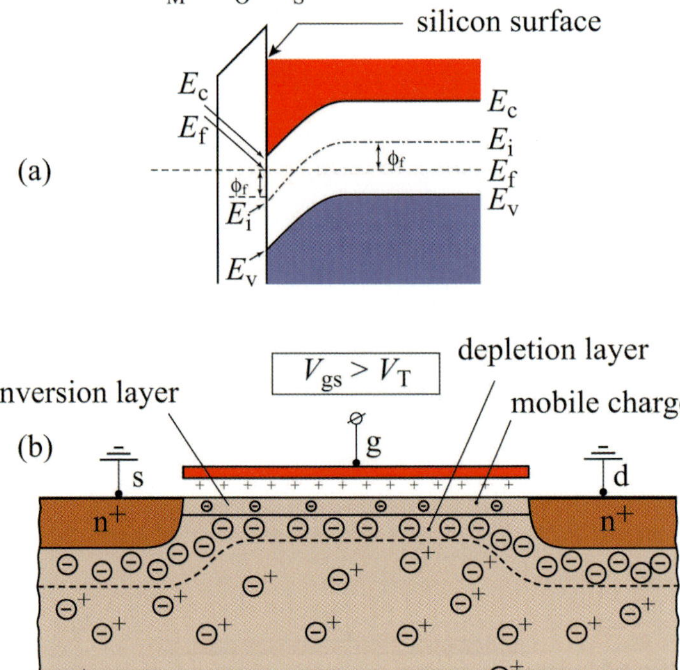

Figure 1.12: *Cross-section of a MOS transistor with $V_{gs}{>}V_T$ ($V_T{>}0$) and $V_{ds}{=}V_{sb}{=}0\,V$ and its corresponding energy band diagram*

Deep in the substrate, however, the energy band corresponds to a p-type semiconductor. A very narrow n-type layer has therefore been created at the surface of a p-type silicon substrate. In addition to the negative acceptor atoms already present, this *inversion layer* contains electrons which act as mobile negative charge carriers. Conduction in the n-type inversion layer is mainly performed by these electrons, which are minority carriers in the p-type substrate. The inversion layer forms a conducting *channel* between the transistor's source and drain. No current flows in this channel if there is no voltage difference between the drain and source terminals, i.e., $I_{ds} = 0\,A$ if $V_{ds} = 0\,V$. The number of electrons in the channel can be controlled by the gate-source voltage V_{gs}.

Assuming that $V_{gs} > V_T$, the effects of increasing V_{ds} from $0\,\text{V}$ are divided into the following regions:

1. $0 < V_{ds} < V_{gs} - V_T$.
   This is called the *linear* or *triode* region of the MOS transistor's operating characteristic.

2. $V_{ds} = V_{gs} - V_T$.
   At this point, a transition takes place from the linear to the so-called saturation region.

3. $V_{ds} > V_{gs} - V_T$.
   This is the *saturation* region of the MOS transistor's operating characteristic.

The three regions are discussed separately on the following pages.

The linear region $\quad \begin{cases} V_{\mathrm{gs}} > V_{\mathrm{T}} > 0 \\ 0 < V_{\mathrm{ds}} < V_{\mathrm{gs}} - V_{\mathrm{T}} \end{cases}$

Figure 1.13 shows the situation in the linear region, in which a current I_{ds} (which flows from drain to source) causes a voltage difference in the channel. The surface potential under the gate decreases from V_{ds} in the drain to $0\,\mathrm{V}$ in the source. The maximum potential difference between the gate and channel is at the source. Therefore, the strongest inversion and the highest concentration of electrons in the inversion layer occur adjacent to the source. The maximum potential difference between the channel and substrate is at the drain. The depletion layer is therefore thickest here. In the linear region, the drain current I_{ds} increases with increasing V_{ds} for a constant V_{gs}.

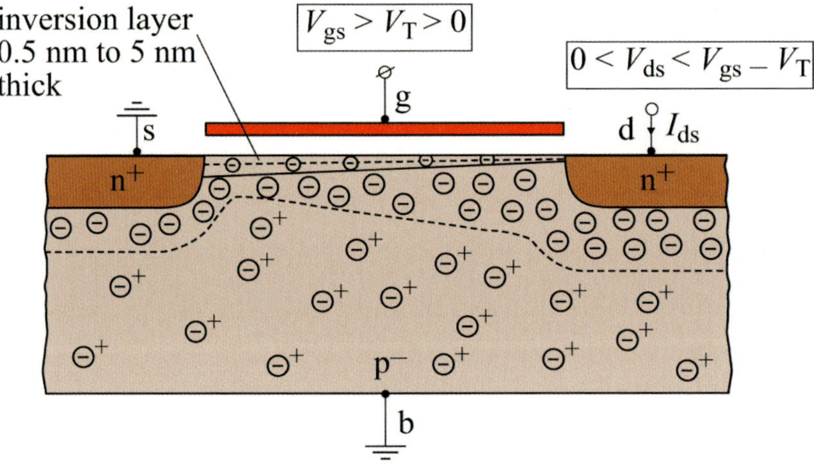

Figure 1.13: *Cross-section of a transistor operating in the linear (triode) region*

$$\textbf{The transition region} \quad \begin{cases} V_{\mathrm{gs}} > V_{\mathrm{T}} > 0 \\ V_{\mathrm{ds}} = V_{\mathrm{gs}} - V_{\mathrm{T}} \end{cases}$$

An increase in V_{ds}, with V_{gs} constant, decreases the voltage difference between the gate and channel at the drain. The inversion layer disappears at the drain when the voltage difference between the gate and channel equals the threshold voltage V_{T}. The channel end then coincides with the drain-substrate junction. This situation occurs when $V_{\mathrm{ds}} = V_{\mathrm{gs}} - V_{\mathrm{T}}$, and is shown in figure 1.14.

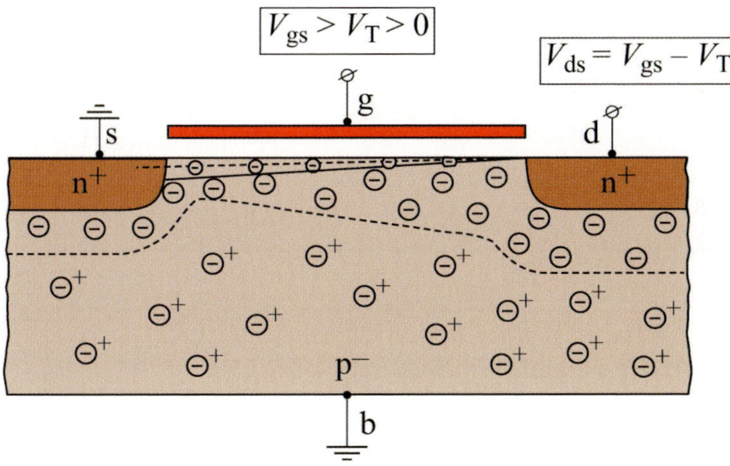

Figure 1.14: *Situation during transition from triode to saturation region, i.e., $V_{\mathrm{ds}}=V_{\mathrm{gs}}-V_{\mathrm{T}}$*

$$\textbf{The saturation region} \quad \begin{cases} V_{\text{gs}} > V_{\text{T}} > 0 \\ V_{\text{ds}} > V_{\text{gs}} - V_{\text{T}} \end{cases}$$

The channel end no longer coincides with the drain when V_{ds} is larger than $V_{\text{gs}} - V_{\text{T}}$. This situation is shown in figure 1.15.

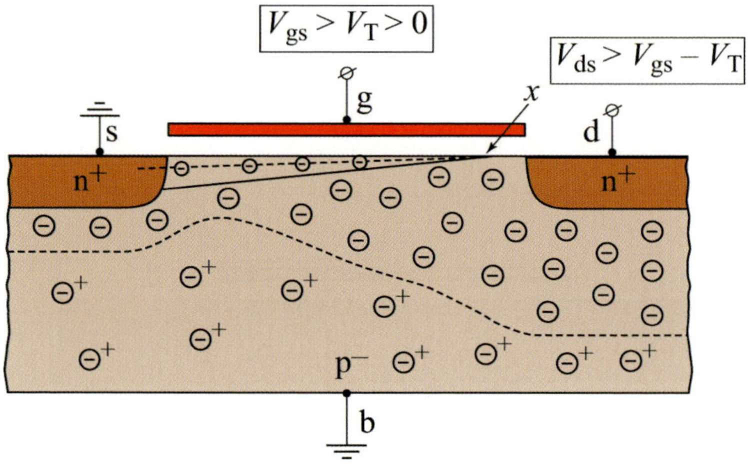

Figure 1.15: *Situation in the saturation region, i.e.,* $V_{\text{ds}} > V_{\text{gs}} - V_{\text{T}}$

The voltage V_x at the end point x of the inversion layer equals $V_{\text{gs}} - V_{\text{T}}$. Therefore, V_{T} is the voltage difference between the gate and channel at position x. If this *pinch-off point* is considered to be the *virtual drain* of the transistor, then I_{ds} is determined by the voltage $V_x = V_{\text{gs}} - V_{\text{T}}$. In other words, the drain current in the saturation region equals the drain current at the transition point between the linear and saturation regions. The value of the *saturation current* is clearly proportional to V_{gs}. Electrons are emitted from the inversion layer into the depletion layer at the pinch-off point. These electrons will be attracted and collected by the drain because $V_{\text{ds}} > V_x$, which builds a large electric field across the very narrow pinch-off region.

Figure 1.16 shows the $I_{\text{ds}} = f(V_{\text{ds}})$ characteristic for various gate voltages. If $V_{\text{ds}} = 0\,\text{V}$, then $I_{\text{ds}} = 0\,\text{A}$. If V_{ds} is less than $V_{\text{gs}} - V_{\text{T}}$, then the transistor operates in the triode region and the current I_{ds} displays an almost linear relationship with V_{ds}. Current I_{ds} increases to its saturation value when $V_{\text{ds}} = V_{\text{gs}} - V_{\text{T}}$. Further increases of V_{ds} above

$V_{gs} - V_T$ no longer cause increases in I_{ds}. The transition between the triode and saturation regions is characterised by the curve $V_{ds} = V_{gs} - V_T$.

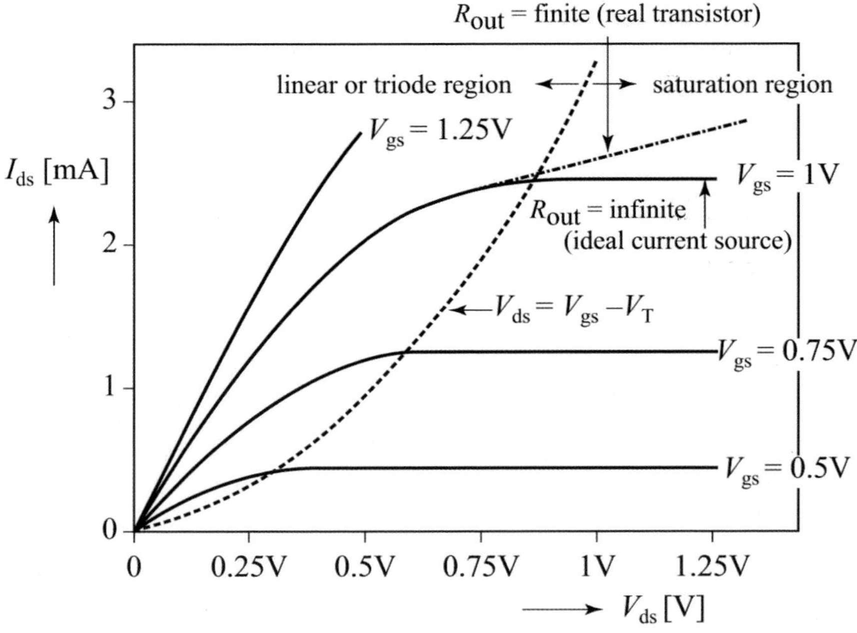

Figure 1.16: *The $I_{ds}{=}f(V_{ds})$ characteristic for various values of V_{gs}*

1.4   Derivation of simple MOS formulae

The inversion layer nMOS transistor shown in figure 1.17 has a width W perpendicular to the plane of the page and an oxide capacitance C_{ox} per unit area. A commonly-used unit for C_{ox} is fF/μm^2, where $1\,\text{fF} = 10^{-15}\,\text{F}$.

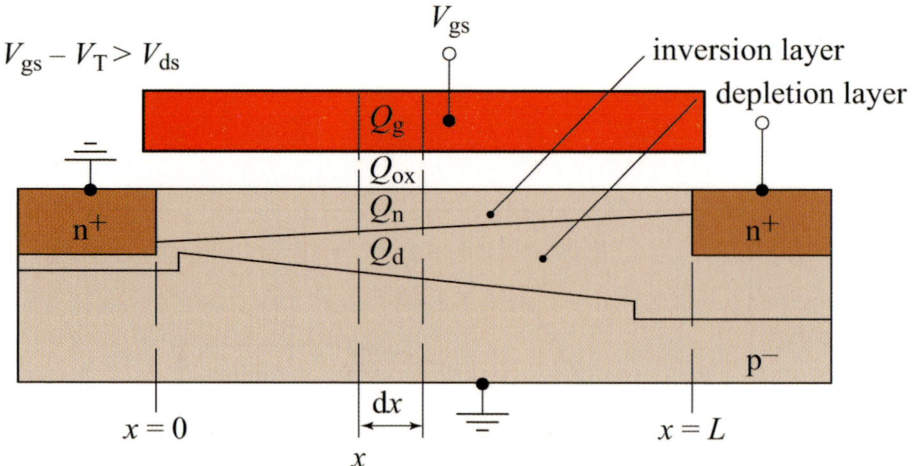

Figure 1.17: *Charges in a MOS transistor operating in the linear region*

Based on the *law for conservation of charge*, the following equality must hold at any position x between the source and drain:

$$Q_g + Q_{ox} + Q_n + Q_d = 0. \tag{1.3}$$

The components in this equation are charges per unit area, specified as follows:

Q_g = the gate charge [C/m^2];
Q_{ox}= primarily a small fixed charge which in practice always appears to be present in the thin gate oxide [C/m^2];
Q_n = the mobile charge in the inversion layer [C/m^2];
Q_d = the fixed charge in the depletion layer [C/m^2].

For gate voltages larger than V_T, the inversion layer shields the depletion layer from the gate. The charge in the depletion layer can then be considered constant:

$$Q_{ox} + Q_d = -C_{ox} \cdot V_T \tag{1.4}$$

The threshold voltage V_T is assumed to be constant. The potential in the channel at a position x is $V(x)$. With $Q_g = C_{ox}[V_{gs} - V(x)]$ and substituting (1.4) into (1.3) yields:

$$Q_n = -C_{ox}[V_{gs} - V_T - V(x)]$$

The total mobile charge $\mathrm{d}Q_\mathrm{m}$ in a section of the channel with length $\mathrm{d}x$ is defined as:

$$\mathrm{d}Q_\mathrm{m} = Q_\mathrm{n} \cdot W \cdot \mathrm{d}x = -W \cdot C_\mathrm{ox}[V_\mathrm{gs} - V_\mathrm{T} - V(x)] \cdot \mathrm{d}x \tag{1.5}$$

$$\Rightarrow \frac{\mathrm{d}Q_\mathrm{m}}{\mathrm{d}x} = -W \cdot C_\mathrm{ox}[V_\mathrm{gs} - V_\mathrm{T} - V(x)] \tag{1.6}$$

The drain current I_ds is expressed as:

$$I_\mathrm{ds} = \frac{\mathrm{d}Q_\mathrm{m}}{\mathrm{d}t} = \frac{\mathrm{d}Q_\mathrm{m}}{\mathrm{d}x} \cdot \frac{\mathrm{d}x}{\mathrm{d}t} \tag{1.7}$$

where $\frac{\mathrm{d}Q_\mathrm{m}}{\mathrm{d}x}$ is defined in equation (1.6) and
$\frac{\mathrm{d}x}{\mathrm{d}t}$ is the velocity v at which the charge Q_m moves from the source to the drain region.

This is the velocity of the electrons in the inversion layer and is expressed as:

$$v = \mu_\mathrm{n} \cdot E = -\mu_\mathrm{n} \cdot \frac{\mathrm{d}V(x)}{\mathrm{d}x} \tag{1.8}$$

where E is the electric field strength and μ_n represents the electron mobility in the inversion layer. The mobility represents the ease in which charge carriers move within a semiconductor.

In practice, the effective mobility appears to be less than one third of the electron mobility in the substrate (see section 2.3). Combining equations (1.6), (1.7) and (1.8) yields:

$$I_\mathrm{ds} = \mu_\mathrm{n} \cdot C_\mathrm{ox} \cdot W \cdot [V_\mathrm{gs} - V_\mathrm{T} - V(x)] \cdot \frac{\mathrm{d}V(x)}{\mathrm{d}x} \tag{1.9}$$

Substituting $\beta_\square = \mu_\mathrm{n} \cdot C_\mathrm{ox}$ yields:

$$I_\mathrm{ds} \cdot \mathrm{d}x = \beta_\square \cdot W \cdot [V_\mathrm{gs} - V_\mathrm{T} - V(x)] \cdot \mathrm{d}V(x) \tag{1.10}$$

Integrating the left-hand side from 0 to L and the right-hand side from 0 to V_ds yields:

$$I_\mathrm{ds} = \frac{W}{L} \cdot \beta_\square \cdot \left(V_\mathrm{gs} - V_\mathrm{T} - \frac{1}{2}V_\mathrm{ds}\right) \cdot V_\mathrm{ds} \tag{1.11}$$

Equation (1.11) has a maximum value when $V_{ds} = V_{gs} - V_T$. In this case, the current I_{ds} is expressed as:

$$I_{ds} = \frac{1}{2} \cdot \frac{W}{L} \cdot \beta_\square \cdot (V_{gs} - V_T)^2 \tag{1.12}$$

If $V_{gs} = V_T$ then $I_{ds} = 0\,\text{A}$. This clearly agrees with the earlier assumption that V_T is positive for an inversion-layer nMOS transistor. The term β is usually used to represent $\frac{W}{L} \cdot \beta_\square$. This factor is called the transistor *gain factor* and depends on geometry. The gain term $\beta_\square$ is a process parameter which depends on such things as the oxide thickness t_{ox}:

$$\beta_\square = \mu_n \cdot C_{ox} = \mu_n \cdot \frac{\epsilon_0 \epsilon_{ox}}{t_{ox}} \tag{1.13}$$

The unit of measurement for both β and $\beta_\square$ is A/V^2. However, $\mu\text{A/V}^2$ and mA/V^2 are the most commonly-used units. For an n-channel MOS transistor, $\beta_\square$ varies from $360\,\mu\text{A/V}^2$ to $750\,\mu\text{A/V}^2$ for oxide thicknesses of $3.2\,\text{nm}$ and $1.6\,\text{nm}$, respectively. Note that these values for $\beta_\square$ resemble the zero-field mobility in the substrate. The effective mobility in the channel, and so the effective gain factor, is much lower due to several second order effects as discussed in chapter 2.

According to equation (1.11), I_{ds} would reach a maximum value and then decrease for increasing V_{ds}. In the discussion concerning figures 1.15 and 1.16, however, it was stated that the current remains constant for an increasing V_{ds} once $V_{ds} > V_{gs} - V_T$. The transistor has two operating regions which are characterised by corresponding expressions for I_{ds}. These regions and their I_{ds} expressions are defined as follows:

1. The linear or triode region. $0 < V_{ds} < V_{gs} - V_T$.

$$I_{ds} = \beta \cdot (V_{gs} - V_T - \frac{1}{2}V_{ds}) \cdot V_{ds} \tag{1.14}$$

2. The saturation region. $V_{ds} \geq V_{gs} - V_T$.

$$I_{ds} = \frac{\beta}{2} \cdot (V_{gs} - V_T)^2 \tag{1.15}$$

According to equation (1.15), I_{ds} is independent of V_{ds} in the saturation region. The output impedance dV_{ds}/dI_{ds} should then be infinite and the transistor should behave like an ideal current source. In practice, however, MOS transistors show a finite output impedance which is dependent on geometry . This is explained in chapter 2. Figure 1.16 shows both the ideal (theoretical) and the real current-voltage characteristics of a transistor with a threshold voltage $V_\mathrm{T} = 0.25\,\mathrm{V}$.

The $I_\mathrm{ds} = f(V_\mathrm{ds})|_{V_\mathrm{gs} = \text{constant}}$ curves in figure 1.16 are joined by the dotted curve $V_\mathrm{ds} = V_\mathrm{gs} - V_\mathrm{T}$ at the points where equation (1.14) yields maximum values for I_ds. This curve divides the $I_\mathrm{ds} - V_\mathrm{ds}$ plane into two regions:

1. Left of the dotted curve: the triode or linear region, which is defined by equation (1.14);

2. Right of the dotted curve: the saturation region, which is defined by equation (1.15).

1.5   The back-bias effect (back-gate effect, body effect) and the effect of forward-bias

The simple MOS formulae derived in section 1.4 appear to be reasonably satisfactory in most cases. The very important *back-bias effect* is, however, not included in these formulae. This effect accounts for the modulation of the threshold voltage by the substrate bias and the subsequent effects on the drain current.

During normal operation (when $V_\mathrm{gs} > V_\mathrm{T}$ and $V_\mathrm{ds} > V_\mathrm{gs} - V_\mathrm{T}$) a depletion layer is formed, as shown in figure 1.15. However, the thickness of the depletion region under the channel increases when a negative *back-bias voltage* (V_sb) is applied to the bulk (b) with respect to the source. This is caused by the increased reverse-bias voltage across the fictive channel-substrate junction. The increased depletion layer requires additional charge. The channel charge therefore decreases if V_gs is held constant. The channel conductivity can only be maintained if V_gs is increased. The threshold voltage is therefore related to the back-bias voltage V_sb. This dependence is expressed as follows:

$$V_\mathrm{T} = V_\mathrm{x} + K\sqrt{V_\mathrm{sb} + 2\phi_\mathrm{f}} \tag{1.16}$$
$$V_\mathrm{T0} = V_\mathrm{x} + K\sqrt{2\phi_\mathrm{f}} \tag{1.17}$$

The terms in these formulae are as follows:

$V_\mathrm{x}\;=\;$ process-related constant threshold voltage term;

$$
\begin{aligned}
V_{\text{T0}} &= \left. V_{\text{T}} \right|_{V_{\text{sb}}=0\text{V}}; \\
K &= \text{process parameter equal to } \tfrac{1}{C_{\text{ox}}}\sqrt{2N_A q \epsilon_0 \epsilon_{\text{si}}}; \\
&\quad \text{also known as the \emph{`body factor'} or } K\text{-factor}; \\
N_A &= \text{substrate (bulk) dope concentration}; \\
V_{\text{sb}} &= \text{source-bulk (back-bias) voltage}; \\
2\phi_{\text{f}} &= \text{band bending where inversion first occurs.}
\end{aligned}
$$

The back-bias effect causes MOS transistors of the same type and dimensions to have different threshold voltages. Assume the nMOS inverter of figure 1.18 is designed in a 1.2 V 65 nm technology. Applying equation (1.16) yields the following equations for transistors T_1 and T_2, respectively:

$$
\begin{aligned}
V_{\text{T}1} &= V_{\text{x}} + K\sqrt{V_{\text{s}_1\text{b}} + 2\phi_{\text{f}}} \\
V_{\text{T}2} &= V_{\text{x}} + K\sqrt{V_{\text{s}_2\text{b}} + 2\phi_{\text{f}}}
\end{aligned}
$$

If the output is 'high' ($\approx 0.7\,\text{V}$), the source-bulk voltages of T_1 and T_2 are $V_{\text{s}_1\text{b}} = V_{\text{ss}} - V_{\text{bb}} = 0\,\text{V}$ and $V_{\text{s}_2\text{b}} = V_{\text{out}} - V_{\text{bb}} = 0.7\,\text{V}$, respectively. The source-bulk voltage can therefore cause the threshold voltage $V_{\text{T}2}$ of the upper transistor to be considerably larger than the threshold voltage $V_{\text{T}1}$ of the lower transistor.

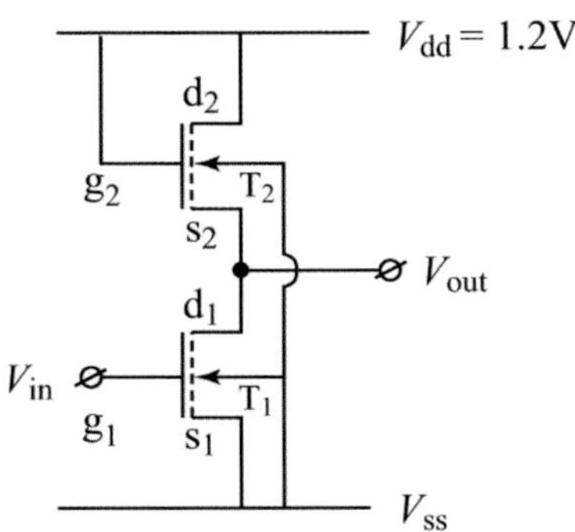

Figure 1.18: *nMOS-inverter with enhancement load*

Figure 1.19 shows the influence of the back-bias effect on different transistor characteristics. Formula (1.16) clearly shows that the threshold voltage V_{T} increases with an increasing back-gate voltage V_{sb}. For a constant V_{gs}, the drain-source current therefore decreases for an increasing V_{sb}. This is illustrated in figure 1.19b.

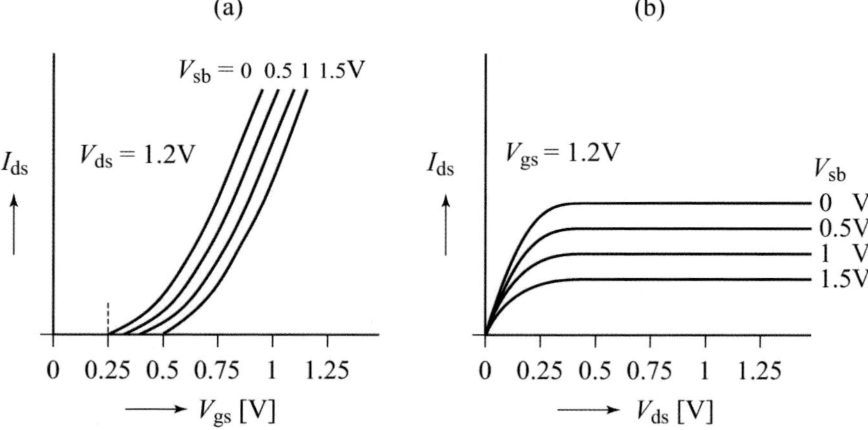

Figure 1.19: *Back-bias effect on MOS transistor characteristics:* (a) $I_{ds}= f(V_{gs})|_{V_{ds}=\text{const}}$ (b) $I_{ds}= f(V_{ds})|_{V_{gs}=\text{const}}$

Figure 1.20 shows the dependence of V_T on V_{sb}. The starting-point of this graph is determined by V_{T0} in equation (1.17) while its curve depends on the K-factor.

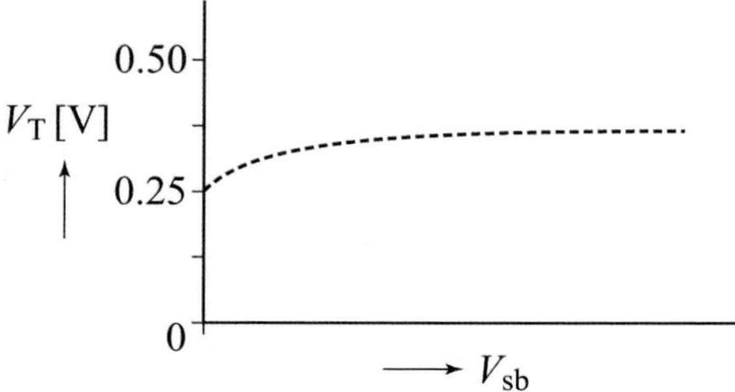

Figure 1.20: $V_T=f(V_{sb})$: *Threshold voltage as a function of source-bulk voltage*

The back-bias effect must be accurately treated when dimensioning MOS circuits. The most important reasons for using a back-bias voltage are as follows:

- Normally, the term V_x in equations (1.16) and (1.17) spreads more than the K-factor. The influence of the K-factor on the threshold voltage is larger when a back-bias voltage is applied. This results in a more stable threshold voltage.

- The depletion layer around the source and drain junctions of the MOS transistor becomes thicker as a result of the increased reverse voltage across these p-n junctions. This reduces the parasitic capacitances of the source and drain.

- Negative voltage pulses which may occur in dynamic MOS logic circuits may forward-bias the p-n diode between the substrate and a source or drain. Application of a negative voltage to the substrate virtually removes this possibility.

- Because an additional back-bias voltage increases the V_T, it reduces the subthreshold leakage current. This usage is described in chapter 8.

Next to the back-bias effect, there also exists a *forward-bias effect*. This effect is sometimes used to reduce the threshold voltage in order to improve the performance of certain logic or memory cores on a chip. In such cases, the source-to-substrate junction is put in forward bias, but with a voltage level below the junction voltage, which is equal to about $0.6\,\mathrm{V}$.

The MOS transistor formulae are summarised as follows:

$$
\begin{aligned}
-\ \text{linear region} \quad &: \quad I_{\mathrm{ds}} = \beta\left(V_{\mathrm{gs}} - V_{\mathrm{T}} - \frac{V_{\mathrm{ds}}}{2}\right)V_{\mathrm{ds}} \\[2ex]
-\ \text{saturation region} \quad &: \quad I_{\mathrm{ds}} = I_{\mathrm{ds_{sat}}} = \frac{\beta}{2}\left(V_{\mathrm{gs}} - V_{\mathrm{T}}\right)^2 \\[2ex]
\text{where} \quad & \quad V_{\mathrm{T}} = V_{\mathrm{x}} + K\sqrt{V_{\mathrm{sb}} + 2\phi_{\mathrm{f}}} \\[2ex]
\text{and} \quad & \quad V_{\mathrm{T0}} = V_{\mathrm{x}} + K\sqrt{2\phi_{\mathrm{f}}}
\end{aligned}
\tag{1.18}
$$

1.6 Factors which characterise the behaviour of the MOS transistor

The previously-discussed current-voltage characteristics represent the relationship between a transistor's current (I_{ds}) and its various applied voltages (V_{gs}, V_{ds} and V_{sb}). A number of important parameters which are frequently used to describe the behaviour of a transistor are explained below.

The *transconductance* g_m describes the relationship between the change ∂I_{ds} in the transistor current caused by a change ∂V_{gs} in the gate voltage:

$$g_m = \frac{\partial I_{ds}}{\partial V_{gs}}\Big|_{V_{ds} = \text{const}} \tag{1.19}$$

Referring to figure 1.16, it is clear that the value of g_m depends on the transistor's operating region:

$$\text{Linear region}: g_m = \beta \cdot V_{ds} \tag{1.20}$$

$$\text{Saturation region}: g_{m\text{sat.}} = \beta \cdot (V_{gs} - V_T) \tag{1.21}$$

Another parameter that characterises conduction in a transistor is its *output conductance*. In the transistor's linear operating region, this conductance (which is also called the channel conductance) is defined as:

$$g_{ds} = \left(\frac{dI_{ds}}{dV_{ds}}\right) = \{\beta(V_{gs} - V_T) - \beta V_{ds}\} \tag{1.22}$$

If V_{ds} is small, then:

$$g_{ds} = \beta(V_{gs} - V_T) \tag{1.23}$$

For an ideal MOS transistor operating in the saturation region, we have $\frac{dI_{ds}}{dV_{ds}} = 0$. The transistor current is then independent of V_{ds}. The output resistance is therefore infinite and the transistor acts as an ideal current source. In practice, however, the MOS transistor always has a finite output resistance and its current remains dependent on V_{ds}. This is illustrated in figure 1.16 and is treated in section 2.4.

We will now briefly discuss two other figure-of-merits, which represent the frequency response of a MOS transistor. If we ignore parasitic effects, the average transit time τ of a carrier across the channel L of an intrinsic device, operating in its linear region is equal to:

$$\tau = \frac{L}{\mu E} = \frac{L^2}{\mu V_{dd}} \tag{1.24}$$

and

$$\tau = \frac{L}{v_{sat}} \tag{1.25}$$

when the device is in velocity saturation (see chapter 2) and where v_{sat} equals the saturation speed of the carriers in the channel.

In most traditional CMOS technologies $v_{\text{sat}} \approx 10^{-11}\,\text{cm/s}$, so that $\tau \approx 10\,\text{ps}$ for a transistor with a channel length $L = 100\,\text{nm}$. This leads to the so-called cut-off (or threshold) frequency f_{T} at which the device can fill and empty the channel as a response to an ac signal:

$$f_{\text{T}} = \frac{1}{2\pi\tau} = \frac{v_{\text{sat}}}{2\pi L} = \frac{g_{\text{m}}}{2\pi C_{\text{gs}}} \tag{1.26}$$

with the average carrier velocity in the channel equal to $g_{\text{m}}/C_{\text{ox}} \approx g_{\text{m}}/C_{\text{gs}}$.

In saturation it is clear that $v_{\text{average}} = v_{\text{sat}}$. Scaling thus leads to a reduction of the transit time and to an increase of the transconductance and cut-off frequency.

1.7 Different types of MOS transistors

1. The previous discussions are all related to *n-channel* MOS transistors. The substrate material of these nMOS transistors is p-type and the drain and gate voltages are *positive* with respect to the source during normal operation. The substrate is the most *negative* electrode of an nMOS transistor.

2. *P-channel* MOS transistors are produced on an n-type substrate. The voltages at the gate and drain of these pMOS transistors are *negative* with respect to the source during normal operation. The substrate is the most *positive* electrode.

Generally, nMOS circuits are faster than those with pMOS transistors. The *power-delay* (τD) *product* of a logic gate is the product of its delay τ and dissipation D. The τD products of nMOS logic gates are lower than those of pMOS logic gates. This is because of the difference between the *mobility* of electrons and holes. Electron mobility is a factor of about 2.5 times higher than hole mobility in both the bulk silicon and inversion layers of the respective devices. Figure 2.1 illustrates this relationship, which is expressed as follows:

$$\mu_{\text{n}} \approx 2.5 \cdot \mu_{\text{p}}$$

The following relationship then follows from equation (1.13):

$$\beta_{\square\mathrm{n}} \approx 2.5 \cdot \beta_{\square\mathrm{p}}$$

An nMOS transistor therefore conducts approximately two and a half times as much current as a pMOS transistor of equal dimensions and with equal absolute voltages. Advanced CMOS technologies, today, apply different channel-stress and crystal-orientation techniques to improve the transistor drive currents and which also drives the performance of the pMOSts closer to that of the nMOSts. These are discussed in chapter 3.

Figure 1.21 shows a schematic overview of transistors which are distinguished on the basis of threshold voltage V_{T}. This distinction applies to both pMOS and nMOS transistors and results in the following types:

- *Enhancement* or *normally-off* transistors:
  No current flows through an enhancement transistor when $V_{\mathrm{gs}} = 0$. $V_{\mathrm{T}} > 0$ for an nMOS enhancement transistor and $V_{\mathrm{T}} < 0$ for a pMOS enhancement transistor.

- *Depletion* or *normally-on* transistors:
  Current flows through a depletion transistor when $V_{\mathrm{gs}} = 0$. $V_{\mathrm{T}} < 0$ for an nMOS depletion transistor and $V_{\mathrm{T}} > 0$ for a pMOS depletion transistor.

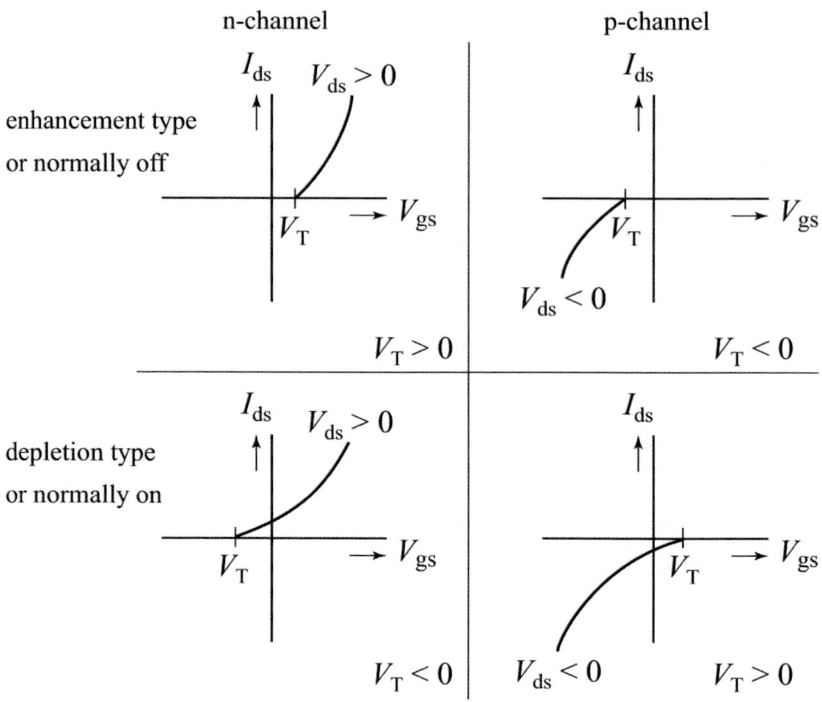

Figure 1.21: *Schematic overview of the different types of MOS transistors*

1.8   Parasitic MOS transistors

MOS (V)LSI circuits comprise many closely-packed transistors. This leads to the presence of parasitic MOS transistors, as illustrated in figure 1.22.

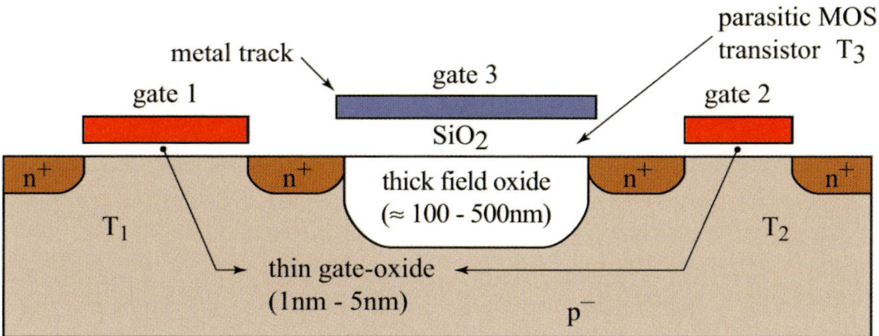

Figure 1.22: *Example of a parasitic MOS transistor*

Transistors T_1 and T_2 are separated by the *field oxide*. Parasitic MOS transistor T_3 is formed by a metal interconnection track on the field oxide and the n^+ areas of transistors T_1 and T_2. This field oxide is thick in comparison with the gate oxide, which ensures that the threshold voltage V_{Tpar} of transistor T_3 is larger than the threshold voltages of transistors T_1 and T_2. The field strength at the silicon surface in T_3 is therefore lower than in T_1 and T_2. Transistor T_3 will never conduct if its gate voltage never exceeds V_{Tpar}.

Many MOS production processes use an extra diffusion or ion implantation to artificially increase the threshold voltage V_{Tpar} of parasitic transistors. For this purpose, boron is used to create a p-type layer beneath the thick oxide in processes that use p^--type substrates. This makes it much more difficult to create an n-type inversion layer in these areas.

Processes that use n^--type substrates use phosphorus to increase $|V_{Tpar}|$. The terms *channel stopper implant* is used to refer to these boron and phosphorous implantations.

Note: Parasitic MOS transistors also appear in bipolar circuits. The absolute value of parasitic threshold voltages is always higher in n-type substrates than in p-type substrates. This is one of the reasons why planar IC technologies were mainly developed on n-epi layers.

1.9 MOS transistor symbols

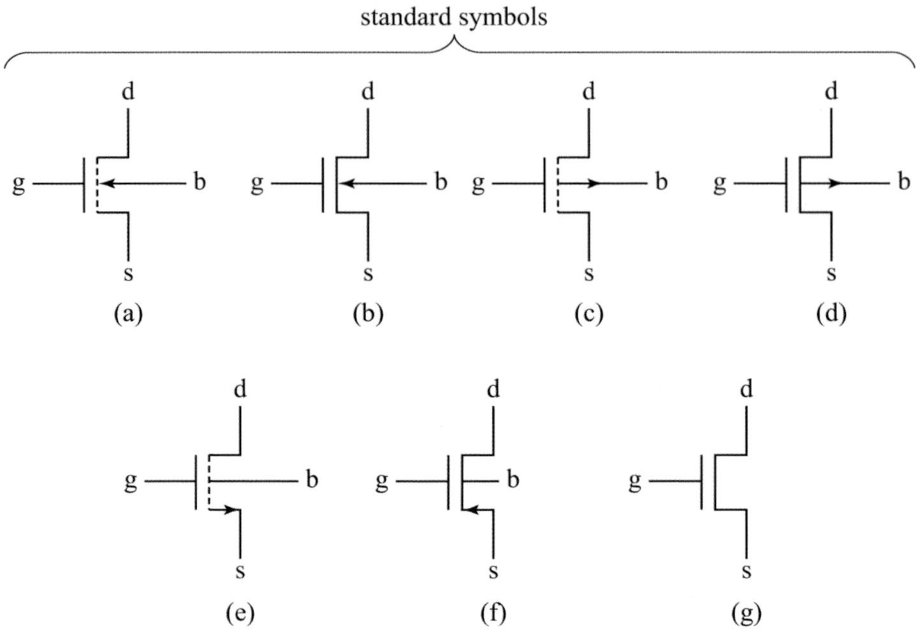

Figure 1.23: *Various transistor symbols*

Figure 1.23 shows various symbols used in literature to represent MOS transistors. Their meanings are as follows:

a) The inward pointing arrow indicates that the transistor is n-channel and the broken line between s and d indicates that it is an enhancement transistor.

b) The solid line from s to d indicates that this n-channel transistor is a depletion device.

c) The outward pointing arrow indicates that the transistor is p-channel and the broken line between s and d indicates that it is an enhancement transistor.

d) The solid line from s to d indicates that this p-channel transistor is a depletion device.

e) This symbol for an n-channel enhancement transistor is analogous to the npn transistor symbol.

f) This p-channel transistor is by definition not necessarily an enhancement type.

g) This general symbol represents a MOS transistor of any type.

Adaptations of the above symbols are also used. MOS symbols must therefore be interpreted with caution. The following rules are generally applied:

1. A transistor symbol with a broken line between its source and drain is always an enhancement or normally-off type;

2. Arrows indicate the forward directions of the substrate-channel 'junctions'.

The symbols in figure 1.24 are used throughout this book.

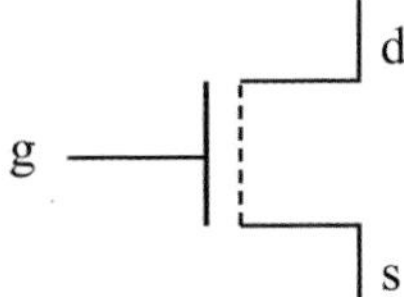

nMOS enhancement transistor

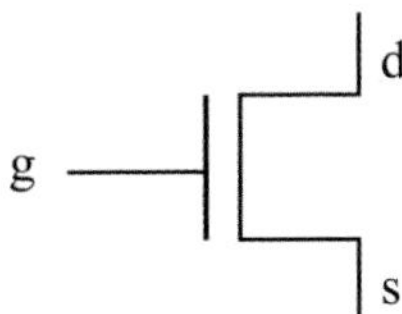

nMOS depletion transistor

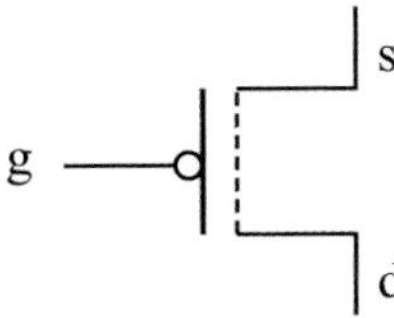

pMOS enhancement transistor

Figure 1.24: *Transistor symbols used throughout this book*

1.10 Capacitances in MOS structures

Figure 1.25 illustrates the MOS capacitance, whose value depends on such things as V_g and the frequency at which it varies. Section 1.3.1 describes the MOS capacitance and presents a qualitative discussion of its related charges, fields and voltages. Figure 1.26 shows a plot of the total capacitance C_t between the gate and ground terminals as a function of their voltage difference.

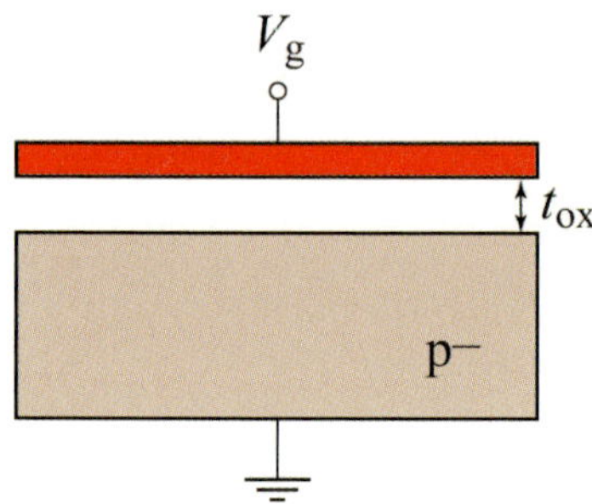

Figure 1.25: *The MOS capacitance*

The various regions of the *C-V curve* in figure 1.26 are explained as follows:

1. $V_g \ll V_T$ for a p-type substrate; $V_g \gg V_T$ for an n-type substrate. Here, the surface potential ϕ_s is highly negative and majority carriers in the p-type substrate will form a surface layer of holes. This accumulation layer is thin in comparison with the oxide thickness and exists as long as V_g is much smaller than V_T. Now, the silicon behaves like a metal plate, and the MOS capacitance is equal to the oxide capacitance C_{ox}. Deviations only appear at very high frequencies ($> 1\,\mathrm{GHz}$), where the *dielectric relaxation time* τ_R is important. For the $10\,\Omega\mathrm{cm}$ silicon, $\tau_R \approx 10\,\mathrm{ps}$ ($=10^{-11}\,\mathrm{s}$).

2. $V_g \approx V_T$, thus $\phi_s \approx 0...2\phi_f$.
   As V_g gradually becomes more positive, the accumulation layer decreases for a p-type substrate. A depletion layer is created under the gate when $\phi_s > 0$. A voltage change ΔV at the gate causes a change ΔQ in the charge at the edge of the depletion layer. In fact, the total capacitance is now determined by the series connection of the gate capacitance and the depletion layer capacitance. The capacitance therefore decreases.

3. $V_{\mathrm{g}} \gg V_{\mathrm{T}}$ for a p-type substrate; $V_{\mathrm{g}} \ll V_{\mathrm{T}}$ for an n-type substrate. Now, ϕ_{s} is highly positive and an inversion layer is created. This layer is thin compared to the oxide thickness. At low frequencies ($< 100\,\mathrm{kHz}$), the capacitance will again be equal to the oxide capacitance C_{ox}. However, the inversion layer for a p-type substrate consists of electrons that are supplied and absorbed by the substrate. This relies on the process of thermal generation and recombination of minorities, i.e., the electrons. With a constant temperature, the speed of the generation/recombination process is limited. This accounts for the lower capacitance shown in figure 1.26 at higher frequencies ($> 1\,\mathrm{MHz}$). At these high frequencies, the capacitance C_{t} will be about equal to the series connection of the gate capacitance and the depletion layer capacitance.

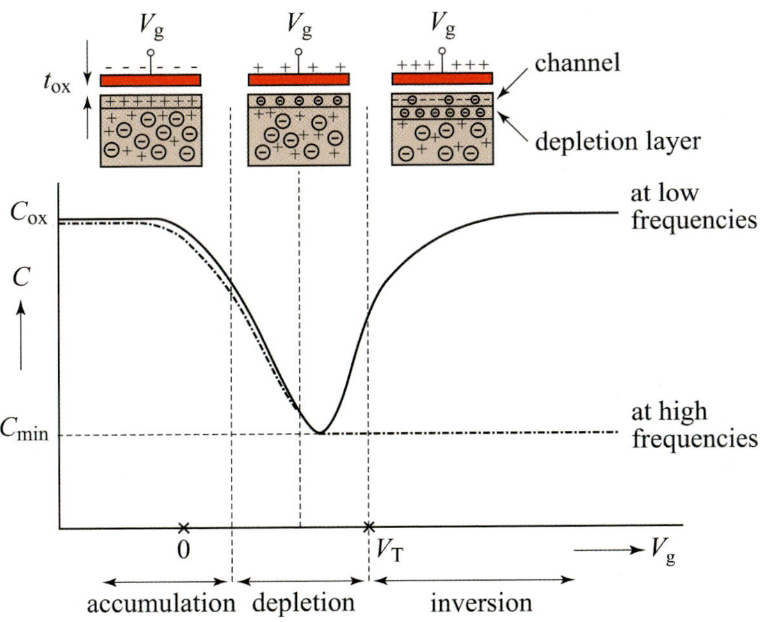

Figure 1.26: *Capacitance behaviour of a MOS structure*

As discussed, the MOS capacitance can be considered as a series connection of two capacitances: the oxide capacitance C_{ox} between the gate and the silicon surface and a capacitance C_{s} between the silicon surface and the substrate interior. This is explained below.

The voltage V_g can be expressed as follows:

$$V_g \;=\; V_{ox} + \phi_{ms} + \phi_s \qquad (1.27)$$

The law for conservation of charge yields the following equation:

$$Q_g + Q_{ox} + Q_n + Q_d \;=\; 0 \qquad (1.28)$$

where:

$$
\begin{aligned}
V_{ox} \;&=\; \text{voltage across the oxide between gate and silicon surfaces;}\\
\phi_{ms} \;&=\; \text{contact potential between gate and substrate;}\\
\phi_s \;&=\; \text{surface potential of the silicon with respect to the}\\
&\qquad \text{substrate interior;}\\
Q_g \;&=\; \text{charge on the gate;}\\
Q_{ox} \;&=\; \text{charge in the oxide;}\\
Q_n \;&=\; \text{charge in the inversion layer;}\\
Q_d \;&=\; \text{charge in the depletion layer.}
\end{aligned}
$$

The following expression for a change ΔV_g in gate voltage can be derived from equation (1.27):

$$\Delta V_g \;=\; \Delta V_{ox} + \Delta\phi_s \qquad (\phi_{ms} \text{ is constant, thus } \Delta\phi_{ms} = 0) \quad (1.29)$$

Substituting $Q_n + Q_d = Q_s$ in equation (1.28) yields:

$$\Delta Q_g \;=\; -\Delta Q_{ox} - \Delta Q_s \qquad (1.30)$$

If Q_{ox} is considered constant, then:

$$\Delta Q_g = -\Delta Q_s \qquad (1.31)$$

Equations (1.29) and (1.31) yield the following expressions:

$$\frac{\Delta V_g}{\Delta Q_g} = \frac{\Delta V_{ox}}{\Delta Q_g} + \frac{\Delta\phi_s}{\Delta Q_g} = \frac{\Delta V_{ox}}{\Delta Q_g} - \frac{\Delta\phi_s}{\Delta Q_s}$$

where:

$$\frac{\Delta Q_g}{\Delta V_g} \;=\; C_t \;=\; \text{the total capacitance of the MOS structure;}$$

$$\frac{\Delta Q_g}{\Delta V_{ox}} \;=\; C_{ox} = \text{oxide capacitance;}$$

$$-\frac{\Delta Q_s}{\Delta\phi_s} \;=\; C_s \;=\; \text{capacitance between the silicon surface and the}$$
$$\text{semiconductor interior (depletion layer capacitance).}$$

C_t can now be expressed as follows:

$$C_t = \left(\frac{1}{C_{\text{ox}}} + \frac{1}{C_s}\right)^{-1} \tag{1.32}$$

Capacitance C_s is responsible for the drop in the $C - V$ curve. The value of C_s is determined by the substrate doping concentration and the potential difference across the depletion layer. The minimum value C_{min} in the $C - V$ curve is also determined by C_{ox}. A smaller C_{ox} leads to a larger $\frac{1}{C_{\text{ox}}}$ and a smaller C_{min}. C_{min} can be as low as $0.1 C_{\text{ox}}$.

The $C - V$ curve is often used during MOS manufacturing processes to get a quick impression of the value of V_{T}.

Figure 1.27 shows a MOS capacitance with an additional n$^+$ area, which causes significant changes in the capacitance behaviour. The structure is in fact equivalent to a MOS transistor without a drain or to a MOS transistor with an external short circuit between its drain and source. This structure is generally called a *MOS capacitance* or a *MOS varactor*. Dynamic MOS circuits, in particular, use this device very often.

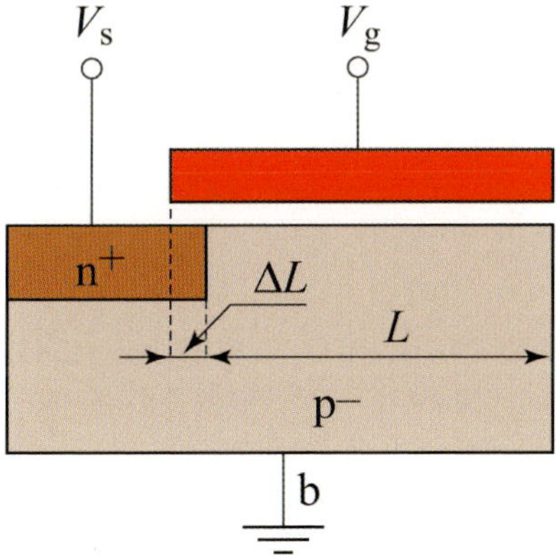

Figure 1.27: *MOS capacitance with source and/or drain area*

While $V_{\text{gs}} < V_{\text{T}}$, there is no inversion layer in a MOS capacitance, and the behaviour of the gate capacitance is unchanged. However, an inversion layer is created when $V_{\text{gs}} > V_{\text{T}}$. The electrons in this inversion layer are supplied by the n$^+$ area instead of by thermal generation/recombination processes of minorities in the substrate. This n$^+$ area can generate and absorb electrons at very high frequencies ($> 1\,\text{GHz}$). Therefore,

C_t will now equal C_{ox} under all normal operating conditions. In this case, C_t represents the capacitance between the gate and source, i.e., $C_t = C_{gs} = C_{ox}(L + \Delta L) \cdot W$.

The dependence of the capacitance C_{gs} on the applied voltage V_{gs} is summarised as follows:

- When $V_{gs} < V_T$, there is no inversion layer. Here, the value of C_{gs} is determined by the channel width W and the gate overlap ΔL on the source/drain area: $C_{gs} = \Delta L \cdot W \cdot C_{ox}$.

- When $V_{gs} > V_T$, there is an inversion layer. Here, C_{gs} is determined by the channel length L: $C_{gs} = (L + \Delta L) \cdot W \cdot C_{ox}$.

The above non-linear behaviour of $C_{gs} = f(V_{gs})$ is shown in figure 1.28.

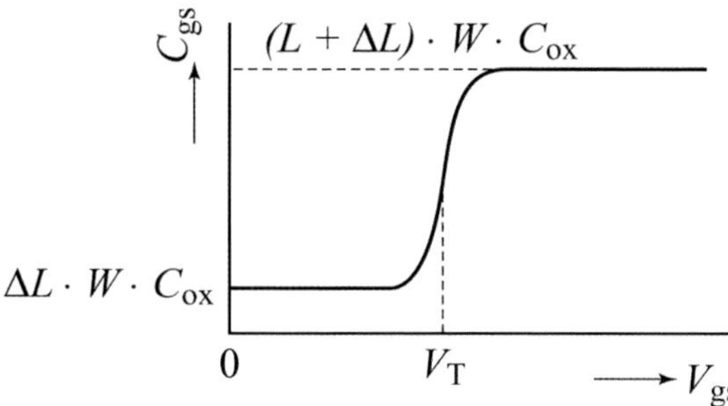

Figure 1.28: *Non-linear behaviour of a MOS capacitance*

Note: There is no inversion layer when $V_{gs} < V_T$. Figure 1.26 shows how the gate-substrate capacitance then behaves.

Figure 1.29 shows the large number of capacitances in a real MOS transistor. These capacitances, which are largely non-linear, are defined as follows:

C_{db}, C_{sb} : drain-substrate and source-substrate capacitances, which are non-linearly dependent on V_{db} and V_{sb}, respectively.

C_{gdo}, C_{gso} : gate-drain and gate-source capacitances, which are voltage-independent.

C_{gd}, C_{gs} : gate-drain and gate-source capacitances (via the inversion layer), which are non-linearly dependent on V_{gs},

V_{gd} and V_{gb}.

C_{gb} : gate-substrate capacitance, which is non-linearly dependent on V_{gb}.

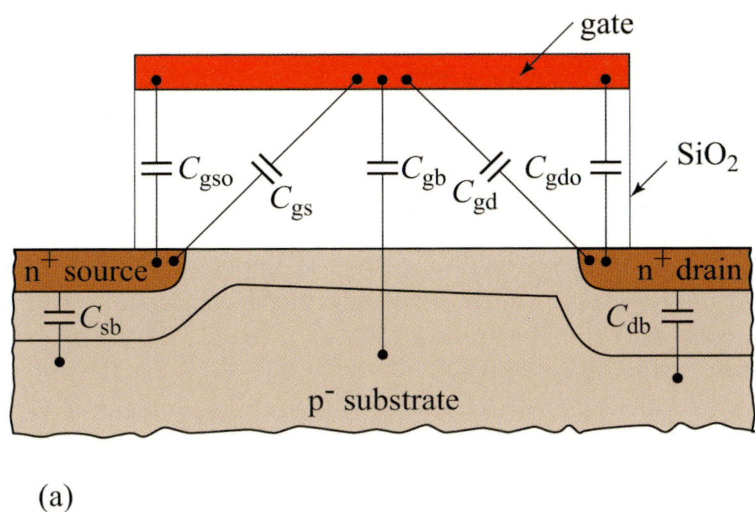

(a)

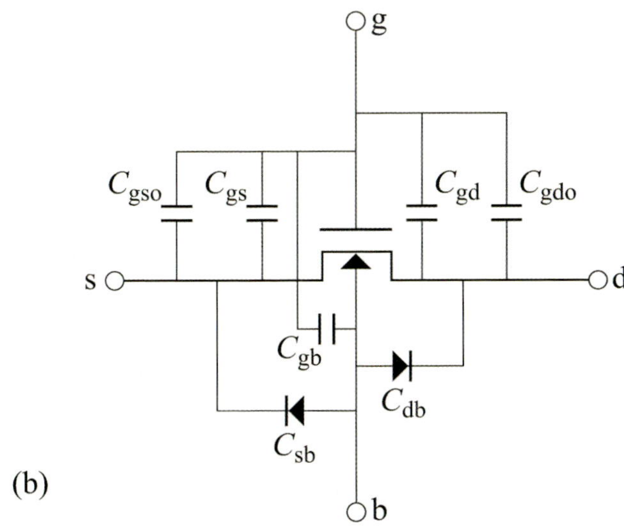

(b)

Figure 1.29: *Capacitances in a MOS transistor*

The values of the C_{db} and C_{sb} diode capacitances in figure 1.29 are expressed as follows:

$$C(V) = \frac{C_o}{(1 + \frac{V}{V_j})^{1/m}} \qquad (1.33)$$

where:

C_o = capacitance when $V=0$;
V_j = junction voltage (0.6 V to 0.9 V);
m = grading factor, $2 \leq m \leq 3$: $m = 2$ for an abrupt junction and $m = 3$ for a linear junction.

Terms C_{gdo} and C_{gso} represent gate overlap capacitances that are determined by the transistor width, the length of the overlap on the drain and source areas, and the thickness of the gate oxide. These capacitances are clearly voltage-independent.

The gate-substrate capacitance C_{gb} is only important if $V_{gs} \ll V_T$. Now, C_{gb} is often expressed as $C_{gb} \approx (0.12 \text{ to } 0.2) \cdot W \cdot L \cdot C_{ox}$. The inversion layer shields the substrate from the gate and $C_{gb}=0$ when $V_{gs} \geq V_T$.

Terms C_{gd} and C_{gs} represent gate-drain and gate-source capacitances, respectively, which are present via the inversion layer (figure 1.28). The values of these capacitances depend strongly on the bias voltage on the terminals of the MOS transistor. The following cases are distinguished:

Case a $V_{gs} < V_T$; no inversion layer, thus $C_{gd}=C_{gs}=0$.
Case b $V_{gs} > V_T$ and $V_{ds}=0$.
    For reasons of symmetry, $C_{gs}=C_{gd}=\frac{1}{2} \cdot W \cdot L \cdot C_{ox}$.
Case c $V_{gs} > V_T$ and $V_{ds} > V_{dsat}$ $(V_{dsat} = V_{gs} - V_T)$.
    The transistor is in saturation and there is no inversion layer at the drain: $C_{gd}=0$ and $C_{gs} = \frac{2}{3} \cdot W \cdot L \cdot C_{ox}$.
    This expression for C_{gs} is derived below.
Case d $V_{gs} > V_T$ and $0 < V_{ds} < V_{dsat}$.
    In this case, a linear interpolation between the values in cases b and c closely corresponds to the actual values, which are shown in figure 1.30.

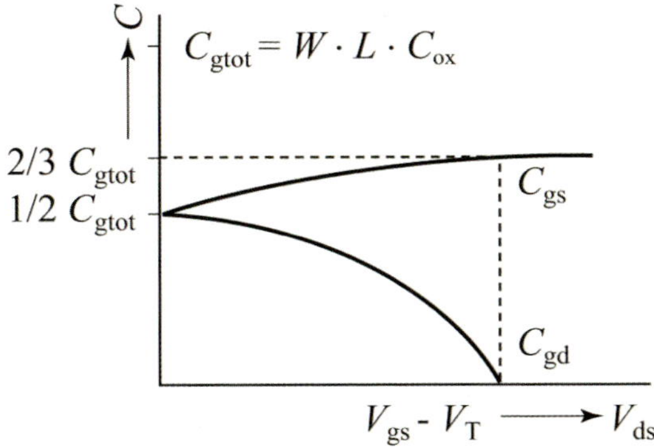

Figure 1.30: *C_{gs} and C_{gd} dependence on V_{ds} for $V_{\mathrm{gs}} > V_{\mathrm{T}}$*

The above expression in case c for the gate-source capacitance C_{gs} of a saturated MOS transistor is explained with the aid of figure 1.31. This figure shows a cross-section of a MOS transistor biased in the saturated region. The channel does not reach the drain area, but stops at a point where the channel potential is exactly $V_{\mathrm{gs}} - V_{\mathrm{T}}$.

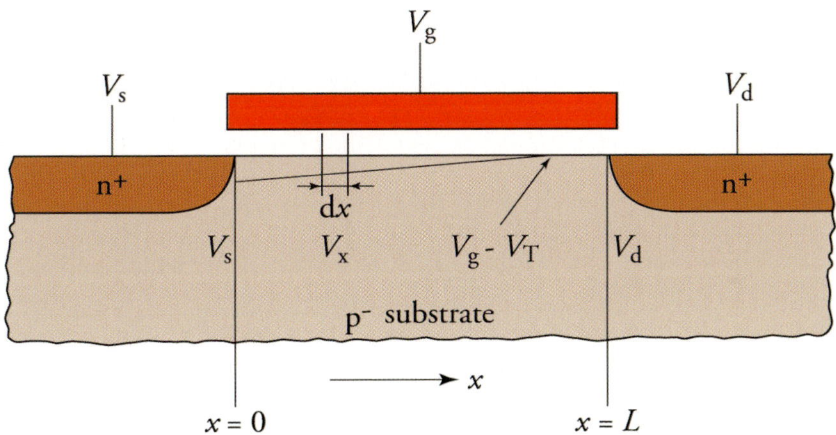

Figure 1.31: *Cross-section of a saturated MOS transistor. $C_{\mathrm{gd}} = 0$ and $C_{\mathrm{gs}} = \frac{2}{3} \cdot W \cdot L \cdot C_{\mathrm{ox}}$.*

Equation (1.5) leads to the following expression for the charge $\mathrm{d}Q$ in a channel section of length $\mathrm{d}x$ at position x:

$$\mathrm{d}Q(x) = Q_{\mathrm{n}} \cdot W \cdot \mathrm{d}x = -W \cdot C_{\mathrm{ox}}[V_{\mathrm{gs}} - V_{\mathrm{T}} - V(x)] \cdot \mathrm{d}x \qquad (1.34)$$

The following expression for $\mathrm{d}x$ is derived from equation (1.9):

$$\mathrm{d}x = \mu_n \cdot C_{ox} \cdot W \cdot [V_{gs} - V_T - V(x)] \cdot \frac{\mathrm{d}V(x)}{I_{ds}} \tag{1.35}$$

Combining equations (1.34) and (1.35) yields the following expression for $\mathrm{d}Q(x)$:

$$\mathrm{d}Q(x) = \frac{\mu_n \cdot C_{ox}^2 \cdot W^2 \cdot [V_{gs} - V_T - V(x)]^2}{I_{ds}} \cdot \mathrm{d}V(x) \tag{1.36}$$

Equation (1.15) yields the following expression for the drain current I_{ds} in a saturated MOS transistor:

$$I_{ds} = \frac{\beta}{2} \cdot (V_{gs} - V_T)^2 = \frac{\mu_n \cdot C_{ox}}{2} \cdot \frac{W}{L} \cdot (V_{gs} - V_T)^2 \tag{1.37}$$

Substituting equation (1.37) in equation (1.36) yields:

$$\mathrm{d}Q(x) = \frac{C_{ox} \cdot W \cdot L \cdot 2 \cdot [V_{gs} - V_T - V(x)]^2}{(V_{gs} - V_T)^2} \cdot \mathrm{d}V(x) \tag{1.38}$$

Integrating equation (1.38) from the source to the imaginary drain gives:

$$Q = \int_{V_s}^{V_{gs} - V_T} \frac{C_{ox} \cdot W \cdot L \cdot 2 \cdot [V_{gs} - V_T - V(x)]^2}{(V_{gs} - V_T)^2} \cdot \mathrm{d}V(x)$$

$$= \frac{C_{ox} \cdot W \cdot L \cdot 2}{(V_{gs} - V_T)^2} \cdot \left[-\frac{1}{3} \cdot [V_{gs} - V_T - V(x)]^3 \right]\Big|_{V_s}^{V_{gs} - V_T}$$

$$\Rightarrow Q = \frac{2}{3} \cdot W \cdot L \cdot C_{ox} \cdot (V_{gs} - V_T) \tag{1.39}$$

The gate-source capacitance C_{gs} can be found by differentiating Q in equation (1.39) with respect to V_{gs}:

$$C_{gs} = \frac{\mathrm{d}Q}{\mathrm{d}V_{gs}} = \frac{2}{3} \cdot W \cdot L \cdot C_{ox} \tag{1.40}$$

The C_{gs} of a saturated MOS transistor is therefore only two thirds of the total value, while the gate-drain capacitance is zero.

In summary:

Most capacitances in a MOS transistor are non-linearly dependent on the terminal voltages. For each capacitance, these dependencies are as follows:

1. The diode capacitances C_{db} and C_{sb}:

$$C(V) = \frac{C_{\mathrm{o}}}{(1+\frac{V}{V_{\mathrm{j}}})^{1/m}}, \quad \text{where } V_{\mathrm{j}} \approx 0.6 \ldots 0.9\,\text{V and } 2 \leq m \leq 3.$$

2. Figure 1.28 shows the voltage dependence of gate-channel capacitances C_{gd} and C_{gs} when the drain and source are short circuited, as is the case in a MOS capacitance. Figure 1.30 shows the voltage dependence of C_{gd} and C_{gs} when the drain and source are at different voltages, i.e., during normal transistor operation.

3. The gate-substrate capacitance C_{gb} is 0 when $V_{\mathrm{gs}} > V_{\mathrm{T}}$ and $C_{\mathrm{gb}} = 0.2 \cdot W \cdot L \cdot C_{\mathrm{ox}}$ if $V_{\mathrm{gs}} < V_{\mathrm{T}}$.

4. The overlap capacitances C_{gdo} and C_{gso} are the only capacitances which are not dependent on the terminal voltages.

1.11   Conclusions

The basic principles of the operation of the MOS transistor can be explained in different ways. The fairly simple approach adopted in this chapter should provide a good fundamental understanding of this operation. The current-voltage characteristics presented are derived by means of the simplest mathematical expressions for MOS transistor behaviour.

Second-order and parasitic effects are not essential to an understanding of the basic principles of MOS transistor operation. They have therefore been neglected in this chapter. However, these effects should be included in accurate descriptions of MOS transistors and are therefore discussed in chapter 2. Most of these effects are included in the MOS transistor models used by commonly-used compact MOS models in circuit simulation programs.

1.12 References

<u>General basic physics</u>

[1] R.S.C. Cobbold,
'Theory and applications of field effect transistors',
Wiley, New York

[2] S.M. Sze,
'Semiconductor Devices: Physics and Technology', 2nd Edition (Illustrated),
John Wiley & Sons Inc., September 2001

[3] Y.P. Tsividis,
'Operation and modelling of the MOS transistor',
WCB Mc Graw-Hill, Boston 1999

[4] C. Kittel,
'Introduction to Solid State Physics (7th edition)',
Wiley, 1996, New York

<u>MOS capacitances</u>

[5] E.W. Greenwich,
'An Analytical Model for the gate Capacity of Small-Geometry MOS structures',
IEEE Transactions on Electron Devices,
ED-30, pp 1838-1839, 1983

[6] J.J.Paulos, D.A. Antoniadis, and Y.P. Tsividis,
'Measurement of Intrinsic Capacitances of MOS Transistors',
ISSCC Digest of technical papers, pp 238-239, 1982

[7] D.E. Ward and R.W. Dutton,
'A Charge-Oriented Model for MOS Transistor Capacitances',
IEEE Journal of Solid-State Circuits, pp 703-707, 1978

[8] H. Kogure, et al.,
'Analysis of CMOS ADC Non-linear Input Capacitance',
IEICE Trans. Electron., Vol. E85-C, No. 5, May 2002

1.13 Exercises

Note: $2\phi_f=1\,\text{V}$ throughout these exercises.

1. What happens to the depletion layer in figure 1.12 when the substrate (b) is connected to a negative voltage ($\approx -1\,\text{V}$) instead of ground?
   What effect does this have on the threshold voltage V_T ?

2. Current I_{ds} in a transistor ($\frac{W}{L}=2$) is $100\,\mu\text{A}$ when its gate-source voltage V is $0.8\,\text{V}$. The current is $324\,\mu\text{A}$ when $V = 1.2\,\text{V}$.

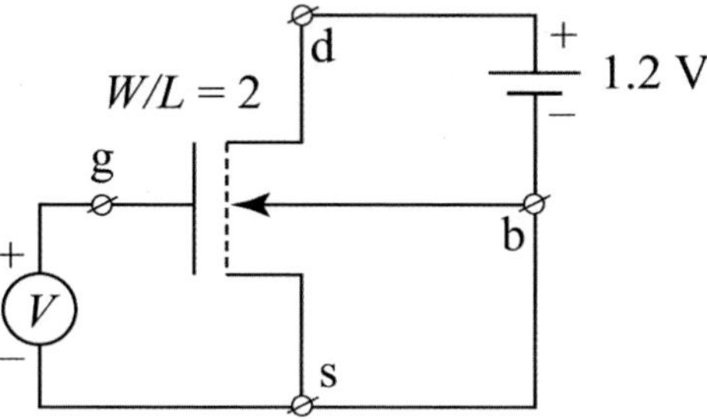

   a) Which transistor operating regions (linear or saturated) do these values of V correspond to?

   b) Calculate $\beta_\square$ and V_T for the given transistor.

3. Given:

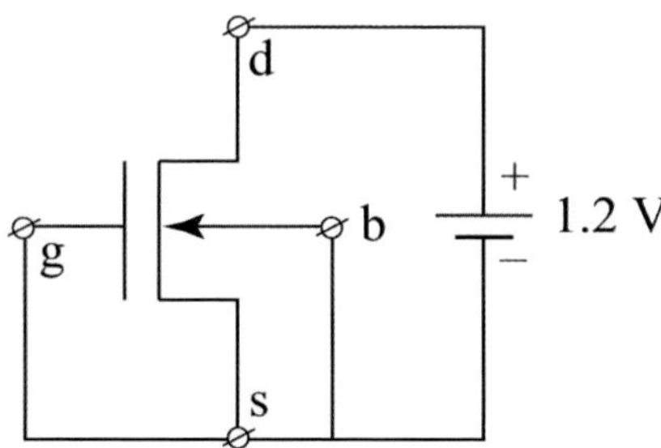

   a) What type is the transistor shown?

   b) Calculate I_{ds} when this transistor has the same β as the transistor in exercise 2 and $V_T= -1\,\text{V}$.

4. Given:

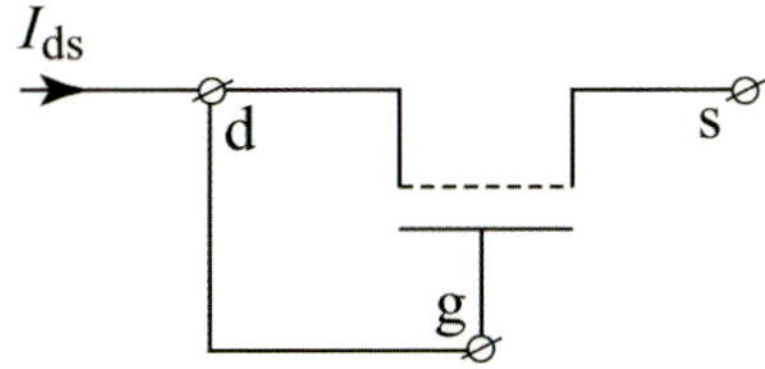

If this is an n-type enhancement MOS transistor and the current $I_{ds} > 0$, explain the following:

a) This transistor is always in its saturation region.

b) This connection is often called a MOS diode.

5. For this exercise, the threshold voltage V_T is 0.25 V. There is *no* thermal generation of electron/hole pairs.

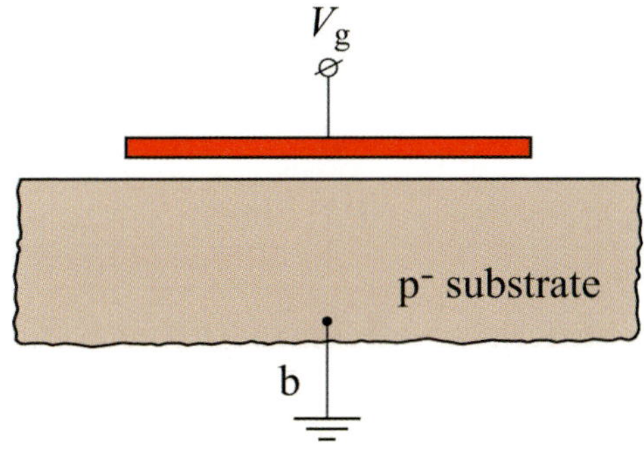

a) The above structure exists when the source and drain areas of an nMOS transistor are excluded.
Copy this structure and include the possible depletion and inversion layers for the following values of V_g: -0.6 V, 0.1 V, 0.6 V and 1.2 V.

b) An n⁺ area is now added to the structure in exercise 5a.

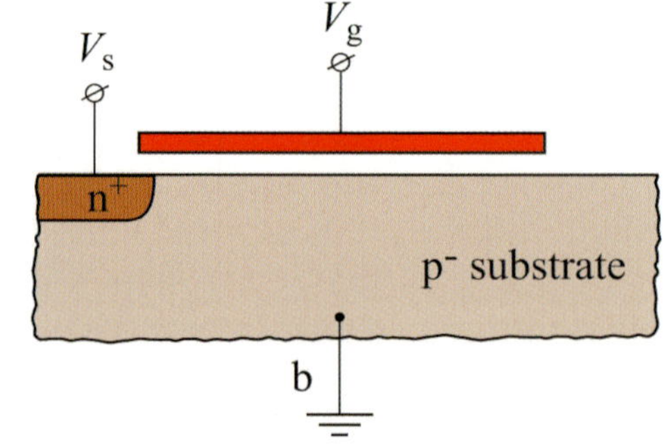

Repeat exercise 5a for $V_s = 0\,\text{V}$ and for $V_s = 0.5\,\text{V}$.

c) The substrate of the structure in exercise 5b is connected to a negative voltage: $V_{bb}=-1\,\text{V}$.
   What happens to the depletion and inversion layers if $V_s = 0\,\text{V}$ and $V_g = 0.5\,\text{V}$?

d) A second n^+ area is added to the structure of exercise 5b to yield the following structure.

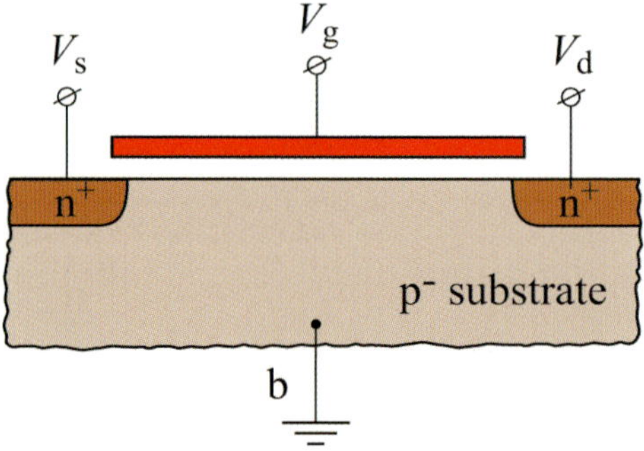

Repeat exercise 5a for $V_s = 0\,\text{V}$ and $V_d = 0.8\,\text{V}$.

e) In practice, there are thermally-generated electron hole pairs in the silicon substrate. The resulting free electrons in the depletion layer move in the opposite direction to the applied external electric field.
   Draw the direction of movement of the thermally-generated electrons and holes for $V_g=1.2\,\text{V}$ in the structure of exercise 5a.
   If this situation continues for a longer period, a new equilibrium is reached and the electrons and holes accumulate in the structure. Draw this situation.

6. The following values apply
in the figure shown:
V_{dd}=1.2 V, β=1 mA/V^2,
V_x=−1 V, V_{bb}=−1 V.

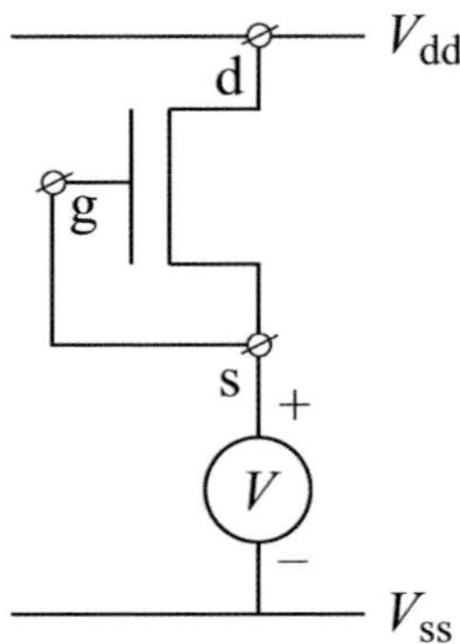

a) What type is the transistor and why?

b) Calculate and draw the graph $I_{ds}=f(V_{ds})$ for $K=0\,V^{1/2}$ and $V_{ds}=0$, 0.2, 0.4, 0.6, 0.8, 1.0 and 1.2 V.

c) Repeat b) for $K=0.2\,V^{1/2}$.

d) Assuming $K=0.2\,V^{1/2}$, calculate the output impedance of the transistor for $V_{ds}=50\,mV$ and for $V_{ds}=0.6\,V$.
(Note: the drain remains at 1.2 V).

7. The following values apply
for the circuit shown:
$V_{dd} = 1.2$ V, $V_{bb} = -1$ V,
$V_{ss} = 0$ V, $K=0.2\,V^{1/2}$,
$\beta_\square = 400\,\mu A/V^2$, $V_{X_L} = -1$ V
and $V_{X_D} = 0.2$ V.

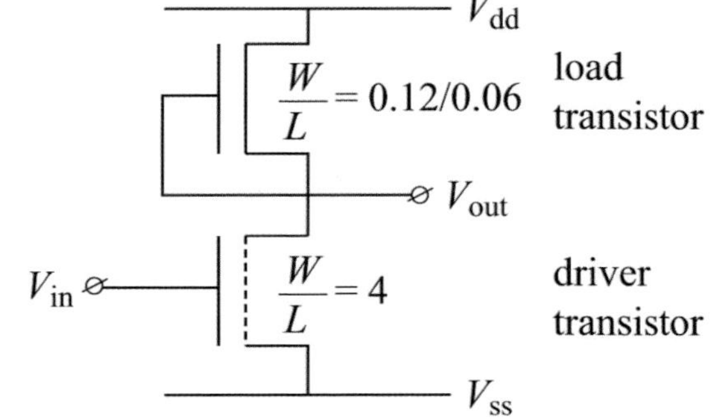

a) Calculate V_{out} for $V_{in}=1.2\,V$.

b) Determine the transconductance of both MOS transistors for this situation.

c) What value does V_{out} reach when $V_{in}=0.1\,V$?

d) The same low output level must be maintained when the load transistor is replaced by an enhancement-type transistor of the same size and with its gate at V_{dd}. Does this require a driver transistor with the same $\frac{W}{L}$ and with a smaller or a larger channel width W? Explain your answer.

8. The aspect ratio of this transistor is $W/L = 200\text{nm}/50\text{nm}$. Results of measurements on it are summarised in the following table:

$V_{sb}[\text{V}]$	$I_{ds}[\mu\text{A}]$	
	$V_{gs}=0.5\text{V}$	$V_{gs}=1\text{V}$
0	40	360
1.25	10	–

   a) Determine V_x, K and $\beta_\square$ for this transistor.

   b) Calculate and draw the graph $V_T=f(V_{sb})$ for at least five V_{sb} values $(0\,\text{V}<V_{sb}<2\,\text{V})$.

9. Define an expression for the transconductance with respect to the substrate voltage V_{sb} when the transconductance with respect to the normal gate voltage is defined as $g_m=\dfrac{\delta I_{ds}}{\delta V_{gs}}$.

10. Assume that we build a decoupling capacitor between V_{dd} and V_{ss} using an nMOS transistor with a gate-oxide thickness $t_{ox} = 1.6\,\text{nm}$.

   a) Draw how this nMOS transistor is connected between the V_{dd} and V_{ss} lines to form this capacitor.

   b) What would be its capacitance value per unit area?

   c) Assuming a pMOS transistor operates fully complementary to an nMOSt, how would you connect such a pMOSt as an additional capacitor in the same circuit as in a).

11. The following values apply in the figure shown: both nMOS transistors are identical, $V_{dd} = 1\,\text{V}$, $V_T = 0.3\,\text{V}$ when $k = 0\,\text{V}^{1/2}$.

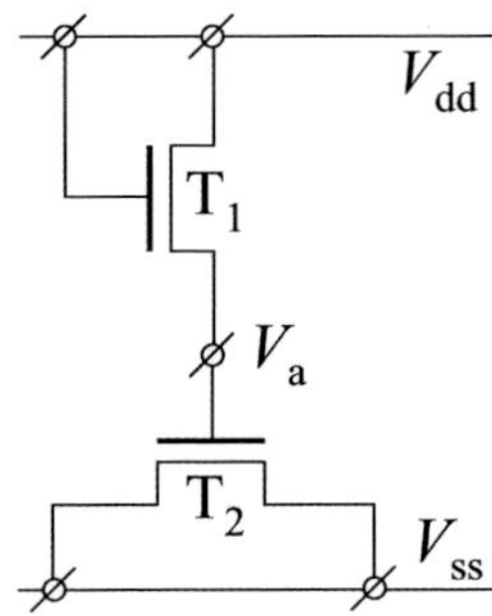

a) With what type of device could you compare T_1?

b) With what type of device could you compare T_2?

c) What would be the voltage V_a when $k = 0\,\text{V}^{1/2}$?

d) What would be the voltage V_a when $k = 0.2\,\text{V}^{1/2}$?

e) If one of the V_{ss} connections of T_2 would be left open (floating), what would be the result in terms of operation of the device T_2 and of the operation of the total circuit?

Chapter 2

Geometrical-, physical- and field-scaling impact on MOS transistor behaviour

2.1　Introduction

The simple formulae derived in sections 1.4 and 1.5 account for the first-order effects which influence the behaviour of MOS transistors. Until the mid-seventies, formulae (1.18) appeared quite adequate for predicting the performance of MOS circuits. However, these transistor formulae ignore several physical and geometrical effects which significantly degrade the behaviour of MOS transistors. The results are therefore considerably more optimistic than the actual performance observed in MOS circuits. The deviation becomes more significant as MOS transistor sizes decrease in VLSI circuits.

This chapter contains a brief overview of the most important effects, in nanometer CMOS technologies, which degrade the performance of MOS devices. The chapter concludes with a detailed discussion on transistor leakage mechanisms.

2.2   The zero field mobility

As discussed in chapter 1, the MOS transistor current is heavily determined by the *gain factor* β of the transistor:

$$\beta = \frac{W}{L} \cdot \beta_\square = \frac{W}{L} \cdot \mu \cdot C_{\text{ox}} \tag{2.1}$$

where W and L represent the transistor channel width and length respectively, C_{ox} represents the gate oxide capacitance per unit of area and μ represents the actual *mobility* of the carriers in the channel. This mobility can be quite different from the zero-field or substrate mobility μ_0, which depends on the doping concentration in the substrate. Figure 2.1 shows zero-field electron and hole mobilities in silicon at room temperature as a function of the doping concentration.

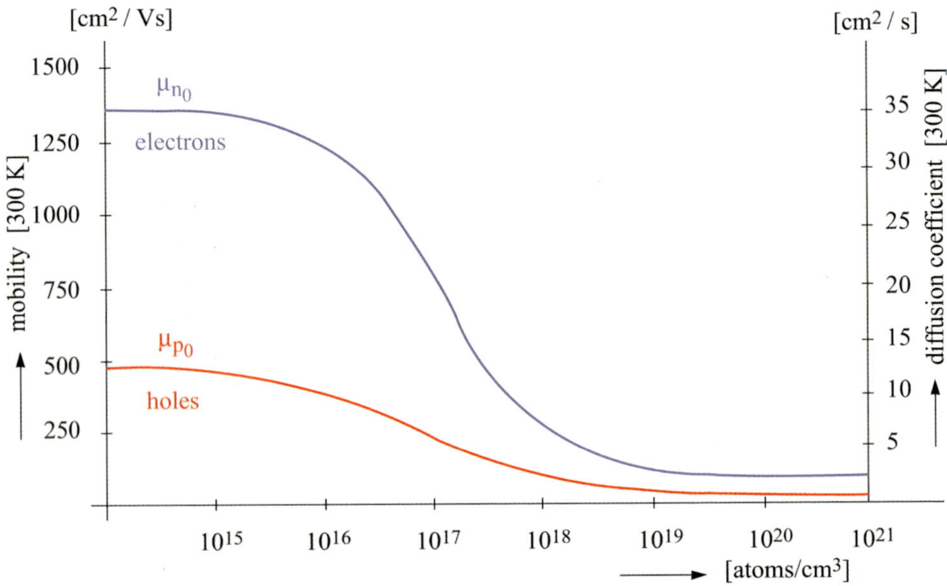

Figure 2.1:   *Zero-field carrier mobility and diffusion coefficient as a function of doping concentration in silicon at room temperature*

For a channel doping concentration of 10^{17} atoms/cm^3, the mobility of electrons (μ_{n_0}) is about three times that of holes (μ_{p_0}), in the absence of an electric field. This is the major reason that the I_{on} current (which is the saturation current when $V_{\text{gs}} = V_{\text{dd}}$) of an nMOS transistor is about two to four times higher than the I_{on} of an equally sized pMOS transistor, depending on the technology node. It also depends on the transistor stress engineering and crystal orientation. However, several other effects dramatically reduce the mobility of the carriers in the channel. These are discussed in section 2.3.

2.3   Carrier mobility reduction

During normal transistor operation, electrical fields are applied in both the lateral (horizontal) and transversal (vertical) directions, which influence the mobility of the carriers in the channel. Moreover, when the chip temperature is increased, either by an increase of the ambient temperature or by the chip's own dissipation, this will have a negative effect on the carrier mobility and thus on the β of each transistor.

2.3.1   Vertical and lateral field carrier mobility reduction

During normal operation, the effective mobility μ of the carriers in the transistor channel is degraded by the mechanisms indicated in figure 2.2. These include the *vertical electric field* E_z, the lateral electric field E_x and the carrier velocity v.

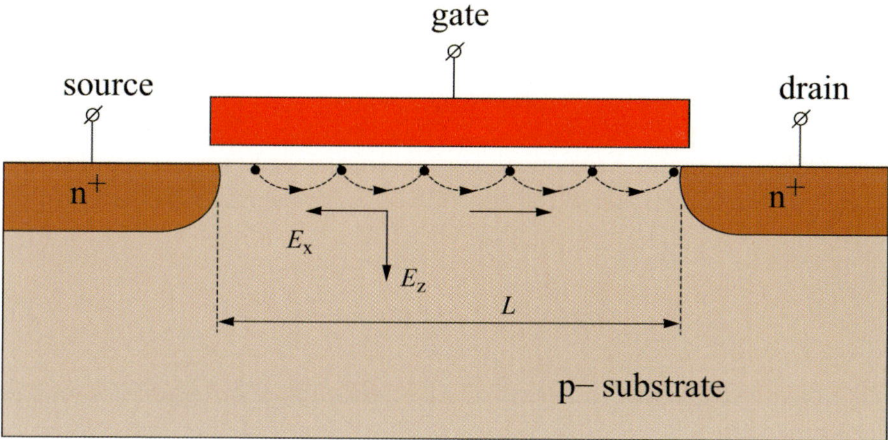

Figure 2.2: *Components which affect carrier mobility in MOS transistors*

When the vertical electric field E_z is high, the minority carriers in an n-channel device are strongly attracted to the silicon surface, where they rebound. The resulting *'surface scattering'* is indicated by the dashed lines in figure 2.2. This causes a reduction of the recombination time and of carrier mobility μ with increasing E_z.

In [1], some experimental results are presented with respect to the vertical field carrier mobility degradation. The vertical electric field depends on the gate voltage and on the substrate voltage. The relationship between these voltages and the mobility can be expressed as follows:

$$\mu = \frac{\mu_0}{1 + \theta_1(V_{gs} - V_T) + \theta_2(\sqrt{V_{sb} + 2\phi_F} - \sqrt{2\phi_F})} \tag{2.2}$$

where, μ_0 represents the zero-field substrate mobility, ϕ_F represents the Fermi level in the substrate and θ_1 and θ_2 are technology defined constants.

The carriers in the transistor channel are accelerated to a maximum velocity when the *lateral electric field* E_x is high. This means that, above a critical field $E_{x_{sat}}$ (figure 2.3), the carrier velocity is no longer related to E_x and reaches a constant level (v_{sat}). A good first-order approximation for this 'velocity saturation' phenomenon is:

$$\mu = \frac{\mu_0}{1 + E_x/E_{x_{sat}}} \tag{2.3}$$

where

$$E_x \approx \frac{V_{ds}}{L} \tag{2.4}$$

Substituting equation (2.4) into equation (2.3) yields:

$$\mu = \frac{\mu_0}{1 + \theta_3 \cdot V_{ds}} \tag{2.5}$$

$$\text{where} \quad \theta_3 = \frac{1}{L \cdot E_{x_{sat}}} \tag{2.6}$$

The above effects are included in the following expression for carrier mobility:

$$\mu = \frac{\mu_0}{(1 + \theta_1(V_{gs} - V_T) + \theta_2(\sqrt{V_{sb} + 2\phi_F} - \sqrt{2\phi_F}))(1 + \theta_3 V_{ds})} \tag{2.7}$$

At high gate voltages, the vertical field influence (represented by the voltage terms containing V_{gs} and V_{sb}) may reduce the transistor current by about 50 %. The lateral field influence may be of the same order of magnitude. Note that this lateral field close to the source dominates the drain-source current. At a level of about $1\,\text{V}/\mu\text{m}$, this lateral field also reduces the electron mobility in the channel of an nMOS transistor by almost 50 %. Thus, the total field-dependent mobility reduction can amount to a factor four.

The actual mobility is equal to the substrate mobility when $E_z = 0$. Some transistor models include the series resistance of the source (R_s) and the drain (R_d) in the surface scattering factor θ_1 and in the velocity-saturation factor θ_3. Moreover, these resistances are weakly dependent on the terminal voltages and are therefore included in the device equations. Their influence can be incorporated in equation (2.7) by replacing θ_1 and θ_3 by $\theta_1{}'$ and $\theta_3{}'$, respectively, where

$$\theta_1' \;=\; \theta_1 + \beta \cdot (R_s + R_d) \tag{2.8}$$

and

$$\theta_3' \;=\; \theta_3 - \beta \cdot R_d \tag{2.9}$$

Because these resistances are in series with the transistor terminals, they have a reducing effect on the transistor current.

Due to the ultra-short channel lengths, today, transistors show electric fields that exceed $10\,\text{V}/\mu\text{m}$. Because $E_{\text{x}_{\text{sat}}}$ is smaller for electrons than for holes, the performance of nMOS transistors (conduction by electrons) is more degraded by velocity saturation than that of pMOS transistors (conduction by holes). Figure 2.3 shows the carrier velocity v as a function of the electric field E_{x} in the channel. The critical field $E_{\text{x}_{\text{sat}}}$ to reach velocity saturation depends on the doping level and is about $3\,\text{V}/\mu\text{m}$ for electrons and about $10\,\text{V}/\mu\text{m}$ for holes. In fact, holes may reach a saturation velocity comparable to the v_{sat} of electrons.

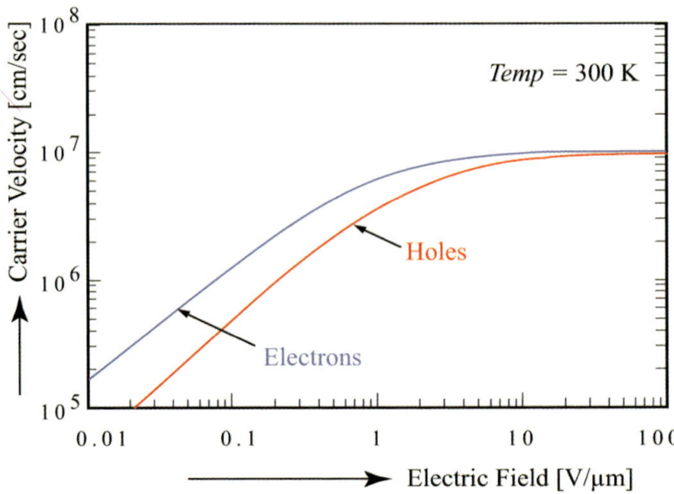

Figure 2.3: *Carrier velocity as a function of the lateral electric field in the channel*

This carrier velocity is defined by:

$$v = \mu \cdot E_\mathrm{x} \approx \mu \cdot V_\mathrm{ds}/L \qquad (2.10)$$

So, in the derivation of the current expression (1.12), we can replace $\mu \cdot V_\mathrm{ds}/L$ by v. Particularly in 90 nm CMOS processes and beyond, most carriers in the channel travel at a maximum *saturation velocity* v_sat. This would lead to a saturation current equal to:

$$I_\mathrm{ds} = \frac{v_\mathrm{sat} C_\mathrm{ox} W}{2} \cdot (V_\mathrm{dd} - V_\mathrm{T}) \qquad (2.11)$$

This reduces the channel length's influence on the current, which is one of the reasons that the transistor's *drive current* has shown negligible increase over the last couple of technology generations. This is not expected to change for future generations; for almost all CMOS processes from 180 nm to 45 nm, the I_on for the nMOS and pMOS transistors remain approximately equal to 650 mA/μm and 270 mA/μm, respectively. As a result, the effective gain factor $\beta_{\square\mathrm{eff}}$ of a transistor with minimum drawn gate length (L_drawn) almost remains at a constant value, which is close to or somewhat below 100 μA/V^2. Figure 2.4 shows this gain factor as a function of the drawn channel length for various technology generations [2]. *LSTP* and *LOP* refer to a *low standby power* and a *low operating power*, respectively.

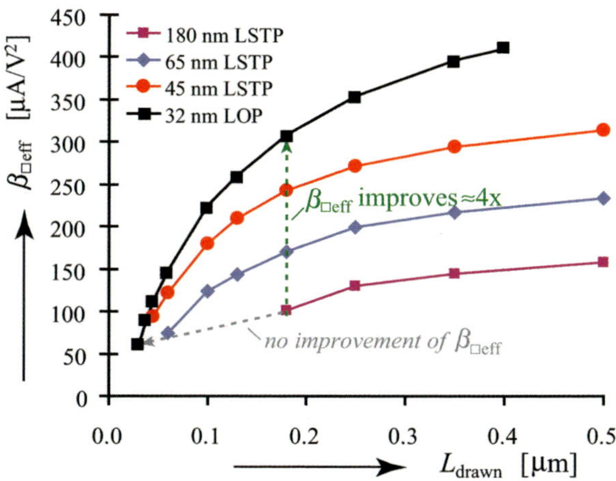

Figure 2.4: *The effective gain factor versus the drawn channel length for various technology generations [2]*

It also shows that $\beta_{\square\text{eff}}$ improves dramatically with increasing L_{drawn}. For example an increase of L_{drawn} in the 45 nm CMOS node from minimum length to about 100 nm, will double the value of $\beta_{\square\text{eff}}$. For digital circuits this channel length increase is certainly not an appropriate solution for performance increase, because it also increases the chip area. However, if we compare a minimum drawn transistor ($L_{\text{drawn}} = 0.18\,\mu\text{m}$) in $0.18\,\mu\text{m}$ technology with a transistor with the same $L_{\text{drawn}} = 0.18\,\mu\text{m}$ in 32 nm technology, the performance improvement is close to a factor of four. So, for analog circuits, which usually take a small portion of the chip area on a mixed analog/digital design, a larger channel length serves two goals. First it improves the transconductance ($g_m \sim \beta_{\square\text{eff}}(V_{\text{gs}} - V_{\text{T}})$) and, secondly, it reduces the threshold voltage variation, which improves the transistor *matching* properties and enables a better performance prediction. These variability-related topics are discussed in chapter 9.

Chapter 3 presents a few mobility enhancement techniques which will improve the performance of both the digital and analog circuits in advanced and future nanometer CMOS ICs.

2.3.2   Stress-induced carrier mobility effects

The mechanical stress induced by shallow-trench isolation (STI) has an increasing effect on the carrier mobility of the nMOS and pMOS devices.

It degrades the mobility of nMOS devices, while it slightly improves it of the pMOS. Chapter 3 discusses mobility enhancement techniques to improve transistor device performance in advanced CMOS technologies.

2.4   Channel length modulation

The ideal I_{ds} vs V_{ds} characteristics illustrated in figure 1.16 do not show the influence of V_{ds} on I_{ds} in the saturation region. In practice, an increase in V_{ds} in the saturation region causes an increase in I_{ds}. This phenomenon is particularly obvious in short-channel devices and is caused by *channel length modulation.*

The distribution of carriers in an nMOS transistor operating in the saturation region ($V_{ds} > V_{dssat} = V_{gs} - V_T$) is illustrated in figure 2.5. The operation of the basic MOS transistor in this region is discussed in section 1.3. Clearly, the end of the inversion layer (which is called the virtual drain) does not reach the actual drain. The effective channel length therefore equals $L - \Delta L_{CLM}$.

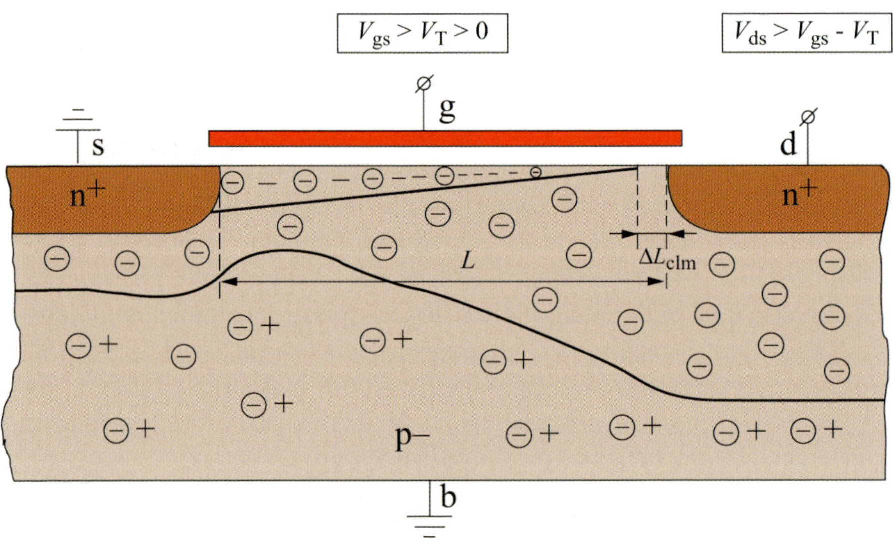

Figure 2.5: *A MOS transistor in the saturation region ($V_{ds} > V_{gs} - V_T$)*

The saturation current specified in equation (1.15) must be changed to account for the effective channel length. The modified expression is as shown in equation (2.12).

$$I_{\text{ds}_{\text{sat}}} = \frac{W}{L - \Delta L_{\text{CLM}}} \cdot \frac{\beta_\square}{2} \cdot (V_{\text{gs}} - V_{\text{T}})^2 \tag{2.12}$$

where ΔL is the length of the depletion region at the silicon surface between the inversion layer and the drain. In the above expression, the total field-dependent mobility degradation, as discussed before, is not included.

The voltage $V_{\text{ds}} - V_{\text{dssat}}$ across this *'pinch-off' region* modulates ΔL_{CLM}. This effect can be modelled by:

$$\frac{\Delta L_{\text{CLM}}}{L} = \alpha \ln \left(1 + \frac{V_{\text{ds}} - V_{\text{dssat}}}{V_{\text{P}}}\right) \tag{2.13}$$

where α and V_{P} are constants, which may vary with the transistor geometry. The expression clearly shows the relation between ΔL_{CLM} and the amount of V_{ds} voltage above V_{dssat}.

The above discussions show that the additional contribution to the drain current of a MOS transistor operating in the saturation region is proportional to $V_{\text{ds}} - V_{\text{dssat}}$. This effect is sometimes approximated by the following modified current expression:

$$I_{\text{ds}} = (1 + \lambda V_{\text{ds}}) \cdot I_{\text{ds0}} \tag{2.14}$$

Where I_{ds0} is the transistor current when the channel length modulation is ignored, and λ is a semi-empirical channel length modulation parameter, whose reciprocal value $(1/\lambda)$ is analogous to the BJT Early voltage. The effect of this channel length modulation on the $I_{\text{ds}} = f(V_{\text{ds}})$ characteristics is shown in figure 2.6, where the extrapolation of the curves in the saturation region would all intersect the x-axis closely to the point $1/\lambda$.

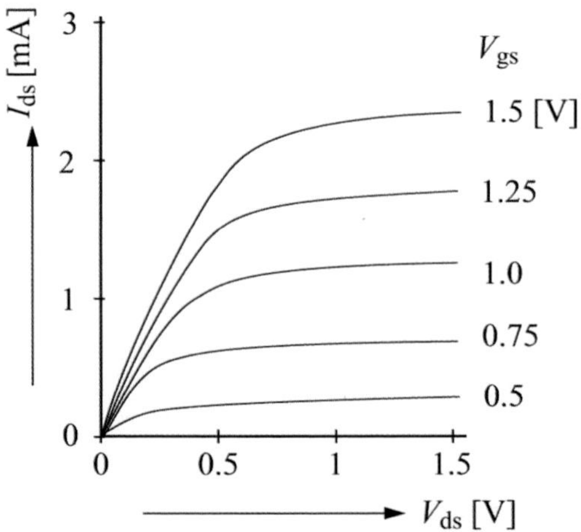

Figure 2.6: *Effect of channel length modulation on the MOS transistor characteristic*

Channel-length modulation is an undesired effect, which is particularly an issue for analog design. Because channel-length modulation is a short-channel effect which rapidly decreases with longer channels, analog designs typically require larger than minimum transistor channel lengths to improve the performance and operating margins.

2.5 Short- and narrow-channel effects

The electrical behaviour of a MOS transistor is primarily determined by its gain factor β, its threshold voltage V_T and its body factor K. Generally, the values of these parameters are largely dependent on the width W and length L of a transistor. The influence of these dependencies increases as transistor dimensions decrease. These small-channel effects, which are discussed below, are particularly significant in deep-submicron and nanometer MOS processes.

2.5.1 Short-channel effects

The cross-section presented in figure 2.7 is used to explain short-channel effects.

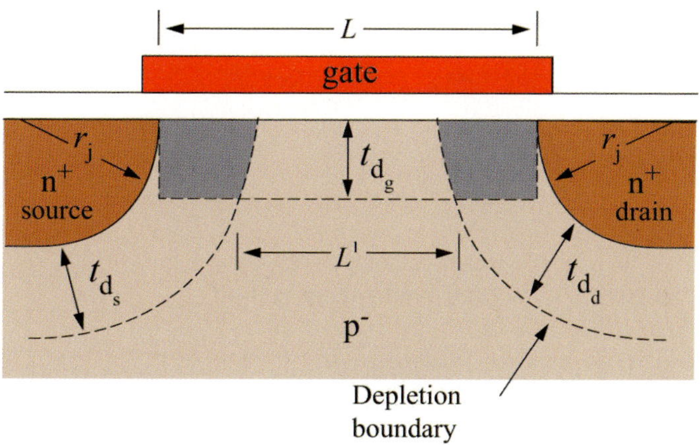

Figure 2.7: *Cross-section of a short-channel transistor, showing several depletion areas that affect each other*

Even in the absence of a gate voltage, the regions under the gate close to the source and drain are inherently depleted of majority carriers, i.e., holes and electrons in nMOS and pMOS transistors, respectively. In a short-channel transistor, the distance between these depletion regions is small. The creation of a complete depletion area under the gate therefore requires a relatively small gate voltage. In other words, the threshold voltage is reduced. This is a typical two-dimensional effect, which can be reduced by shallow source and drain diffusions. However, the associated smaller diffusion edge radii cause a higher electric field near the drain edge in the channel when $V_{ds} > V_{gs} > V_T$. One way to overcome this problem is to reduce the supply voltage. This short-channel effect on the threshold voltage occurs at shorter gate lengths and causes *threshold voltage roll-off*, see figure 2.8).

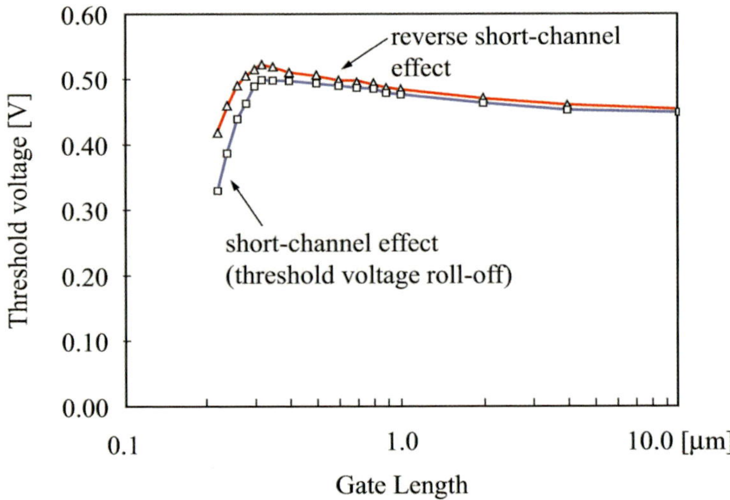

Figure 2.8: *Short-channel and reverse short-channel effect on the threshold voltage V_T of an nMOS transistor*

The use of shallow source and drain extensions (see figure 3.39), with less doping than the real source and drain implants, in combination with local higher doped channel regions (so-called *halo* or *pocket implants*) suppresses the depletion-layer width in the channel and contributes to a reduction of the *short-channel effect* (SCE). The implant is optimised for transistors with the smallest channel lengths in a given process. These transistors will have the nominal threshold voltage while transistors with longer channels will have higher threshold voltages.

A second effect that depends on the channel length is the *reverse short-channel effect (RSCE)*. In conventional CMOS devices, this effect, which involves increasing threshold voltages V_T with decreasing gate length, is caused by a lateral non-uniform channel doping induced by locally enhanced diffusion.

As described before, current devices use so-called halo implants to suppress short-channel effects. Figure 2.9 shows a possible dope profile in a device with halos. In devices with relatively long channels, these halos occupy a smaller region of the channel. When the channel becomes shorter, these halos get closer to one another and will also cause V_T roll-up. In 180 nm CMOS technologies and beyond, these halos intentionally cause roll-up and suppress the onset of roll-off.

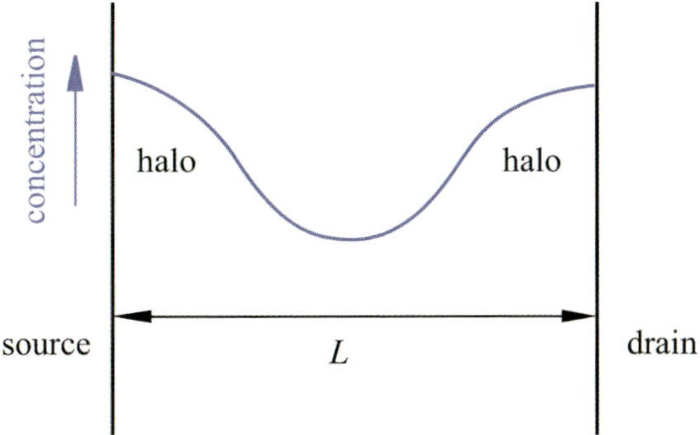

Figure 2.9: *Potential doping profile in the channel of a MOS device including the halos*

2.5.2 Narrow-channel effect

Also, the width of an active device influences the threshold voltage. The depletion layer extends under the edges of the gate, where the gate electrode crosses the field oxide. With a LOCOS type of field isolation, see figure 2.10, this effect is primarily caused by the encroachment of the channel stop dopant at the edge of the field isolation.

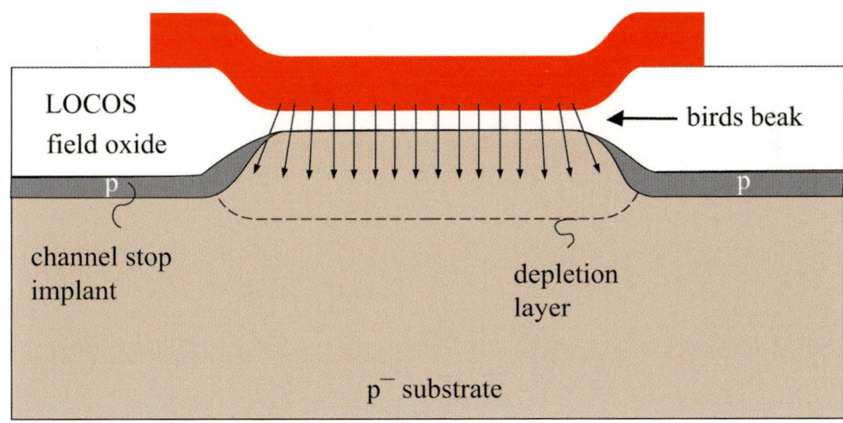

Figure 2.10: *Cross-section of a narrow-channel transistor showing the distribution of electric field lines under the gate*

The additional depletion region charge has to be compensated by an additional gate voltage. This results in an increase of the threshold voltage at reduced width of the device. The encroachment of channel stop dopant is especially pronounced for a conventional diffused well technology. The channel stop dopants are implanted prior to the high-temperature LOCOS oxidation and cause a large shift in V_T. In a retrograde implanted well process, the field oxidation is performed prior to the well implants and less encroachment of dopant atoms occurs under the gate edge. However, the threshold voltage is still increased as a result of the bird's beak and two-dimensional spreading of the field lines at the edge.

Figure 2.11 shows this *narrow-channel effect*, together with the influence of the channel width on the threshold voltage in a Shallow-Trench Isolation (see chapter 3) scheme. In contrast to the conventional narrow-width effect, the threshold voltage is even decreased at very narrow channel widths of around $0.2\,\mu$m. This *Inverse Narrow-Width Effect (INWE)* is attributed to a sharp corner at the top of the shallow-trench isolation. The fringing field at this corner results in an increased electrical field strength and reduces the threshold voltage. Also, the quality of the oxide used to fill the trench is not as good as the thermally grown LOCOS field oxide. A positive fixed oxide charge is present in the oxide and, in nMOS devices, it contributes to the decreased threshold voltage. This contribution of the fixed oxide charge is less severe than the fringing field compound and depends also on the deposition method used to fill the trench.

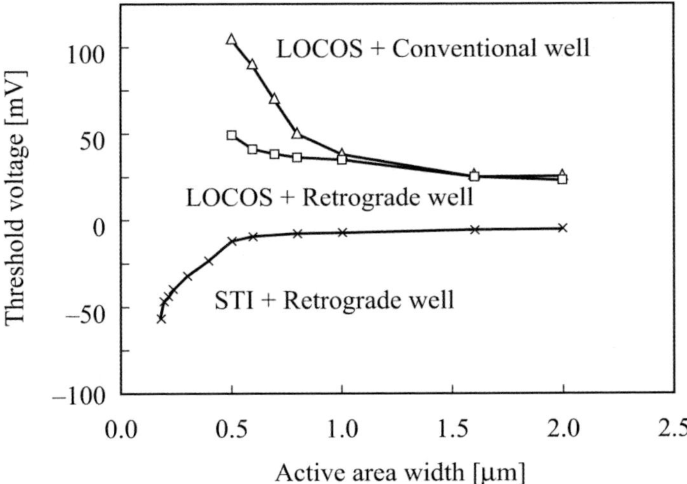

Figure 2.11: *Shift of threshold voltage of nMOS devices as a function of the active area width for different well technology and field isolation schemes*

2.6 Temperature influence on carrier mobility and threshold voltage

An increase in the operating temperature of a MOS transistor affects its behaviour in two different ways:

1. The mobility of the majority carriers, e.g., electrons in an nMOS transistor, in the channel decreases. Consequently, the transistor gain factor $\beta_\square$ also decreases. Its temperature dependence is expressed as follows [3]:

$$\beta_\square \left(Temp\right) = \beta_\square \left(298\,\mathrm{K}\right) \cdot \left(\frac{298}{Temp}\right)^{3/2} \tag{2.15}$$

The exponent 3/2 in this expression is more applicable to the electron mobility. For holes this exponent is closer to 1. PMOS transistor currents are therefore less temperature dependent than those of nMOS transistors.

2. The threshold voltage V_T of both nMOS and pMOS transistors decreases slightly. The magnitude of the influence of temperature

71

change on threshold voltage variation ΔV_T depends on the substrate doping level. A variation of -0.7 mV/°C is quite typical.

Both effects have different consequences for the speed of an IC. This speed is determined by the speed τ of a logic gate, which is defined as:

$$\tau = \frac{CV}{I} = 2 \cdot \frac{CV}{\beta(V_\mathrm{gs} - V_\mathrm{T})^2} \tag{2.16}$$

In conventional CMOS processes the overall circuit performance reduces with increasing temperatures, because its effect on the mobility reduction in the transistor current was traditionally larger than the effect on the reduction of the threshold voltage. This was one of the reasons to keep high-speed processors cool, by using a fan. Also worst-case corner simulations were usually done at high temperatures. However, today's CMOS technologies offer several different threshold voltages to support both high-speed and low-leakage applications. For general-purpose and high-speed processes, V_T is relatively low and a further reduction with 0.7 mV/°C has less influence on this speed than the reduction in the β. For low-leakage processes, with a relatively large V_T, both effects partly compensate each other, because of the increasing competition between mobility and threshold voltage, so that there is a reduced influence on the speed. At a certain supply voltage the above two mechanisms fully cancel each others contribution to the transistor current, such that the circuit speed has no longer a relation with the temperature. This is the so-called *zero-temperature-coefficient (ZTC) voltage* [4,5]. This reducing temperature dependence, which is expected to continue with further scaling of the supply voltage, has serious consequences for the static timing analysis, as it may invalidate the approach of defining PVT (process, voltage and temperature) corners, by independently varying voltage and temperature [6]. Figure 2.12 shows the frequency response of a high-V_T ring oscillator as a function of the supply voltage, for different operating temperatures.

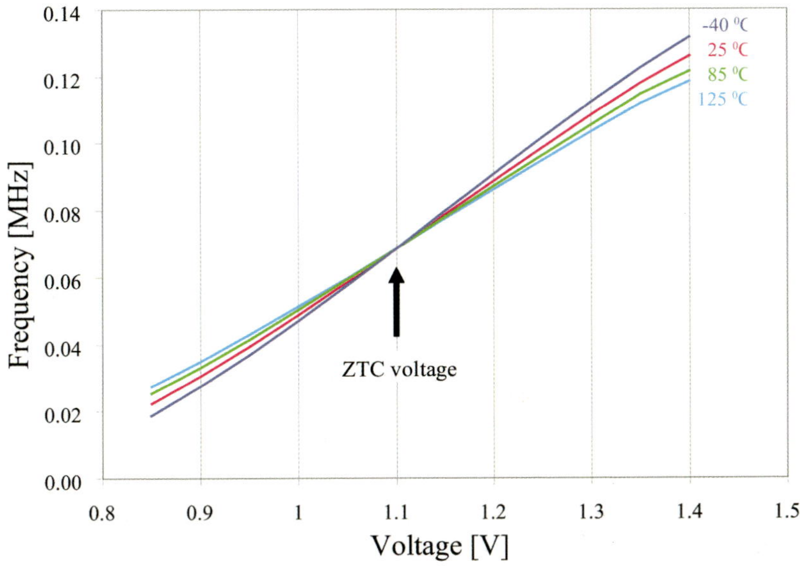

Figure 2.12: *Ring oscillator frequency response as a function of the supply voltage at different temperatures*

Above the ZTC voltage of 1.1 V, which is close to the nominal supply voltage of 1.2 V in this technology, the frequency reduces with increasing temperature, while below this voltage the effect is opposite. For the same ring oscillator fabricated with a standard V_T, this ZTC is reduced to 0.95 V.

As a result of this varying temperature behaviour, the worst-case and best-case corners for simulation need to be reconsidered, since for modern CMOS technologies a higher temperature not automatically corresponds to a lower performance! For the 45 nm technology node and beyond, the temperature effect will diminish further, because of an increasing compensation of the β and V_T contributions to the transistor current [7].

ZTC also has consequences for certain failure analysis methods (see chapter 10) that use local heating to detect changes in circuit behaviour, because these changes will become smaller and less visible in modern technologies.

2.7 MOS transistor leakage mechanisms

Due to the continuous scaling of the physical MOS device dimensions, such as channel length and gate-oxide thickness, and increasing doping levels to suppress short-channel effects (SCEs), MOS devices will increasingly drift away from an ideal switching behaviour. As a result, an increasing number of leakage mechanisms is influencing their performance, particularly during off-state. Figure 2.13 shows the major contributions to the total transistor leakage current.

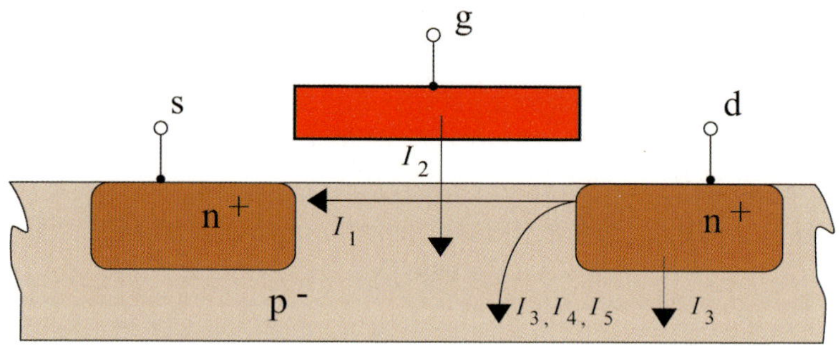

I_1 = Sub-threshold Leakage (incl. Drain-Induced Barrier Lowering (DIBL))
I_2 = Gate Oxide Tunneling
I_3 = Reverse Bias Junction Leakage
I_4 = Gate-Induced Barrier Drain Leakage (GIDL)
I_5 = Impact Ionisation current

Figure 2.13: *Leakage mechanisms in nanometer CMOS technologies*

The reduction of the supply voltage (V_{dd}) for a series of technology generations has caused the threshold voltage (V_T) to reduce accordingly, in order to achieve sufficient performance increase. A lower V_T leads to a larger off-current: a larger drain-to-source leakage current when the gate-to-source voltage is zero: subthreshold leakage current.

The reduction of the gate-oxide thickness below 2.5 nm has resulted in charge carriers tunnelling through the oxide from the channel to the gate or vice versa and causes a gate-leakage current. This tunnelling is not caused by the field only, but mainly by the penetration of the carriers into the oxide. A potential difference of 2 V across an oxide thickness of 2 nm will cause tunnelling, while a 5 V potential difference across 5 nm oxide won't.

Finally, the increasing number of SCEs, in particular the threshold voltage roll-off effect, requires additional local pocket implants in the channel region close to the source and drain. This leads to a larger reverse-bias drain junction leakage current.

A few other leakage mechanisms contribute to the total leakage, such as gate-induced drain leakage (GIDL) and impact ionisation current. The next subsections will discuss the mechanisms behind these leakage currents in more detail. Particularly during the standby mode of most portable electronic devices these leakage currents cause a relatively large standby power, thereby limiting the battery lifetime. Design methods to reduce these leakage power components are presented in chapter 8.

2.7.1 Weak-inversion (subthreshold) behaviour of the MOS transistor

An nMOS transistor operates in the 'weak-inversion' region when its gate-source voltage (V_{gs}) is below its threshold voltage (V_T), see figure 2.14.

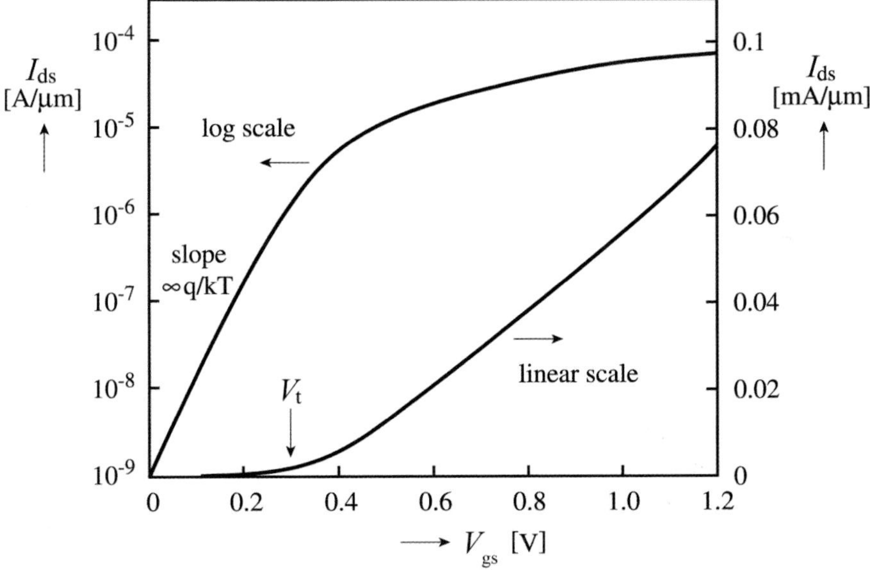

Figure 2.14: *MOSFET current representation on linear scale (right axis) and log scale (left axis)*

Below the threshold voltage, the current decreases exponentially. On a logarithmic scale, the slope (for gate voltages below the threshold voltage this slope is called the *subthreshold slope*) is inversely proportional to the thermal energy kT. Some electrons still have enough thermal energy to cross the gate-controlled potential barrier (figure 2.15) and travel to the drain. At (very) low threshold voltages, the resulting *subthreshold leakage current* may lead to an unacceptably high power consumption. This leakage current should not exceed a few tens of nano-amperes for a one hundred million transistor chip in standby mode (no circuit activity and at zero gate voltage). This operating region is also called the 'subthreshold region'. The *subthreshold slope* (s_{subthr}) depends on the technology and is expressed in mV/dec. The lower the value, the steeper the slope. It ranges from $s_{\mathrm{subthr}} \approx 63\,\mathrm{mV/dec}$ for SOI processes to $s_{\mathrm{subthr}} \approx 80\,\mathrm{mV/dec}$ for bulk CMOS processes. This means that the subthreshold leakage current increases about 18 times for every $100\,\mathrm{mV}$ reduction in V_{T}, for bulk CMOS transistors.

The normal strong-inversion equations (that apply above the threshold voltage) do not apply to the weak-inversion region. The drain-source current in a transistor with a long channel and a constant drain-source voltage operating in the weak-inversion region is expressed as follows:

$$I_{\mathrm{dssub}} = \frac{W}{L} \cdot C\, I_{\mathrm{ds0}}\, \mathrm{e}^{V_{\mathrm{gs}}/mU_{\mathrm{T}}} \qquad (2.17)$$

The terms in equation (2.17) are defined as follows:

$$
\begin{aligned}
C &= 1 - \mathrm{e}^{-V_{\mathrm{ds}}/U_{\mathrm{T}}} \\
U_{\mathrm{T}} &= \frac{kT}{q} \approx 25\,\mathrm{mV} \text{ at room temperature} \\
I_{\mathrm{ds0}} &= \text{characteristic current at } V_{\mathrm{gs}} = 0\,\mathrm{V};\ I_{\mathrm{ds0}} \propto \mathrm{e}^{-V_{\mathrm{T}}/mU_{\mathrm{T}}} \\
m &= \text{slope} \approx 1.5 \text{ and depends on the channel length}
\end{aligned}
$$

Equation (2.17) applies when V_{gs} is not larger than a few U_{T} below V_{T}. The subthreshold transistor current I_{dssub} can lead to a considerable *standby current* in transistors that are supposedly inactive. Expression (1.16) in chapter 1 shows that when we apply a negative (positive) voltage to the substrate or back gate of an nMOS (pMOS) transistor, we can increase the threshold voltage, thereby reducing the subthreshold leakage. Methods to use this back-bias effect in reducing the standby power of CMOS ICs, are discussed in chapter 8.

An accurate description of the behaviour of a transistor operating in the weak-inversion region is contained in references [8,9]. The following statements briefly summarise this operating region:

1. At low V_T, there is a significant subthreshold current when $V_{gs} = 0\,V$. This has the following consequences:

    (a) There is a considerable standby current in (C)MOS VLSI and memory circuits;

    (b) The minimum required clock frequency of dynamic circuits is increased as a result of leakage currents. DRAMs are among the circuits affected.

2. In long-channel transistors, the source and drain depletion regions have a limited influence on the electric field in the channel. These devices show a much lower threshold-voltage dependence on channel length and drain voltage. In short-channel transistors, on the contrary, the source and drain depletion layers cover a much larger part of the channel (see figure 2.7). They interact with each other, such, that the drain voltage reduces the potential barrier at the source. Figure 2.15 shows the influence of the drain voltage and channel length on the barrier height. An increase of the drain-source voltage in short-channel transistors will reduce the barrier height. This *Drain-Induced Barrier Lowering effect (DIBL)* leads to a reduction of the threshold voltage V_T of the transistor.

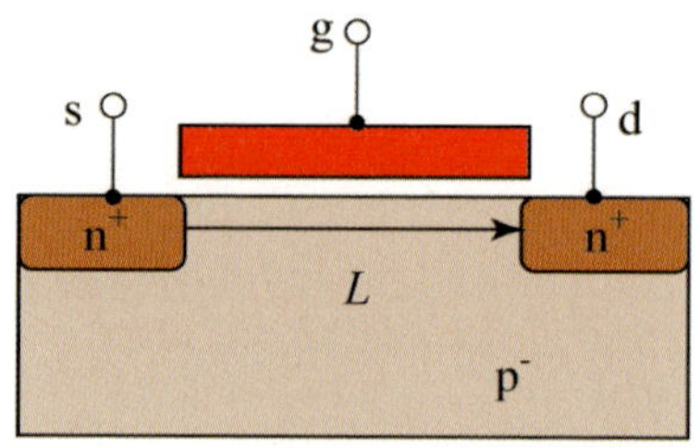

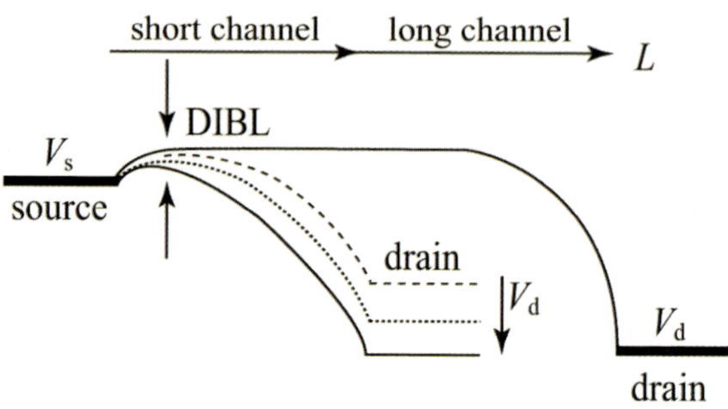

Figure 2.15: *The effect of the drain voltage on lowering the barrier*

In a 65nm CMOS technology, for example, a transistor with a channel length of 60 nm can carry a 6 times larger subthreshold leakage current than one with a channel length of 90 nm at the same operating voltages! The total subthreshold leakage current is also dependent on the temperature: it almost doubles for every $25°C$ increase in temperature. It is clear that these effects are very important in the development and the modelling of deep-submicron technologies.

3. Analogue circuit techniques use weak-inversion behaviour in low-current applications. The voltage gain of a MOS transistor operating in the weak-inversion region is relatively high and comparable to the voltage gain of bipolar transistors.

2.7.2   Gate-oxide tunnelling

The continuous scaling of MOS devices over the past four decades has caused a reduction of their lateral dimensions with an average scaling factor $s \approx 0.7$. In order to achieve a higher speed, the transistor current needs to be increased. From the current expressions, derived in chapter 1, it can be seen, that we need to increase the transistor gain factor β, in order to achieve a higher current drive. Because β is inversely proportional to the gate-oxide thickness t_{ox}. This thickness was required to be reduced with the same factor, leading to an increasing probability of direct electron tunnelling through it. However, when t_{ox} becomes less than 2.5 nm, this tunnelling becomes visible in the total transistor leakage picture. Both this probability and the resulting leakage current are strong exponential functions of t_{ox}, as well as a function of the voltage across the gate oxide [11]. It is not only the electrical field across the oxide that determines the amount of tunnelling, but also, and even more dominantly, the penetration of carriers into the oxide, e.g., 2 V across 2 nm oxide causes tunnelling, while 5 V across 5 nm does not. Figure 2.16 shows these relationships [8]. The gate leakage increases roughly by a factor of ten for every 0.2 nm reduction in oxide thickness. It also increases by about a factor of ten with the doubling of the voltage across the oxide.

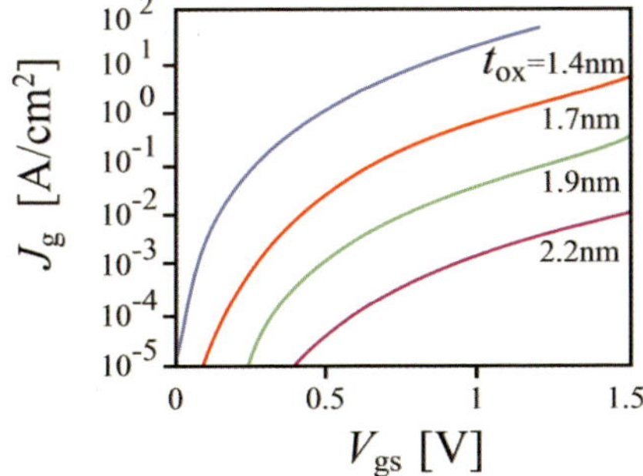

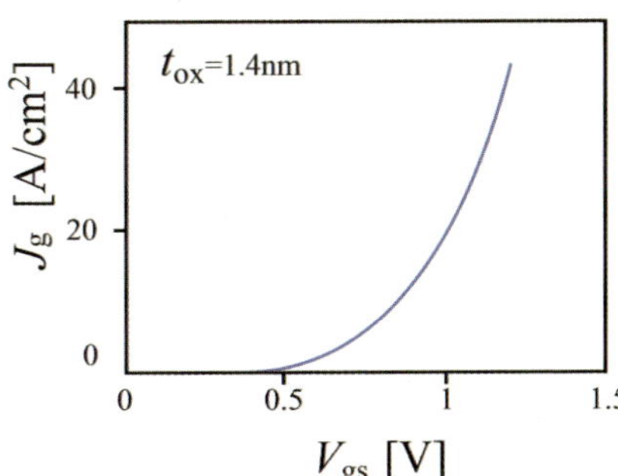

Figure 2.16: *Gate-oxide leakage current density as a function of the oxide thickness and oxide voltage*

The total *gate-oxide leakage* is dependent on the state of the transistor. In the transistor on-state, the whole channel area contributes to the gate-oxide leakage current. In the off-state, only the gate-to-source and/or gate-to-drain overlap area contribute, depending on the voltage on these terminals.

It should be noted that gate-oxide (SiO_2) leakage will be dominated by the n-MOS device, because the p-MOS device shows an increased barrier for hole tunneling. As a result, a pMOS transistor will exhibit roughly an order of magnitude lower gate-oxide leakage current than its nMOS counterpart [9].

The gate-oxide leakage can be reduced by applying an isolating material with a relatively high dielectric constant, so that the dielectric thickness can be increased and the gate-oxide leakage reduced. Researchers are currently exploring the potential of hafnium-based dielectrics (e.g., HfO_2, HfSiO, HfSiON), but most of these materials show poor electrical stability, low charge-trapping capability, yield and reliability. Further research is still required to improve the quality of a new gate dielectric.

2.7.3 Reverse-bias junction leakage

In commonly used CMOS circuits, when the source and drain junctions are biased, they are reverse biased. The larger reverse bias across the drain junction is particularly responsible for the increasing junction leakage. There are several mechanisms that contribute to this junction leakage. Two of them, which are commonly known from the basic conventional diode operation, are diffusion and drift of minority carriers and electron-hole pair (Shockley-Read-Hall; SRH) generation in the depletion region of the reverse biased junction. For the lower temperature range, the generation mechanism dominates and is proportional to the intrinsic carrier concentration n_i. The diffusion mechanism is dependent on the thermal energy kT, which causes the carriers to move at random even when no field is applied. At high temperatures, this dominates the leakage contribution, which is then more proportional to $n_i{}^2$.

If the reverse electrical field increases but is still in the relatively low region, the reverse leakage current is mainly assisted by interface traps. This so-called *trap-assisted tunnelling (TAT)* increases with the density of traps and can be increased by electrical stress.

If no special technology measures had been taken, four decades of transistor channel length scaling would have resulted in the source and drain depletion layers touching each other, causing the SCE as discussed in section 2.5.1. To suppress SCE, close to the source and drain regions in the channel, halos (pockets) are implanted with increasing peak doping levels ($\approx 1 - 2 \cdot 10^{19}$ atoms/cm^3), depending on the technology node. Because of this, the depletion layers become so thin, that also here direct tunnelling of carriers will occur, just like through the thin gate oxide.

A high field across the reverse-biased p-n junction at the drain causes significant tunnelling of electrons through this junction, from the valence band of the p-halo to the conduction band of the n-drain (figure 2.17). This so-called *sub-surface band-to-band tunnelling* (sub-surface BTBT) is a major contribution to the total *reverse-bias junction leakage current* at high reverse electrical fields.

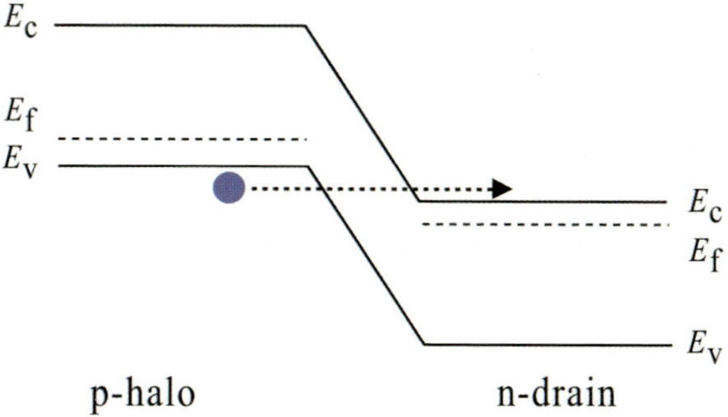

Figure 2.17: *Sub-surface BTBT through an nMOS drain junction*

This BTBT-leakage current is exponentially dependant on the reverse-bias electric field across the junction and on the doping levels at both sides of the junction.

The use of a negative back-bias voltage increases the reverse-bias potential and may lead to a dramatic increase of this BTBT leakage. So, the use of a reverse body-bias to reduce the overall transistor leakage becomes questionable, since it influences the various leakage components in a different way: it will reduce the subthreshold leakage, but it will increase the reverse-bias junction leakage. Section 2.7.6 presents a discussion on the combined leakage behaviour in relation with the temperature and the use of a reverse body bias.

2.7.4 Gate-induced drain leakage (GIDL)

When a large drain-to-gate bias is applied, a deep depletion condition is created in the drain region underneath the gate. This can create a sufficient energy-band bending (greater than the silicon bandgap) at the Si-SiO$_2$ interface in the drain for valence-band electrons to tunnel into the conduction band. In other words: when, in case of an nMOS transistor, $V_{gs} \leq 0\,\mathrm{V}$ while $V_{ds} = V_{dd}$, electrons may tunnel through this deep-depletion layer and flow to the drain terminal, while the remaining holes flow to the substrate. Figure 2.18 shows the effect by the energy band diagram.

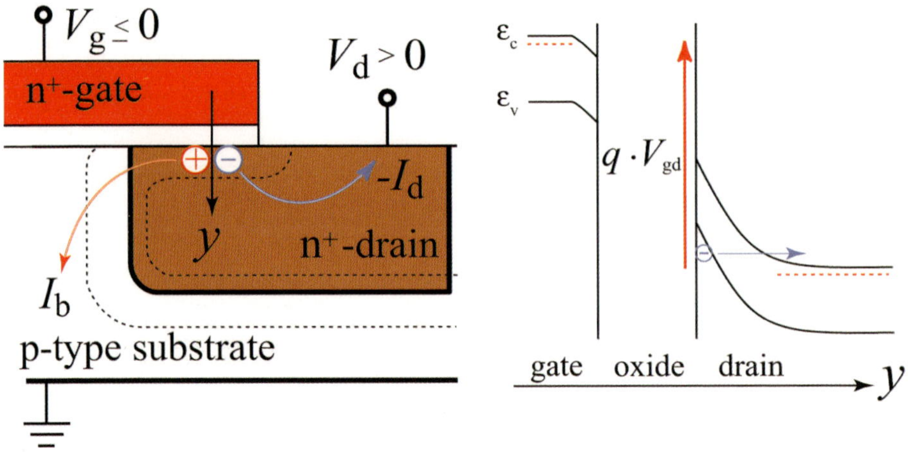

Figure 2.18: *Cross-section and energy-band diagram used to explain the GIDL effect*

This *surface BTBT* together with trap-assisted tunnelling current at the interface are the main causes of the so-called *gate-induced drain leakage (GIDL)* current [10]. This current only depends upon the conditions near the gate to drain overlap. A decreasing gate-oxide thickness causes an increase of the drain-to gate electrical field and results in an increase of the GIDL current. The resulting drain and substrate currents increase exponentially with V_{dg}. The effects of GIDL remain almost constant with constant-field scaling (see chapter 11). It is not expected that the use of high-ϵ dielectrics will have much impact on GIDL.

2.7.5   Impact Ionisation

As a result of the continuous scaling of the devices, the electric field near the drain has reached extraordinary large values in short-channel devices. Consequently, a carrier can acquire sufficient energy that it can cause *impact ionisation* upon impact with an atom, e.g., such a *hot carrier* can cause the transition of an electron from the valence band to the conduction band, leaving a free hole behind. This produces an extra conduction electron and a hole. These generated carriers face the same large electric field that has caused their formation. Therefore, in an nMOS device, both electrons will flow to the drain, while the hole drifts to the substrate. In other words: impact ionisation occurs, when the excess of energy that a carrier has collected, from passing a high electric field, is used to create other carriers. This gives rise to an increase in

the drain-source current I_{ds} and to a substrate current I_b. The degree
to which impact ionisation occurs and the magnitude of I_b are directly
proportional to the square of the maximum electric field in the pinch-off
region near the drain. Every measure to reduce this maximum causes a
reduction in I_b.

In an nMOS device, a carrier impact may generate high-energy electrons which may be scattered toward the substrate-to-gate oxide interface. Electrons with the highest energy can even be injected into the gate
oxide. This so-called *hot-carrier effect* will degrade the device performance and it may lead to reliability problems (see chapter 9). Impact
ionisation and the occurrence of hot carriers are more pronounced in
short-channel devices operating at the maximum drain source voltage
(equal to the supply voltage), when the gate source voltage is equal to
half of the supply voltage.

Because analog circuits tend to be designed with non-minimum channel length devices, they will suffer less from these effects. Because the
bandgap of silicon is $1.15\,\text{V}$, the trend of reducing the supply voltages
further below $1.2\,\text{V}$, will cause a reduction of occurrence of impact ionisation in future CMOS devices.

2.7.6 Overall leakage interactions and considerations

The previous subsections presented the most important individual leakage components in an MOS transistor. However, these leakage components are interdependent and respond differently to a change in one or
more of the parameters (e.g., oxide thickness, channel length, channel
doping, source and drain doping, gate material/doping, terminal voltages, chip temperature).

Figure 2.19 (top) shows an example of the gate, drain, and bulk
currents of an nMOS transistor in a $65\,\text{nm}$ low-leakage (LSTP) CMOS
technology as a function of the gate voltage at $25\,^{\circ}\text{C}$ and $125\,^{\circ}\text{C}$. Figure 2.19 (bottom) shows the influence of an additional reverse substrate-bias voltage on the current-to-voltage behaviour.

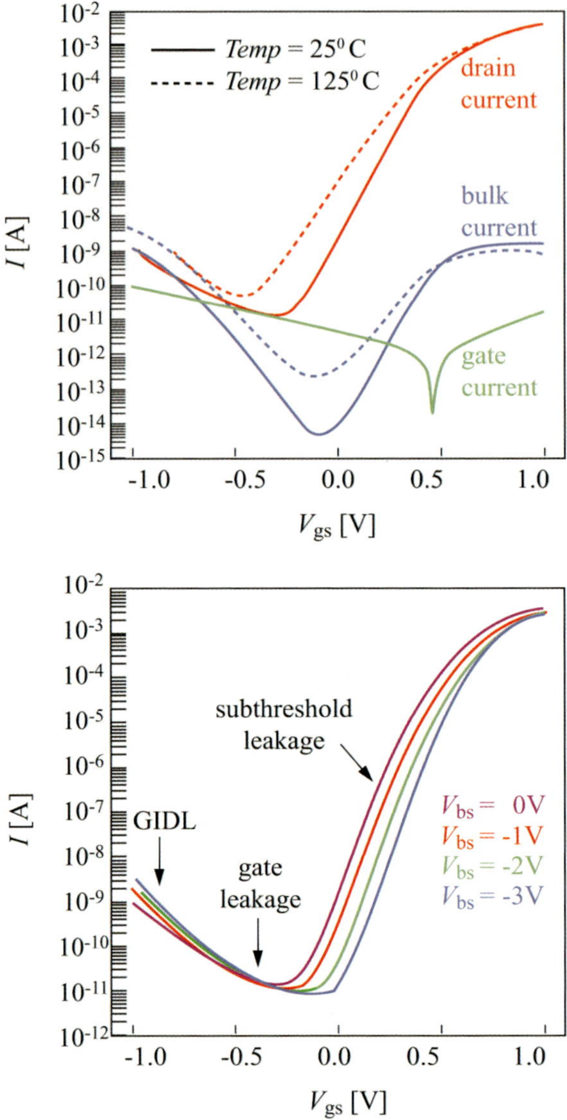

Figure 2.19: *Most dominant contributions to the total leakage current in an nMOS transistor in a low-leakage 65 nm CMOS technology at 25°C and 125°C (top), and at different back bias voltages (bottom)*

The subthreshold current is exponentially proportional with the temperature, while the gate tunnelling current is almost completely independent of it. For an nMOS transistor in a general-purpose 65 nm CMOS technology, the contributions of the gate and subthreshold currents will dramatically increase, because such a process has a much thinner gate oxide and a smaller V_T.

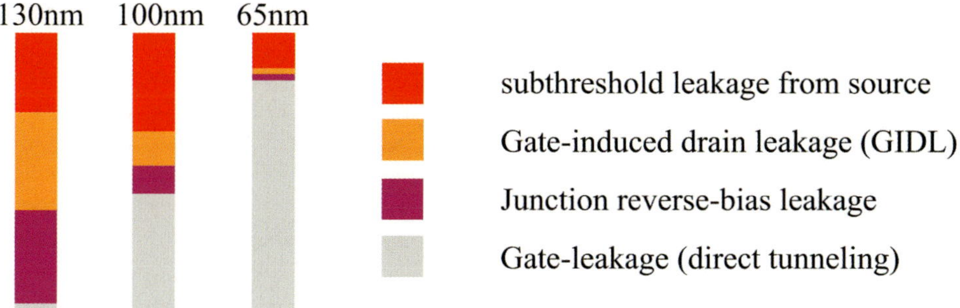

Figure 2.20: *Relative contributions of the various leakage mechanisms to the total transistor leakage current [12]*

Figure 2.20 shows an example of the relative contributions of the various leakage mechnisms to the total transistor leakage current [12], which itself increases exponentially (Figure 2.21; [13]) with further technology scaling. It reflects a process with very thin gate oxide. However, these contributions may vary dramatically between different technology nodes and between low-power and general-purpose processes from different foundries.

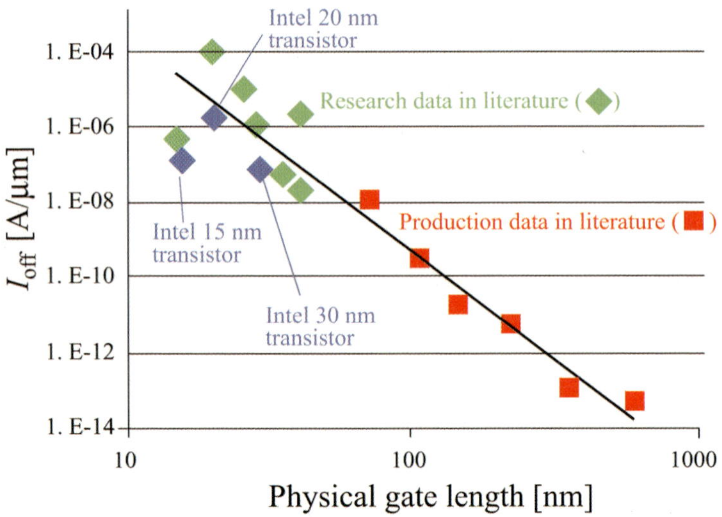

Figure 2.21: *Exponential increase of the total leakage current vs channel length (technology generation) [13]*

More on leakage current mechanisms can be found in [14]. Alternative technology and design solutions to reduce leakage currents are presented in chapter 8.

2.8   MOS transistor models

All previously discussed physical mechanisms, combined with an accurate description of the (overlap) capacitances, are included in today's (compact) MOS models. A MOS model gives a description of the physical behaviour of a MOS device from weak-inversion to strong-inversion operation and calculates the device currents, charges and noise as a function of the terminal voltages. Particularly the analog and RF applications of the MOS transistor require an accurate description of their operation over the full range of operating voltages. A compact model is based on only a limited number of parameters. In combination with a circuit simulator, a compact model allows full simulation of the electrical behaviour of rather complex analog and RF circuits and predicts their behaviour, before they are integrated on real silicon.

The majority of recently developed MOS models are based on so-called surface-potential (SP) models, in which the surface potential equations at both ends of the channel are solved. In December 2005 the Compact MOS Council (CMC) has selected the PSP model, which was a merge of two existing SP models, to replace BSIM as the industrial standard MOSFET model for future CMOS technologies [10,15]. More details about the physical and mathematical basics of MOS modeling are beyond the scope of this book, but are largely available through the internet.

2.9 Conclusions

The formulae derived in chapter 1 provide a good insight into the fundamental behaviour of MOS devices. These formulae were used to predict circuit behaviour with reasonable accuracy until the mid 1980-ies. The continuous drive for higher circuit densities with smaller transistors, however, has given rise to an increased contribution from physical and geometrical effects. These effects cause deviations from the ideal transistor behaviour assumed in chapter 1. In addition, the magnitude of these deviations increases as transistor dimensions shrink. These effects combine to reduce the ideal transistor current by more than a factor four for channel lengths below $0.25\,\mu$m. There are also effects that permanently degrade the performance of a MOS transistor. Particularly its behaviour in off-state has a great impact on the standby power consumption of CMOS ICs. Therefore weak-inversion behaviour and transistor leakage-current mechanisms have been discussed in detail, to allow better understanding of the low-standby power solutions presented in chapter 8. The continuous scaling of both devices and interconnects also has severe consequences for the reliability of the IC and may dramatically reduce its lifetime. These consequences and potential technology and design solutions are discussed in chapter 9.

2.10 References

[1] A.J. Walker and P.H. Woerlee,
'A mobility model for MOSFET device simulations',
Journal de Physique, colloque C4, Vol. 49, No. 9, Sept. 1988, p. 256

[2] M.Vertregt,
'The Analog Challenge in Nanometer CMOS',
IEDM Digest of Technical Papers, pp. 11-18, December 2006

[3] R.S.C. Cobbold,
'Theory and applications of field effect transistors',
John Wiley & Sons, Inc. New York

[4] I.M. Filanovsky, A Allam,
'Mutual Compensation of Mobility and Threshold Voltage Temperature Effects with Applications in CMOS Circuits',
IEEE Transactions on Circuits and Systems: Fundamental Theory and applications, vol. 48, no. 7, pp. 876-884, July 2001

[5] E. Long, et al.
'Detection of Temperature Sensitive Defects Using ZTC',
Proceedings of 22nd IEEE VLSI Test Symposium (VTS 2004)

[6] A. Dasnan, et al.,
'Handling Inverted Temperature Dependance in Static Timing Analysis',
ACM Transactions on Design Automation of Electronic Systems, Vol. 11, No. 2, April 2006, pp. 306-324

[7] R.Kumar, et al.,
'Reversed Temperature-Dependent Propagation Delay Characteristics in Nanometer CMOS Circuits',
IEEE Transactions on Circuits and Systems-II: Express Briefs, Vol. 53, No. 10, October 2006, pp. 1078-1082

[8] R. van Langevelde, et al.,
'Gate current: modelling, ΔL extraction and impact on RF performance'
IEDM Technical Digest, pp. 289-292, 2001

[9] F.Hamzaoglu, et al.,
'Circuit-Level Techniques to Control Gate Leakage for Sub-100nm

CMOS'
Proceedings of the 2002 ISLPED Symposium, pp. 60-63

[10] G. Gildenblat, et al.,
'PSP: An Advanced Surface-Potential-Based MOSFET Model for Circuit Simulation',
IEEE Transactions on Electron Devices, Vol. 53, No. 9,
pp.1979-1993, September 2006

[11] D. Lee, et al.,
'Gate Oxide Leakage Current Analysis and Reduction for VLSI Circuits',
IEEE Transactions on VLSI Systems, Vol. 12, No. 2, February 2004,
pp. 155-166

[12] J. Assenmacher,
'BSIM4 modelling and Parameter Extraction',
http://www.ieee.org/r5/denver/sscs/References/2003_03_Assenmacher.pdf

[13] G. Marcyk, et al.,
'New Transistors for 2005 and Beyond',
http://www.eas.asu.edu/ vasilesk/EEE531/TeraHertzlong.pdf

[14] A. Scholten, et al.,
'The Physical Background of JUNCAP2',
IEEE Transactions on Electron Devices, Vol. 53, No. 9,
pp.2098-2107, September 2006

[15] R.Woltjer, et al.,
'An industrial view on compact modeling',
Proceedings of the 36th European Solid-State Device Research Conference, Sept. 2006, Page(s):41 - 48

General basic physics

[16] S.M. Sze,
'Very Large Scale Integration Technology',
Mc Graw-Hill, 2nd edition, 1998

2.11 Exercises

1. At 25°C the magnitude of an nMOS transistor's gain factor β is $240\,\mu\text{A}/\text{V}^2$ and its threshold voltage V_T is 0.4 V.

   a) Calculate the gain factor β when the transistor is operating at 65°C.

   b) Calculate the threshold voltage for the temperature in a).

   c) What would be the consequences of this reduced threshold voltage for the standby current in an SRAM, for instance?

2. Assume the transistor in exercise 1 is saturated with its gate connected to its drain. At what V_gs would the influence of the temperature difference in exercise 1 on the gain factor β and on the threshold voltage V_T fully compensate each other (in other words: what would be the ZTC voltage value of V_gs)?

3. a) What is the effect on the gain factor β of a pMOS transistor with $L = 60\,\text{nm}$ when the mobility is only influenced by velocity saturation caused by a very large horizontal electric field, $E_\text{x} = 0.9 \cdot E_\text{xsat}$?

   b) Calculate the drain-source voltage at which the relevant reduction in mobility occurs if $\theta_3 = 0.45\,V^{-1}$.

4. Assume equation (2.7) can be approximated by:

$$\mu = \frac{\mu_0}{(1 + \theta_1'(V_\text{gs} - V_\text{T}) + \theta_3' \cdot V_\text{ds})}$$

   What can you say about the relation between the mobility and the series resistance of the drain (R_d), when the transistor operates in the saturation region?

Chapter 3

Manufacture of MOS devices

3.1 Introduction

Until the mid-eighties, the nMOS silicon-gate process was the most commonly-used process for MOS LSI and VLSI circuits. However, nearly all modern VLSI and memory circuits are made in CMOS processes. CMOS circuits are explained in chapter 4; the technology used for their manufacture is discussed in this chapter.

Modern nanometer CMOS processes, with channel lengths below 100 nm, have emerged from the numerous manufacturing processes which have evolved since the introduction of the MOS transistor in integrated circuits. Differences between the processes were mainly characterised by the following features:

- The minimum feature sizes that can be produced.

- The gate oxide thickness.

- The number of interconnection levels.

- The type of substrate material. Alternatives include n-type and p-type, high-resistive or low-resistive, bulk silicon, epitaxial or SOI wafers.

- The choice of the gate material. Initially, the gate material was the aluminium implied in the acronym MOS (Metal Oxide Semiconductor). Molybdenum has also been used. Until 120 nm MOS processes and above, however, nearly all use *polycrystalline* silicon (polysilicon) as gate material. One of the main reasons is that a polysilicon gate facilitates the creation of self-aligned source and drain areas. Another reason for using polysilicon as gate material is that it allows accurate control of the formation of the gate oxide. From 90 nm onwards, a stack of W-WN-polysilicon and SiO_xN_y is used. A combination of a metal gate with high-ϵ dielectrics is first introduced in the 45 nm node, but will certainly be an option in the 32 nm node.

- The method to isolate transistors. Conventional CMOS processes used the so-called LOCOS isolation while most of today's processes use Shallow-Trench Isolation (STI), see section 3.5.

- The type of transistors used: nMOS, pMOS, enhancement and/or depletion, etc.

Many of the transistor parameters, in terms of performance, power consumption, and reliability, are determined by the substrate that is used as starting material. A short summary on the properties and use of the different substrate materials will therefore be presented first.

Modern manufacturing processes consist of numerous photolithographic, etching, oxidation, deposition, implantation, diffusion and planarisation steps. These steps are frequently repeated throughout the process and they currently total more than 500. The IC fabrication discussion starts with a brief description of each step. Most processes use masks to define the required patterns in all or most of the IC diffusion and interconnect layers. Modern CMOS manufacturing processes use between 25 and 50 masks. However, the initial discussion of IC manufacturing processes in this chapter focuses on a basic nMOS process with just five masks.

Subsequently, a basic CMOS process flow is briefly examined. Fundamental differences between various CMOS processes are then highlighted.

Finally, a sample nanometer CMOS process is explained. Many of the associated additional processing steps are an extension of those in the basic CMOS process flow. Therefore, only the most fundamental deviations from the conventional steps are explained. The quality and reliability of packaged dies are important issues in the IC manufacture industry. An insight into the associated tests concludes the chapter.

3.2 Different substrates (wafers) as starting material

A very critical element in the operation of an integrated circuit is the electrical isolation between the individual devices. Unintended electrical interference can dramatically affect their performance. Smaller minimum feature sizes reduce the distance between devices and increase their sensitivity at the same time. An important factor in the isolation properties is the substrate on which the devices are built. In all discussions, so far, we have assumed a *bulk silicon* substrate (wafer) as the starting material for our (C)MOS processes. However, most standard CMOS technologies today use epitaxial wafers, while several high-performance microprocessors are made on SOI wafers. The properties and use of these substrates (wafers) will be discussed next.

3.2.1 Wafer sizes

From an economical perspective, larger wafers have lead to reduced IC manufacturing costs. This rule drove the wafer diameter from about 1 inch ($\approx$ 25 mm), about four decades ago, to 12 inches (= 300 mm) today. This has put severe pressure on maintaining the wafer flatness, its resistivity and low crystal defect density homogeneous across a rapidly increasing wafer area. However, the introduction of a new wafer diameter generation requires a huge amount of development costs. A possible move from 300 mm wafer fabs to 450 mm wafer fabs may need \$15 billion to \$20 billion of development costs [1]. According to the International Technology Roadmap for Semiconductors (ITRS), a 450 mm fab is projected to emerge around 2012 or so. This huge investment can only be earned back by those companies that run expensive products in very high volumes. All semiconductor tool vendors currently have no investments in the development of 450 mm tools, so it is questionable whether the first 450 mm waferfab will be built before 2015.

3.2.2 Standard CMOS Epi

Epitaxial wafers consist of a thin, mono-crystalline silicon layer grown on the polished surface of a bulk silicon substrate [2]. This so-called *epi layer* is defined to meet the specific requirements of the devices in terms of performance, isolation and reliability. This layer must be free of surface imperfections to guarantee a low defect density and limit the number of device failures. Since the carriers in a transistor channel only travel in the surface region of the device, the epi layer thickness is only defined by the transistor architecture (source/drain and STI depths) and ranges from one to a few microns. Usually the total wafer thickness is typically $750\,\mu$m, but may range between $400\,\mu$m and $1\,$mm, depending on the wafer size and technology node. It means that the top epi layer forms only less than one percent of the total wafer and that the major part of the wafer mainly serves as a substrate carrier for the ICs made on it.

Although the resistance of this substrate hardly affects the performance of digital circuits it has influence on the robustness of the ICs built on it. Most conventional CMOS processes, including the 180 nm node, use/used low-resistivity (5-10 mΩcm at doping levels between $5 \cdot 10^{18}$ and $1 \cdot 10^{19}$ atoms/cm^3) wafers, in order to reduce the chance of latch-up occurrence (see chapter 9). With reducing supply voltages the chance for triggering the parasitic transistor to initiate latch-up, is also diminishing. This, combined with the increasing integration of GHz RF functions, has made the use of high-resistivity (10-50 Ωcm at doping levels between $1 \cdot 10^{15}$ and $1 \cdot 5^{15}$ atoms/cm^3) substrates very popular from the 120 nm CMOS node onwards. It leads to performance increase of passive components, such as inductors, but also to a better electrical isolation between the noisy digital circuits and the sensitive RF and analog ones (less substrate noise; chapter 9).

Because the full device operation occurs within this thin top epi layer, it puts severe demands on the homogeneity of the layer thickness, of the resistivity and of the crystal defectivity. When growing single crystal silicon, either for creating bulk silicon wafers or for creating thin epi layers, a few typical defects in the silicon may show up. *Point defects* may originate from single empty locations (*vacancies*) in the monocrystalline atomic structure (figure 3.1), while *micro defects* or *crystal-oriented particles (COP)* can be the result of a cluster of voids.

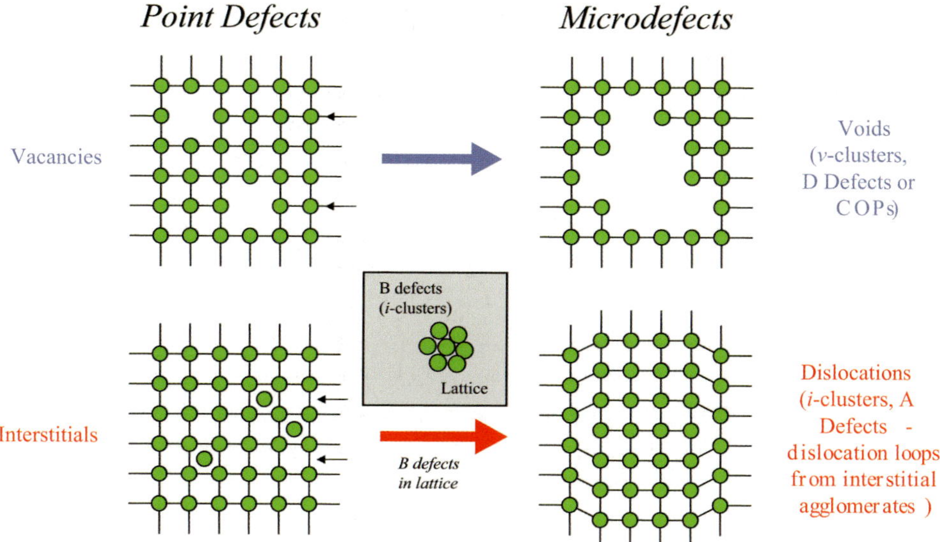

Figure 3.1: *Defects in silicon (Source: MEMC)*

Interstitials are atoms located in between the atoms of the crystal, while *dislocations* may be caused by clusters of interstitials. The average atomic spacing is also dependant on the covalent atomic radius of the specific material: Silicon (Si) 1.17 Å, Boron (B) 0.88 Å, Phosphorous (P) 1.10 Å, Arsenic (As) 1.18 Å, Stibnite (Sn) 1.36 Å. So, B is a smaller atom than Si. Doping Si with B (or P) reduces the average atomic spacing of the Si crystal. Another result of this is that the average atomic spacing in the p^- epi layer is larger than that in the p^+ substrate, because the substrate contains a higher concentration of smaller atoms. Large differences in the atomic spacing of different layers may lead to so-called misfit dislocations. To prevent *misfit dislocations* in a thin epi layer on a resistive substrate a simple rule of thumb is applied [3,4]:

epi thickness in $\mu m \leq$ *substrate resistivity in* $m\Omega$

Stand-alone memories may require their own substrates. Trench capacitor DRAMs (chapter 6), for instance, in which the trench capacitor cells are fabricated inside the substrate using depths over $7\,\mu m$, have much more interaction with the silicon (defects, leakage, stress) than stacked-capacitor DRAMs (chapter 6), where the capacitor is located fully above the silicon. They therefore need a low-COP or defect-free top layer with a thickness larger than the trench depth, to prevent substrate defect generation during the deep-trench fabrication.

These examples show that not all ICs can be made on the same substrate. The following subjects discuss substrates that enhance the device performance.

3.2.3   Crystalline orientation of the silicon wafer

As discussed in chapter 2, the effective mobility of the carriers in the channel has reduced dramatically over time, due to the continuous scaling of the transistors. Suppressing short-channel effects by increasing the channel doping has led to an increased density of charged impurity scattering sites, thereby reducing the mobility of the carriers in the channel. The intrinsic speed of a logic gate, in first approximation, is proportional to the mobility. Therefore, a lot of research is currently performed in a variety of ways to improve carrier mobility. In this respect also the crystalline orientation of the silicon substrate plays an important role.

Traditionally, CMOS has been fabricated on wafers with a (100) crystalline orientation, mainly due to the high electron mobility and low interface trap density. However, the pMOS transistors on this substrate suffer from a low mobility. By moving away from the (100) orientation, electron mobility is degraded, while hole mobility is improved. Compared to a traditional (100) wafer, a (110) wafer can show hole mobility improvements up to 30% in practice, while electron mobility may have degraded by about 5-10%. An optimum technology, with a much better balance between nMOS and pMOS device performance would be a hybrid-orientation technology: the (100) plane for NMOSts and the (110) plane for the pMOSts [5,6], see also section 3.9.4.

If the pMOS channel is oriented along the <100> direction on a (100) wafer, its mobility and performance may be increased by about 15%, with almost no degradation of the nMOSt performance. Another advantage is that the pMOS transistor will also exhibit a reduced variability. This is only a minor change in the starting wafer, with no further consequences for the device technology and layout (figure 3.2).

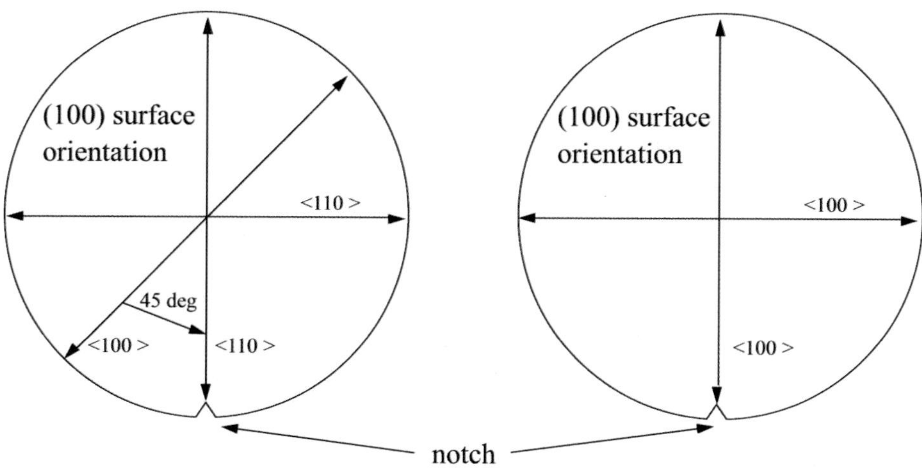

Figure 3.2: *a) traditional notch grinding and b) grinding the notch in the <100> direction (Source: MEMC)*

The only difference is that the wafer flat alignment or notch is changed from the standard <110> direction to the <100> direction. Traditionally, the notch is cut during crystal grinding in the <110> direction (figure 3.2.a). To orient the channel direction along <100>, requires a crystal rotation of 45° to grind the notch in <100> direction (figure 3.2.b). This orientation change is a low cost solution to enhance the pMOS device, logic gate and memory cell performance with no risk or consequences for the integration process. This wafer option is already in use in high volume production

3.2.4   Silicon-on-insulator (SOI)

Bulk-CMOS devices show relatively large source/drain capacitances. This can be avoided with the *SOI-CMOS* devices illustrated in figure 3.3. The complete isolation of nMOS and pMOS transistors associated with this process also completely removes the possibility of latch-up.

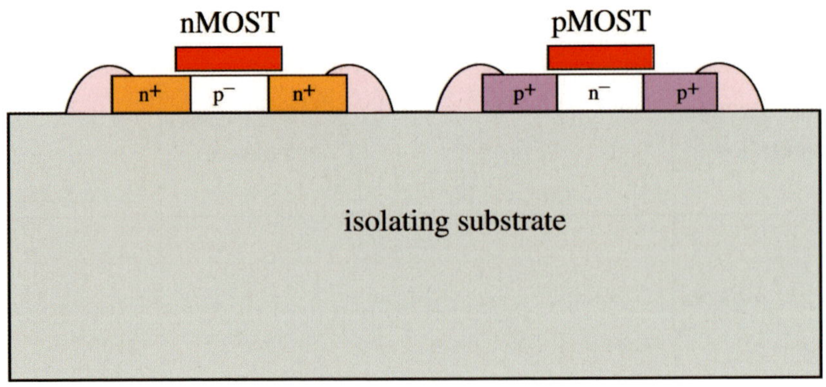

Figure 3.3: *Cross-section of a basic SOI-CMOS process*

Neither the nMOS nor pMOS transistor channels require over-compensating impurity dopes. Very small body effects and source/drain capacitances are therefore possible for both types of transistor. In addition, the n^+ and p^+ source and drain regions do not have bottom junctions. Consequently, the parasitic capacitances are much less than those of the bulk-CMOS processes. This makes the SOI-CMOS process particularly suitable for high-speed and/or low-power circuits. Murphy's law, however, ensures that there are also several disadvantages associated with SOI-CMOS processes. The absence of substrate diodes, for example, complicates the protection of inputs and outputs against the *ESD* pulses discussed in chapter 9.

Sapphire was originally used as the isolating substrate in SOI-circuits, despite the fact that it is substantially more expensive than silicon. The *SIMOX* ('*Separation by IMplantation of OXygen*') process provides a cheap alternative for these *silicon-on-sapphire* or 'SOS-CMOS' processes. Several modern SOI-CMOS processes are based on SIMOX. These processes use a retrograde implantation of oxide atoms to obtain a highly concentrated oxygen layer beneath the surface of a bare silicon wafer. The resulting damage to the wafer's crystalline structure is corrected in an annealing step. The result is shown in figure 3.4.

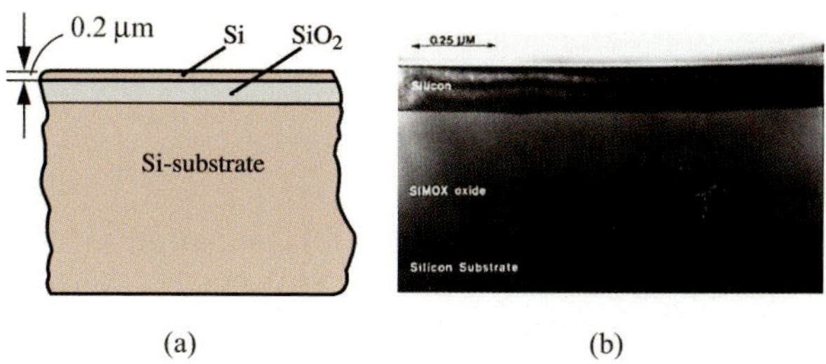

Figure 3.4: *(a) Cross-section of a SIMOX wafer and (b) SEM photograph of such a cross-section*

SIMOX wafers are delivered with a *buried-oxide layer (BOX layer)* (SiO2) varying from less than 50 nm to 150 nm, with a top silicon layer varying from less than 10 nm to 100 nm [7]. This is done to reduce the consequences of damage on the wafer surface. Fully depleted devices can be realised by reducing the thickness of the top layer to below 50 nm, for example, during processing. An alternative to the SIMOX process flow to create SOI, is the Smart Cut process flow (figure 3.5)

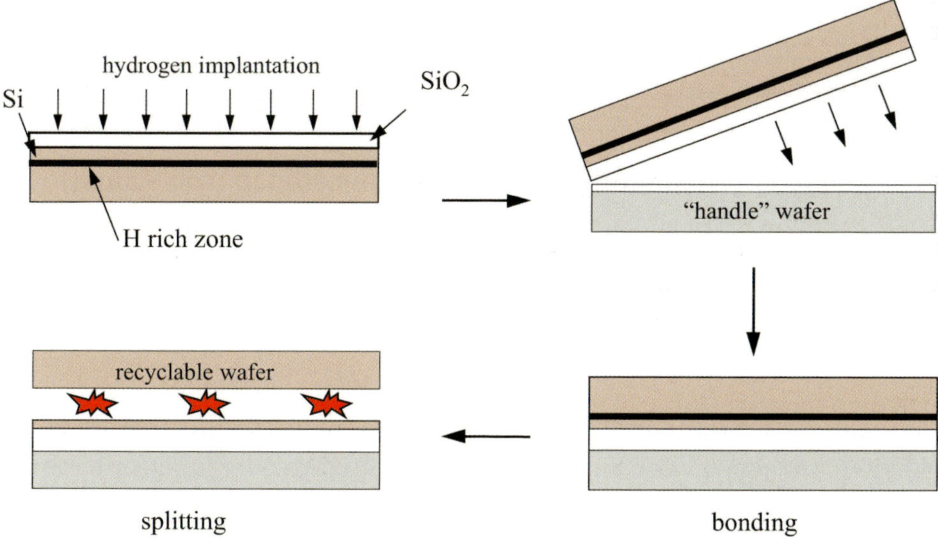

Figure 3.5: *Smart-Cut process flow (Source: SOITEC)*

After the original wafer is first oxidized to create an isolating layer, H$^+$ ions are implanted to form a "weak" layer at a certain distance below the surface. The thickness of the top layer is determined by the implantation energy. Next the wafer is cleaned and bonded upside-down to another wafer for further handling. During the "smart cut" step, the wafer is heated, such that the wafer is split exactly at the implanted weak H^+ layer. The remaining part of the wafer is reused again as original wafer, or as carrier for a new SOI wafer, and the process cycle starts again. Finally, the SOI wafer needs an annealing step to recover the atomic structure, which was damaged during the implantation step. After a CMP planarisation step, the SOI wafer is ready. This smart-cut technology can be used for a wide range of SOI and BOX thickness.

In an SOI device with a thick top silicon layer (figure 3.6.a), this layer can only become partially depleted *(PD-SOI)* during operation, showing such parasitic effects as the floating-body and Kink effect. A thin-body device (<50 nm) (figure 3.6.b) will become fully depleted *(FD-SOI)* and does not show these effects.

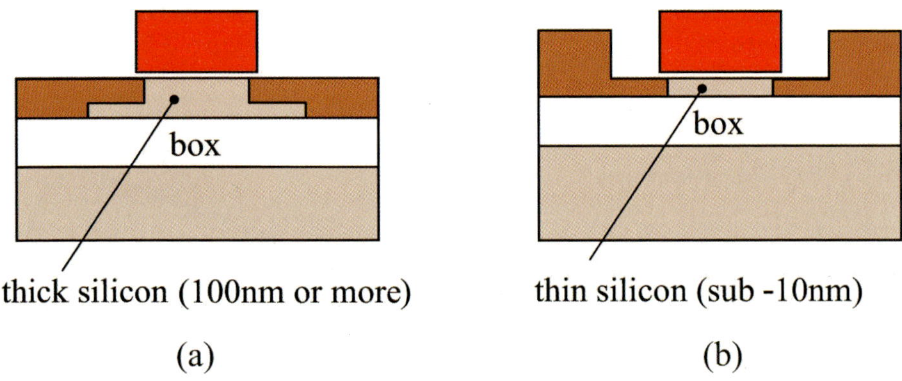

Figure 3.6: *Cross section of a (a) partially-depleted SOI device and (b) a fully-depleted SOI device*

For advanced low-voltage CMOS (≤ 1 V) system-on-chip designs with digital, analogue and RF parts, SOI is expected to offer a better performance than bulk CMOS technology [8,9]. SOI is said to deliver more speed at the same power consumption, or to consume less power at the same speed. Furthermore, SOI realises better isolation between digital, analogue and RF parts on the IC. Those circuits will therefore be less affected by substrate noise. Additionally an SOI transistor has lower parasitic capacitances and consequently exhibits a better RF performance. SOI devices are thermally insulated from the substrate by the buried-oxide layer. This leads to a substantial elevation of temperature *(self-heating)* within the SOI device, which consequently modifies the output *IV*-characteristics of the device, showing negative conductance. These effects, which are considerably larger in SOI than in bulk devices under similar conditions, must be taken into account by device technology engineers, model builders and designers. Since the body is isolated, SOI circuits show several advantages, compared to bulk-CMOS:

- smaller junction capacitances

- no deep well required (this is especially an advantage for FD-SOI)

- less n^+ to p^+ spacing, due to absence of wells

- significant reduction in substrate noise (questionable at high frequencies $> 1.5\,\text{GHz}$)

- no manifestation of latch-up

- reduced soft-error rate (SER), because the electron-hole pairs generated in the substrate can not reach the transistors

- steeper subthreshold slope, which can be close to the theoretical limit of $63\,\text{mV/decade}$, compared to around $80\,\text{mV/decade}$ for bulk CMOS devices

The future for planar partially depleted SOI devices is not completely clear. The relative performance benefit due to the smaller junction capacitances of SOI will gradually reduce because this advantage diminishes with scaling. Junction area capacitance decreases with the square of the scaling factor while gate and perimeter capacitances decrease only linearly. Next to this, the increasing impacts of interconnect capacitances and delays will also reduce the performance benefits of SOI.

For the 45 nm node most semiconductor manufacturers still use bulk CMOS as their main process technology. However, beyond this node, FD-SOI may become a good alternative to bulk-CMOS. Since the channel region is fully depleted, it largely eliminates the neutral body. It therefore hardly exhibits the floating-body, history and kink effects. Moreover, it is expected to show improved short-channel effects (SCE) and drain-induced barrier lowering (DIBL). FD-SOI requires a reduced channel-doping concentration, leading to a higher mobility and a much steeper *subthreshold slope*, which almost matches the ideal value of $\approx$

$63\,\mathrm{mV/decade}$ (figure 3.7), compared to the $\approx 80\,\mathrm{mV/decade}$ for a bulk-CMOS process.

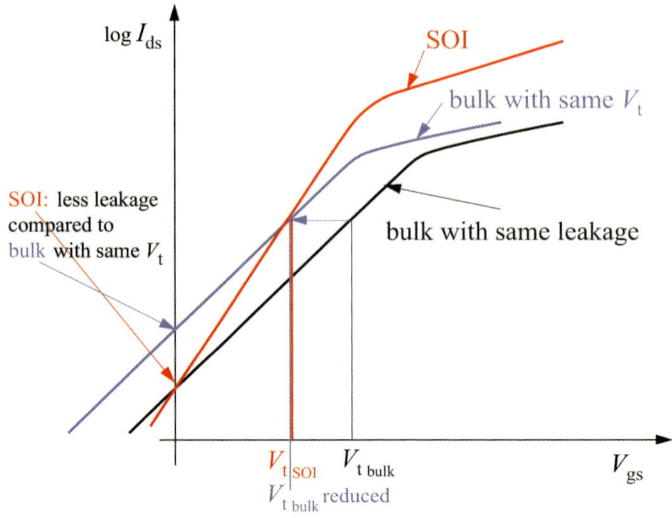

Figure 3.7: *Schematic illustration of current characteristics and sub-threshold behaviour of bulk-CMOS and FD-SOI*

The diagram shows that in an SOI process, a transistor may have a lower V_{T} than in a bulk-CMOS process, while carrying the same subthreshold leakage current. This advantage can either be used for speed improvement, when running SOI at the same supply voltage as bulk-CMOS, or for power reduction, when running SOI at a lower supply voltage but at the same speed. FD-SOI allows sub-1V RF circuits, with improved F_t and F_{max} and reduced noise levels.

The transistors in such a nanometer FD-SOI process are fabricated in a thin film, with a thickness $\approx$ 10-20 nm. Because the body between source and drain is fully depleted, the V_{T}-spread in these devices is much less dominated by the doping levels. Instead, it now depends heavily on the film thickness, whose uniformity across an 8 inch or 12 inch wafer has become a major criterion in the success of FD-SOI. This uniformity is therefore likely to have a more global (inter-chip) than local (intra-chip) impact on the variability in device operation. Below the 22nm node planar SOI devices are expected to show device current degradation due to "quantum confinement" [10].

104

Many other alternative device and process options are seen as potential candidates for technologies beyond the 45 nm node. A flavour of technology options in both the devices and interconnects is presented in section 3.9.4.

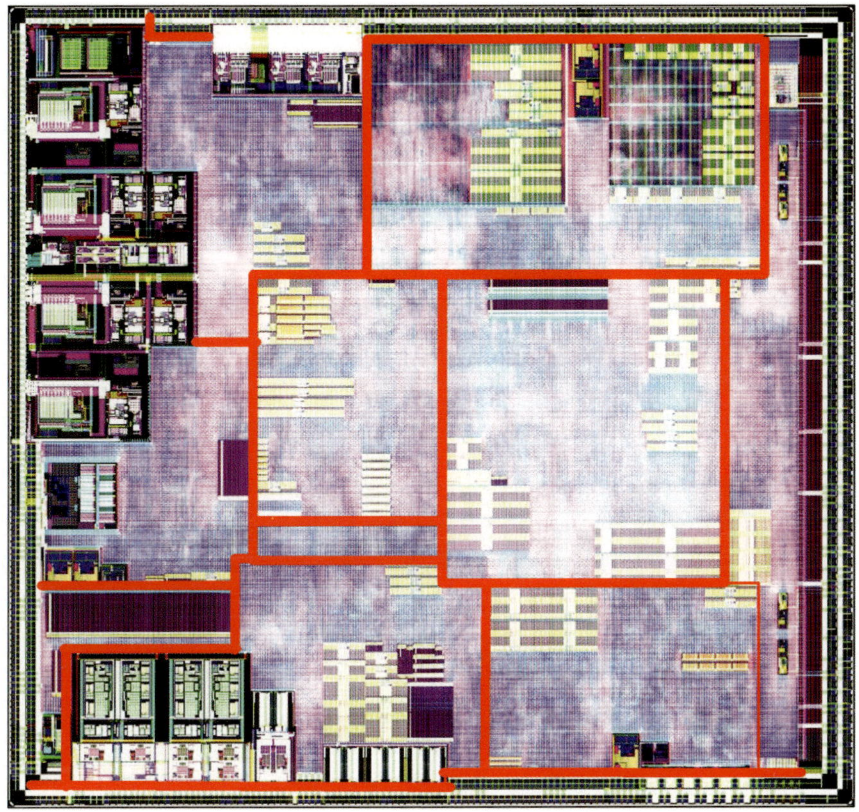

Figure 3.8: *Example of a complex signal processor chip, containing several synthesized functional blocks (Source: NXP Semiconductors)*

3.3 Lithography in MOS processes

3.3.1 Lithography basics

The integration of a circuit requires a translation of its specifications into a description of the layers necessary for IC manufacture. Usually, these layers are represented in a *layout*. The generation of such a layout is usually done via an interactive graphics display for handcrafted layouts, or by means of synthesis and place-and-route tools, as discussed in chapter 7. Figure 3.8 shows an example of a complex IC containing several synthesized functional blocks.

A complete design is subjected to functional, electrical and layout design rule checks. If these checks prove satisfactory, then the layout is stored in a computer file (gds2 file). A software program (post-processor) is used to convert this database to a series of commands. These commands control an *Electron-Beam Pattern Generator (EBPG)* or a *Laser-Beam Pattern Generator (LBPG)*, which creates an image of each mask on a photographic plate called a *reticle*. Such a reticle contains a magnified copy of the mask patterns. The reticle pattern is thus demagnified as it passes through the projection optics. During the printing process, often pellicles are used to protect the reticle from harmful particles. A *pellicle* is a very thin transparent membrane adhered to a metal frame, which keeps particles out of focus during the lithographic process, so it will not image onto the wafer and reduces the possibility of printing defects. Particularly with the introduction of 193 nm, the light transmission loss in the pellicles increases with the number of exposures, such that they frequently need to be replaced.

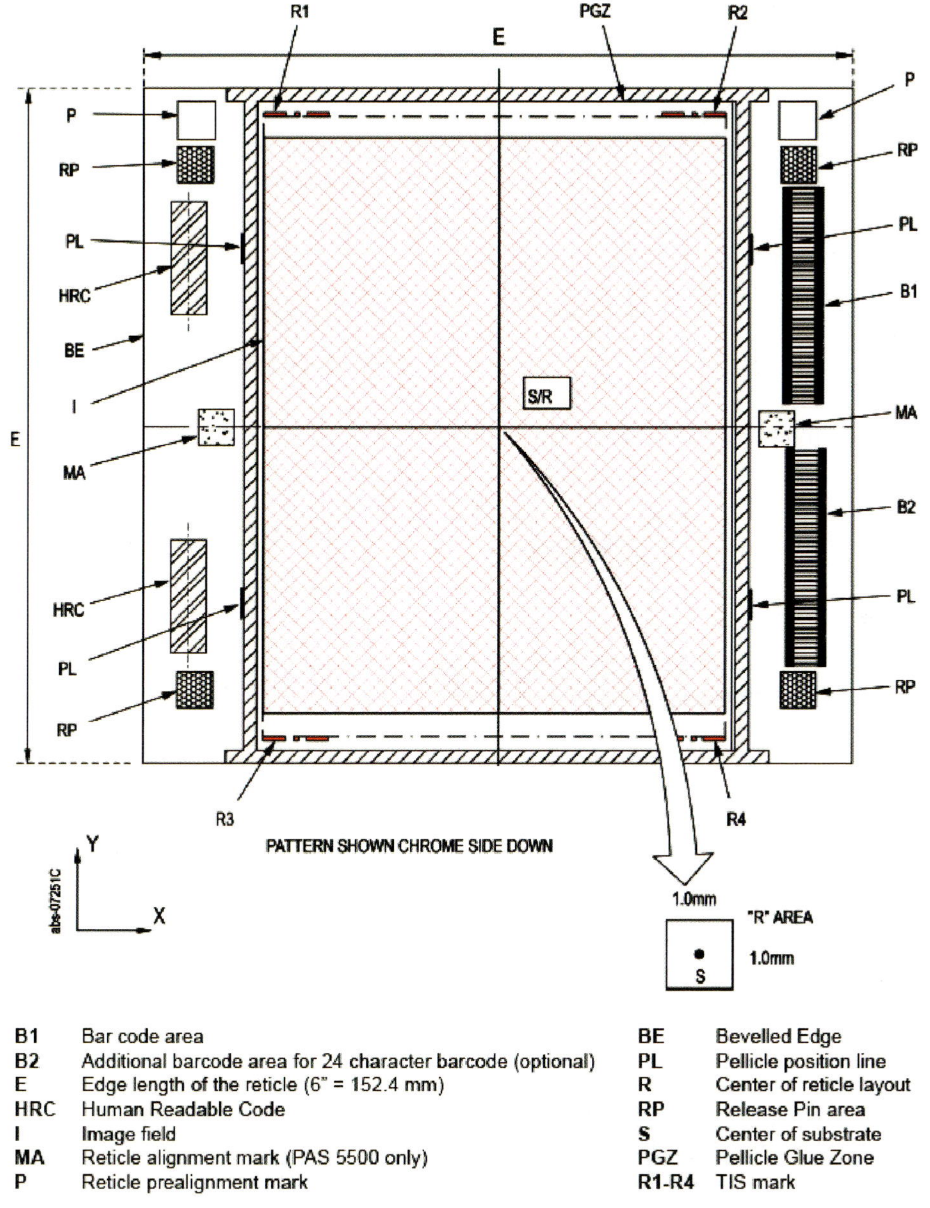

Figure 3.9: *Schematic layout of a 4x reduction reticle for step and scan systems (Source: ASML)*

Needles to say that F should be minimized and DOF should be maximized. In fact, a trade off has to be made. Whereas the resolution of the imaging system is improving (reducing F) by increasing NA, its depth of focus will be reduced. This reduction can be compensated by a highly planarized top surface for exposure and by DOF enhancement techniques, which are outside the scope of of this text.

For many technology generations in the past, the values for k_1 and NA were about the same, resulting in minimum feature sizes, which were about equal to the wavelength of the used light source. $0.35\,\mu$m feature sizes were mostly printed on i-line (365 nm) steppers. From a cost perspective, there is a strong drive to extend the wavelength of the light source to smaller technologies. The 248 nm Deep-UV (DUV) steppers, with a krypton-fluoride (KrF) light source, are even used for 90 nm feature sizes, while the argon-fluoride (ArF) 193 nm DUV can potentially be used for feature sizes until 60 nm with dry lithography and until 40 nm with immersion lithography. Steppers (scanners) with shorter wavelengths will become very expensive and need many workarounds, as traditional optical lithography will no longer be viable at much shorter wavelengths.

When creating smaller feature sizes with the same wave length, we need to compensate for non-ideal patterning, such as: lens aberrations, variations in exposure dose, pattern sensitivity, die distribution across the reticle and the field (reticle) size. The extension of the use of the 193 nm wavelength to sub 100 nm technologies can not be done without the use of several additional *Resolution Enhancement Techniques (RET)*: Optical-Proximity Correction (OPC), Off-Axis illumination (OAI), Phase-Shift Masks (PSM), better resist technologies, immersion lithography and design support. In the following these techniques are discussed in some detail to present the reader a flavour of the increasing complexity and costs of the lithographic process, starting with the basic conventional binary mask.

The conventional binary mask is used in combination with the 193 nm light source to depict features with half pitch (HP) sizes as small as 90 nm. A *binary (photo) mask* is composed of quartz and chrome features (figure 3.11) [11]. Light passes through the clear quartz areas and is blocked by the opaque chrome areas. Where the light reaches the wafer, the photo-resist is exposed, and those areas are later removed in the develop process, leaving the unexposed areas as features on the wafer. Binary masks are relatively cheap and they show long life times, because they can be cleaned an almost infinite number of times. Moreover, they use the lowest exposure dose and enable high throughput rates. Preferably all masks should be binary masks since it would reduce the overall production costs.

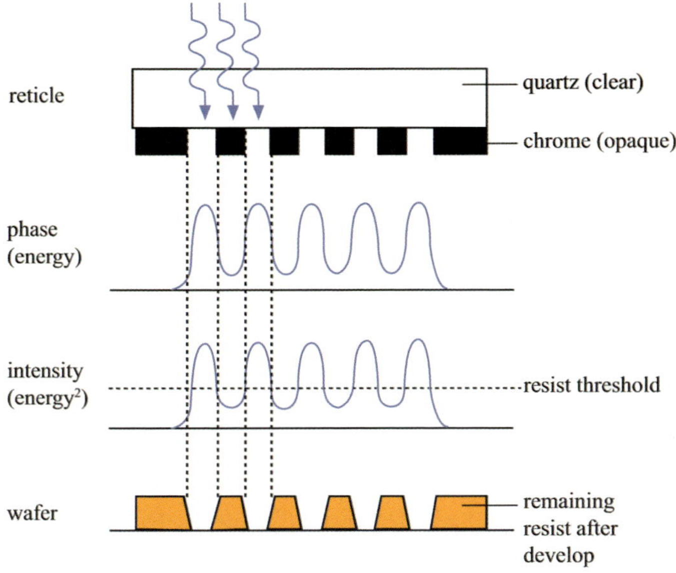

Figure 3.11: *Basic use of a binary photo mask (Source: ASML)*

As feature sizes and pitches shrink, the resolution of the projection optics begins to limit the quality of the resist image. In the example above, there is significant energy (and intensity, which is proportional to the square of the energy) even below the opaque chrome areas, due to the very close proximity of the neighbouring clear quartz areas. This "unwanted" energy influences the quality of the resist profiles, which are ideally vertical.

A conventional binary mask with a dense pattern of lines will produce a pattern of discrete light diffraction orders (-n, -(n-1),.., -2, -1, 0, 1, 2,.., n-1, n). The example in figure 3.12 shows a so-called *three-beam imaging* system. Here a binary mask is used in combination with a projection lens that acts as a first order ray filter. This prevents the capture of higher order rays.

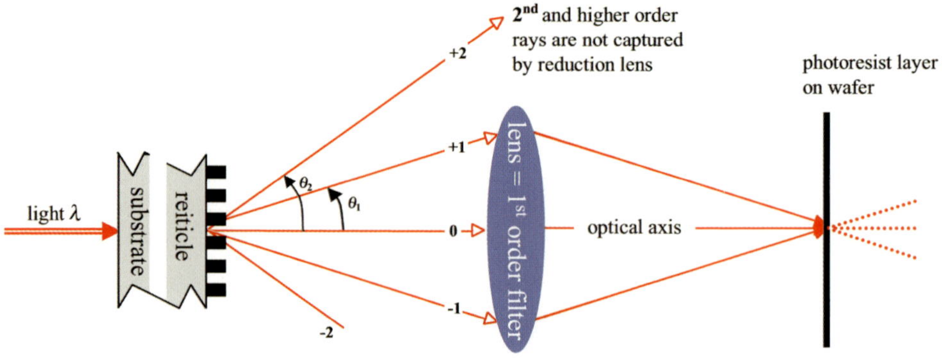

Figure 3.12: *Three-beam imaging concept*

The interference of the zero-order diffracted light beam with the two first-order diffracted light beams produces a reduced (4:1) image of the pattern. If the line pitch in the pattern becomes smaller, the first-order light beam diffracts with an angle, which is too large to be captured by the lens, which is then incapable of producing the right image. Therefore phase-shift techniques, such as off-axis illumination and PSM, are designed to "sharpen" the intensity profile, and thus the resist profile, which allows smaller features to be printed. When a binary mask is illuminated at a different from normal angle, this angle can be chosen such that one of the first-order diffracted light beams can no longer be captured by the lens and the image is produced by only two diffracted beams (the zero and remaining first-order). This so-called *off-axis illumination (OAI)* technique (figure 3.13) is therefore an example of *two-beam imaging*. A further optimisation of this imaging technique can be achieved by choosing the angle of illumination such that the remaining beams are symmetric with respect to the center of the lens. An OAI system can improve the resolution limit of a dense line pattern with a factor of two.

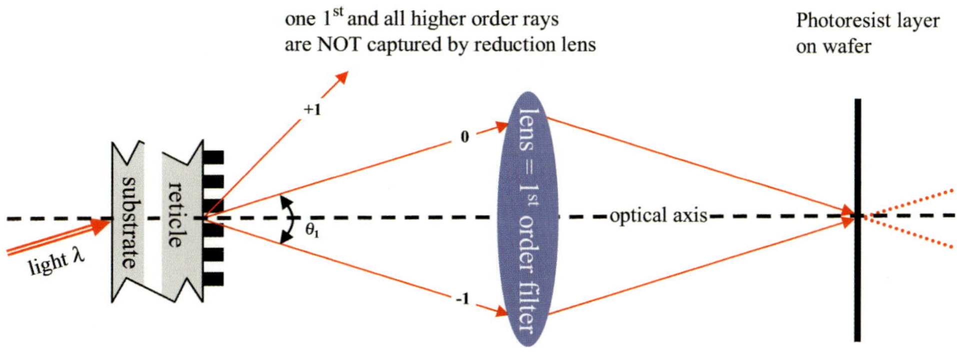

Figure 3.13: *Off-axis illumination (two-beam imaging concept)*

However, another benefit from a two-beam imaging system comes from the enhanced *depth-of-focus (DOF)*. It can be seen that in a three-beam imaging system (figure 3.12), the first-order diffracted beams travel across a different path than the zero-order beam, before arriving at the wafer surface. It can therefore provide only a very narrow range, in which the zero and first diffraction orders remain in phase (basically only in the focal plane), limiting its depth of focus. Outside this range it creates a phase error. A minor displacement of the wafer out of the focal plane causes an increase of this phase error and leads to a degraded image at the wafer surface. In a two-beam imaging system (figure 3.13), assuming full spatial symmetry, the diffraction patterns are in phase and will interfere properly. The same wafer displacement in such a system will result in a satisfactory image over a longer range, thereby increasing its depth of focus.

An alternative to off-axis illumination is the *Phase-Shift Mask (PSM)* technology, which has been pioneered in recent years to extend the limits of optical lithography. PSM technology is divided into two categories: attenuated PSM and alternating PSM.

Attenuated Phase Shift Masks (AttPSM) form their patterns through adjacent areas of quartz and, for example, molybdenum silicide (MoSi). Unlike chrome, MoSi allows a small percentage of the light to pass through (typically 6% or 18%). However, the thickness of the MoSi is chosen so that the transmitted light is 180° out of phase with the light that passes through the neighboring clear quartz areas (figure 3.14) [11]. The light that passes through the MoSi areas is too weak to expose the resist, and its 180° phase shift reduces the intensity in these areas such that they appear to be "darker" than similar features in chrome. The result is a sharper intensity profile which allows smaller features to be printed on the wafer. The 180° phase shift is only achieved for light at a given fixed wave length. AttPSM masks can therefore only be used for one type of scanners only, while binary masks can be used for scanners with different wavelengths.

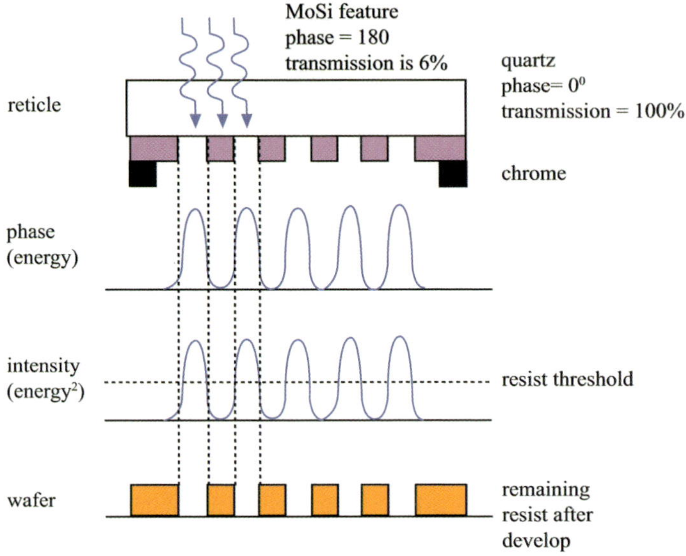

Figure 3.14: *Basic use of an attenuated phase-shift mask (attPSM) (Source: ASML)*

In fact, the use of attPSM filters one of the first order diffracted light beams of a three-beam imaging system (figure 3.12), which makes it a two-beam imaging system, similar to OAI imaging (figure 3.13). Figure 3.15 shows a comparison of the three different imaging systems. It clearly shows the improvement of the DOF in the two-beam imaging systems.

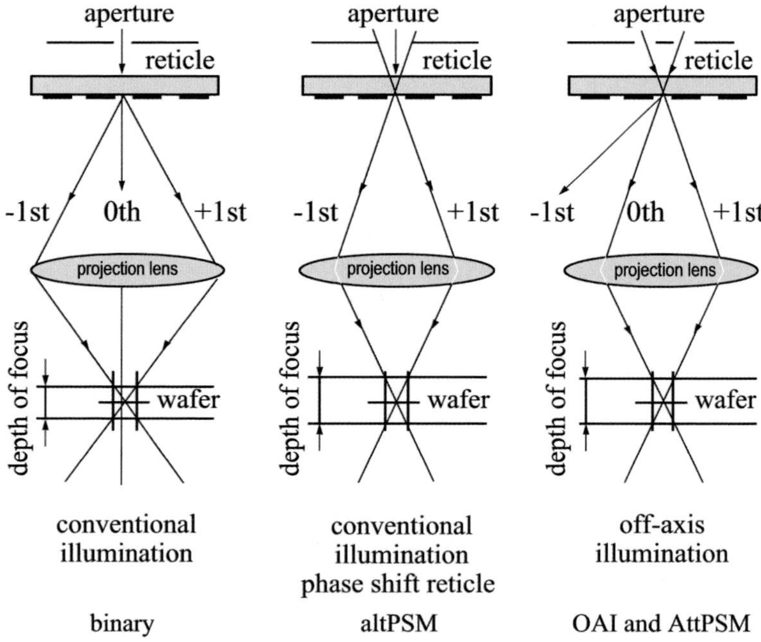

Figure 3.15: *Comparison of the three different imaging systems (Source: ASML)*

OAI systems and attenuated phase-shift masks are used for critical patterns that require higher resolution than photolithography systems that employ binary masks only. An alternative powerful but complex two-beam illumination system is the *alternating phase-shift mask (altPSM)* concept (figure 3.16). Such masks employ alternating areas of chrome, $0°$ phase quartz and $180°$ phase-shifted quartz to form features on the wafer [11]. The pattern is etched into the quartz on the reticle causing a $180°$ phase shift compared to the unetched areas ($0°$ phase). As the phase goes from positive to negative, it passes through 0. The intensity (proportional to the square of the phase) also goes through 0, making a very dark and sharp line on the wafer. The process of manufacturing the mask is considerably more demanding and expensive than that for binary masks. Furthermore, the AltPSM requires an additional binary "trim" mask and exposure step, resulting in extra costs and decreased stepper/scanner throughput, however it enables excellent CD control.

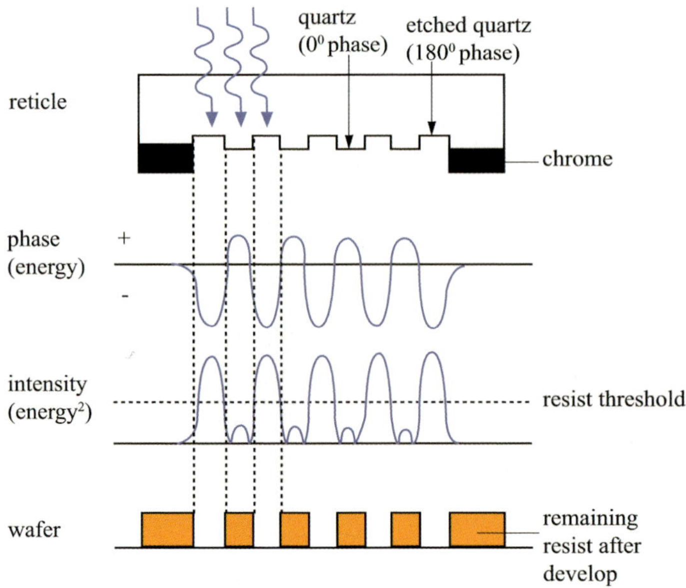

Figure 3.16: *Basic use of an alternating phase-shift mask (altPSM) (Source: ASML)*

AltPSM is used for the production of high-performance ICs that only allow extremely limited variations in line width, such as high-speed microprocessors.

As explained, the above presented lithographic techniques are basically applied to increase the resolution and/or depth of focus of the total illumination system. Another technique, which is currently already applied to enhance the lithographic properties is called *immersion lithography*. If we immerse the photolithographic process in water ($n = 1.43$) and if we assume that $\sin\alpha$ in the expression (3.1) can reach a maximum value of 0.95, then this "water-immersion lithography" can yield an NA close to 1.37. Only the lower part of the optics is immersed in water (figure 3.17).

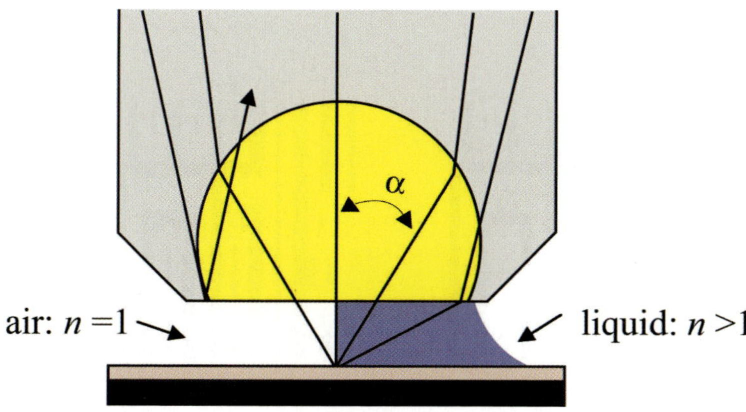

Figure 3.17: *Basic principle of immersion lithography (Source: ASML)*

The left half in the figure shows the diffraction of the light beams in air, with a diffraction index $n = 1$ and some of the beams being reflected. The right halve uses an immersion liquid with $n > 1$, which reduces the amount of reflected light, increasing the resolving power and allowing finer feature sizes. Immersion lithography also improves the DOF, which may resolve some of the related topography problems.

Compared to an air-based system, immersion lithography shows a number of additional problems. To achieve a high throughput, the stage has to step quickly from one chip position to the next, which may create *bubbles* into the water, deteriorating the imaging capability of the system. There are several solutions to this problem, but these are beyond the scope of this text.

Using one of the above described resolution enhancement techniques (RETs) is a prerequisite to create lithographic images with a satisfactory resolution and *DOF*. But it is not sufficient. When printing patterns with sub-wavelength resolution they need to be compensated for the aberrations in the patterning. In other words: the fabricated IC patterns are no longer accurate replica of the originally designed patterns. So, we need already to compensate (make corrections) for these shortcomings in the mask. Figure 3.18 shows how *optical proximity correction (OPC)* is applied in the mask-definition process. The right mask pattern is used during lithography, to get left (original layout) pattern image on the chip. More optimal imaging results can be achieved by using so-called *subresolution assist features (SRAFs)*, such as scattering bars and hammerheads, which are not printed onto the wafer, but help to reduce resolution enhancement variations across the mask.

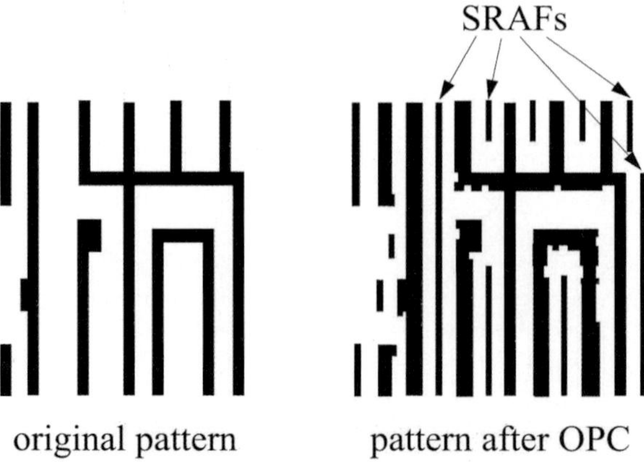

Figure 3.18: *OPC (including SRAFs) applied in the mask-definition process (Source: ASML)*

This has several consequences for the layout designer: he should leave enough space to add OPC features and/or he should draw the patterns with constant proximity and/or he should leave enough space to add SRAFs. It will certainly make the design process more complex.

While the above described RETs improve the resolution of the imaging system, the use of OPC masks will make them work. Mask costs, however, very much depend on the applied technology. When normalising the costs of a binary mask to 1, then an attPSM (without OPC) mask would cost 1.2 times as much and an attPSM (with OPC) mask 2.5 times. The use of altPSM is much more costly (6 times and 10 times more for altPSM without and with OPC, respectively), since it requires an additional binary trim mask and thus needs double exposure.

For the time being, we still have to rely on innovations that extend the use of photolithography beyond the 40 nm node. Therefore support from the design side might alleviate some of the expected problems when extending the use of 193 nm lithography into the sub 50 nm CMOS technologies. To improve yield, complex Design for Manufacturability (DfM) design rules have already been used in many technology nodes.

For the 45 nm node this is certainly not enough. It will also require strict *Design for Lithography (DfL)* design rules. DfL, also called *litho-friendly design*, litho-driven design, or litho-centric DfM, is focused on more regular layout structures. It will simplify the lithographic process, it supports SRAFs and might reduce the mask costs. It may also lead to a more aggressive scaling and to yield improvement due to a smaller variety of patterns to be printed. Moreover, more regularity in the standard cells may also lead to a better portability to the next technology node. Figure 3.19 shows two layout versions of a standard cell: the original layout with a plot of simulated line widths and the litho-friendly layout with a plot of simulated line widths, showing more regularity.

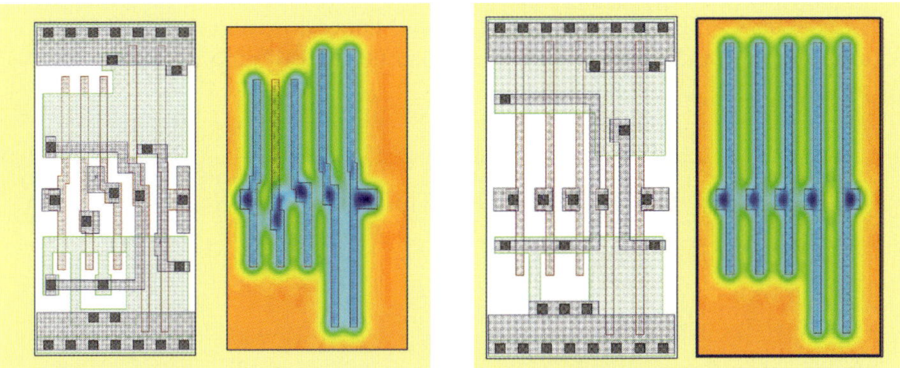

Figure 3.19: *Comparison of an original (a) and a litho-friendly layout (b) with more regularity (Source: NXP Semiconductors)*

For this particular cell, litho-friendly design shows a relatively large impact on the cell area. For an average library, however, the area increase can be limited to just a few percent. Next to the already discussed implications of RET and DfL for layout design, these techniques are increasingly being supported by the design flow and get more and more attention from Design for Yield (DfY) EDA-tools and tool vendors. An overview of EDA-vendor DfY activities is presented in [12].

Litho-friendly design usually uses a limited number of poly pitches. Such a fixed-pitch litho-friendly library design is a step towards a *fully-regular library architecture*. Next to the process spread caused by lithographic imperfections, such an architecture may also reduce the influence of other process-spread mechanisms, by using only one size NMOS and one size pMOS transistor. The high-density gate-array architecture shown in figure 7.34 is an example of such an architecture, which can also be used as standard-cell template.

Before we continue our discussions, it is good to present some typical sizes and dimensions which are characteristic for a 45 nm CMOS process (table 3.1). It shows that there are different definitions for critical dimensions, feature sizes and pitches. They not only depend on the type of circuit, but also on the particular phase during the lithographic and manufacturing process.

Table 3.1: Various definitions for critical dimensions (CD), feature sizes (F), pitches and channel lengths (L, L_{eff}) depending on the lithographic and manufacturing process step and on the type of circuit, for a 45 nm process

Dimension	LOGIC		Stand-alone memory
	Low cost	High performance	(excluding SRAM)
pitch	180	150	90
litho print	$65 = CD$	55-70	45
resist trim	50	45	-
poly etch	$45 = L = F$	40	45
out diffusion	$40 = L_{\text{eff}}$	35	45

Let's summarize the individual contributions of the above-described RETs: the combination of PSM and OPC may lead to a minimum k_1 of about 0.25, while water immersion can lead to a maximum NA of approximately 1.37. Using these values, for a 193 nm lithography, in expression (3.1) for F, leads to a minimum feature size ($\equiv$ half pitch; most common for memories) of around 40 nm. Smaller line widths, e.g., 30 nm, can be obtained, but with larger spacings, e.g., 90 nm, at the cost of additional area. This will most commonly be used for the manufacture of logic ICs (ASICs).

It is thus expected that the lifetime of 193 nm optical lithography can be extended for one or more technology nodes. Potentially, the feature size can be smaller by using a light-transparent liquid with a larger refraction index than water. It is clear that all additional lithographic solutions to enable smaller feature sizes, will reduce pattern aberrations, but not prevent them. After reaching the limits of 193 nm immersion lithography, so after the year 2010, 157 nm DUV (deep ultra violet)-line (from a fluorine laser source) lithography could have been an option. However, other immersion liquids have to be explored (e.g., Fomblin oil), as the transparency of water is lost at wavelengths below 185 nm. According to today's knowledge, 193 nm lithography allows immersion liquids with a larger refraction index (n) than that of suitable liquids in 157 nm immersion lithography. This latter lithography is expected to only extend the lifetime of photolithography for one more technology node. Industry has therefore decided to skip the 157 nm lithography because the development costs are not expected to be recovered. Below 157 nm optically transparent materials used for lens and mask will become increasingly absorbent. A potential work-around for this absorption problem would be to build a completely reflective optical system, but the reflection efficiency of even the best-polished optical devices is limited to only a few percent. Extreme-UV lithography, which will discussed below, uses reflective optics.

Finally, particularly the semiconductor memory vendors have found a way to increase bit density without the use of very advanced and expensive lithography tools. By using multiple layers of silicon *(3-D stacked silicon)*, memory capacity can be increased dramatically, without increasing the footprint of the memory chip. Some SRAM products use cells with three vertically stacked transistors, while some flash memories are currently being fabricated using two layers of memory cells. OTPs may be built from four layers already. NAND-flashes with eight layers of silicon are in development. 3-D technologies are currently only economically viable when the complexity of the devices fabricated in these stacked layers is very limited. Because non-volatile memories use only one type of transistor in the cells (see chapter 6) they are particularly suited for 3-D stacking. Therefore these layers are only used to fabricate arrays of memory cells and require only three masks per layer, which can be fabricated by existing photolithography tools. These arrays use the peripheral address selection and sense amplifier circuits of the original first memory array.

3.3.2 Lithographic alternatives beyond 40 nm

Beyond 40 nm there are a couple of alternative solutions:

- Use of an immersion liquid with higher refraction index, creating a so-called *super-fluid NA* in the range of 1.55 to 1.75. This would also require different lenses.

- Use of a *Double Patterning Technology (DPT)*. When the pitch of two lines in a dense pattern is less than 80 nm, it becomes a sub-resolution pitch, which can no longer be imaged correctly, with current lithographic techniques. Therefore this can be done with an image split: first image the odd lines with twice the minimum feature pitch (figure 3.20) and then image the even lines, also with twice the pitch.

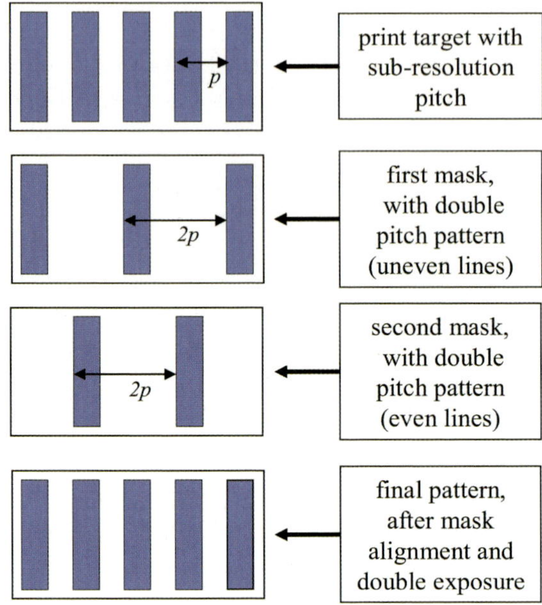

Figure 3.20: *Example of double patterning*

This procedure requires two masks and two exposures. The biggest challenge is the high accuracy of the alignment of the masks during exposure.

- Use of *spacer lithography*.

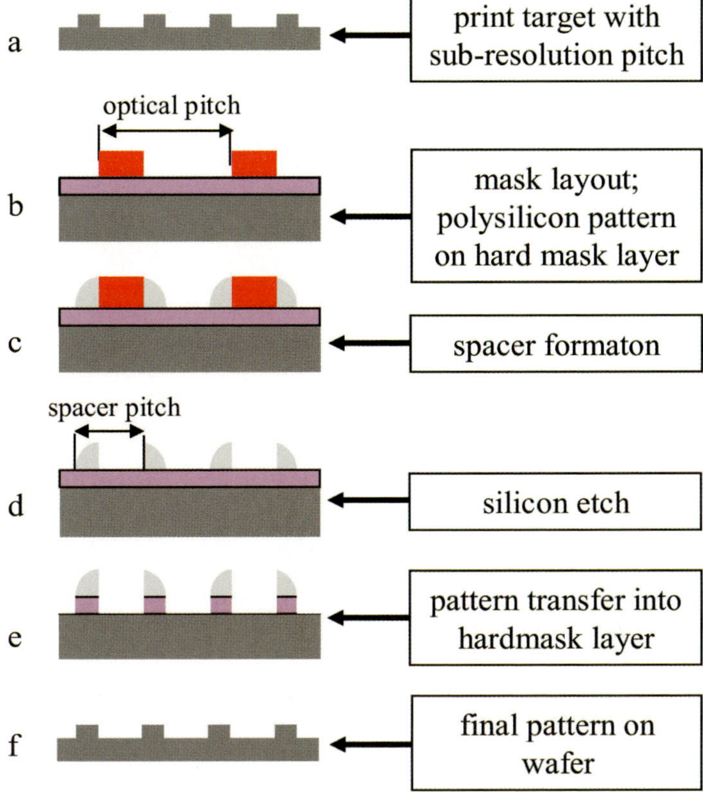

Figure 3.21: *Basic steps in spacer lithography*

In this technology the final pattern on the wafer is created by the formation of sub-resolution features during semiconductor process steps, rather than by sub-resolution lithography. The process flow in this technology is as follows (figure 3.21). The print target is shown in (a). As a first step, a hard mask layer is deposited or grown on the wafer. To support the formation of sub-resolution spacers a sacrificial polysilicon layer is deposited on the wafer and patterned with a relatively large optical lithography pitch (b). Next, an oxide (or nitride or other) layer is deposited on top of the structure and then etched back until sub-resolution sidewall spacers are left (c). Then the sacrificial polysilicon is removed (etched) (d), followed by a pattern transfer from spacer to hard mask (e). Finally the pattern in the hard mask is used to create the final pattern on the wafer (f). This spacer technology is a convenient approach to achieve sub-resolution patterning with relatively large optical resolution pitches, avoiding problems of e.g., overlay between successive exposures in a double patterning technology. Another advantage of this technique is that the printed *critical dimension uniformity (CDU)* is independent of the *line-edge roughness (LER)*. LER is caused by the diffusion of resist during a heat step after the exposure (post-exposure bake at 200-220°C), but before the development of the resist. This diffusion is random and may lead to diffusion lengths of 40 nm, which causes intra-line variations leading to frayed lines. In spacer technology, however, the pattern transfer is done through spacers and not through resists, showing almost no LER. A disadvantage of the spacer lithography is that it is only applicable for mono *CD (critical dimension)* which reflects the smallest geometrical features (contacts, metal width, trenches, etc. which can be fabricated), so, for patterns with only one width. Patterns with features that also have two times the line width can be produced by the formation of two spacers directly positioned next to each other. For printing patterns with different widths, DPT needs to be used. An example of nanofabrication spacer lithography can be found in [13].

3.3.3 Next generation lithography

- Use of *Extreme-UV (EUV) lithography*. With a light source wave length of 13.5 nm, EUV is often regarded as the most probable potential lithography solution for technology nodes beyond 30 nm. However, EUV "light" is absorbed by all materials, including air. Therefore mirrors have to be used in a vacuum-based system with reflective instead of refractive optics and reticles. Still a lot of problems need to be solved before it can be used in high-volume production. A few of them will be mentioned here. First, there is no suitable resist for high-volume production available, yet. Second, because the light needs to propagate through a system of six to seven mirrors, and one mirror absorbs about 70% of the light intensity at the EUV-wavelength, extremely powerfull light sources are needed in combination with relatively long exposure times. To generate 1 W of EUV power, the RF power needed to activate the plasma light source may be as high as 500 kW. The lack of appropriate power sources and resists is a major bottleneck for bringing this lithography quickly to the market. This explains the need for an improved light-transmission system to improve the throughput time and reduce the power consumption.

In 2006 the first EUV lithography tools (demo tool: US$ 65 million!!) have already been shipped. It is not meant for production but it will support R&D programs at IMEC (Leuven, Belgium) and at CNSE (University of Albany, New York) [14]. Production EUV tools are not expected to enter the market before 2010.

- Use of alternative techniques to fabricate image-critical patterns in sub 30 nm technologies. For many years, *X-ray lithography (XRL)* has been a potential candidate for *next-generation lithography (NGL)*. It uses x-rays, which generate photons with a wavelength of roughly 1nm to expose the resist film deposited on the wafer, enabling much finer features than current optical lithography tools. However, it has some major disadvantages. Conventional lenses are unable to focus x-rays and, consequently, XRL tools can not use a lens to shrink a mask's features. Therefore its 1:1 pattern transfer methodology requires mask patterns with only one-fourth of the feature sizes used in the 4:1 photo-lithography masks. In addition, it requires an extremely expensive synchrotron, which converts an electron beam into an x-ray beam. It is therefore expected that the use of XRL will be limited to fabrication processes that create niche devices, such as MEMS.

- An alternative to photolithography is the *nano-imprint lithography (NIL)*, which may also be an option for the 32 nm node and/or beyond. This 1:1 technology is based on physically pressing a hard mold (typically identical to the quartz/chrome material commonly used for optical lithography) with a pattern of nano structures onto a thin blanket of thermal plastic monomer or polymer resist layer on the sample substrate, to which the structure needs to be replicated. This imprinting step is usually done with the resist heated, such that it becomes liquid and can be deformed by the pattern on the mold. After cooling down, the mold is separated from the sample, which now contains a copy of the original pattern. Its mayor advantage is that it can replicate features with nanometer dimensions [15]. This process is already used in volume production in electrical, optical and biological applications. For semiconductor applications, the "step-and-flash" imprint (SFIL) seems to be the most viable one. It allows imprinting at room temperature with only a little pressure using a low-viscosity UV curing solution instead of the resist layer. The higher the sensitivity to UV, the less exposure time the solution needs and the higher the throughput. In this imprint technology some of the wafer process complexity has moved to the fabrication of the mold. Still a lot of key issues, particularly related to overlay and defects, need to be solved, but the results of this disruptive technology, so far, are promising. A potential barrier for using the imprint lithography is that it requires very advanced lithographic processes to create the patterns on the mold. Because it is a 1:1 pattern transfer process, the pattern dimensions are only one-fourth of those printed on a photo mask, which is one of its major challenges. Moreover, low throughput may become the real show stopper for this technology. Reference [15] discusses the process and potentials of nano-imprint in more detail.

Moore's law is driven by the economical requirements of the semiconductor markets. This means that all semiconductor technologies (litho, diffusion, packaging, testing, etc.) are cost driven. For the lithography it means that there is a constant drive to make masks cheaper or to use cheaper masks for certain low-resolution process steps. Binary masks are relatively simple and cheap, but guarantee high throughput and can be non-destructively cleaned. Attenuated PSM masks are immersed in a chemical liquid for cleaning, which is a destructive process, such, that they can only be cleaned about three times and are therefore much more expensive.

To minimise mask costs during the fabrication process, the more expensive masks are only used to image those patterns that really need the smallest feature sizes. For the production of one type of memory for example, different mask categories can be used. To reduce the production costs of a flash memory process of 22 masks, it may use 4 ArF (attPSM + OPC) masks, 12 KrF (6 binary and 6 attPSM) and 6 I-line (binary) masks.

3.3.4 Mask cost reduction techniques for low-volume production

On so-called *multi-project wafers (MPW)* several products are included on the same maskset to reduce overal mask costs (figure 3.22).

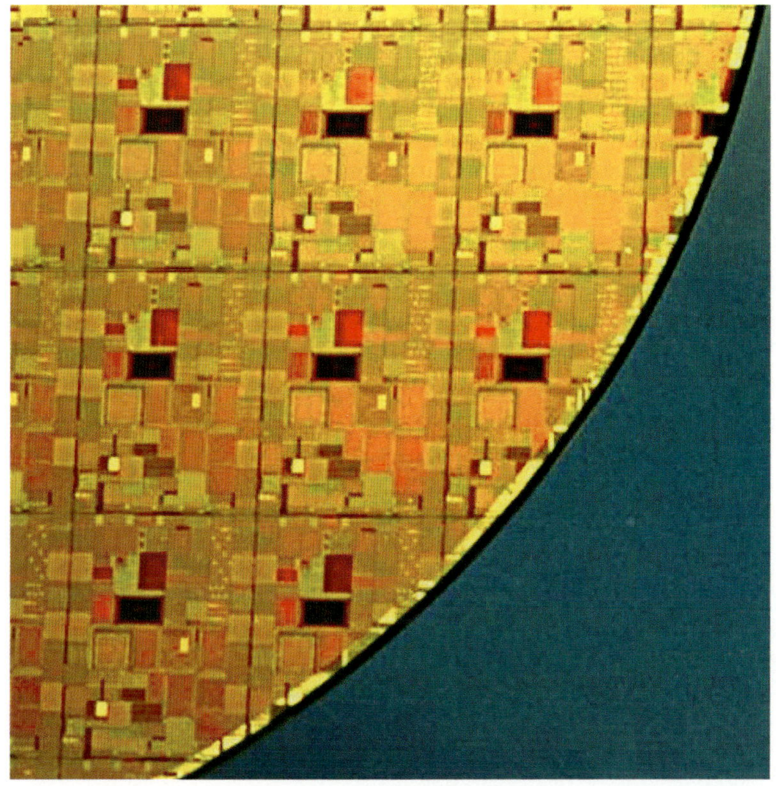

Figure 3.22: *Example of a multi-project wafer (MPW)*

Another way to share the mask costs is the *multi-layer reticle (MLR)*, on which several mask layers of the same product are grouped together to reduce the physical number of masks. These MLRs do not combine designs of different products. Both techniques are particularly used for small-volume designs, for prototyping, and for educational purposes. To save mask costs completely, *direct writing techniques* use an *electron-beam (e-beam)* or *laser-beam* system, which writes the layout pattern directly onto a wafer resist layer, without using a mask. It requires the deposition of an additional conductive layer on the resist layer, to prevent damage by electron charging during the patterning process. The resolution yielded by an e-beam machine is better than $5\,\text{nm}$, but at a lower throughput, because it writes every feature individually. It is free of wavelength aberration. Laser-beam systems are gaining market share at the cost of e-beam systems, because they are cheaper since they do not require a vacuum environment. Because of their low throughput, both e-beam and laser-beam systems usage, today, is limited to fabricate low-volume samples, such as MPWs, prototyping products and test silicon for process development. Next to that these techniques are used to fabricate the physical glass-chrome masks (reticles) for use in photolithography processes. These direct-writing techniques are also called *mask-less lithography (MLL or ML2)* and are currently also being explored as an alternative for, or successor of the conventional photolithography, even for high volume production. The main reason is the rapidly increasing costs of an optical mask set, which reaches the $2 million mark for the 65 nm node, although these costs will reduce when the process is getting more mature. Over the last decade, a lot of progress has already been made to improve throughput. The potentials of mask-less e-beam lithography are further discussed in [16].

More information on future lithography techniques can be found in numerous publications and also on the internet and is beyond the scope of this book.

To summarise the evolution of the wafer stepper/scanner, table 3.2 presents several key parameters which reflect the improvements made over different generations of steppers/scanners.

Table 3.2: The evolution of the wafer scanner (Source: ASML)

Status @ 2007 of most-advanced litho-tools					
Name		**I-line**	**DUV**	**193**	**193i**
Illumination source		Hg lamp	KrF laser	ArF laser	ArF laser
Reduction		4x	4x	4x	4x
Wavelength	nm	365	248	193	193
NA_{max} projection lens		0.65	0.93	0.93	1.35
k_{1_min}		0.5	0.25	0.25	0.25
Minimum pitch	nm	562	133	104	71
Depth of Focus	Dense Lines	600	300	150	150
Overlay	SMO	12	6	6	7
	MMO	20	10	10	12
Stage Repeatability	nm	10	6	4	4
Lens distortion	nm	n.a.	3.5	1.7	1.4
Wafer size	inch	8"/12"	8"/12"	8"/12"	8"/12"
Throughput	wph	165/135	165/135	150/122	165/122
Cost (MillionUS$)		5	13	19	> 36

Pattern imaging

The photolithographic steps involved in the transfer of a mask pattern to a wafer are explained with the aid of figure 3.23. Usually, the first step is oxidation and comprises the growth of a 30 to 50 nm thick silicon-dioxide (SiO_2) layer on the wafer. Subsequently, a nitride (Si_3N_4) layer is deposited (figure 3.23.a). Next, this nitride layer is covered with a *photoresist layer* (3.23.b). The mask is used to selectively expose the photoresist layer to light (3.23.c+d). The photoresist is then developed, which leads to the removal of the exposed areas if the photoresist is positive. The resulting pattern in the resist after development (3.23.e) acts as an etch barrier in the subsequent nitride etching step (3.23.f), in which the unprotected nitride is removed (stripped). Finally, the remaining resist is removed and an image of the mask pattern remains in the nitride layer (3.23.g). This nitride pattern acts as a barrier for a subsequent processing step.

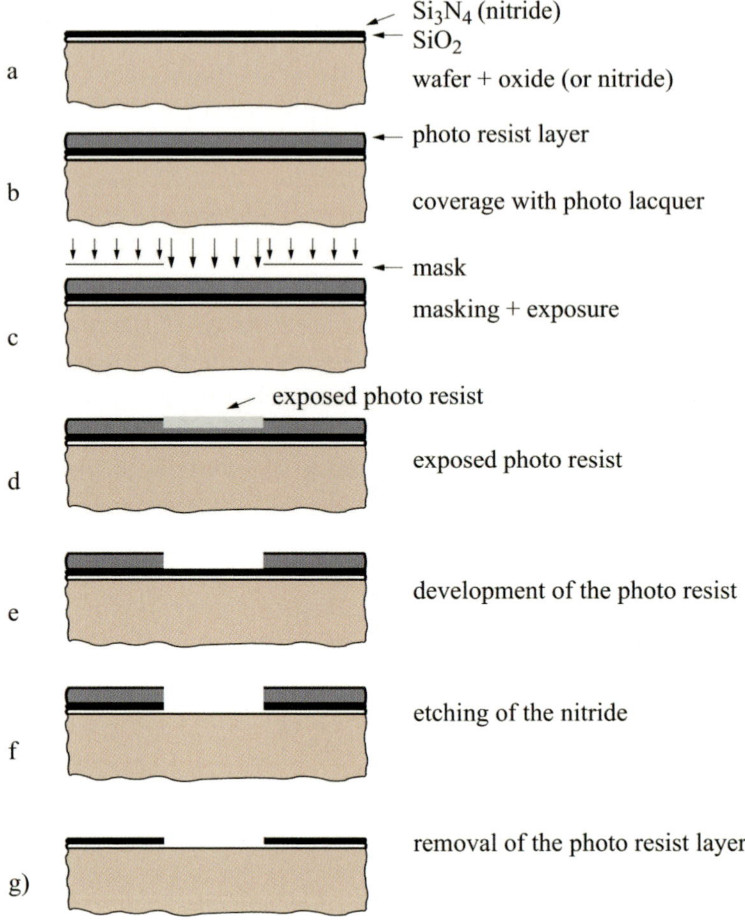

Figure 3.23: *Pattern transfer from mask to wafer*

Both positive and negative resists exist. The differences in physical properties of these resist materials result in inverting images, see figure 3.24.

The combination of pattern transfer and one or more processing steps is repeated for all masks required to manufacture the IC. The types of layers used for the pattern transfer may differ from the silicon-dioxide and silicon-nitride layers described above.

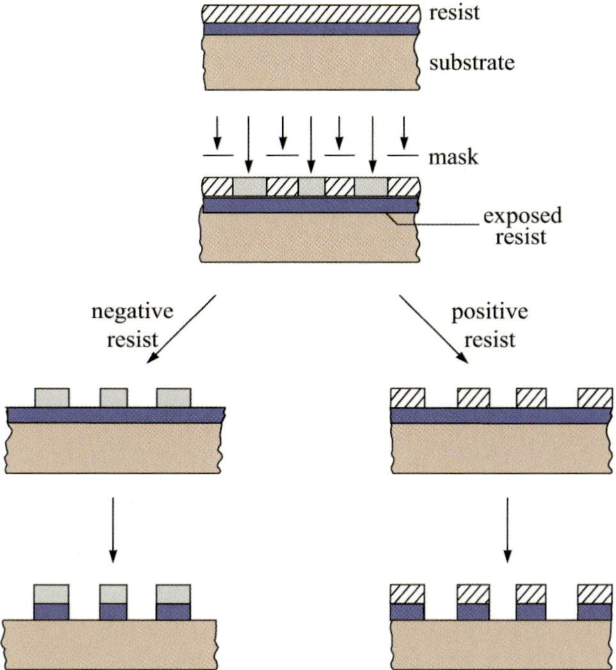

Figure 3.24: *The use of positive and negative resist for pattern imaging*

The principle, however, remains the same. The processing steps that follow pattern transfer may comprise etching, oxidation, implantation or diffusion and planarisation. Deposition is also an important processing step. These steps are described in detail in the following sections.

3.4   Etching

The previously described photolithographic steps produce a pattern in a nitride or equivalent barrier layer. This pattern acts as a protection while its image is duplicated on its underlying layer by means of *etching* processes. There are several different etching techniques. The etching process must fulfil the following requirements: a high degree of anisotropy, good dimensional control, a high etching rate to minimize processing time, a high selectivity for different materials, a perfect homogeneity and reproducability (e.g., 1 billion trenches in a Gb DRAM) and a limited damage or contamination to satisfy reliability standards. The degree of anisotropy depends on the requirements of the process step, e.g., during the STI etch an extremely vertical and sharp profile may increase stress and the occurrence of defects.

With *wet etching*, the wafer is immersed in a chemical etching liquid. The wet-etching methods are *isotropic*, i.e., the etching rate is the same in all directions. The associated *'under-etch'* problem illustrated in figure 3.25(a) becomes serious when the minimum line width of the etched layer approaches its thickness.

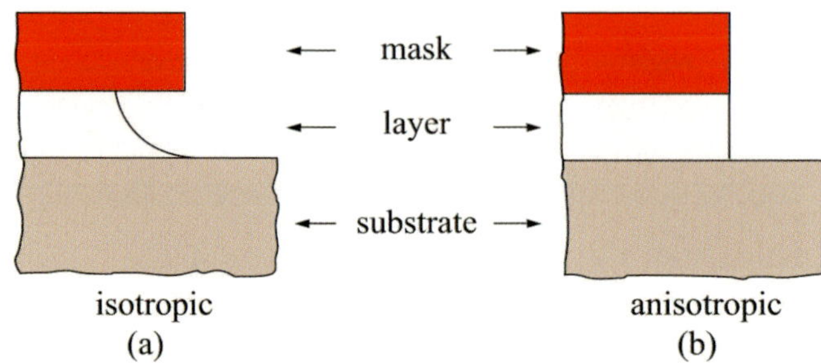

Figure 3.25: *The results of different etching methods*

Dry etching methods may consist of both physical and chemical processes (anisotropic) or of a chemical process only (isotropic). Dry-etching methods, which use a plasma, allow *anisotropic* etching, i.e., the etching process is limited to one direction by the perpendicular trajectory of the ions used at the wafer surface. The result, shown in figure 3.25(b), is an accurate copy of the mask pattern on the underlying layer.

With *plasma etching* techniques [17], the wafers are immersed in a plasma containing chlorine or fluorine ions that etch, e.g., Al and SiO_2 respectively. It comprises a plasma chamber, which contains a certain process gas. To transfer from the gas state into the plasma state, the chamber is pumped to the required pressure and energy is supplied to produce a glow-discharge plasma by a radio frequency (RF) electromagnetic field. This causes ionisation of the low-temperature plasma: after collision with molecules, they create many different gaseous species: free radicals, electrons, ions, neutrals, photons and by-products. These are then accelerated by an electrical field towards the surface material, which can then be etched quickly and selectively. The etching process depends on the gas pressure and flux and on the applied RF field. In *sputter etching* techniques, the wafer is bombarded by gas ions such as argon (Ar^+). As a result, the atoms at the wafer surface are physically dislodged and removed.

Finally, a combination of plasma and sputter etching techniques is used in *Reactive Ion Etching (RIE)*. During RIE ionised gaseous molecules from the plasma are accelerated by an electric field toward the surface and react with the surface atoms forming new electrically neutral molecules which then floats away.

Satisfactory etching processes have been developed for most materials that are currently used in IC manufacturing processes. New process generations, however, require improved selectivity, uniformity, reproducibility and process control. Selectivity can be improved by the compound of the gaseous plasma or by the creation of polymers at the underlying layer. The use of an additional carbonaceous substance such as CHF_3 during etching enhances its anisotropic properties. The use of this substance creates a thin layer close to the side wall of a contact hole, for example, which improves the anisotropy of the etching process. A second advantage is that carbon reacts with oxygen. It therefore increases the selectivity of the etching process because, when used in the etching of a contact-to-silicon, the reaction is stopped immediately on arrival at the silicon surface. Carbon does not react with silicon.

For critical anisotropic etching steps, both low-pressure etching techniques and *High-Density Plasma (HDP)* techniques are used. In HDP, energy is coupled into the plasma inductively to increase the number of free electrons. HDP is operated at low (some mtorr) pressure. This in turn results in a higher plasma density and a higher degree of ionisation. HDP is used to provide high-aspect ratios.

The focus on new etching techniques does not preclude further development of existing techniques such as high-pressure etching and RIE.

Many process steps use plasma or sputter-etching techniques, in which charged particles are collected on conducting surface materials (polysilicon, metals). Also during ion implantation, charge can be built up. These techniques can create significant electrical fields across the thin gate oxides; this is called the *antenna effect*. The gate oxide can be stressed to such an extend that it can be damaged (so-called process or *plasma-induced damage: PID*) and the transistor's reliability can no longer be guaranteed. The antenna effect can also cause a V_T-shift, which affects matching of transistors in analog functions. It is industry practice to introduce additional "antenna design rules" to limit the ratio of antenna area to gate oxide area. There are different rules for polysilicon, contact, via and metal-antenna ratioes. These ratioes may vary e.g., from 10 (contact-on-poly area to poly-gate area) to 5000 (accumulated-

metal area to poly-gate area). Also, protection diodes are used to shunt
the gate. Each input to a logic gate in a standard-cell library then
contains a protection diode.

3.5   Oxidation

The dielectrics used in the manufacture of nanometer CMOS circuits
must fulfil several important requirements [18]:

- high breakdown voltage

- low dielectric constant of inter metal dielectrics

- high dielectric constant for gate dielectric

- no built-in charge

- good adhesion to other process materials

- low defect density (no pinholes)

- easy to be etched

- permeable to hydrogen.

One of the materials that incorporates most of these properties is silicon dioxide (SiO_2). SiO_2 can be created by different processes: thermal
oxidation or deposition. A *thermal oxide* was used to isolate the transistor areas in conventional MOS ICs. In these isolation areas, the oxide
must be relatively thick to allow low capacitive values for signals (tracks)
which cross these areas. This *thick oxide* was created by exposing the
monocrystalline silicon substrate to pure oxygen or water vapour at a
high temperature of 900°C to 1200°C. The oxygen and water vapour
molecules can easily diffuse through the resulting silicon dioxide at these
temperatures. The following respective chemical reactions occur when
the oxygen and water vapour reach the silicon surface:

Dry oxidation : Si (solid) + O_2 (vapour) $\longrightarrow$ SiO_2 (solid)
Wet oxidation : Si (solid) + $2H_2O$ (vapour) $\longrightarrow$ SiO_2 (solid) + $2 H_2$

The *Local Oxidation of Silicon (LOCOS)* process is an oxidation technique which has found universal acceptance in MOS processes with gate

lengths down to $0.5\,\mu$m. Silicon is substantially consumed at the wafer surface during this process. The resulting silicon-dioxide layer extends about 46% below the original wafer surface and about 54% above it. The exact percentages are determined by the concentration of the oxide, which contains about $2.3 \cdot 10^{22}$ atoms/cm^3, while silicon contains about $5 \cdot 10^{22}$ atoms/cm^3. A disadvantage of the LOCOS process is the associated rounded thick oxide edge. This *bird's beak* is shown in figure 3.26(a).

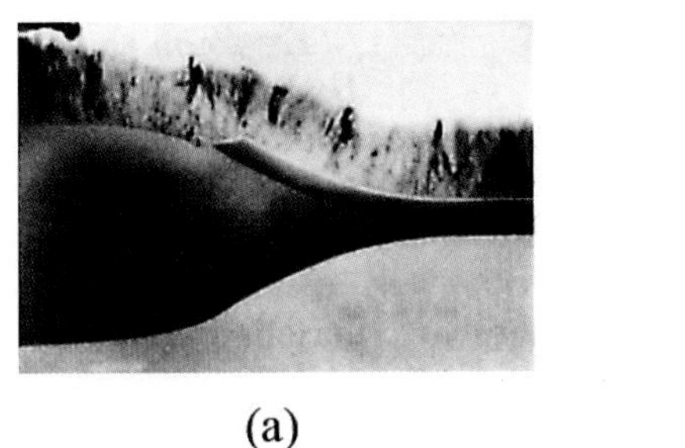 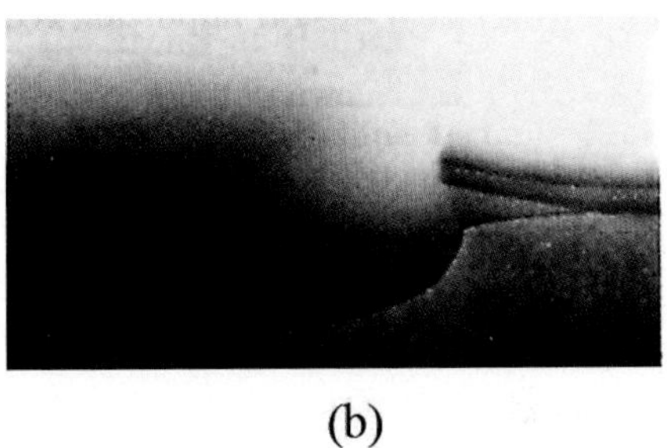

(a)   (b)

Figure 3.26: *Comparison of (a) a conventional LOCOS process and (b) a new oxide formation process which yields a suppressed bird's beak*

The formation of the bird's beak causes a loss of geometric control, which becomes considerable as transistor sizes shrink. Intensive research efforts aimed at suppression of bird's beak formation have resulted in lengths of just 0.1-$0.15\,\mu$m for an oxide thickness of $0.5\,\mu$m. Such a bird's beak is shown in figure 3.26(b).

Even with a suppressed bird's beak, the use of LOCOS is limited to the isolation of over-$0.25\,\mu$m transistors.

An important alternative to these LOCOS techniques, already used in $0.35\,\mu$m CMOS technologies and below, is the *Shallow-Trench Isolation (STI)*. STI uses deposited dielectrics to fill trenches which are etched in the silicon between active areas. The use of STI for nanometer technologies is discussed later in this chapter (section 3.9.3).

Another important application of thermally grown oxide is the oxide layer between a transistor gate and the substrate in conventional CMOS processes. This *'gate oxide'* must be of high quality and very reliable.

135

Defects such as pinholes and oxide charges have a negative effect on electrical performance and transistor lifetime. Because the gate oxide is only a few atoms thick, it is particularly a challenge for the industry to scale it further and/or find alternative ways to increase its capacitance. Figure 3.27 shows a cross section of a MOS transistor.

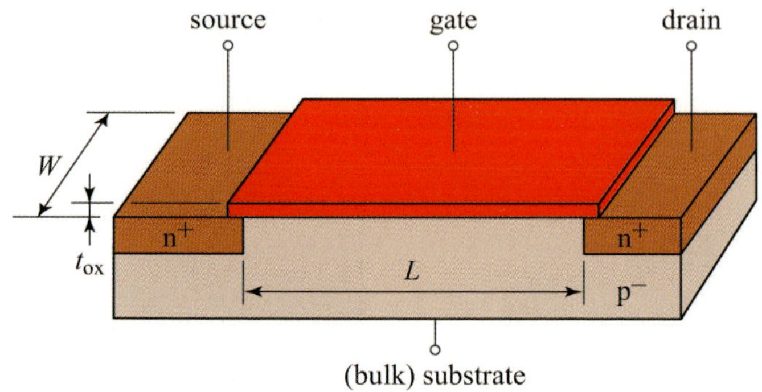

Figure 3.27: *Schematic cross section of a MOS transistor*

The *gate-oxide thickness* must be sufficiently uniform across the die, die to die, wafer to wafer, and from run to run. It scales with the technology node according to table 3.3, which shows representative values for various technology nodes. L represents the physical gate length.

Table 3.3: Trends in gate-oxide thickness and threshold voltage

Technology	L [nm]	t_{ox} [nm]	V_{dd} [V]	V_t [V]
0.35 μm	350	7	3.3	0.6
0.25 μm	250	5	2.5	0.55
0.18 μm	180	3.2	1.8	0.55/0.45
0.13 μm	120	2	1.2	0.45/0.35/0.2
90 nm	80	2.2/1.6	1.2/1.0	0.45/0.4/0.35/0.3/0.2
65 nm	60	1.8/1.2	1.2/1.0	0.5/0.4/0.3/0.2
45 nm	40	1.8/1	1.1/0.9	0.5/0.4/0.3/0.1

The table also shows the divergence in gate oxide thicknesses, supply and threshold voltages. This is due to the fact that today's semiconductor technologies must support applications with a wide range of performance requirements: high-density, low active power, low standby power, high speed, etc. In each technology node, the input- and output (I/O) transistors usually operate at a larger voltage (1.8 V, 2.5 V and/or 3.3 V) and require an additional oxide thickness and threshold voltage. The simultaneous use of more oxide thicknesses and threshold voltages in one process is of course at the cost of more mask, lithograhpy and processing steps. Technology nodes, today, offer different process versions, e.g., a *general-purpose (GP)* process and a *low-leakage (LL)*, also called *low-standby power (LSTP)*, process. Each of the processes offers usually only two or three different threshold voltages, to limit the number of masks, lithography and processing steps.

The use of dielectric SiO_2 layers below about 2 nm thickness causes gate oxide direct *tunnelling*, resulting in currents which may exceed a level of $1 \, A/cm^2$. At these gate-oxide thicknesses, pMOS transistors with heavily boron-doped polysilicon suffer from boron penetration into the gate oxide, causing an undesirable positive threshold-voltage shift and a performance and reliability degradation. The quality of the gate oxide is greatly improved with nitrided gate oxide. It reduces boron penetration and improves gate oxide charge-to-breakdown [19]. On the other hand, too much nitrogen close to the gate-oxide/Si-substrate interface enhances Negative Bias Temperature Instability (NBTI; see also chapter 9) [20]. Moreover, the combination of thinner gate oxide and increased channel doping also causes depletion of the bottom region of the gate material and this effect becomes more pronounced with further scaling of the oxide thickness. This is called *gate depletion*. As a result of these effects, the currently-used *double-flavoured polysilicon* (n^+ doped gate for nMOS transistors and p^+ doped gate for pMOS transistors) is most likely to be replaced by a metal. Other alternatives, which are currently under research and also prevent gate depletion, include *fully-silicided (FUSI)* polysilicon gates. Most advanced CMOS processes use atomic-layer deposition (ALD) to fabricate the very thin gate-oxide layer. This is discussed in the next subsection.

3.6 Deposition

The *deposition* of thin layers of dielectrical material, polysilicon and metal is an important aspect of IC production.

The growth of an *epitaxial film* (layer) is the result of a deposition step combined with a chemical reaction between the deposited and substrate material. If the deposited layer is the same material as the substrate, it is called *homo-epitaxy* or epi-layer for short. Silicon on sapphire is an example of *hetero-epitaxy*, in which the deposited and substrate materials differ [21]. Epitaxial deposition is created by a *Chemical Vapour Deposition (CVD)* process. This is a process during which vapour-phase reactants are transported to and react with the substrate surface, thereby creating a film and some by-products. These by-products are then removed from the surface. Normally, the actual film created by a CVD process is the result of a sequence of chemical reactions. However, a different overall reaction can generally be given for each of the silicon sources. The hydrogen reduction of silicon tetrachloride ($SiCl_4$), for example, can be represented as:

$$SiCl_4 + 2H_2 \longrightarrow Si + 4\,HCl$$

Several parameters determine the growth rate of a film, including the source material and deposition temperature. Usually, high temperatures ($> 1000\,°C$) are used for the depositions because the growth rate is then less dependent on the temperature and thus shows fewer thickness variations. The overall reaction for the deposition of polysilicon is:

$$SiH_4 \text{ (vapour)} \longrightarrow Si \text{ (solid)} + 2\,H_2 \text{ (vapour)}$$

This reaction can take place at lower temperatures, because SiH_4 decomposes at a higher rate. The creation of dielectric layers during IC manufacture is also performed by some form of CVD process. The most commonly-used dielectric materials are silicon dioxide (SiO_2) and silicon nitride (Si_3N_4). In an *Atmospheric-Pressure CVD (APCVD)* process, the material is deposited by gas-phase reactions. This deposition generally results in overhangs and a poor step coverage (figure 3.29). APCVD is currently used to deposit Boron PhosphoSilicate Glass *(BPSG)* epitaxial layers and form the scratch-protection layer (PSG). BPSG is a dielectric which is deposited on top of polysilicon (between polysilicon and first metal). BPSG contains boron and phosphorus for a better flow (spread) of the dielectric. The phosphorus also serves to improve internal passivation. The following reactions apply for the deposition of SiO_2 and Si_3N_4, respectively:

LPCVD:	$Si(OC_2H_5)_4$	$\longrightarrow SiO_2 + $ by-products
PECVD:	$Si(OC_2H_5)_4 + O_2$	$\longrightarrow SiO_2 + $ by-products
LPCVD:	$3SiCl_2H_2 \quad + 4NH_3$	$\longrightarrow Si_3N_4 + 6HCL + 6H_2$

Two versions of CVD have been introduced by the above reactions: LPCVD and PECVD. *LPCVD* is a low-pressure CVD process, usually performed in a vacuum chamber at medium vacuum (0.25-2.0 torr) and at temperatures between 550 and 750 °C. Under these conditions, the vapour-phase reactions are suppressed, while the decomposition now occurs at the surface, leading to a much better step coverage. In the previously-discussed CVD process, the chemical reactions are initiated and sustained only by thermal energy. *PECVD* is a plasma-enhanced CVD process. A *plasma* is defined to be a partially ionised gas which contains ions, electrons and neutrals. The plasma is generated by applying an RF field to a low-pressure gas, thereby creating free electrons within the discharge regions [21]. The electrons gain sufficient energy so that they collide with gas molecules, thereby causing gas-phase dissociation and ionisation of the reactant gases. At room temperature, a plasma therefore already contains high-energy electrons. Thus, even at low temperatures, a PECVD process can generate reactive particles; it therefore has a higher deposition rate than other CVD processes.

If we compare the previous reactions to depositing SiO_2, we see that the LPCVD which occurs at high temperature therefore needs no additional oxygen, while the PECVD process needs additional oxygen because the oxygen cannot be dissociated from the *TEOS* (tetra ethylorthosilicate: $Si(OC_2H_5)_4$) at low temperatures. A *Sub-Atmospheric CVD (SACVD)* process occurs at temperatures around 700 to 800 °C. Because of the high pressure ($\approx 1/2$ atmosphere instead of a few torr), the deposition speed will be higher, resulting in a higher throughput. This form of CVD is particularly used for BPSG.

Metal layers are deposited by both physical and chemical methods. In Physical Vapour Deposition (PVD) methods, such as *evaporation* and *sputtering*, the material is physically moved onto the substrate. PVD-Evaporation is a deposition process, in which a vapour of the material to be deposited is transported to the wafer in a low-pressure environment. After condensation at the wafer surface, it forms a thin film on it. When using the PVD-sputtering technique for the deposition of aluminium, for instance, an aluminium target is bombarded with argon ions, which physically dislodge aluminium molecules from the target, causing a flux of aluminium to flow from the target to the wafer surface. The aluminium was alloyed with 0.5% copper to improve elctromigration behaviour. After deposition of the aluminium photolithographic and etching steps are used to create the required metal pattern.

Copper cannot be deposited and etched as easy as aluminium. Potential etching plasmas create non-volatile residuals that remain on the wafer. Moreover, Copper defuses through oxides leading to transistor threshold voltage shifts and reliability problems. Therefore, a copper back-end technology is quite different from a conventional aluminium one. In the latter, the aluminium deposition step is followed by a dry etching step to etch the metal away according to the mask pattern and then filling the gaps with a dielectric. A copper back-end uses a so-called damascene process flow, in which the conventional subtractive metal etching process flow is replaced by a metal inlay process flow. Figure 3.28 shows a comparison of both flows.

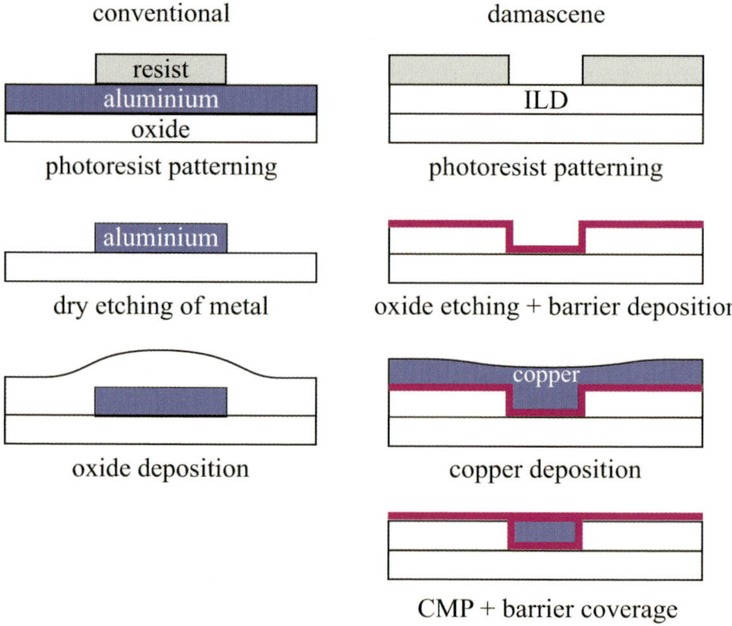

Figure 3.28: *Comparison of conventional and damascene processing*

In a damascene process, first trenches are etched in the inter-level dielectric (ILD) layer, most commonly an oxide layer. Then a thin barrier layer is deposited by an atomic layer deposition (ALD) step on top of the ILD layer and prevents the diffusion of copper. Next, a seed layer is deposited to provide a conductive layer, which is required for the electroplate-deposition process of the copper, to improve copper adhesion and coverage. Then, copper deposition is done, using an electrochemical process: *electroplating*, in which the wafer is immersed in a (salt/acid) solution of copper sulfate (and some other acids and/or additives to enhance the filling capabilities) and connected to a negative terminal of the power supply. The positive supply terminal is connected to a copper body, which creates copper ions into the salt solution. These positively charged copper ions are attracted to the negative wafer surface and form a thick copper blanket across the total wafer. Then a planarisation step, called CMP (section 3.8) polishes the wafer until it has reached the bottom of the barrier layer (copper and barrier are removed in one step!). Copper tracks are then remaining as a metal inlay in the trenches (Damascene processing), similar to the metal inlay in swords, made in ancient times in Damascus, Syria. Then again a barrier layer is deposited to cover the top of the copper inlays, such that copper is fully encapsulated within the barrier material. Today, most fabs use a dual-damascene backend, in which both the *vias* (also called *studs*, or *pillars*, which are contacts between two metal layers) and trenches are simultaneously etched into the ILD layer. Also in the next sequence of deposition steps for the barrier, the seed layer and the electroplate copper, respectively, the vias and tracks are simultaneously filled, thereby reducing processing costs.

Although the resistance of copper is 40% less than that of aluminium, this advantage can not fully be exploited, because part of the available track volume is occupied by the barrier material, which has a much higher resistance value. The use of copper instead of aluminium for interconnection resulted in only a limited reduction of the effective interconnect resistivity by 25 to 30%. In combination with the use of low-ϵ dielectrics, the interconnect capacitance is reduced and leads to faster or less-power circuits. Copper can also withstand higher current densities, resulting in a reduced chance of electromigration (see chapter 9).

CVD methods form the chemical alternative for the deposition of metals. Tungsten (W), for example, may yield the following CVD reaction:

$$WF_6 + 3H_2 \longrightarrow W + 6HF$$

The choice of deposition method is determined by a number of factors, of which *step coverage* is the most important. Figure 3.29 shows an example of bad aluminium step coverage on a contact hole in a conventional CMOS process. Such a step coverage can dramatically reduce the lifetime of an IC. It also causes problems during further processing steps and the associated temperature variations can lead to voids in the aluminium.

Moreover, the local narrowings cannot withstand high current densities. *Current densities* of $\approx 10^5$ A/cm^2 are not exceptional in modern integrated circuits. Excessive current densities in metal tracks cause *electromigration*. This leads to the physical destruction of metal tracks and is another phenomenon that reduces the reliability of ICs. This topic is examined more closely in chapter 9.

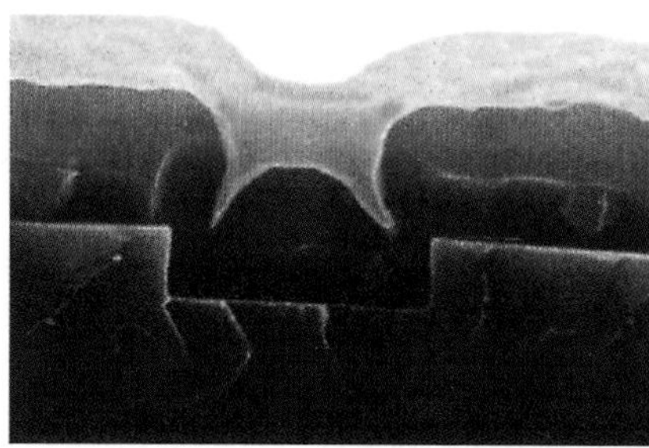

Figure 3.29: *Example of poor step coverage in a conventional CMOS process*

One deposition step that got a lot of attention over the last decade is the so called *atomic layer deposition (ALD)*, particularly for its potential applications in advanced (high-ϵ) gate dielectrics, DRAM capacitor dielectrics and copper diffusion barriers in advanced CMOS and memory processes. Without going deep into the chemical and physical reactions, ALD basically uses pulses of gas, creating one atomic layer at a time. So, the deposited film thickness is only dependent on the number of deposition cycles providing extremely high uniformity and thickness control. It is therefore also of interest in all (sub) nanometer applications that benefit from accurate control of (ultra-) thin films. More details on ALD can be found in [22].

3.7   Diffusion and ion implantation

Diffusion and *ion implantation* are the two most commonly-used methods to force impurities or dopants into the silicon.

Diffusion

Diffusion is the process by which the impurities are spread as a result of the existing gradient in the concentration of the chemical. Diffusion is often a two-step process.

The first step is called *pre-deposition* and comprises the deposition of a high concentration of the required impurity. The impurities penetrate some tenths of a micrometer into the silicon, generally at temperatures between 700 to 900°C. Assuming that the impurities flow in one direction, then the flux is expressed as:

$$J = -D \cdot \frac{\delta C(x,t)}{\delta x}$$

where D represents the *diffusion coefficient* of the impurity in [cm^2/s] and $\frac{\delta C}{\delta x}$ is the impurity concentration gradient.

As the diffusion strongly depends on temperature, each different diffusion process requires individual calibration for different processing conditions. During the diffusion process, silicon atoms in the lattice are then substituted by impurity atoms.

The second step is called *drive-in diffusion*. This high-temperature ($> 1000\,°C$) step decreases the surface impurity concentration, forces the impurity deeper into the wafer, creates a better homogeneous distribution of the impurities and activates the dopants. This drive-in diffusion also causes an identical lateral diffusion.

As a result of the increased requirements of accurate doping and doping profiles, diffusion techniques are losing favour and ion implantation has become the most popular method for introducing impurities into silicon.

Ion Implantation

The ion implantation process is quite different from the diffusion process. It takes place in an *ion implanter*, which comprises a vacuum chamber and an ion source that can supply phosphorus, arsenic or boron ions, for example. The silicon wafers are placed in the vacuum chamber and the ions are accelerated towards the silicon under the influence of electric and magnetic fields. The *penetration depth* in the silicon depends on the ion energy. This is determined by the mass and electrical charge of the ion and the value of the accelerating voltage. Ion implanters are

- throughput is lower than in diffusion process

- complex and expensive implanters

- initial cost of equipment: 2 to 5 M\$.

The use of ion implantation in the formation of source/drain regions becomes increasingly challenging as these junctions become very shallow in scaled processes. The doping concentration does not increase with scaling. Only the energy during implantation must be adjusted to create those shallow junctions. *Silicidation* of sources and drains becomes a problem in that silicide can penetrate through the shallow junctions. This is called *junction spiking*. Unsilicided sources and drains show a five to ten times higher sheet and contact resistance, affecting the electrical properties of the transistors. Because of this, all modern CMOS processes today use silicided sources and drains.

3.8 Planarisation

The increase in the number of processing steps, combined with a decrease in feature sizes, results in an increasingly uneven surface. For example: after completing the transistors, an isolation layer is deposited before the metal layers are deposited and patterned. The step height of the underlying surface is replicated into this isolation layer. This introduces two potential problems in the fabrication process. When the first metal is directly deposited onto this layer, its thickness can dramatically reduce at these steps, causing an increase in metal resistance and an increase in the occurrence of electromigration. Secondly, as already discussed in the lithography section, new lithography tools allow a smaller depth-of-focus (DOF), tolerating only very small height variations. During imaging, these variations can introduce focus problems at the high and low areas. Therefore, all current CMOS processes use several planarisation steps. These steps flatten or 'planarise' the surface before the next processing step is performed.

In conventional CMOS processes, *planarisation* was used during the back-end of the process, i.e., in between the formation of successive metal layers to flatten the surface before the next metal layer was defined. In such a *Spin-On-Glass (SOG)* formation, the surface was coated with a liquid at room temperature. After this, the wafer was rotated (spun), such that the liquid flowed all over the wafer to equalise the surface.

Next, the wafer undergoes a high-temperature curing process to form a
hard silicate or siloxane film. To prevent cracking, phosphorus was often
incorporated in the film. The resulting dielectric layer was planarised
to a certain extent. An advantage of SOG is that very small gaps are
easy to fill. However, with SOG, the surface is locally, but not globally,
planarised, see figure 3.31. On locally rough areas (A and B), the surface
is reasonably planarised.

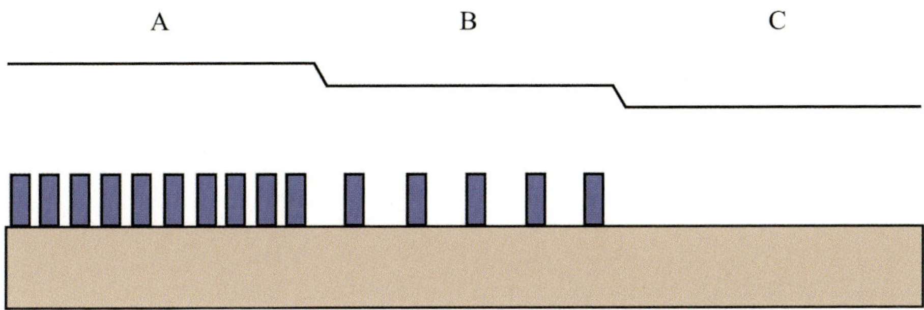

Figure 3.31: *SOG planarisation results*

There is still a global height difference after SOG planarisation, depending on the local pattern densities (area A, B and C). In a multilevel
metal chip, this effect would be much worse and would lead to etching
problems and problems with the *DOF* of the stepper. In all CMOS technologies below $0.25\,\mu$m, a very good alternative planarisation technique
is used: *Chemical Mechanical Polishing (CMP)*.

CMP is based on the combination of mechanical action and the simultaneous use of a chemical liquid (slurry) and actually polishes the
surface, see figure 3.32.

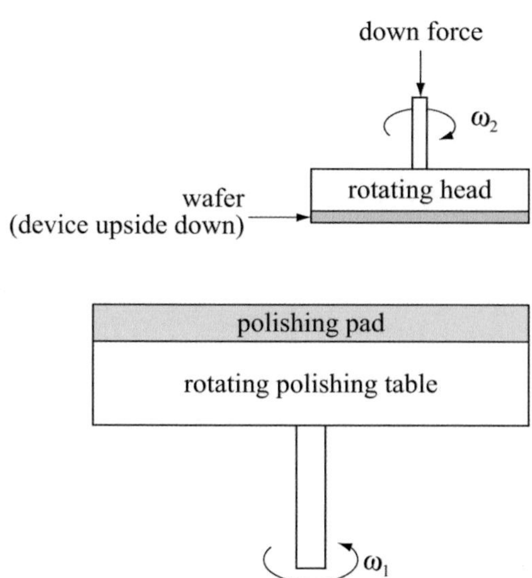

Figure 3.32: *Schematic overview of the CMP polishing process*

The *slurry* contains polishing particles (e.g., silica or alumina) and an etching substance (KOH or NH_4OH (e.g., ammonia)). A polishing pad together with the slurry planarises the wafer surface. Because CMP is also based on a mechanical action, it is much better suited for the local and global planarisation of rough areas, by offering a reduced topography for a more consistent focus across the field of exposure. It is particularly used for the creation and oxide filling of trenches (STI; section 3.9.3) and during the metallisation (back-end) part of a multi-layer metal process.

From the previous text the reader might conclude that CMP leads to an ideal planarisation result. However, there are several issues related to differences in pattern densities and differences in polishing rates of the various materials. Figure 3.33 shows the polishing results at three different phases of the CMP process.

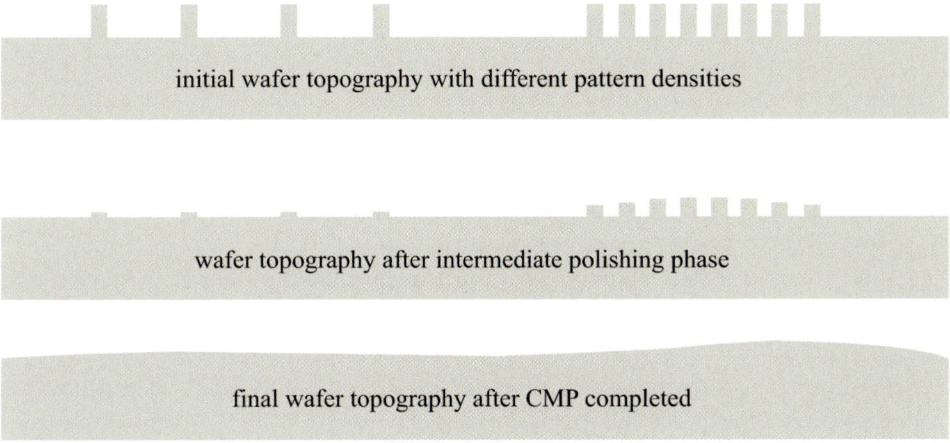

initial wafer topography with different pattern densities

wafer topography after intermediate polishing phase

final wafer topography after CMP completed

Figure 3.33: *Changing wafer topography after different CMP polishing phases*

The forces, exhibited during the polishing process, cause a higher pressure on the individual features in sparsely dense areas than in high dense areas. As a result, an increased polishing rate is observed on areas with very sparse patterns, compared to areas with the high-density patterns. This may lead to problems with the *DOF* during the lithography process and to reliability problems because of different contact heights.

Figure 3.34: *Potential problems of copper CMP*

As discussed in section 3.6, the copper CMP process includes the simultaneous removal of copper and barrier. The soft center of relatively large copper areas (wide copper lines and pads) polishes faster than the barrier/dielectric interface. This so-called *dishing* effect (figure 3.34) increases the resistance of these lines and reduces pad reliability. Also due to the difference in polishing rates, areas with dense copper patterns will polish faster than areas with only sparse copper patterns. This so-called *erosion* will also lead to thinner copper lines with higher resistance.

These polishing problems, in combination with the increased porosity of the inter-metal dielectrics, require constant monitoring through test structures for maintaining or improving both yield and reliability.

Particularly the mechanical degradation of the pads may lead to problems as cracking and peeling-off during packaging.

Measures to prevent planarisation problems in the back-end metallisation process include the creation of dummy metal patterns in scarcely-filled areas. The idea is to create metal patterns with as uniform a density as possible. These *dummy metal patterns*, sometimes also called *tiles*, should be automatically defined during chip finishing. Figure 3.35 shows an example of the use of tiling to achieve an improved metal distribution for optimised planarisation.

Figure 3.35: *Improved homogenous metal distribution by the use of tiles (purple)*

The use of tiles improves the quality of global planarisation and also results in a better charge distribution (reduced *antenna effect*) during back-end processing (deposition and etching of the successive metal layers). The shape of the individual tiles should be chosen such that it hardly affects the yield, performance, and signal integrity of a logic block.

A disadvantage of CMP is the mechanical wear of the polishing pad. As a result, the speed of polishing is reduced and, sometimes after each wafer, a diamond-brush step is performed to recondition the pad. After about 500 wafers, the polishing pad must be completely replaced. Figure 3.36 shows the result of the CMP planarisation technique in a multi-metal layer process.

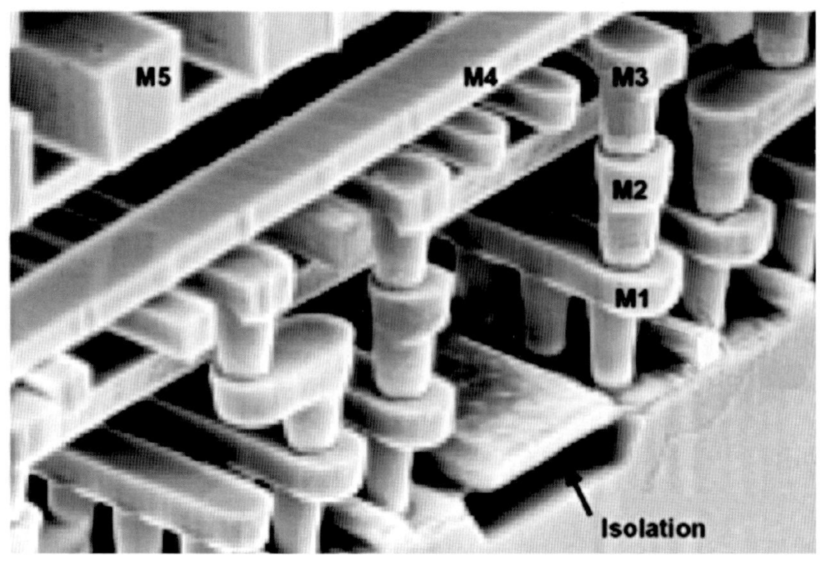

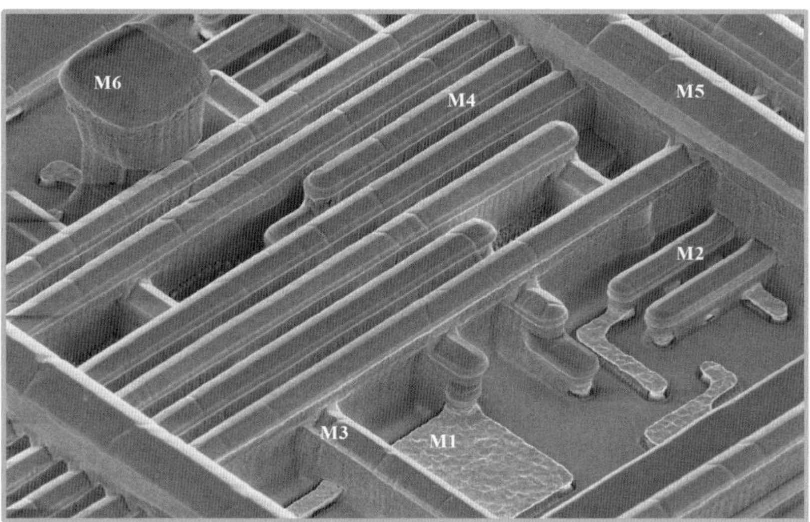

Figure 3.36: *Cross sections of CMOS back end, showing the potentials of CMP planarisation (Source: NXP Semiconductors)*

3.9 Basic MOS technologies

Sections 3.3 to 3.8 illustrate that MOS processes mainly consist of several basic actions that are repeated. In modern CMOS processes, the total number of actions has increased to several hundreds.

In this section, a basic nMOS process with just five masks is discussed. A good understanding of this *silicon-gate nMOS* process enables a smooth transition to the complex modern CMOS processes. With the exception of some new steps, these CMOS processes are just an extension of the basic nMOS process presented here. A good insight into both technology types is a prerequisite when comparing the advantages and disadvantages of nMOS and CMOS.

Finally, a nanometer CMOS process is presented and the associated fundamentally new steps are discussed. The section is concluded with a quantitative discussion of CMOS technology options beyond 45 nm.

3.9.1 The basic silicon-gate nMOS process

An *nMOS process* which uses a mere five masks is explained with the aid of figure 3.37. First, an oxide is grown on the base silicon wafer. Next, the oxidised silicon wafer is coated with a silicon nitride (Si_3N_4) layer, as shown in figure 3.37(a).

The first mask is the ACTIVE mask, which is used to define nitride areas corresponding to substrate regions where transistors should be formed. After the nitride is etched, boron is implanted through the resulting holes to produce the channel stopper, discussed in section 1.8 and indicated in figure 3.37(b). The wafer is then oxidised to produce the LOCOS areas in figure 3.37(c). The resulting thick oxide only exists at places that were not covered by the nitride. The channel stopper is thus automatically present everywhere beneath the LOCOS oxide. This is a great advantage of the LOCOS process. The removal of the remaining nitride reveals the areas in which transistors will be created. Now, the oxide is removed by a wet HF dip. The next step is the growth of a thin oxide in these areas.

The thickness of this oxide varies from a few to a few tens of nanometers in most MOS processes. The threshold voltage adjustment implantation which follows this oxidation damages the thin oxide. The implantation is therefore done through this *sacrificial pad oxide*. Low-energy impurity atoms such as iron (Fe) and/or copper (Cu) from the ion implanter may be caught in and/or masked by the sacrificial gate oxide during the implantation. This sacrificial pad oxide is subsequently removed and the actual thin gate oxide is grown. In some processes, however, impurities are implanted through the sacrificial pad oxide, e.g., during a threshold voltage (correction) implant. The properties of a MOS transistor are largely determined by the gate oxide. Gate oxidation is therefore one of the most critical processing steps. Its thickness is between 1 and 7 nm (see table 3.3).

After this, a polysilicon layer of about 0.1 to 0.4 μm thickness is deposited. A subsequent phosphorus diffusion, used to dope the polysilicon, is followed by photolithographic and etching steps, which yield polysilicon of the required pattern on the wafer. The POLY mask is the second mask step in this process and is used to define the pattern in the polysilicon layer. This step corresponds to figure 3.37(d). The polysilicon is used both as MOS transistor gate material, where it lies on thin oxide, and as an interconnection layer, where it lies on thick oxide (LOCOS). The *sheet resistance* of polysilicon interconnections lies between 20 and 300 $\Omega/\square$. Polysilicon can therefore only be used for very short interconnections (inside library cells).

Phosphorus (P) or arsenic (As) are mainly used to create the source and drain areas. The sheet resistance of these areas is about the same as that of polysilicon. Today's polysilicon source and drain areas are silicided to reduce the resistance values to about 50 $\Omega/\square$ (see section 3.9.3). The edges of the n^+ areas are defined by the LOCOS and the polysilicon gate. Source and drain areas are thus not defined by a mask but are *self-aligned* , according to the location of the gate. The overlap of the gate on the source and drain areas is therefore determined by the *lateral diffusion* of the source and drain under the gate. In the nMOS processes that used diffusion to create sources and drains, the length of the lateral diffusion is about 60% of the diffusion depth of the drain and source.

Currently, lower doped thin drain extensions are used which show a lateral diffusion of about 40% of their depth, see also section 3.9.3. With a *drain extension* of 20 nm, the lateral diffusion is only about 8 nm in a 45 nm process. The *effective transistor channel length* is therefore equal to the polysilicon width minus twice the lateral diffusion.

The wafer is then covered with a new oxide layer, deposited by an LPCVD step. The resulting SILOX layer indicated in figure 3.37(e) is about 300 to 800 nm thick. The CONTACT mask is the third mask step in this process and is used to define contact holes in the SILOX layer, see also figure 3.37(e). The metal layer is then deposited by means of sputtering, see section 3.6. The METAL mask is the fourth mask in this sample process. It is used to define the pattern in the aluminium or tungsten layer.

Basically, the processing is now completed, see figure 3.37(f). However, as a final step, the entire wafer is covered with a plasma-nitride *passivation* layer. This *scratch-protection* layer protects the integrated circuit from external influences. Figure 3.37(f) shows the situation before deposition of the scratch protection. With a final mask step, the scratch protection is etched away at the bonding pad positions to be able to make wiring connections from the chip to the package. This mask and the associated processing steps are not included in the figure.

In summary, the mask sequence for the considered basic silicon-gate nMOS process is as follows:

1. ACTIVE   definition of active areas
2. POLY   polysilicon pattern definition
3. CONTACT   definition of contact holes between aluminium and monocrystalline silicon or polysilicon
4. METAL   interconnection pattern definition in aluminium.

Finally, the NITRIDE mask is used to etch openings in the nitride passivation layer, to be able to connect bonding pads with package leads.

Note: The temperatures used for the source and drain diffusion exceed 900°C. Aluminium evaporates at these temperatures. Self-aligned source/drain formation is therefore impossible in an aluminium-gate process. Molybdenum gates have also been experimented with. However, they have never been industrially applied. In current CMOS technologies the sources and drains are implanted rather than diffused, due to the very high accuracy of the channel length definition.

The silicon-gate nMOS process has the following properties:

- Small gate-source and gate-drain overlap capacitances, caused by the self-aligned implantations.

- A relatively low number of masks, i.e., basically five to six.

- Three interconnection layers, i.e., n^+ diffusion, polysilicon and aluminium. However, intersections of n^+ and polysilicon interconnections are not possible as these result in the formation of a transistor. Chapter 4 presents a basic summary on the properties of nMOS circuits.

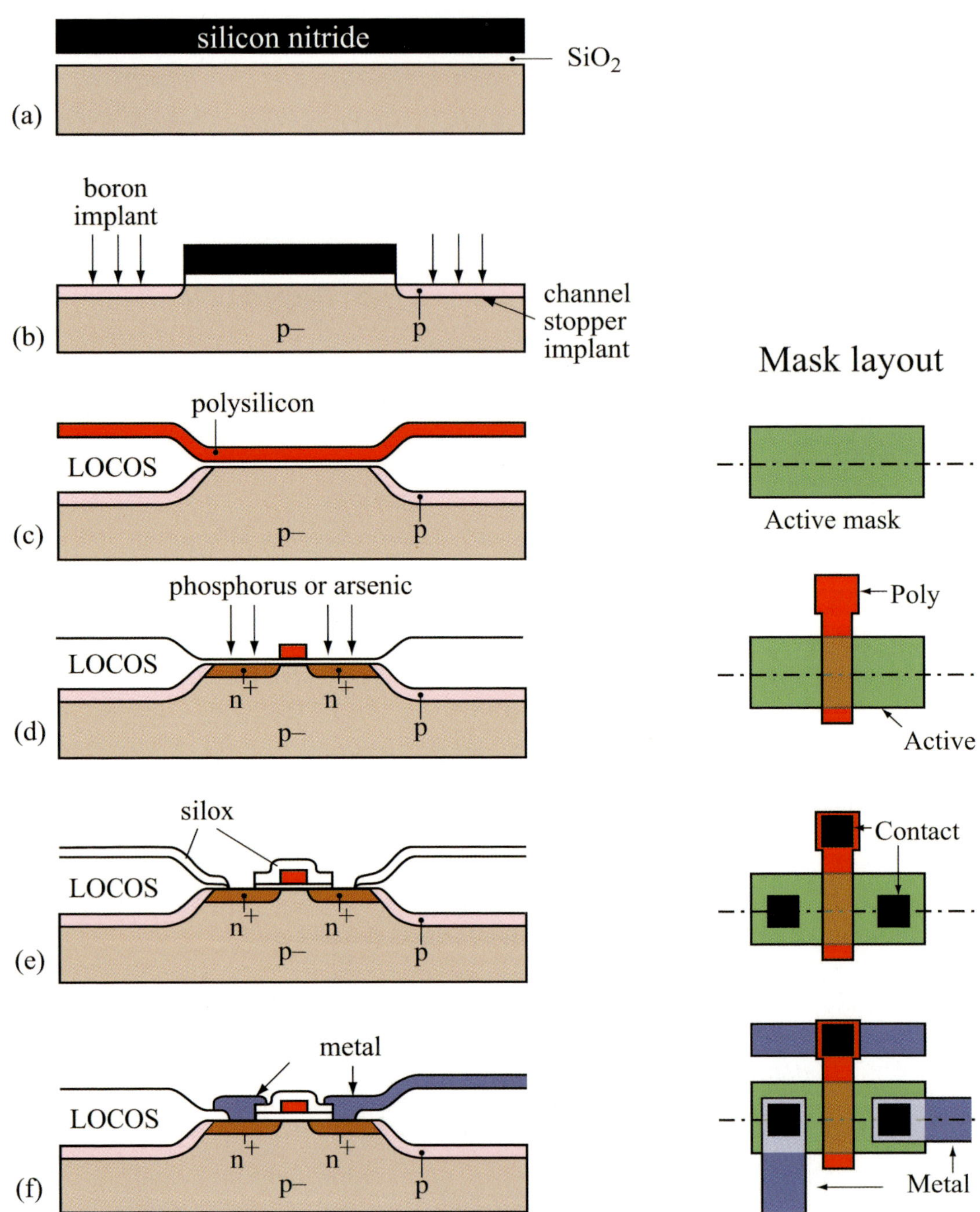

Figure 3.37: *The basic silicon-gate nMOS process with LOCOS isolation*

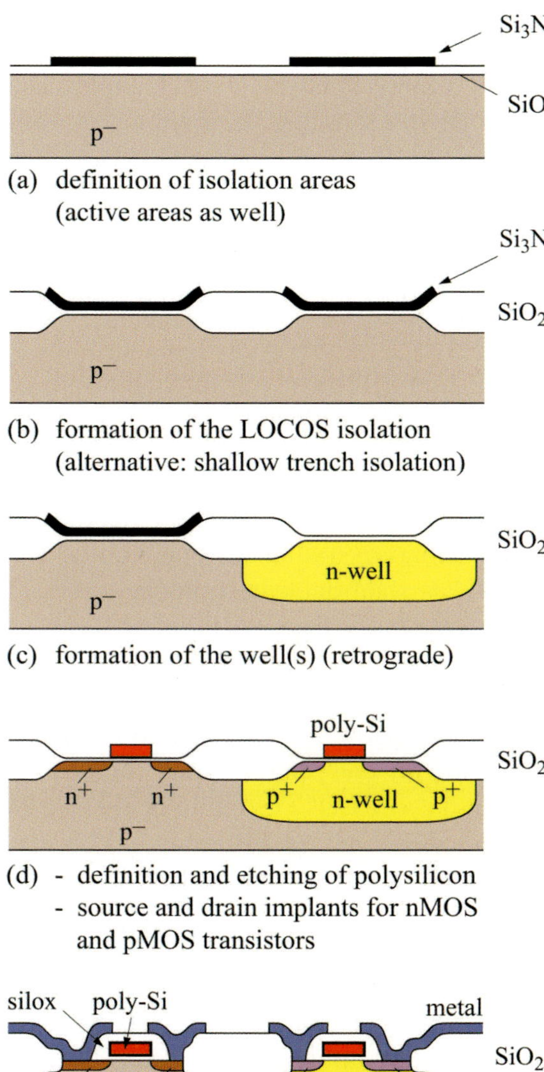

Figure 3.38: *The basic CMOS process with LOCOS isolation*

3.9.2   The basic Complementary MOS (CMOS) process

CMOS circuits and technologies are more complex than their nMOS counterparts. In addition, a static CMOS circuit contains more transistors than its nMOS equivalent and occupies a larger area in the same process generation. However, CMOS circuits dissipate less power than their nMOS equivalents. This is an important consideration when circuit complexity is limited by the 1 W maximum power dissipation associated with cheap plastic IC packages. In fact, reduced dissipation is the main reason for using CMOS instead of nMOS.

Both n-type and p-type transistors are integrated in CMOS processes. Figure 3.38 illustrates the flow of a simple CMOS process with an *n-well*, or *n-tub*, in which the pMOS transistors are implemented. This process serves as an example for the many existing CMOS technologies.

The basic CMOS process begins with the oxidation, to some tens of nanometers, of a monocrystalline p-type silicon wafer. A layer of silicon nitride (Si_3N_4) is then deposited on the wafer. This is followed by a photoresist layer. A mask is used to produce a pattern in the photoresist layer corresponding to *active areas*. Circuit elements will be created in these areas.

The defined pattern determines which silicon nitride remains during a subsequent etching step. The photoresist is then completely removed, as shown in figure 3.38(a). LOCOS oxide is then grown by exposing the wafer to oxygen at a high temperature. This oxide will not be grown on the areas that are still covered by the nitride. The LOCOS oxide separates active areas, see figure 3.38(b) for an indication of the result. Instead of LOCOS, STI is used in deep-submicron and nanometer CMOS processes to separate active areas (see next subsection). A new photoresist layer is then deposited and the p-type transistor areas are 'opened' during photolithographic steps. In conventional processes, the n-well was created by depositing a high concentration of donors (mostly phosphorous) in these areas, as shown in figure 3.38(c). Initially, these ions collect at the silicon surface but they diffuse more deeply during a subsequent high temperature step. Today, the n-well (and p-well) are implanted (see next subsection). A layer of polysilicon is then deposited on the wafer, which now consists of n-type n-well areas with a limited submicrometer depth and p-type substrate areas.

Polysilicon doping reveals either n-type polysilicon for both nMOS and pMOS transistor gates, or *double-flavoured polysilicon* (n-type and p-type polysilicon for nMOS and pMOS transistor gates, respectively).

This is also sometimes referred to as n^+/p^+ *dual polysilicon*.

A photolithographic step follows and the polysilicon pattern is etched. The resulting polysilicon is used for short interconnections and for transistor gates.

Separate masks are used for the self-aligned source/drain implantations: nplus and pplus for the nMOS and pMOS transistors in the substrate and n-well, respectively. The result is shown in figure 3.38(d).

The first step in the creation of interconnections between the different transistor areas is to deposit an SiO_2 (SILOX) layer on the wafer. Contact holes are etched in this layer to allow connections to the gates, drains and sources of the transistors. A metal layer is then deposited, in which the final interconnect pattern is created by means of photolithographic and etching steps. Figure 3.38(e) shows the final result.

Modern CMOS processes use 25 to 35 masks. Basically, these processes are all extensions of the simple CMOS process described above. VLSI and memory processes now use channel (gate) lengths of 45 nm to 0.35 μm and offer several levels of polysilicon and/or metal. These multiple interconnection layers facilitate higher *circuit densities*. The next section discusses a state-of-the-art nanometer CMOS process.

3.9.3 An advanced nanometer CMOS process

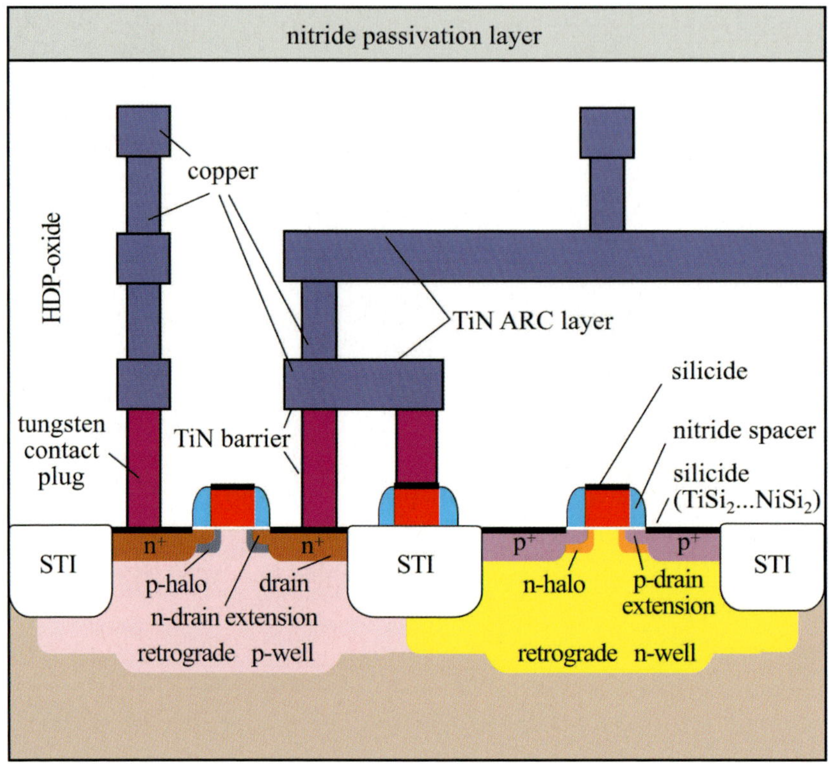

Figure 3.39: *An advanced nanometer process with STI isolation*

Compared to the basic CMOS process discussed before, an advanced
nanometer CMOS process, with channel lengths below 100 nm, incorporates several major different processing steps. These differences will
now be discussed in some detail.

Shallow-trench isolation

Actually, LOCOS is thick SiO_2 that is thermally grown between the
active areas. In contrast, *Shallow-Trench Isolation (STI)* is implemented
at significantly lower temperatures, preventing many warpage and stress
problems associated with a high-temperature step. The STI process
starts with a thermally-grown oxide with a thickness between 10 nm to
14 nm. This is followed by an LPCVD deposition of 100 nm to 160 nm

nitride. Next, the active areas are masked and a dry etch step is applied
to create the trenches, which have a typical depth between 250 nm and
500 nm. The corners at the bottom and the top of the trench are rounded
by a thermally-grown oxide layer (between 20 nm and 50 nm) along the
side walls of the trench, see figure 3.40.

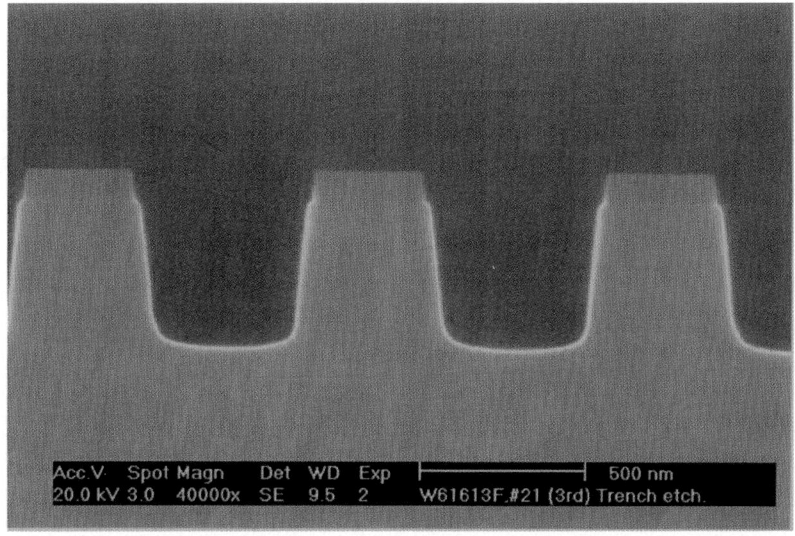

Figure 3.40: *Cross-section after etching the trenches in the silicon*

After removing the resist, a thick oxide High-Density Plasma (HDP),
typically 700 nm to 1100 nm, is deposited. HDP is capable of filling
the high aspect ratio of the trenches, which includes the pad oxide and
nitride layer thicknesses. As shown in figure 3.41, the step coverage of
the oxide is dependent on the geometry of the active area mask.

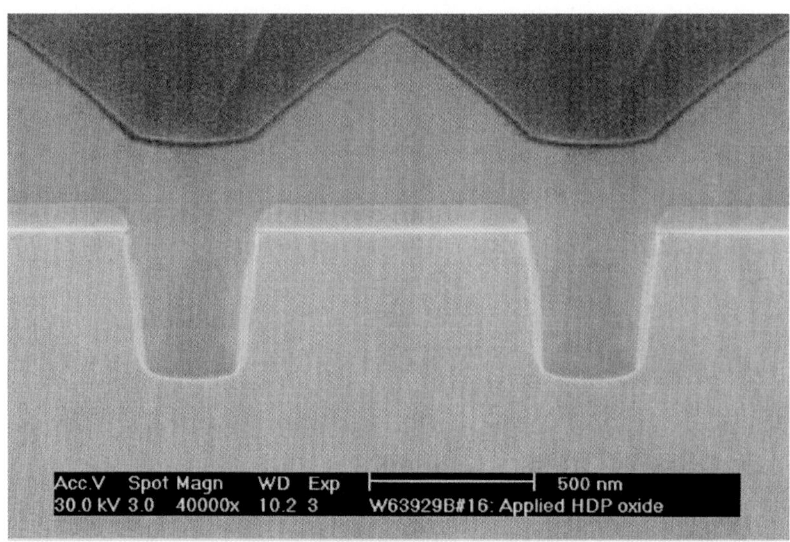

Figure 3.41: *STI process cross-section after thick oxide deposition*

In dense areas, the oxide level is well above the silicon nitride, while the oxide thickness equals the deposited oxide thickness in large open areas. The remaining topology is planarised using CMP, see section 3.8. The nitride layer is used as chemical etch stop, see figure 3.42.

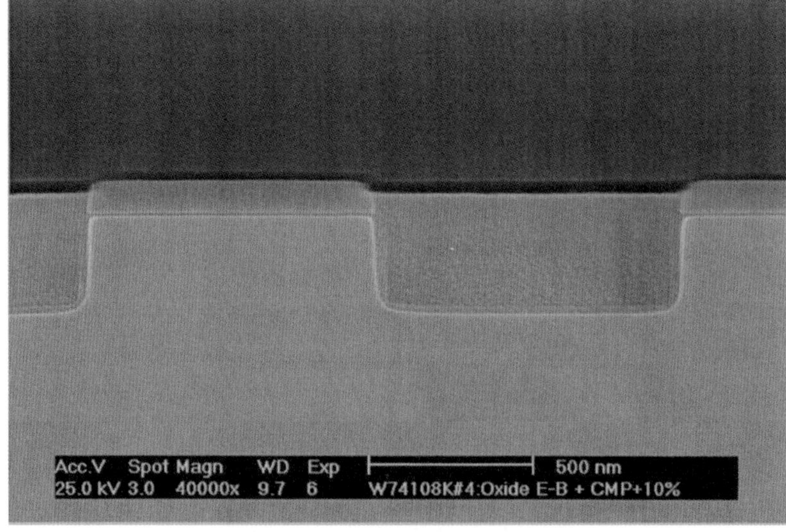

Figure 3.42: *SEM cross-section after CMP*

Next, the nitride masking layer is removed, using a wet etch and subsequently sacrificial oxide, gate oxide (by ALD) and polysilicon is deposited, etc. Figure 3.43 shows a cross-section through the width of the
device. The gate oxide between the polysilicon layer and the monocrystalline silicon substrate can be as thin as 1 nm in very advanced nanometer CMOS ICs.

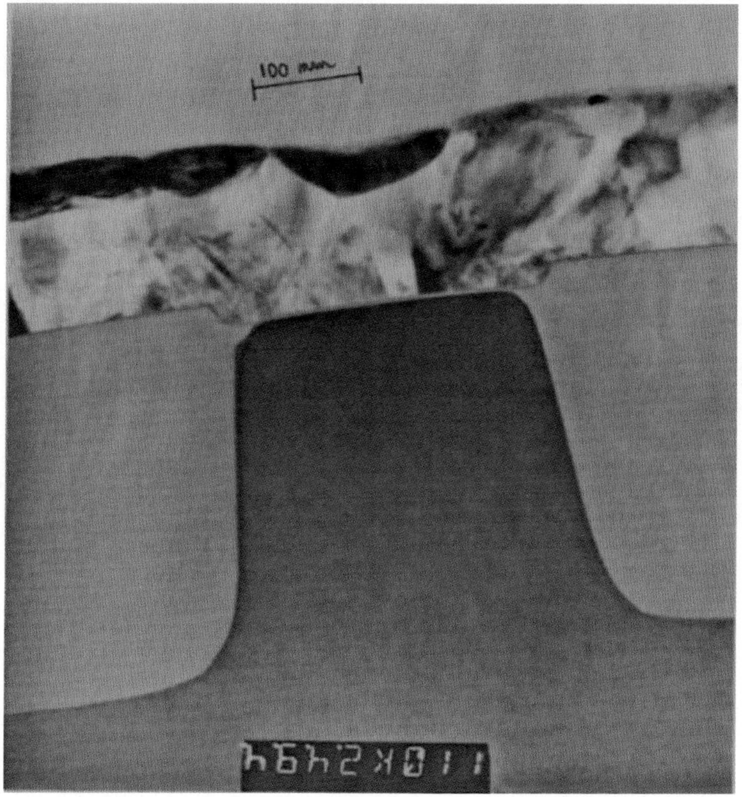

Figure 3.43: *TEM cross-section through the width of the device*

In this way, device widths far below 100 nm can be well defined. Figure 3.44 shows a comparison between LOCOS and STI field isolation
techniques. It is clear that the STI is much more accurately defined
and enables the creation of high aspect-ratio field-oxide isolation areas
to improve the circuit density in nanometer CMOS ICs.

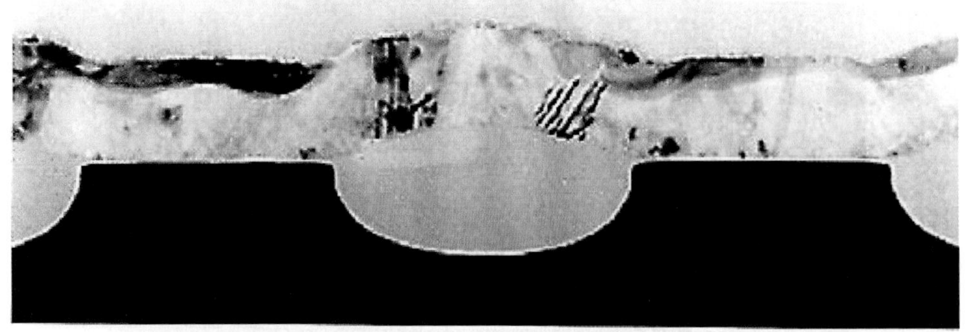

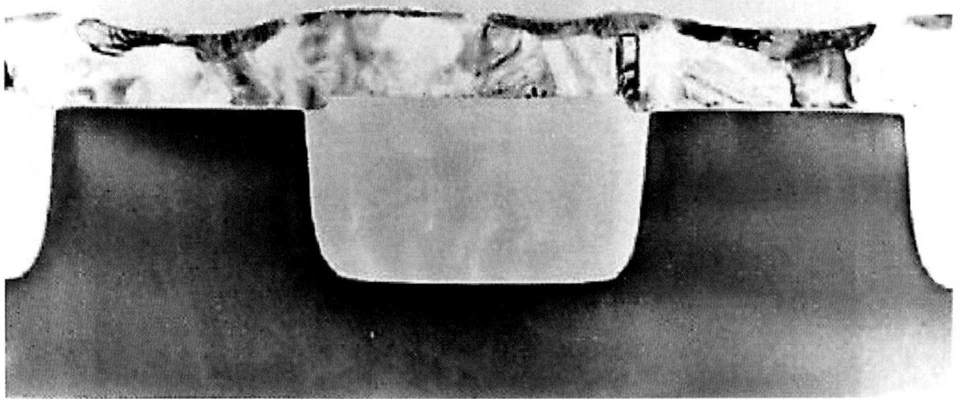

Figure 3.44: *Comparison between LOCOS (top) and STI field isolation (bottom) techniques*

Retrograde-well formation

A *retrograde-well* process (figure 3.39) uses both n-wells and p-wells, and is also called a twin-well process. These wells form the substrate for p-type and n-type devices, respectively. High-energy implantation of the wells yields doping profiles with maxima between 250 and 600 nm beneath the wafer surface in active areas. The maximum dope level beneath thick oxide areas (STI areas) is only a short distance below the bottom of these oxides. The implantation therefore acts as a very effective *channel stopper* for parasitic devices in these areas.

Only a limited temperature is required to drive the well implants to appropriate depths, which results in limited lateral diffusion. Conse-

164

quently, the wells can be accurately defined and their separation from source and drain areas of their own type (e.g., n-well to n^+ source/drain regions and p-well to p^+ source/drain regions) can be relatively small. This is the most important reason for applying retrograde-well processing.

Each well can be optimised to yield the highest performance for both types of transistors. This can be done by minimising source/drain junction capacitances and body effect or by using an *'anti-punch-through' (APT)* implant. Another advantage is the associated feasible symmetrical electrical behaviour. In addition, the two wells are usually each other's complement and can be formed by defining only a single mask during the design, while the other one is defined during the post processing or chip finishing. Also the throughput time for a retrograde well is shorter than that of a diffused-well. Finally, another significant advantage of twin-well CMOS processes is formed by the better scaling properties, which facilitate the rapid transfer of a design from one process generation to another. The consequences of scaling are extensively discussed in chapter 11.

Optimizing technologies for high-speed digital designs generally degrades analogue circuit performance of long-channel devices. Careful optimisation of the front-end process (including the wells) is required to improve mixed analogue/digital circuit performance [24].

Drain extension

The *hot-carrier effect*, which will be discussed in chapter 9, manifests itself more when carriers acquire more kinetic energy than about 3.2 eV. In 1.2 V processes and below, it becomes almost impossible for the charge carriers to penetrate into the gate oxide (energy equals $q \cdot V = 1.2\,\text{eV}$ in a 1.2 V process). Carriers can only acquire such energies after a lot of collisions in the pinch-off region. As the pinch-off regions are very narrow for nanometer CMOS technologies, this is becoming very unlikely to happen.

The LDD (chapter 9) implants, as used in processes of $0.35\,\mu\text{m}$ and larger to reduce the probability of occurence of hot carriers, are thus replaced by a more highly doped source/drain extension (figure 3.39). This source and drain extension is produced similar to the LDD. However, the peak doping concentration ($\approx 1 \cdot 10^{20} - 2 \cdot 10^{20}\,\text{atoms/cm}^3$), today, is much higher than usually applied in an LDD and almost equals the peak dope in the highly doped source and drain regions. It results in a lower series resistance. Moreover, oxide spacers have been mostly replaced by nitride spacers and a lot more doping-profile engineering has been performed, to create smooth junctions tot reduce junction leakage (band-to-band tunnelling). This is achieved by a combination of three different implants: a very thin off-axis As implant for the source/drain extension, a much deeper As n^+ implant for the source/drain formation, followed by an even deeper Phosphorous implant with a reduced doping, to create the smooth junction. This source/drain extension implant is much less deep (10-20 nm) than the actual source/drain junctions, which allows a better control of the channel length and reduces the short-channel effects. Actually, such an extension acts as a hard mini-drain. In some cases in literature, only one implant is used to create the drain. This is then without extension implant, and called *Highly-Doped Drain (HDD)*. The phosphorous halo with increased dope in the channel around the drain, reduces the depletion layer thickness and suppresses short-channel effects such as threshold roll-off and punch-through.

Silicides, polycides and salicides

Silicidation is the process of creating a surface layer of a refractory metal silicide on silicon. *Silicides* may be formed by the use of $TiSi_2$, WSi_2, $CoSi_2$, $NiSi$ or other metal silicides. When, for example, a titanium film is deposited directly on a silicon surface, after the definition of the polysilicon and the formation of the source/drain junctions, the titanium and the silicon react to form a silicide layer during a subsequent heating step. Titanium (and some other metals) react with exposed polysilicon and source/drain regions to form $TiSi_2$ silicide (or other silicides). A layer of titanium nitride (TiN) is formed simultaneously on the silicon dioxide. This will be selectively etched away. Silicidation yields low-ohmic silicide top layers in polysilicon and source/drain regions to reduce RC delays by five to ten times, and improve circuit performance. Because the silicidation step is maskless, it is also called *self-aligned silicide* or *salicide*. In a *polycide* process only the polysilicon is silicided. Sheet resistance values for silicided and unsilicided source, drain, and polysilicon regions are presented in table 4.2 in chapter 4.

Ti/TiN film

Titanium (Ti) is used in the contact holes to remove oxides and to create a better contact with the underlying silicide. A *titanium nitride* (TiN) film is used in the contacts, as well as on top of the PETEOS (plasma-enhanced tetra-ethyl orthosilicate) oxide, because of its good adhesive properties. When the tungsten is being etched away with a plasma, TiN is used as an etch stop. The TiN is also responsible for an increased resistance of the contact plugs.

Anti-Reflective Coating (ARC)

Reflections during exposure of a metal mask may cause local narrowing in the resist pattern and, consequently, in the underlying metal pattern, which is to be defined. A titanium nitride film is often deposited on top of the metal layer and serves as an *Anti-Reflective Coating (ARC)*. Today, organic ARC is used during all lithographic steps in nanometer technologies. This film is highly absorbent at the exposure wavelength. It absorbs most ($\approx 75\%$) of the radiation that penetrates the resist. It also suppresses scattering from topographical features.

Contact (re)fill

In many processes, particularly those which include planarisation steps, oxide thickness may vary significantly. Deep contact holes with high aspect ratios require special techniques to guarantee good filling of such contacts. This *contact filling* is often done by tungsten, called (tungsten) plugs, pillars or studs. As these aspect ratios become more aggressive with scaling, poor step coverage and voids in the contact plug become apparent. To fill the plugs void-free, very thin Ti and TiN films are used as a low resistance glue layer for better adhesion to the dielectric.

Damascene metal patterning

In $0.18\,\mu$m CMOS processes and above, metal patterning is done by depositing an aluminum layer, followed by a dry etching step to etch the aluminum away according to a mask pattern. In the damascene process, copper patterns are created by etching trenches in the dielectric, overfilling these trenches with copper and then polishing the overfill away using CMP, until the polishing pad lands on the dielectric. Damascene copper processing is discussed in some detail in section 3.6.

Damascene patterning is used, particularly in $120\,$nm and below, to form copper wires. In a *dual-damascene* process, plugs (studs, pillars) and wires are deposited simultaneously. This process replaces the deposition of the plug and its etching, thereby reducing processing costs.

The damascene process is mainly used to pattern copper, which cannot be etched like aluminium in plasma reactors. The copper will create too many by-products which remain on the surface and cannot be removed. The use of copper instead of aluminium for interconnection results in a reduction of the interconnection resistivity by 25 to 30%. This advantage is mainly exploited by a reduction of the metal height, so that about the same track resistance is achieved, but at a reduced mutual wire capacitance. This serves two goals: power reduction due to the reduced load capacitance of the driving gate and cross-talk reduction due to the smaller mutual wire capacitance to neighbouring wires. In combination with the use of low-ϵ dielectrics, the speed can be improved even more, or the power can be reduced further. Copper can also withstand higher current densities (reduced chance of electromigration, see also chapter 9).

3.9.4 CMOS technology options beyond 45nm

Approaching the end of Moore's law, by reaching the physical limits of scaling planar CMOS devices, has challenged both process and design engineers to create solutions to extend CMOS technology scaling towards 10nm feature sizes. Local circuit speed is dominated by the devices (transistors' driving currents) while the global speed is dominated by a combination of the devices and interconnects (signal propagation). There are several issues related to the continuous scaling of the devices and interconnects.

Devices

The transistor's driving current depends heavily on its threshold voltage and carrier mobility. Scaling introduces several mechanisms that reduce this mobility, directly or indirectly. First of all, the carrier velocity saturation and surface scattering affects, introduced in chapter 2, are responsible for a two to six times mobility reduction. Apart from this, there is an increased depletion of the bottom side of the polysilicon gate *(gate depletion; gate inversion)*, due to the increased levels of halo implants for suppression of short-channel effects. Because mainly this bottom side of the gate is responsible for the drive current of the transistor, this gate depletion will dramatically reduce it. Current R&D focus is on the potentials of fully-silicided *(FUSI gate)* and metal gates. It has proven very difficult to replace polysilicon gates with an appropriate metal-gate material. This is due to the fact that the metal workfunction (which also determines the V_T) is affected by the metal-gate composition, the gate dielectric and heat cycles. Few metal gates have been identified giving a correct V_T after integration in a manufacturable CMOS process flow.

In a FUSI gate the chemical reaction during silicidation continues until the gate is siliced all the way down to the bottom of the gate. Its operation then resembles that of a metal gate, and does not show bottom depletion. It is expected that FUSI or *metal gate* may be introduced in the 45 nm or 32 nm CMOS node.

The conventional way of increasing the transistor current is to reduce the gate-oxide thickness. But with oxide thickness values (far) below 2 nm the transistor exhibits relatively large gate leakage currents, which increase with a factor of ten for every 0.2 nm further reduction of the oxide thickness. A high-ϵ gate dielectric (hafnium oxide, zirconium oxide and others) is therefore a must to continue device scaling with an affordable leakage budget. The search for the right combination of high-ϵ gate dielectric with the right gate electrode with the right work function and tolerance to high-temperature process steps is very difficult. Intel has developed a so-called *gate-last CMOS process*, in which the sources and drains are created before the gate electrode, and has developed the Penryn dual-core processor with 410 million transistors in 45 nm CMOS with high-ϵ gate dielectrics and metal gate [25].

Another way of increasing the transistor current is to improve the channel mobility. The use of *strained silicon* is one of the alternatives to achieve this. To achieve the best mobility improvements, the strain should be compressive for the pMOS transistors and tensile for the nMOS transistors. In unstrained nanometer CMOS processes the average hole mobility in the silicon is about two times lower than the electron mobility. Therefore, in many cases, the improvement of the pMOS transistor mobility has been given more priority.

In a *strain-relaxed buffer (SRB)* technology, a SiGe layer is grown on a silicon substrate. Germanium atoms physically take more space than silicon.

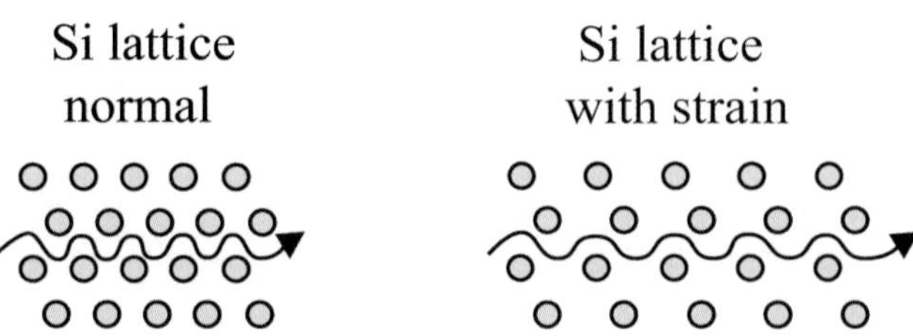

Figure 3.45: *Strained Si shows a reduced atom density, allowing improved carrier mobility*

Next, a thin (about 10 nm thick) silicon layer is grown on top of the thicker SiGe layer. This top layer's atomic structure adapts itself to the atomic structure of the SiGe layer below. This creates strain in this silicon top layer, introducing physical stress in it, thereby increasing the channel mobility. The left picture in figure 3.46 shows a cross section of such a transistor. Experimental SiGe strained silicon showed 20% improvement in channel mobility [26]. To achieve a sufficient improvement in mobility, about 20% to 30% of the silicon atoms must be replaced by germanium. Germanium, however, exhibits a much larger thermal resistance than silicon, leading to self-heating problems comparable to SOI. A second problem related to this type of strained SiGe is the fact that germanium oxide is dissolvent to water, which is used during wafer cleaning to remove residual material from previous processing steps.

A third problem is that the SRB technology implicitly creates threading dislocations from the top of the SiGe layer into the strained silicon top layer [27]. These may have severe impact on the junction leakage and yield. Other SiGe methods have replaced the SRB technology. An alternative means of introducing strain to enhance the mobility is to embed an epitaxially grown strained $Si_{1-x}Ge_x$ (embedded silicon germanium; eSiGe) film in the source and drain areas *(recessed source/drain)*. Germanium atoms are slightly larger than silicon atoms (5.66 Å vs 5.43 Å, which generates a *compressive strain* in the transistor channel, which results in an enhanced hole mobility (figure 3.46 right picture) in pMOS transistors [28]. However, it puts severe demands to the transistor engineering, in particular with the alignment (overlay) of the gate with respect to the STI isolation areas. In order to fabricate a device with symmetrical behaviour, the self-aligned source and drain must be of equal size to induce the same amount of stress into the channel. *Tensile strain*, as opposed to compressive strain, can be created by using Carbon (3.56 Å) which has a smaller lattice constant to substitute some silicon atoms. nMOS and pMOS transistors react differently under the influence of strain. As a result, the introduction of tensile strain improves the performance of nMOS devices while it degrades the performance of pMOS devices and vice versa. nMOS and pMOS devices are therefore built with built-in tensile and compressive strain, respectively.

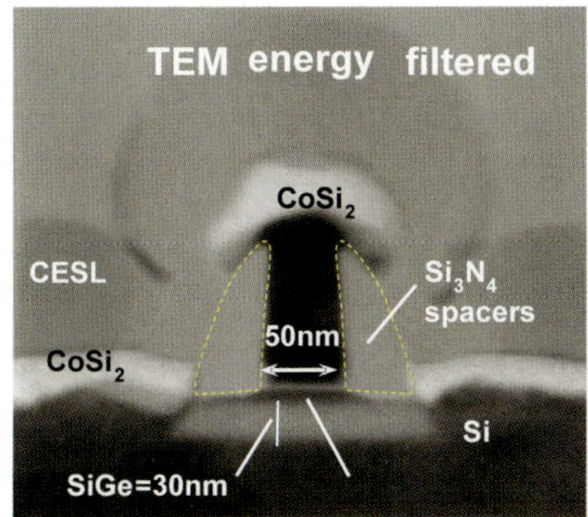

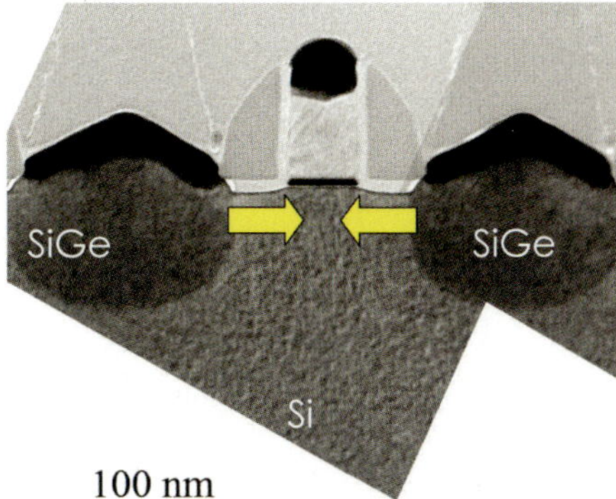

Figure 3.46: *Use of process-induced strain to enhance mobility in an nMOS transistor (left) (Source: ST Microelectronics) and a strained $Si_{1-x}Ge_x$ film in the source and drain areas of a pMOS transitor (right) (Source: NXP Semiconductors)*

The carrier mobility in the channel is also related to their physical crystal orientation (see also section 3.2). It is known that the mobility of holes in a (110) silicon substrate with a current flow along the <110> direction is about two times higher than in conventional (100) silicon. A combination of (110) oriented crystal lattice for the pMOS transistors with a (100) lattice for nMOS provides a much better balance between nMOS and pMOS transistor performance. The (110) orientation for the pMOS could lead to a 45% increase in drive current [29]. Figure 3.47 shows a cross section of a potential nMOS and pMOS device architecture built with different crystal orientations.

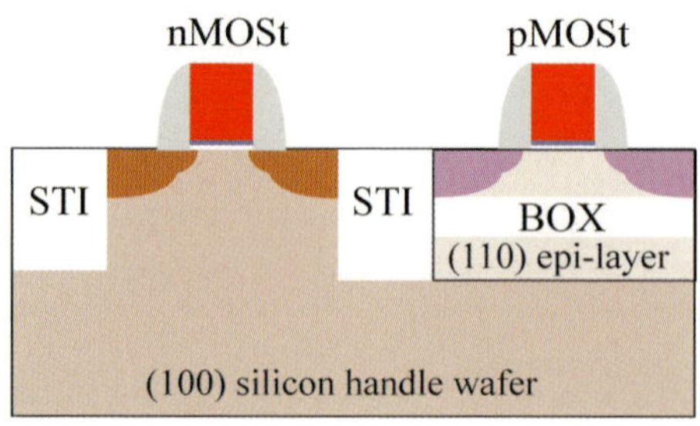

Figure 3.47: *Hybrid-substrate architecture with nMOSt on (100) and pMOSt on (110) crystal orientation*

Figure 3.48 shows a summary of a potential technology options to boost the intrisic device speed.

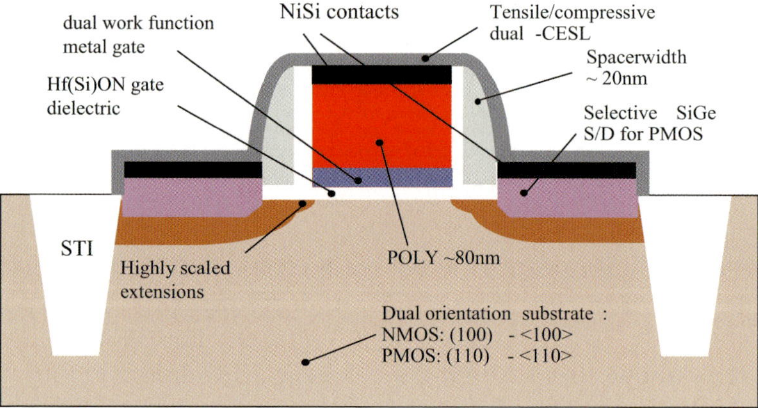

Figure 3.48: *Potential technology options for performance boost of MOS devices (Source: NXP Semiconductors)*

The optimum combination of stress and device orientations has driven and will still further drive the Ion to much higher values than available in today's high-volume CMOS processes as discussed in chapter 2.3.1. Figure 3.49 shows the relative improvements of the Ion currents for nMOS and pMOS transistors, respectively, relative to the year of mass production [30].

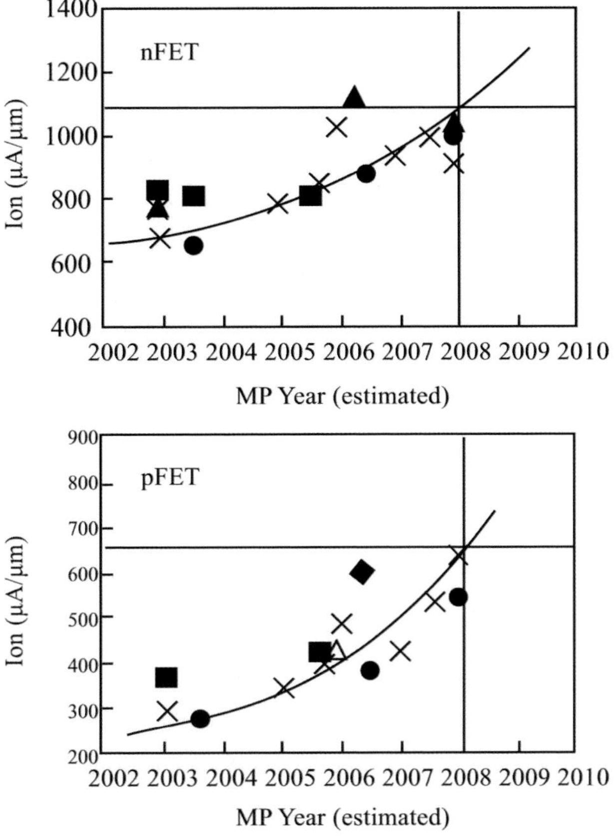

Figure 3.49: *MOSFET performance trend relative to estimated mass production year.* $I_{\text{off}} = 100\,nA/\mu m$ *and* $vdd = 1.0$ *V.* [30]

However, it is not only the real value of I_{on} that counts, but it is more the total $I_{\text{ds}} = f(V_{\text{ds}})$ characteristic that counts, because during switching the transistor cycles through the whole current to voltage characteristic.

A fourth alternative to increase the transistor current is to use a double-gate or FinFET transistor. In a *double-gate transistor* (figure 3.50.a), the transistor body is still lateral, but embedded in between two gates, a bottom gate and a top gate. Above a certain thickness of the body, there are two parallel channels contributing to the total current of the device, which now behave as two parallel fully-depleted SOI transistors.

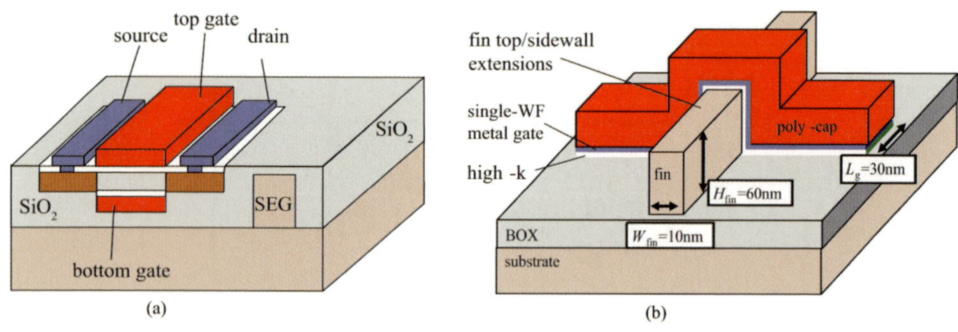

Figure 3.50: *a) Double-gate transistor and b) cross section of a FinFET (Source: NXP Semiconductors)*

In a *FinFET* architecture, a narrow vertical substrate, about 10 to 30 nm thick (figure 3.50.b), is located on top of a BOX (burried-oxide) layer and then covered with a thin gate-oxide layer. Then a thin metal layer with a poly silicon cap is formed, covering the gate-oxide areas at all sides: left, top and right side. If the fin (or body) is very thin, this device will operate as a fully-depleted SOI transistor with a higher driving current, due to the parallel current channels. The width of the transistor is determined by the height of the thin substrate, meaning that only one-size (width) transistors can be fabricated. In this example device the transistor width is equal to the width of the fin + two times its height, resulting in a transistor width of 130 nm. The double-gate and FinFET devices are also called multi-gate FET or MuGFET. These devices help to control leakage currents and reduce short-channel effects. Because they do not exhibit doping fluctuations, their matching properies are expected to be much better. However still a lot of innovations from both the technologists and the designers are required to economically build complex ICs with them at reasonable yield.

Interconnects

There are several reasons why future CMOS ICs still need an increasing number of interconnect layers. Every new technology node offers us more transistors at a two times higher density. This requires more metal resources to support the increasing need for connecting these transistors. Secondly, they require a more dense power distribution network to be able to supply the increasing current needs. Since the introduction of 120 nm CMOS technologies, the aluminium back-end has been

174

replaced by a copper back-end. Due to the required use of a barrier layer in the copper (section 3.6) formation process, the effective copper metal track resistance has only reduced by about 25% compared to aluminium. This has been exploited by reducing the metal height, so that metal tracks show resistances comparable to aluminium, but show less mutual capacitance to neighbouring signals, while maintaining the signal propagation across them. However, further reductions of the metal heights are limited by the increasing current densities and the chance of electromigration. There is also an issue in the scaling of the contacts and vias. Since their number and aspect ratio (height/width ratio) increase with scaling, while their sizes decrease, they are becoming a very important part in the determination of the global chip performance, reliability and yield. Because of the increasing currents, the contacts and vias show an increasing amount of voltage drop, particularly when the signal line switches many times from one metal layer to another. Another result of the increasing current is the increased possibility of electromigration occurrence, thereby threatening the reliability. Finally, due to the high aspect ratios, there is an increased chance for bad contacts or opens, which will affect the yield. Already today, but certainly in the future, *design for manufacturabilty (DfM)* becomes an integral part of the design flow to support yield-improving measures (see also chapter 10). A few examples are: 1) *wire spreading*, where wires are routed at larger pitches (spreaded) because there is more area available than needed by minimum pitch routing and 2) *via doubling*, where more vias are used for the same connection to improve yield.

Most of the further improvements of the interconnect network has to come from further reduction of the dielectric constant (low-ϵ dielectrics) of the *inter-level dielectric (ILD)* layers between the metal layers and between the metal lines within one layer. During the last two decades, this dielectric constant has gradually reduced from 4 to 2.5. It is expected that it will reduce to close to 2, but it still needs many innovations to guarantee sufficient reliability. Some research is currently focused on *air-gaps*, in which the dielectric material between metal lines in the same layer is replaced by air only. This will reduce the dielectric constant to even below 2 (the effective dielectric constant will not be equal to 1 (of air), because there are also mutual electric-field lines from the top and bottom areas of neighbouring metal lines. The reliability of these air gaps is an even bigger challenge.

The combined move from aluminium to copper wiring and from oxide to low-ϵ dielectrics required a change in the bonding process because the adhesion and stability are different. Low-ϵ dielectrics are more porous and include more air, so they become less robust and more sensitive to plasma damage during damascene processing and to pressure during test (probing) and bonding. Particularly when *bond-over-active* techniques are used, where pads are not only located at the chip's periphery but also on top of circuits, these low-ϵ dielectrics must guarantee sufficient reliability. So, changing pad-related design and technology concepts also influences the reliability of the bonding process. Poor bond pad surface contamination may lead to a bond pad metal peel-off which leads to wedge bond or ball bond lifting.

Finally, the continuous process scaling also affects the copper resistivity level. Further scaling leads to an increase of the copper resistivity due to side-wall, grain-boundary and impurity scattering effects which reduce the electron mean free path to 40nm. It also drives the need for ultra-thin, high conductivity barriers and the exploration of "barrier-less" approaches. Figure 3.51 shows the expected trend according to the 2006 ITRS roadmap [31]. A further discussion on copper resistance and its modelling can be found in [32].

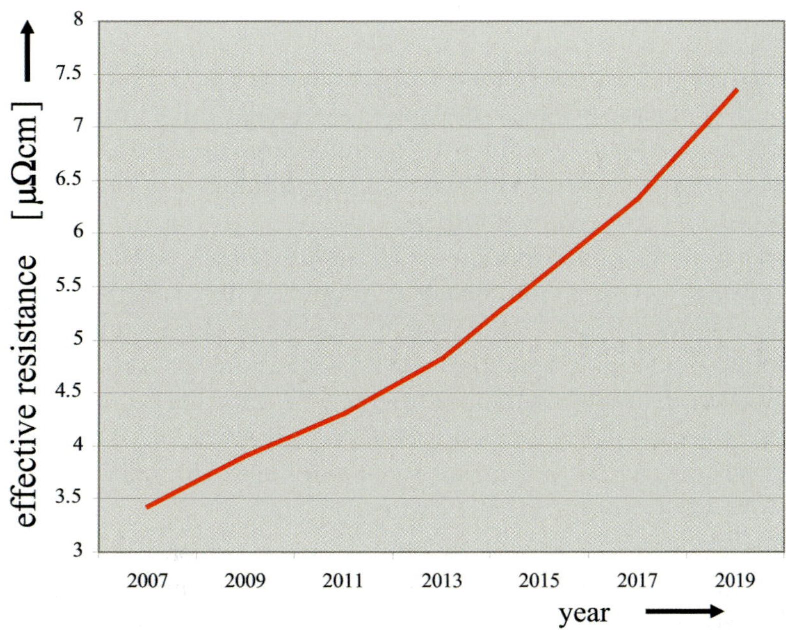

Figure 3.51: *Expected increase of the effective copper resistance of the first metal layer [ITRS 2006]*

3.10 Conclusions

It is clear that the realisation of optimum electronic systems is based on a perfect match between the substrate (wafer), the transistors, and the interconnections. The increasing number of application areas have lead to a large variety of substrate and technology options to support high-speed and low-power products.

So, the processing requirements for different types of circuits can be quite diverse. RAMs, for example, require a technology that allows very high bit densities. CMOS static RAMs therefore require tight n^+-diffusion to n-well spacings. This can be achieved when a retrograde-well implantation is used to minimise lateral well diffusion.

The discussions have started with a basic description of the most important processing steps that are repeatedly used throughout the fabrication of a CMOS chip. For educational purposes, the complexity of the described processes gradually increased from a simple five-mask nMOS process, to a complex over-thirty-masks nanometer CMOS process.

Finally, several trends are discussed which focus on future technology requirements. Chapters 9 and 11 focus on the physical and electrical design consequences of the continuous scaling process.

Finally the increasing complexity of both the lithographic and manufacturing process is reflected by the growing cost of a fab. To ramp up a fab to volume production in a 65 nm requires a time frame of about two and a half years and a budget of $ 3.5 billion. This has prompted many semiconductor companies to become *'fab-lite'* or maybe even totally *fabless*. This trend will certainly be continued toward the 32 and 22 nm technology nodes.

3.11 References

[1] I. Hadar,
'450mm Wafers: Not the Best Solution',
Electronic News, 7/11/2006

[2] www.memc.com

[3] M. Porrini,
'Growing Ingots of Single Crystal Si',
MEMC Silicon Workshop at IMEC, Leuven, Belgium, June 22, 2006

[4] G. Vaccari,
'Silicon Epitaxi for CMOS and Power Applications',
MEMC Silicon Workshop at IMEC, Leueven, Belgium, June 22, 2006

[5] L. Chang, et al.,
'CMOS Circuit Performance Enhancement by Surface Orietation Optimization',
IEEE Transactions on Electron Devices, Vol. 51, No. 10, October 2004

[6] M. Yang, et al.,
'Hybrid-Orientation Technology (HOT): Opportunities and Challenges',
IEEE Transactions on Electron Devices, Vol. 53, No. 5, May 2006

[7] Eric Neyret,
'Ultra-thin SOI: Roadmap and Manufacturing Technology',
June 2002, www-lpm2c.grenoble.cnrs.fr.nanosciences/CIRP/NEYRET.PDF

[8] T. Buchholtz, et al.,
'A 660 MHz 64b SOI Processor with Cu Interconnects',
ISSCC, Digest of Technical Papers, February 2000

[9] J.L. Pelloie, et al.,
'SOI Technology Performance and Modelling',
ISSCC, Digest of Technical Papers, 1999, pp 428-429

[10] H. Majima, et al.,
IEEE Electron Device Letter, pp. 396-398, 2000

[11] Copied with permission from ASML from the following website:
http://www.asml.com/asmldotcom/show.do?ctx=10448&rid=10131

[12] Tom Lecklider,
  'Yield: The Key to Nanometer Profits',
  Evaluation Engineering, March 2005
  www.evaluationengineering.com/archive/articles/0305/0305yield.asp

[13] Y.K. Choi, et al.
  'Sublithographic nanofabrication technology for nanocatalysts and
  DNA chips',
  J. Vac. Sci. Technol. B21(6), Nov/Dec 2003, pp. 2951-2955

[14] Mark LaPedus,
  'ASML ships world's first EUV tool',
  www.eetimes.com 08/28/2006

[15] Peter Singer,
  'Nanoimprint Lithography: A Contender for 32 nm?'
  Semiconductor International, Issue August 1, 2006

[16] Hans C. Pfeiffer, et al.,
  'Microlithography World - The history and potential of maskless E-
  beam lithography',
  Solid State Technology, February 2005
  http://sst.pennnet.com/Articles/Article_Display.cfm?Section=ARTCL&
  ARTICLE_ID=221612&VERSION_NUM=4&p=28

[17] Keizo Suzuki and Naoshi Itabashi,
  'Future prospects for dry etching',
  fure&App/. Chem., Vol. 68, No. 5, pp. 1011-1015, 1996

[18] Dipankar Pramanik,
  'Challenges for intermetal dielectrics',
  Future Fab International, 1997

[19] Bryant Mann,
  'Development of Thin Gate Oxides for Advanced CMOS Applications',
  22nd Annual Microelectronic Engineering Conference, May 2004

[20] Y. Mitani, et al.,
  'NBTI Mechanism in Ultra-thin Gate Dielectric-Nitrogen-originated
  Mechanism in SiON-',
  International Electron Devices Meeting Technical Digest, p.509-512,
  2002

[21] S. Wolf and R.N. Tauber,
'Silicon Processing for the VLSI Era',
Volume-1, Process Technology, Lattice Press, 1986

[22] Booyons S. Lim et al.,
'Atomic layer deposition of transition metals',
nature materials, VOL 2, November 2003,
www.nature.com/naturematerials

[23] Leonard Rubin and John Poate,
'Ion Implantation in Silicon Technology',
The Industrial Physicist, June/July 2003, pp. 12-15

[24] R.F.M. Roes, et al.,
'Implications of pocket optimisation on analog performance in deep
sub-micron CMOS',
ESSDERC, digest of technical papers, 1999, pp. 176-179

[25] M. Bohr, et al.,
'The High-k Solution',
IEEE Spectrum, October 2007, pp. 23-29

[26] S. Thompson, et al.,
'A 90nm logic technology featuring 50nm strained silicon channel
transistors, 7 layers of Cu interconnects, low-k ILD, and 1 mm SRAM
cell',
IEEE International Electron Devices Meeting, 2002

[27] G. Eneman, et al.,
'N+/P and P+/N Junctions in Strained Si on Strain Relaxed SiGe
Buffers: the Effect of Defect Density and Layer Structure',
Mater. Res. Soc. Symp. Proc. Vol. 864 © 2005 Materials Research
Society, pp. E3.7.1-E3.7.6

[28] P.R. Chidambaram,
'35% Drive Current Improvement from Recessed-SiGe Drain Extensions on 37 nm Gate Length PMOS',
2004 Symposium on VLSI Technology Digest of Technical Papers,
pp. 48-49

[29] M. Yang et al.,
'High Performance CMOS Fabricated on Hybrid Substrate With Dif-

ferent Crystal Orientations'
Electron Devices Meeting, 2003. IEDM '03 Technical Digest

[30] K.Ishimaru,
'45 nm/32 nm CMOS; Challenge and Perspective',
ESSCIRC 2007, Digest of Technical Papers, pp. 32-35

[31] International Technology Roadmap for Semiconductors, 2005 edition and 2006 update
www.itrs.net/reports.html

[32] Pawan Kapur, et al.,
'Technology and Reliability Constrained Future Copper Interconnects - Part I: Resistance Modelling'
IEEE Transactions on Electron Devices, Vol. 49, No. 4, April 2002, pp. 590-597

3.12 Exercises

1. Why is the formation of the gate oxide a very important and accurate process step?

2. Briefly explain the major differences between the diffusion process and the ion-implantation process. What are the corresponding advantages and disadvantages?

3. What are the possible consequences of an aluminium track with a bad step coverage?

4. Describe the main differences between the formation of LOCOS and STI.

5. What are the major advantages of self-aligned sources and drains?

6. Why is planarisation increasingly important in modern deep-sub-micron technologies?

7. Assume that the sixth metal layer in a $0.25\,\mu m$ CMOS process is optional. In which designs would you use the sixth metal and why? What is/are the advantage(s)/disadvantage(s) of using the sixth metal layer?

8. Why was copper not used earlier in the metallisation part of a CMOS process?

9. What are the disadvantages of plasma etching?

10. What are 'tiles', as meant in the manufacture of a deep-submicron chip? Why may they be needed in such a design?

11. For which type of circuits would SOI be particularly beneficial in terms of speed and power?

12. Summarise all potential (technological as well as electronic) solutions to increase the I_{on} current of a transistor. Distinguish between nMOS and pMOS solutions.

Chapter 4

CMOS circuits

4.1 Introduction

Although it existed in the seventies, it took until the mid-eighties before CMOS became the leading technology for VLSI circuits. Prior to that time, only a few circuits were designed in CMOS. These early designs were generally limited to analogue circuits and digital circuits that dissipated little power. Examples include chips for calculators, watches and remote controls. CMOS offers both n-type and p-type MOS transistors. Initially, this meant that CMOS circuits were more costly than their nMOS equivalents.

The majority carriers in pMOS and nMOS transistors are holes and electrons, respectively. The mobility of holes is about three times lower than electron mobility. This makes pMOS circuits significantly slower than nMOS circuits of equal chip area. The continuous drive for increased integrated circuit performance therefore led to the early disappearance of pMOS technologies. The demand for higher packing densities and performance led to an increase in the complexity of nMOS processes.

In particular, the quest for a lower τD product (power delay product) necessitated the availability of several different transistor threshold voltages in a single nMOS process. These included a few enhancement threshold voltages ($V_\mathrm{T} > 0$) and different depletion threshold voltages ($V_\mathrm{T} < 0$). Even threshold voltages of zero volts had to be available. These threshold voltages were provided at the cost of additional masks and extra processing steps, which rapidly elevated the complexity of nMOS processes to about the level of CMOS processes. A few advantages afforded by CMOS processes therefore led to their domination of the MOS IC world.

Modern manufacturing processes make it possible to integrate increasingly complex circuits and even complete systems on a single chip. The resulting number of transistors per chip may reach hundreds of millions to billions. The associated power dissipation can easily exceed the critical 1 W maximum limit for cheap small plastic IC packages. Circuits that are manufactured in CMOS processes generally consume less than one tenth of the power dissipated by an nMOS equivalent. Moreover, CMOS circuits have better noise margins. These advantages have led to the use of CMOS for the integration of most modern VLSI circuits. These include memories, digital signal processors, microprocessors, speech synthesizers, data communication chips and complete Systems On Chip (SOC).

The various CMOS processes and their characteristic properties are extensively treated in section 3.9. This chapter starts with a discussion on basic nMOS circuits to be able to understand CMOS circuit properties more easily. Basic design principles and problems associated with CMOS are subjects of this chapter. Several different types of both static and dynamic CMOS circuits are discussed. Related reliability issues, such as latch-up, are discussed in chapter 9, together with other topics that improve the IC's robustness. The chapter ends with a section on CMOS layout design. A layout design method is illustrated by means of an example.

Finally, it should be noted that many examples are based on an n-well CMOS process. Initially, this process was chosen because of its compatibility with the conventional nMOS process. In addition, many dynamic CMOS circuits are 'nMOS-mostly'. Currently, most processes are twin-well CMOS processes, in which the nMOS and pMOS transistors can both be realised with optimum performance.

4.2   The basic nMOS inverter

4.2.1   Introduction

Generally, the electrical properties of a static nMOS circuit are completely determined by its DC behaviour and transient response. These will be explained with the aid of one of the most elementary MOS circuits, i.e., the inverter. In the following we treat the nMOS-only circuits as if they were designed in a 45 nm technology node, along with their supply and threshold voltages.

Figure 4.1 shows schematics of an inverter and its different types of 'load elements'.

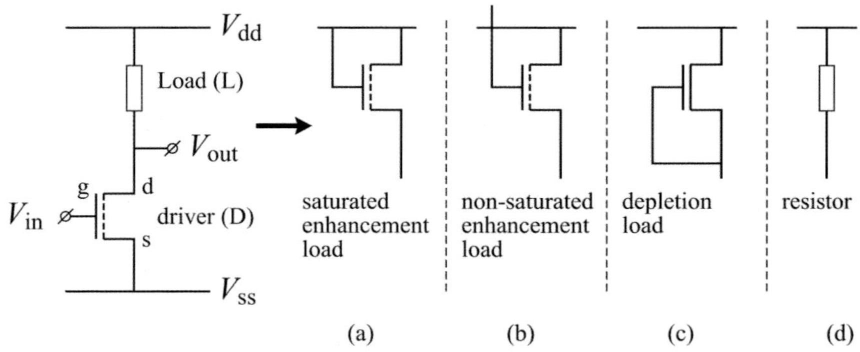

Figure 4.1: *An inverter and its different types of load elements*

The inverter's DC behaviour and transient response are discussed for its different types of load elements. The discussions are based on formulae (1.18) which express the current in a transistor as follows:

$$\text{Linear region}: \quad I_{ds} = \beta(V_{gs} - V_T - V_{ds}/2)V_{ds} \quad (V_{ds} < V_{gs} - V_T)$$

$$\text{Saturation region}: \quad I_{ds} = \beta/2(V_{gs} - V_T)^2 \quad (V_{ds} \geq V_{gs} - V_T)$$

$$\text{Where}: \quad V_T = V_x + k\sqrt{V_{sb} + 2\phi_f}$$

Two criteria are important when determining the dimensions of transistors in MOS logic gates:

- The location of the operating points. These are the output voltages V_L and V_H, which correspond to the logic values '0' and '1', respectively. Output voltage V_L, for example, must be a *'noise margin'* less than the threshold voltage V_{T_D} of the n-type enhancement driver transistor. The noise margin ensures that subsequent logic gates always interpret V_L correctly. V_{T_D} is about 0.3 V and a noise margin of about 0.15 V is normally used. This implies that $V_L \leq 0.15$ V in nMOS circuit design.

- The transient response. This implicitly refers to the rise and fall
  times associated with changes in the output's logic levels.

In the next sections, these criteria are discussed for the four types of
inverters shown in figure 4.1.

4.2.2 The DC behaviour

The DC behaviour of inverters with different types of load elements are
explained separately below with the aid of figure 4.2. This figure shows
the '*driver transistor*' characteristic $I_{ds}= f(V_{ds})|_{V_{gs}=V_H}$ together with
the '*load lines*' of the different load elements in figure 4.1. The shapes
of the load lines are characteristic of the respective load elements.

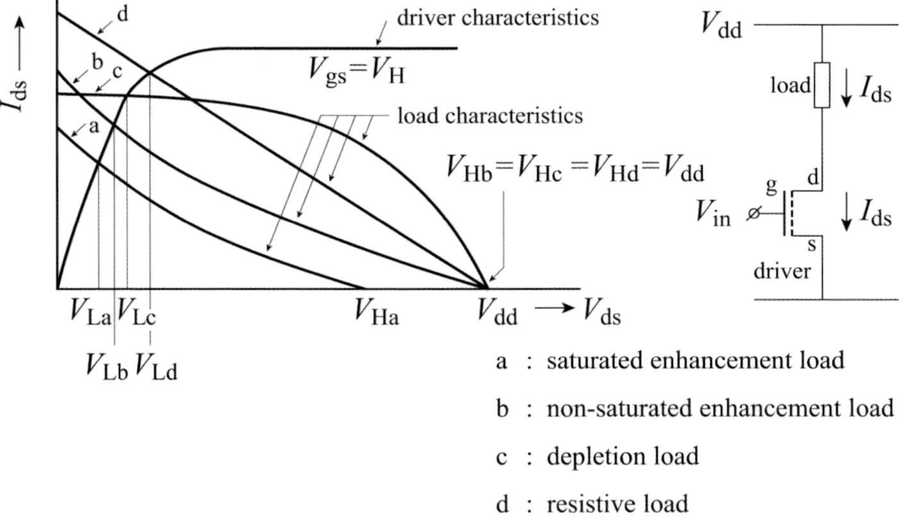

Figure 4.2: *Inverter characteristics for different types of load elements*

The output voltage of an inverter is 'low' ($V_{out}= V_L$) if its input voltage is
'high' ($V_{in}= V_H$) and vice versa. The output low level values corresponding to the different load elements are determined by the intersection of
the driver characteristic and the relevant load line. These values are
indicated by V_{La}, V_{Lb}, etc. in figure 4.2. The indicated positions are
chosen for clarity and are not typical for the various load elements. The
point of intersection between a load line and the driver characteristic
is in fact chosen by the designer. For inverters that use transistors as

188

load elements, this point is determined by the *'aspect ratio'* A, which is expressed as follows:

$$A = \frac{(\frac{W}{L})_{\text{D}}}{(\frac{W}{L})_{\text{L}}}$$

Achieving a correct 'low' level in static nMOS logic clearly requires a minimum ratio between the driver and load transistor sizes. This type of circuit is therefore called *ratioed logic*.

Saturated enhancement load transistor

The DC behaviour of an inverter with a *saturated enhancement load transistor* is explained with the aid of figure 4.3, which shows a schematic diagram of the inverter. The load line and four driver characteristics, for different values of V_{in}, are also shown.

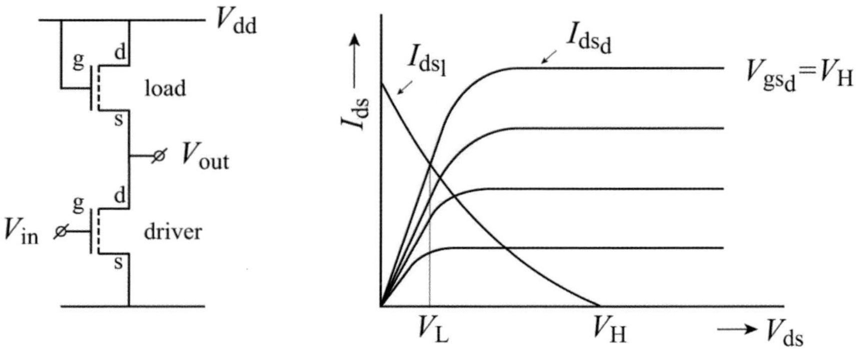

Figure 4.3: *An inverter with a saturated enhancement load transistor*

The minimum drain-source voltage of the load transistor is a threshold voltage, i.e., as $V_{\text{ds}}=V_{\text{gs}}$, $V_{\text{ds}_{\text{L}}} > V_{\text{gs}_{\text{L}}} - V_{\text{T}_{\text{L}}}$ always applies. Therefore, the load transistor always operates in the saturation region. The application of formulae (1.18) yields the following expression for the current in the load transistor:

$$I_{\text{ds}_{\text{L}}} = \frac{\beta_{\text{L}}}{2}(V_{\text{ds}_{\text{L}}} - V_{\text{T}_{\text{L}}})^2$$

The DC operation of an inverter with a saturated enhancement load transistor is described as follows:

- If $V_{\text{in}} = V_{\text{L}} < V_{\text{T}_{\text{D}}}$, then the driver transistor is 'off' and $I_{\text{ds}_{\text{D}}} = I_{\text{ds}_{\text{L}}} = 0$.
  According to the above expression for $I_{\text{ds}_{\text{L}}}$, the output voltage is then: $V_{\text{out}} = V_{\text{H}} = V_{\text{dd}} - V_{\text{T}_{\text{L}}}$.

- If $V_{\text{in}} = V_{\text{H}} \gg V_{\text{T}_{\text{D}}}$ then $V_{\text{out}} = V_{\text{L}}$. The driver current $I_{\text{ds}_{\text{D}}}$ and the load transistor current $I_{\text{ds}_{\text{L}}}$ will then be equal:

$$I_{\text{ds}_{\text{D}}} = I_{\text{ds}_{\text{L}}}$$

$$\Rightarrow \underbrace{\beta_{\text{D}} \cdot \left(V_{\text{H}} - V_{\text{T}_{\text{D}}} - \frac{V_{\text{L}}}{2} \right) \cdot V_{\text{L}}}_{\text{driver transistor in \ linear region}} = \underbrace{\frac{\beta_{\text{L}}}{2} \left((V_{\text{dd}} - V_{\text{L}}) - V_{\text{T}_{\text{L}}} \right)^2}_{\text{load transistor always saturated}}$$

Assuming $V_{\text{L}} \ll V_{\text{dd}}$ and $V_{\text{L}}/2 \ll V_{\text{H}} - V_{\text{T}_{\text{D}}}$ yields:

$$\left(\frac{W}{L}\right)_{\text{D}} \cdot (V_{\text{H}} - V_{\text{T}_{\text{D}}}) V_{\text{L}} = \left(\frac{W}{L}\right)_{\text{L}} \cdot \frac{1}{2} \cdot (V_{\text{dd}} - V_{\text{T}_{\text{L}}})^2$$

With $V_{\text{dd}} - V_{\text{T}_{\text{L}}} = V_{\text{H}}$, this reduces to the following expression for the aspect ratio A of this inverter:

$$A = \frac{\left(\frac{W}{L}\right)_{\text{D}}}{\left(\frac{W}{L}\right)_{\text{L}}} \geq \frac{V_{\text{H}}^2}{2(V_{\text{H}} - V_{\text{T}_{\text{D}}}) V_{\text{L}}} \tag{4.1}$$

The use of a saturated enhancement load transistor is disadvantaged by the associated 'threshold loss', which produces a high level V_{H}, and this is only $V_{\text{dd}} - V_{\text{T}_{\text{L}}}$ rather than V_{dd}. The corresponding relatively low input voltage applied to a subsequent logic gate results in a lower speed. The use of a non-saturated enhancement or depletion load transistor overcomes this problem and produces a V_{H} equal to V_{dd}.

The non-saturated enhancement load transistor

An inverter with a *non-saturated enhancement load* transistor is illustrated in figure 4.4.

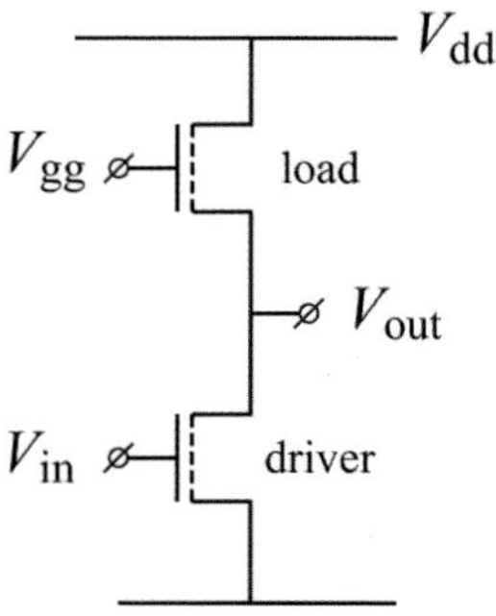

Figure 4.4: *An inverter with a non-saturated enhancement load transistor*

The gate of the load transistor is connected to an extra supply voltage V_{gg} instead of the supply voltage V_{dd}. The extra supply voltage is large enough to ensure that the load transistor always operates in the non-saturated region, i.e., $V_{gg} > V_{dd} + V_{T_L}$.

The DC operation of the above inverter is described as follows:

- $V_{in} = V_L < V_{T_D} \quad \Rightarrow \quad I_{ds_L} = 0\,mA$ and $V_{out} = V_H = V_{dd}$.

- $V_{in} = V_H \gg V_{T_D} \Rightarrow V_{out} = V_L$.

The driver now operates in the linear region. The driver and load transistor currents are equal:

$$
\begin{aligned}
I_{ds_D} &= (\frac{W}{L})_D \cdot \beta_\square \left(V_H - V_{T_D} - \frac{V_L}{2} \right) V_L \\
&= (\frac{W}{L})_L \cdot \beta_\square \left(V_{gg} - V_L - V_{T_L} - \frac{V_{dd} - V_L}{2} \right) (V_{dd} - V_L) \\
&= I_{ds_L}
\end{aligned}
$$

Assuming $V_L \ll V_{dd}$, $\frac{V_L}{2} \ll V_H - V_{T_D}$ and $V_{gg} - V_{T_L} \gg V_L$ yields the following expression for the inverter's aspect ratio A:

$$
A = \frac{(\frac{W}{L})_D}{(\frac{W}{L})_L} = \frac{\left(V_{gg} - V_{T_L} - \frac{V_{dd}}{2} \right) \cdot V_{dd}}{(V_H - V_{T_D}) \cdot V_L}
$$

Since $V_H - V_{T_D} < V_{dd}$, the aspect ratio A is expressed as follows:

$$
A = \frac{(\frac{W}{L})_D}{(\frac{W}{L})_L} \geq \frac{V_{gg} - V_{T_L} - \frac{V_{dd}}{2}}{V_L} \tag{4.2}
$$

The use of a non-saturated enhancement transistor as load element has the following advantages:

- High V_H ($=V_\mathrm{dd}$);

- Large noise margin;

- Fast logic.

The most significant disadvantage is the extra supply voltage required V_gg ($V_\mathrm{gg} \geq V_\mathrm{dd} + V_\mathrm{T_L}$), which may necessitate an extra pin on the chip package. Alternatively, V_gg can be electronically generated on the chip. This results in a *'bootstrapped load'* element, as shown in figure 4.5.

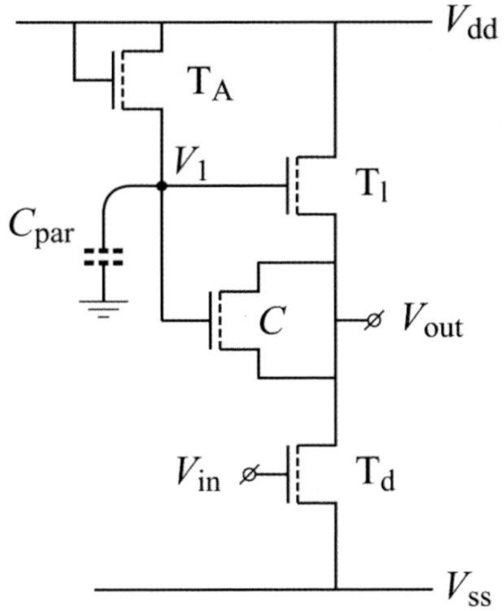

Figure 4.5: *The bootstrapped inverter*

The DC operation of the bootstrapped inverter is explained as follows:

- If $V_\mathrm{in} = V_\mathrm{H}$, then $V_\mathrm{out} = V_\mathrm{L}$ and $V_1 = V_\mathrm{dd} - V_\mathrm{T_A}$. The MOS *'bootstrap' capacitance* C therefore charges.

- When V_in switches from V_H to V_L, then V_out increases by ΔV_out and V_1 increases by ΔV_1. The magnitude of ΔV_1 is determined by the values of the bootstrap capacitance C and the parasitic capacitance C_par such that:

$$\Delta V_1 = \frac{C}{C + C_\mathrm{par}} \cdot \Delta V_\mathrm{out}$$

This means that V_1 immediately passes the V_{dd}–V_{T_A} level and transistor T_A therefore no longer conducts. The voltage V_1 can then further increase to a voltage greater than V_{dd}. The maximum value of V_1 is determined by the capacitance ratio:

$$a = \frac{C}{C + C_{par}}$$

The value of a required to produce a 'high' output voltage is: $V_H = V_{dd}$ and is derived as follows:
$V_H = V_{dd}$ when $V_1 \geq V_{dd} + V_{T_L}$.
$\Delta V_1 = a \cdot \Delta V_{out}$ and $V_1 = V_{dd} - V_{T_A} + a \cdot \Delta V_{out}$.

The load transistor T_L must remain in the linear operating region. The following equation therefore applies:
$V_1 - V_{T_L} > V_{dd}$

$$\Rightarrow V_{dd} - V_{T_A} - V_{T_L} + a \cdot \Delta V_{out} > V_{dd}$$

$$\Rightarrow \Delta V_{out} > \frac{V_{T_A} + V_{T_L}}{a}$$

The output high level must be equal to the supply voltage, i.e., $V_{out} = V_H = V_{dd}$. Therefore, $\Delta V_{out} = V_{dd} - V_L$. Assuming $V_{T_A} \approx V_{T_L}$ yields the following expression for a:

$$a > \frac{2V_{T_L}}{V_{dd} - V_L} \tag{4.3}$$

- If $V_{in} = V_H$, then $V_{out} = V_L$ and the gate voltage of the load transistor T_L is $V_{dd} - V_{T_A} \approx V_{dd} - V_{T_L}$. Load transistor T_L therefore operates in the saturation region when $V_{out} = V_L$. The aspect ratio A of the bootstrapped inverter is therefore identical to that given in equation (4.1) for the inverter with a saturated enhancement load transistor.

The bootstrapped inverter has the following advantages:

1. There is no threshold loss when the bootstrap capacitance C is correctly dimensioned.

2. There is no extra supply voltage required, because the voltage V_1 is pumped to more than a threshold voltage above V_{dd}.

3. This basic bootstrap mechanism is also called a *charge-pump*, which is used in many E(E)PROMs and flash memories to generate the much higher programming and/or erasing voltages. To achieve such high voltages (≥ 10 V), several of these charge pumps are put in series.

The depletion load transistor

The manufacture of depletion transistors requires an extra mask (DI) and additional processing steps. There are, however, considerable advantages associated with the use of a depletion transistor as load element. These include the following:

- The output high level equals V_{dd}, i.e., $V_H = V_{dd}$;

- There is no extra supply voltage required;

- Circuit complexity is minimal and bootstrapping is unnecessary;

- Noise margins are high.

For these reasons, before the move to CMOS, most nMOS processes were '*E/D technologies*' and contain both enhancement and depletion transistors. Some manufacturers, today, even include depletion transistors in their CMOS technologies.

Figure 4.6 shows an inverter with a *depletion load* transistor.

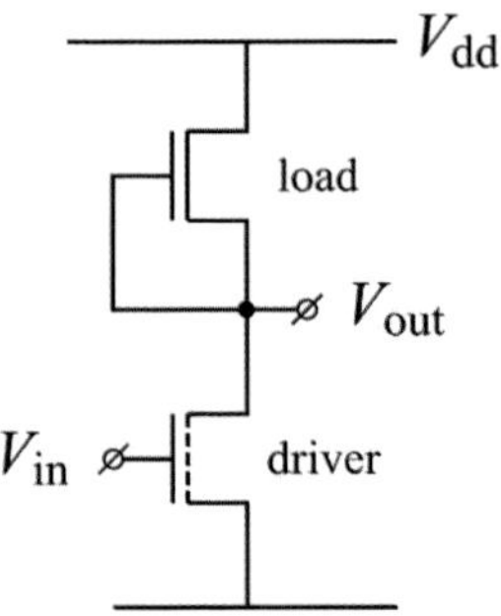

Figure 4.6: *An inverter with a depletion load transistor*

The DC operation of the inverter with a depletion load transistor is described as follows:

- The depletion load transistor has a negative threshold voltage which was usually between $-1\,\text{V}$ and $-3\,\text{V}$. Therefore, $V_{\text{out}}=V_{\text{H}}=V_{\text{dd}}$ when $V_{\text{in}}=V_{\text{L}}<V_{\text{T}_{\text{D}}}$.

- When $V_{\text{in}} = V_{\text{H}} > V_{\text{T}_{\text{D}}}$, then $V_{\text{out}}=V_{\text{L}}$ and $V_{\text{gs}_{\text{L}}}(=0\,\text{V}) < V_{\text{ds}_{\text{L}}}+V_{\text{T}_{\text{L}}}$. In this case, the load transistor operates in the saturation region while the driver transistor operates in the triode region. Equating the currents in the load and driver transistors yields:

$$I_{\text{ds}_{\text{D}}} \quad = \quad I_{\text{ds}_{\text{L}}}$$

$$\Rightarrow (\frac{W}{L})_{\text{D}} \cdot \beta_{\square} \cdot \left(V_{\text{H}} - V_{\text{T}_{\text{D}}} - \frac{V_{\text{L}}}{2} \right) \cdot V_{\text{L}} = (\frac{W}{L})_{\text{L}} \cdot \frac{\beta_{\square}}{2} \cdot V_{\text{T}_{\text{L}}}{}^2$$

If $\frac{V_{\text{L}}}{2} \ll V_{\text{H}} - V_{\text{T}_{\text{D}}}$, then the aspect ratio A of the depletion-load inverter can be expressed as follows:

$$A = \frac{(\frac{W}{L})_{\text{D}}}{(\frac{W}{L})_{\text{L}}} \geq \frac{V_{\text{T}_{\text{L}}}{}^2}{2V_{\text{L}} \cdot (V_{\text{H}} - V_{\text{T}_{\text{D}}})} \tag{4.4}$$

The resistive load

VLSI circuits may consist of tens to hundreds of millions of logic gates which may dissipate no more than 1 to $100\,\text{nW}$ each. A supply voltage of $1.2\,\text{V}$ therefore requires a *resistive load* of several tens of $\text{M}\Omega$ per logic gate. Both diffusion and polysilicon have a *sheet resistance* of about $100\,\Omega/\square$. Realisation of a $10\,\text{M}\Omega$ resistance in a $50\,\text{nm}$ wide polysilicon track therefore requires a length of $5\,\text{mm}$. At the cost of extra processing complexity, however, large resistances can be realised on small chip areas. For random-access memories (RAMs), the disadvantages of complex processing were justified by very large production quantities (1.5 billion 4M-DRAMs in 1996). The addition of a second polysilicon layer with very high resistivity in static RAM processes facilitated the realisation of memory cells that were considerably smaller than the *full-CMOS* cells. The use of resistive load elements (figure 4.7) was therefore mainly limited to application in static memories and was not normally encountered in VLSI circuits. Because this circuit's operation resembles that of an inverter with a depletion load transistor, we will no longer focus on this type of load.

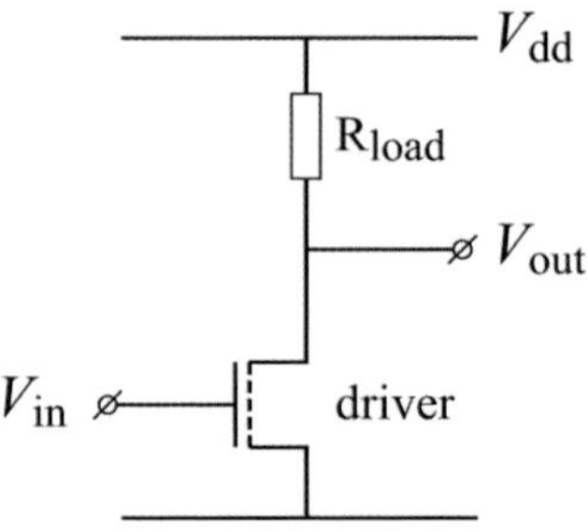

Figure 4.7: *An inverter with a resistive load*

4.2.3 Comparison of the different nMOS inverters

NMOS inverters with different load elements are now compared.

Adopting a 1 pF load capacitance, a *circuit analysis* program was used to simulate the charging and discharging characteristics that correspond to these load and driver transistors, respectively. The charging characteristic associated with the load resistance was also simulated. The results are shown in figure 4.8.

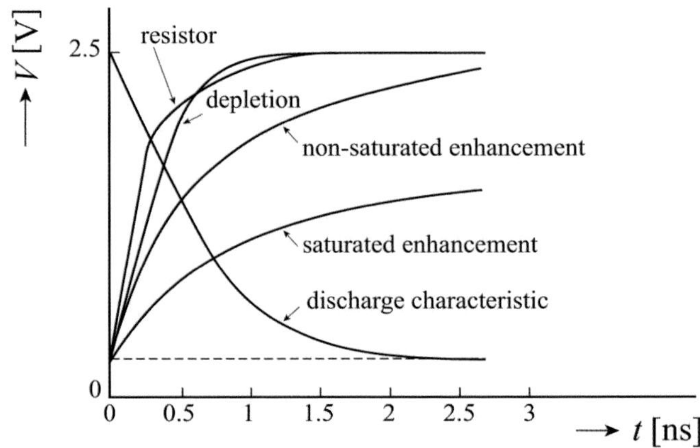

Figure 4.8: *Charging characteristics of nMOS inverters with different types of load, identical load capacitances and the same initial current I_0*

The performance of logic gates, built with the previously presented different types of load elements was quite different. Particularly the saturated enhancement load, because of the relatively large voltage drop at high level, made this logic much slower than logic with a non-saturated enhancement load or with the depletion load (figure 4.8). Logic with the non-saturated enhancement load required either an additional supply voltage to keep the load always in its saturation region, or it required bootstrapping techniques to do so. Both solutions were not so well appreciated and have made E/D nMOS logic (so, logic with the depletion load) the most popular technique to realise VLSI chips, until the mid eighties when CMOS became the most dominant IC technology. For this reason the next paragraph uses the depletion load for further evaluation of nMOS logic gates.

4.2.4 Transforming a logic function into an nMOS transistor circuit

An inverter is transformed into a logic gate by replacing the driver transistor by a combination of MOS transistors. The combination may comprise series and/or parallel transistors. Each transistor gate is controlled by a logic signal. A complex logic function can therefore be implemented in a single logic gate with an associated propagation delay. The following transformation rules apply:

1. An *AND* function is realised by a *series connection* of transistors.

2. An *OR* function is realised by a *parallel connection* of transistors.

Because logic gates are an adaptation of the basic inverter, the output signal is always the inverse of the function that is derived when the transistors in the driver section are interpreted according to the above rules. In fact, implementations always comprise NAND, NOR or AND-OR-NOT functions.

Example: A 'full adder' is described by the following logic functions (see also section 7.3.5):

$$S \ = x\,\bar{y}\,\bar{z} + \bar{x}\,\bar{y}\,z + \bar{x}\,y\,\bar{z} + x\,y\,z$$
$$C_o = x\,y + x\,z + y\,z$$

Symbols x and y represent two bits which must be added. Symbol z represents the 'carry-in'. S represents the binary sum of x, y and z while C_o represents the 'carry-out'.

The logic function S can also be written as:

$$S = x\,(y\,z + \overline{y}\,\overline{z}) + \overline{x}\,(y\,\overline{z} + \overline{y}\,z)$$

This function corresponds to the implementation in figure 4.9, which realises the inverse $(\overline{S})$ of the sum function.

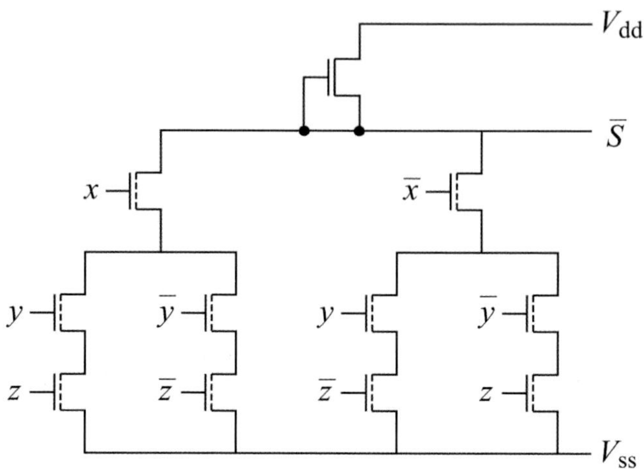

Figure 4.9: *An implementation of the function $\overline{S}$*

Figure 4.10 shows a realisation of the inverse of the carry function.

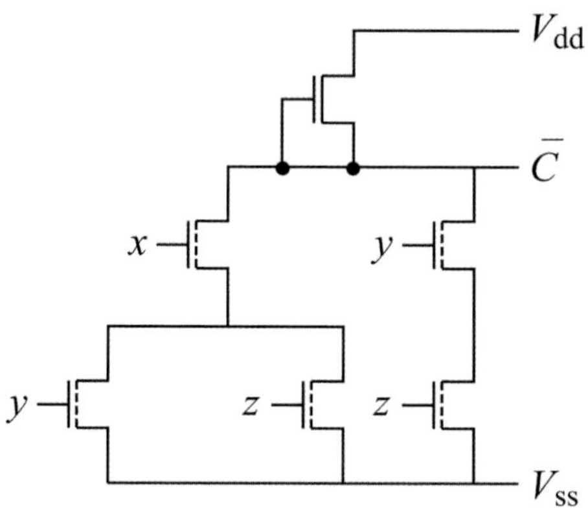

Figure 4.10: *Implementation of the full adder inverse 'carry-out' function*

An nMOS transistor's gain factor β equals $\beta_\square \cdot \frac{W}{L}$. The gain factor β_{total} of n transistors connected in series is expressed as follows:

$$\beta_{\text{total}} = \left(\frac{1}{\beta_1} + \frac{1}{\beta_2} + \ldots \frac{1}{\beta_n}\right)^{-1}$$

If all the transistors have equal dimensions, then:

$$\beta_{\text{total}} = \beta/n$$

The discharge time constant associated with these n transistors is then directly proportional to n. In fact, the speed of a logic gate is largely determined by the number of transistors that are connected in series in the driver section. It is thus generally advisable to keep this number to a minimum. Figure 4.11, for example, shows a NAND gate with n driver transistors in series. The effective $(\frac{W}{L})$ ratio of these n transistors is expressed as follows:

$$\left(\frac{W}{L}\right)_{\text{total}} = \frac{1}{(\frac{W}{L})_1^{-1} + (\frac{W}{L})_2^{-1} + \ldots + (\frac{W}{L})_n^{-1}} \tag{4.5}$$

The $(\frac{W}{L})$ aspect ratio of the driver transistor in an inverter can be calculated using the formulae in sections 4.2.2. For a NAND gate with n inputs, the inverter's driver transistor (D) must be replaced by n transistors in series. The NAND gate will be as fast as the inverter if its transistors each have an aspect ratio $n \cdot (\frac{W_i}{L_i})$, where W_i and L_i are the width and length, respectively, of the inverter's driver transistor.

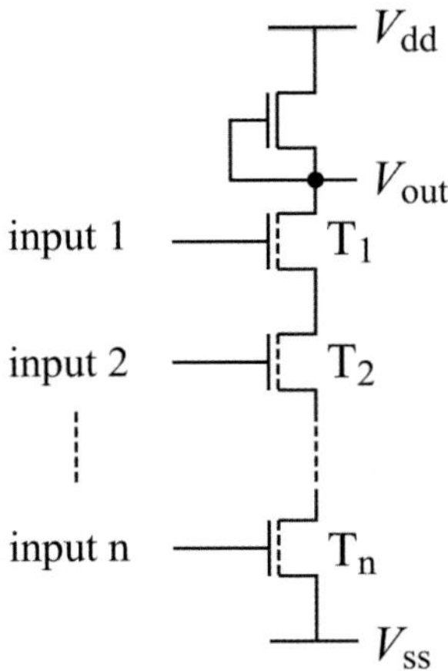

Figure 4.11: *NAND gate with n inputs and thus n transistors in series*

The number of parallel sections in a logic gate is also a critical implementation factor. The circuit area and hence the parasitic capacitances associated with the logic gate increase with the number of parallel sections. This causes an increase in the gate's propagation delay.

This section presented an overview on the electrical design of nMOS circuits and the creation of basic nMOS logic gates. A major disadvantage of nMOS logic is the associated power consumption. Each logic gate with a low level at its output consumes DC power. Therefore, even when a large logic nMOS chip has no signal transitions, there is a large DC power consumption. CMOS circuits, which require more complex technologies than nMOS circuits, do not consume DC power when there is no activity. This is the most important reason for the domination of CMOS circuits in the integrated circuit market.

4.3 Electrical design of CMOS circuits

4.3.1 Introduction

The acronym CMOS stands for Complementary Metal Oxide Semiconductor'. The word 'complementary' indicates that transistors of different types can be manufactured in CMOS processes. The types are n-channel and p-channel, or 'nMOS and 'pMOS'. The nMOS transistor and its operation have been extensively treated before. The pMOS transistor has been briefly mentioned. Where necessary, additional details about its operation are provided in this chapter. The nMOS and pMOS transistors used in CMOS processes are both of the enhancement type. Section 1.7 reveals that the threshold voltage of the nMOS transistor is therefore positive while that of the pMOS transistor is negative. This is shown in figure 4.12.

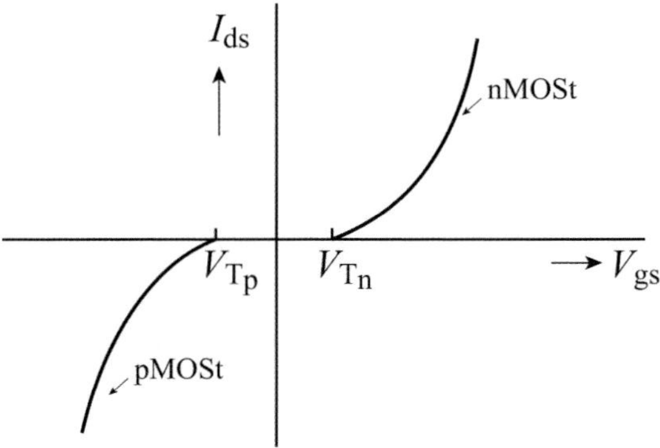

Figure 4.12: *The $I_{ds}=f(V_{gs})$ characteristics of nMOS ($V_{T_n}>0$) and pMOS ($V_{T_p}<0$) enhancement transistors*

The formulae discussed in section 1.5, which describe the back-bias effect on the threshold voltages of nMOS and pMOS transistors, are as follows:

$$V_{T_n} = V_{X_n} + K_n\sqrt{V_{sb} + 2\phi_f} \quad \text{(enhancement type}: V_{X_n} > 0,\ K_n > 0)$$

$$V_{T_p} = V_{X_p} + K_p\sqrt{V_{ws} + 2\,|\phi_f|} \quad \text{(enhancement type}: V_{X_p} < 0,\ K_p < 0)$$

In the CMOS process that is considered in this section, the pMOS transistor is integrated in an n-well. Voltage V_{ws} in the above expression for the threshold voltage V_{T_p} of a pMOS transistor represents the voltage between the n-well and the source of the transistor.

The above expressions and figure show that the operation of the pMOS transistor is the exact complement of the nMOS transistor's operation. The electrical operation of the nMOS and pMOS transistors can be summarised as follows: the pMOS transistor's behaviour with respect to the supply voltage is identical to the nMOS transistor's behaviour with respect to ground and vice versa.

4.3.2 The CMOS inverter

A basic *CMOS inverter* consists of an nMOS transistor and a pMOS transistor connected as shown in figure 4.13. The n-well serves as a substrate for the pMOS transistor. It is formed by the diffusion or ion implantation techniques discussed in chapter 3.

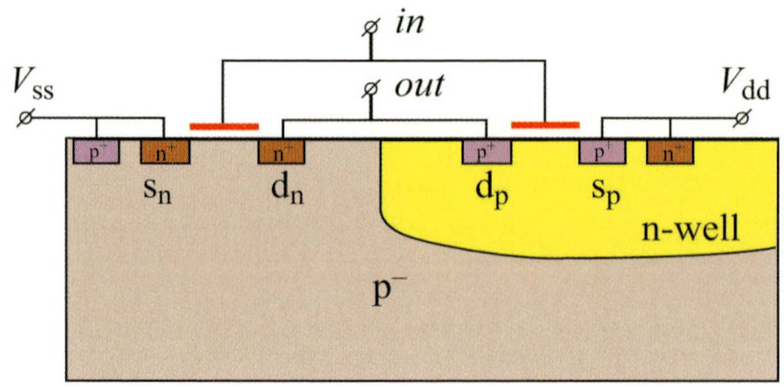

Figure 4.13: *Transistor connections for a CMOS inverter*

Figure 4.14 shows the circuit diagram of a CMOS inverter.

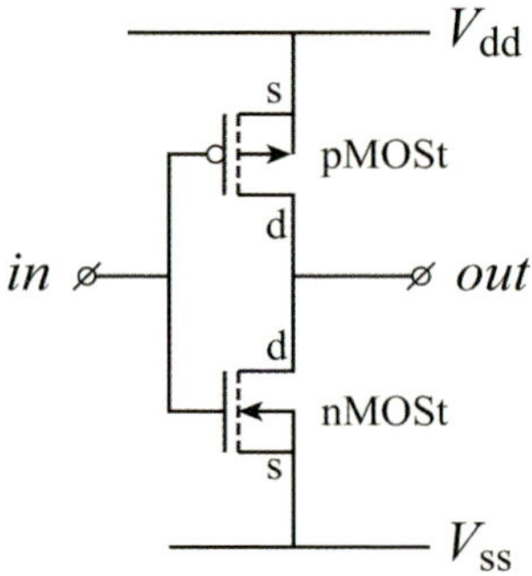

Figure 4.14: *Circuit diagram of a CMOS inverter*

The influence of substrate voltage on the threshold voltage of a transistor is discussed in section 1.5. This back-gate effect is proportional to the square root of the channel dope of the transistor and is represented by the K-factor. The K-factor of the pMOS transistors in a retrograde twin well process can be of the same order as that of the nMOS transistors.

In technologies with channel lengths above 100 nm, the performance of the pMOS transistor is hampered by the mobility of holes, which is approximately two to four times lower than the mobility of electrons. This leads to the following relationship between the effective $\beta_\square$ factors of nMOS and pMOS transistors (including second order effects):

$$\beta_{\square n} \approx 2.5 \cdot \beta_{\square p}$$

For equal absolute threshold voltage values, the pMOS transistor in the layout of an inverter with symmetrical behaviour will therefore be about

2.5 times the size of the nMOS transistor. This size ratio is expressed in the 'aspect ratio' A of the CMOS inverter as follows:

$$A = \frac{\left(\frac{W}{L}\right)_{\mathrm{p}}}{\left(\frac{W}{L}\right)_{\mathrm{n}}} = \frac{A_{T_{\mathrm{p}}}}{A_{T_{\mathrm{n}}}} = \frac{\beta_{\square\mathrm{n}}}{\beta_{\square\mathrm{p}}} \tag{4.6}$$

In many processes, all polysilicon areas and the sources and drains of nMOS transistors in an n-well CMOS process are n^{+} areas. The sources and drains of the pMOS transistors are p^{+} areas. It should be clear from figure 4.13 that p^{+} and n^{+} areas may *never* be directly connected, not even in a stick diagram. Such an interconnection would produce a pn diode which only conducts in one direction. Connections between n^{+} and p^{+} areas must therefore always be made in metal. Many CMOS processes currently include *double-flavoured polysilicon*, or *dual-dope polysilicon*: n^{+} polysilicon gate for the nMOS transistor and p^{+} polysilicon for the pMOS transistor.

In advanced nanometer CMOS technologies, the difference between the effective $\beta_{\square}$ factors of nMOS and pMOS transistors is reducing to between $A = 1$ and $A = 2$, due to mobility reduction effects, such as discussed in chapter 2, and mobility enhancement techniques, such as discussed in chapter 3. In this textbook we will use the value $A = 1.5$ in examples and exercises.

The electrical behaviour of the CMOS inverter

An nMOS inverter comprises a driver and a load transistor. However, the pMOS and nMOS transistors in a CMOS inverter are both driver transistors. Figure 4.15 shows a CMOS inverter and its transfer characteristic $V_{\mathrm{out}} = f(V_{\mathrm{in}})$. The gates of the pMOS (T_{p}) and nMOS (T_{n}) transistors are connected to form the inverter input. It is important to remember that $V_{T_{\mathrm{p}}} < 0$ and $V_{T_{\mathrm{n}}} > 0$.

The transfer characteristic is explained as follows:

T_{n} is 'off' and T_{p} is 'on' for $V_{\mathrm{in}} < V_{T_{\mathrm{n}}}$.
    The output voltage V_{out} then equals V_{dd}.

T_{p} is 'off' and T_{n} is 'on' for $V_{\mathrm{in}} > V_{\mathrm{dd}} + V_{T_{\mathrm{p}}}$.
    V_{out} then equals V_{ss}.

In both of the above stable situations, one transistor is always 'off' and no DC current can flow from supply to ground. The current characteristic $I = f(V_{in})$ in figure 4.15b reflects this fact. The absence of DC current in the two stable situations is the most important advantage of CMOS when compared with nMOS. A current only flows from supply to ground during an input voltage transition, for which the following conditions apply:

$$V_{T_n} < V_{in} < V_{dd} + V_{T_p}$$

Figure 4.15b shows the trajectory of the transient current associated with the input voltage transition from V_{ss} to V_{dd}. The areas where T_n and T_p operate in their respective saturation and triode regions are indicated in figure 4.15a.

The saturation areas are described for the two transistors as follows:

T_n: $V_{ds}|_{T_n} > V_{gs}-V_{T_n}$ i.e., $V_{out}> V_{in}-V_{T_n}$.   This is the area above the dotted line $V_{out}=V_{in}-V_{T_n}$ in the transfer characteristic.

T_p: $V_{ds}|_{T_p} < V_{gs}-V_{T_p}$ i.e., $V_{out}-V_{dd}< V_{in}-V_{dd}-V_{T_p}$.   This is the area below the dotted line $V_{out}=V_{in}-V_{T_p}$ in the transfer characteristic.

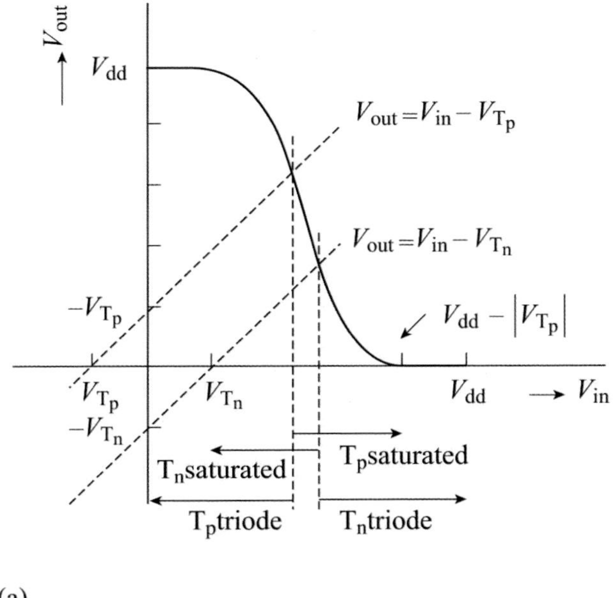

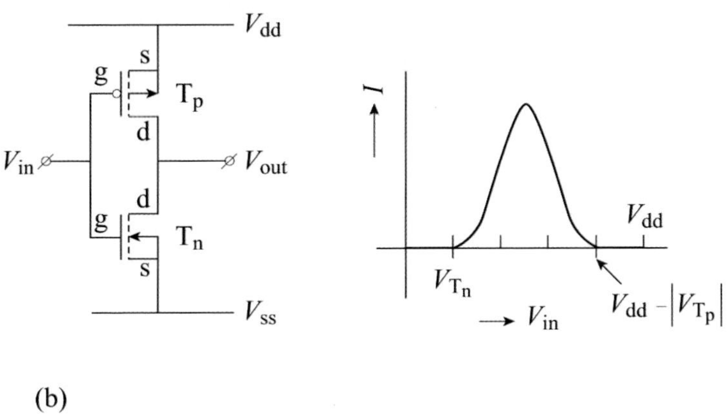

Figure 4.15: *Transfer characteristic (a) and current characteristic (b) of a MOS inverter*

Figure 4.15 shows that the transistors in an inverter are both saturated during transitions between logic levels. Theoretically, their output impedances are then infinite. Application of Ohm's Law reveals that a finite current should then cause an infinitely large change in the output voltage. In practice, the output impedances are always finite and the maximum voltage change is limited. However, the transfer characteristic of the CMOS inverter is still very steep.

It must be noted that figure 4.15 is drawn on the basis of the assumptions that $V_{T_n}=-V_{T_p}$ and $V_{dd}>V_{T_n}+|V_{T_p}|$. The reader should verify that the transfer characteristic of the inverter displays hysteresis when $V_{T_n}+|V_{T_p}|>V_{dd}$.

The charging and discharging behaviour of a CMOS inverter can also be described by means of the static characteristic $I = f(V_{out})$ shown in figure 4.16. This characteristic is obtained when a pulse V_{in} with rise and fall times of 0 ns is applied at the inverter input. Capacitance C is the load capacitance present at the transistor's output. The currents through the pMOS and nMOS transistors are I_p and I_n, respectively.

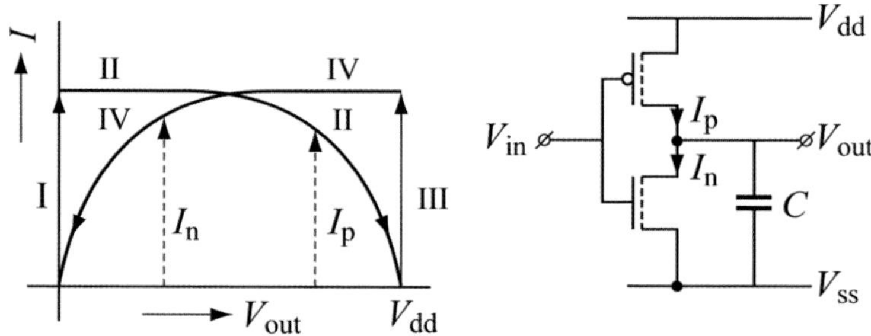

Figure 4.16: *Static CMOS-inverter characteristic*

The curves in figure 4.16 are explained as follows:

Trajectory I   : I_p rises from 0 to $I_{p_{max}}$ when V_{in} falls from V_{dd} to V_{ss}.
Trajectory II : C charges to V_{dd} and I_p decreases to 0.
Trajectory III : I_n rises from 0 to $I_{n_{max}}$ when V_{in} rises from V_{ss} to V_{dd}.
Trajectory IV : C discharges to V_{ss} and I_n decreases to 0.

In figure 4.16, it is assumed that the βs and the V_Ts of the nMOS and pMOS transistors are equal. The current characteristics are therefore symmetrical with respect to $V_{out} = \frac{1}{2}V_{dd}$.

Designing a CMOS inverter

A true CMOS logic gate contains a pMOS transistor for every nMOS transistor. A *pseudo-nMOS* version, however, uses just one active pull-up pMOS transistor with its gate connected to ground. Here, a DC current flows from supply to ground when the output is 'low'. The complementary behaviour of the transistors in true CMOS circuits ensures the absence of DC current at both the low and high stable operating points. This type of CMOS logic is therefore 'ratioless' and the voltages V_H and V_L associated with the respective 'high' and 'low' output levels are independent of the transistor sizes. In fact, V_H equals the supply voltage V_dd while V_L equals $0\,\mathrm{V}$. The dynamic discharge characteristic of a CMOS inverter is obtained when a step voltage (which rises from $0\,\mathrm{V}$ to V_dd in $0\,\mathrm{ns}$) is applied to its input. This is illustrated in figure 4.17. As shown in figure 4.18, the dynamic charge characteristic is obtained when the input step voltage falls from V_dd to $0\,\mathrm{V}$ in $0\,\mathrm{ns}$.

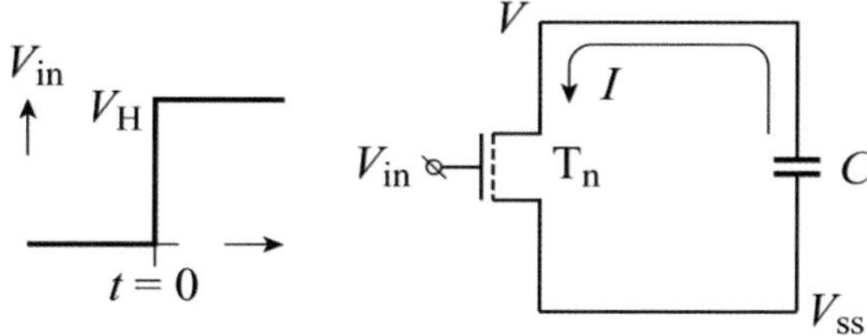

Figure 4.17: *Discharging a load capacitance through an nMOS transistor*

A simple expression, derived from equalising the current expressions for the transistor and the capacitor, defines the gain factor β_n of an nMOS transistor which will discharge a capacitance C from V_dd to V in time t when a step voltage with amplitude V_dd is applied to its gate at $t = 0$:

$$\beta_\mathrm{n} = \frac{4 \cdot C}{V_\mathrm{dd} \cdot t} \tag{4.7}$$

The required dimensions of the nMOS transistor are obtained by equating the gain factor β_n to $\beta_{\square\mathrm{n,eff}} \cdot A_{\mathrm{T_n}}$, where $A_{\mathrm{T_n}}$ is the aspect ratio of the transistor and equals $(W/L)_{\mathrm{T_n}}$. $\beta_{\square\mathrm{n,eff}}$ includes the second-order effects described in chapter 2, which reduces the effective mobility, and, as a consequence it reduces $\beta_{\square\mathrm{n}}$ to $\beta_{\square\mathrm{n,eff}}$.

Example:
Given: A 65 nm CMOS process with $\beta_{\square n,eff}=150\,\mu A/V^2$ and $V_{dd}=1.2\,V$.

Problem: Determine the aspect ratio A_{T_n} of an nMOS transistor T_n which will discharge a load capacitance $C=20\,fF$ from V_{dd} to $0.1\cdot V_{dd}$ in $100\,ps$ when a voltage V_{dd} is applied to its gate.

Solution: Substituting in (4.7) yields:

$$\beta_n = \frac{4\cdot 20\cdot 10^{-15}}{1.2\cdot 10^{-10}} = 660\mu A/V^2$$

Equating β_n to $\beta_{\square n,eff}\cdot A_{T_n}$ and substituting $\beta_{\square n,eff}=150\,\mu A/V^2$ yields:

$$A_{T_n} = \left(\frac{W}{L}\right)_{T_n} \approx 4.4$$

In this example, the mobility-reduction effects (chapter 2), are included in the simple basic current equations, and are represented by the reduced value of $\beta_{\square n,eff}$, which is about four to five times less than the $\beta_{\square n}$ calculated from the zero field mobility.

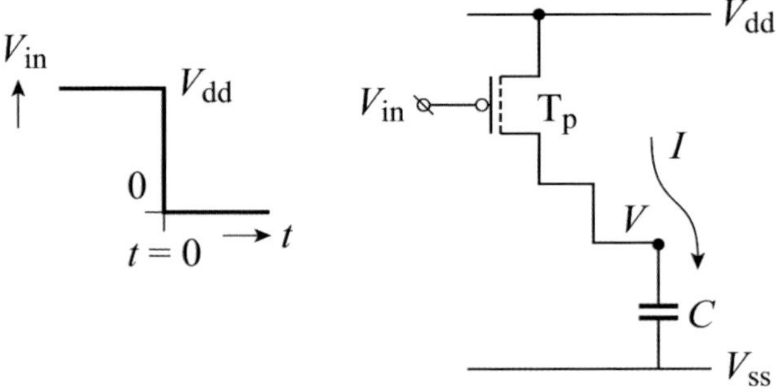

Figure 4.18: *Charging a load capacitance through a pMOS transistor*

The charging of a load capacitance through a pMOS transistor illustrated in figure 4.18 is analogous to discharging through an nMOS transistor. The expression for the gain factor β_p of a pMOS transistor, which will charge a capacitance C from $0\,V$ to a voltage $V = 0.9\cdot V_{dd}$ in time t when its gate voltage falls from V_{dd} to $0\,V$ in $0\,ns$ is, therefore simply obtained by the same equation (4.7).

Example:

Given: The information in the previous example plus $\beta_{\square\mathrm{p,eff}}=100\,\mu\mathrm{A}/\mathrm{V}^2$.

Problem: Determine the aspect ratio $A_{\mathrm{T_p}}$ of a pMOS transistor $\mathrm{T_p}$ which will charge the load capacitance C from $0\,\mathrm{V}$ to $0.9\cdot V_{\mathrm{dd}}$ in $100\,\mathrm{ps}$ when $0\,\mathrm{V}$ is applied to its gate.

Solution: This problem is the complement of the previous example.

Therefore, the following expression applies (see equation (4.6)):

$$A_{\mathrm{T_p}} \;=\; A_{\mathrm{T_n}}\cdot A = 4.4\cdot 1.5 = 6.6$$

The rise and fall times of *buffer circuits* must be equal. These circuits must therefore use the previously-mentioned value of about 1.5 to 2 for the aspect ratio A expressed in formula (4.6). For CMOS logic, however, values for A of around 1.5 are currently used. Larger values yield larger pMOS transistors and thus increase the load capacitance presented to previous logic gates. For CMOS circuits other than inverters, factors $(\frac{W}{L})_{\mathrm{p}}$ and $(\frac{W}{L})_{\mathrm{n}}$ in formula (4.6) are the effective values which apply to the transistors in the p and n sections, respectively. The dimensions of these transistors must be selected so that the value for A is optimal. For technologies beyond $65\,\mathrm{nm}$, different stress techniques and device orientations are used to improve the mobility of nMOS and pMOS transistors. The pMOSt mobility however, will benefit most from these techniques, such that it is expected that in the near future, pMOS and nMOS transistors will exhibit almost equal performance at the same W/L ratio.

Dissipation of a CMOS inverter

During the last two decades, CMOS technology has become the most dominant technology for VLSI circuits. The most important reason for this is its low static power consumption. This is because of the absence of DC currents during periods when no signal transients occur in static CMOS circuits. However, a *short-circuit current* flows from supply to ground when a change in a logic circuit's input voltage causes the output voltage V_{out} to change. This short-circuit current leads to additional power dissipation [12].

The *power dissipation* of a basic CMOS inverter is explained with the aid of figure 4.19.

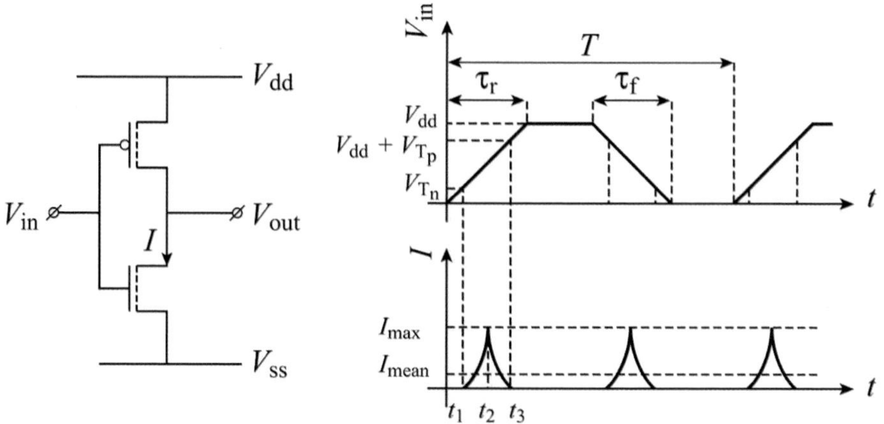

Figure 4.19: *Current through an unloaded inverter*

Only the nMOS transistor conducts when the input voltage V_{in} of this static CMOS inverter is 'high' (V_{dd}). Similarly, only the pMOS transistor conducts when the input voltage V_{in} is 'low' (V_{ss}). Therefore, the inverter does not dissipate power when the input is in either of the above stable states. However, during a transient at the input, there is a period when both the nMOS and pMOS transistors conduct. A short-circuit current then flows from supply to ground while the input voltage is between V_{T_n} and $V_{dd} - |V_{T_p}|$. This current I is shown in figure 4.19. If a load capacitance C_L is connected to the inverter output, then the dissipation consists of two components:

1. *Dynamic power dissipation:*

$$P_1 \;=\; C_L \cdot V^2 \cdot f \qquad (4.8)$$

2. *Short-circuit power dissipation:*

$$P_2 = I_{\mathrm{mean}} \cdot V \qquad (4.9)$$

In the above equations, f $(= 1/T)$ is the frequency at which the voltage change V occurs on C_L and I_{mean} is the average short-circuit current. Clearly, the dynamic component P_1 is independent of transistor dimensions when parasitic capacitances at the output, such as pn-junction capacitances, are neglected. It is expressed in equation (4.8) and is explained with the aid of figure 4.20.

210

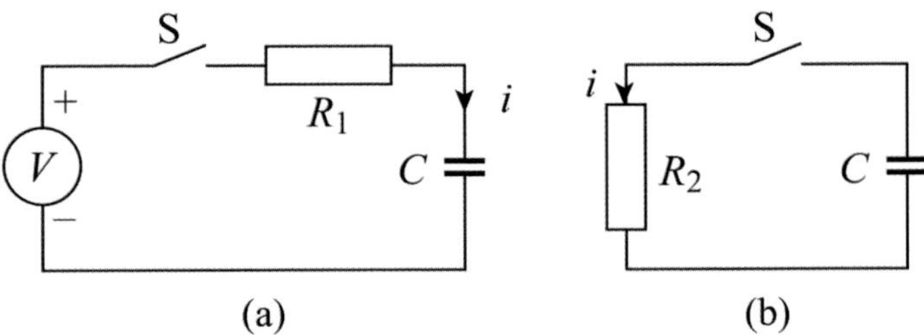

Figure 4.20: *Charging and discharging a capacitance*

Capacitance C is charged and discharged via resistors with values R_1 and R_2, respectively. During charging, the power dissipation in R_1 equals:

$$P_{R_1} = \int_0^\infty i^2(t) \cdot R_1 \cdot dt \qquad \text{with } i(t) = \frac{V}{R_1} \cdot e^{-t/(R_1 C)}$$

The solution to this integral is as follows:

$$P_{R_1} = \frac{1}{2} \cdot C \cdot V^2$$

P_{R_1} is thus independent of R_1. Similarly, the power dissipation P_{R_2} during discharging is independent of the value of R_2 and also equals $C \cdot V^2/2$.

The total power P supplied by the voltage source V during a complete charge-discharge cycle is the sum of P_{R_1} and P_{R_2}, i.e., $P = C \cdot V^2$. For f cycles per second the total power dissipation is:

$$P = C \cdot V^2 \cdot f$$

This dynamic power dissipation appears in all types of logic, including static MOS circuits, bipolar circuits, TTL circuits, etc.

The short-circuit component P_2, however, is proportional to transistor dimensions; it also depends on the size of the load capacitance. An expression for I_{mean} in formula (4.9) is derived on the assumption that the inverter's load capacitance is zero [12]. Although an asymmetric inverter is not fundamentally different, the inverter is also assumed to be symmetric. In this case, the following equations apply:

$$\beta_n = \beta_p = \beta \quad \text{and} \quad V_{T_n} = -V_{T_p} = V_T$$

During the period t_1 to t_2 in figure 4.19, the short-circuit current I increases from 0 to I_{max}. Throughout this period, the output voltage

V_{out} is more than a threshold voltage V_{Tn} larger than the input voltage V_{in}. The nMOS transistor is therefore saturated and application of the simple MOS formulae (1.15) yields the following expression for I during this period of time:

$$I = \frac{\beta}{2}(V_{\text{in}} - V_{\text{Tn}})^2 \quad \text{for } 0 \leq I \leq I_{\text{max}}$$

The symmetry of the inverter produces a maximum value for I when V_{in} equals $V_{\text{dd}}/2$. In addition, the current transient during the period t_1 to t_3 is symmetrical with respect to the time t_2. The mean current I_{mean} (i.e., the effective current which flows during one cycle period T of the input signal) can therefore be expressed as follows:

$$I_{\text{mean}} = 2 \cdot \frac{2}{T} \int_{t_1}^{t_2} I(t)\mathrm{d}t = \frac{4}{T} \int_{t_1}^{t_2} \frac{\beta}{2}(V_{\text{in}}(t) - V_{\text{T}})^2 \mathrm{d}t \tag{4.10}$$

The input voltage V_{in} is assumed to have a symmetrical shape and linear edges, with rise and fall times equal to τ. The value of V_{in} as a function of time t during an edge is therefore expressed as follows:

$$V_{\text{in}}(t) = \frac{V_{\text{dd}}}{\tau} \cdot t$$

The following expressions for t_1 and t_2 can be derived from figure 4.19:

$$t_1 = \frac{V_{\text{T}}}{V_{\text{dd}}} \cdot \tau \quad \text{and} \quad t_2 = \frac{\tau}{2}$$

Substituting these expressions for $V_{\text{in}}(t)$, t_1 and t_2 in equation (4.10) yields:

$$I_{\text{mean}} = \frac{2\beta}{T} \cdot \int_{\tau/2}^{\frac{V_{\text{T}}}{V_{\text{dd}}} \cdot \tau} \left(\frac{V_{\text{dd}}}{\tau} \cdot t - V_{\text{T}}\right)^2 \mathrm{d}\left(\frac{V_{\text{dd}}}{\tau} \cdot t - V_{\text{T}}\right)$$

The solution to this equation is:

$$I_{\text{mean}} = \frac{1}{12} \cdot \frac{\beta}{V_{\text{dd}}} \cdot (V_{\text{dd}} - 2V_{\text{T}})^3 \cdot \frac{\tau}{T}$$

Substituting this expression for I_{mean} into formula (4.9) yields the following expression for the short-circuit dissipation of a CMOS inverter with no load capacitance:

$$P_2 = \frac{\beta}{12} \cdot (V_{\text{dd}} - 2V_{\text{T}})^3 \cdot \frac{\tau}{T} \tag{4.11}$$

Formula (4.11) clearly illustrates that the short-circuit dissipation is proportional to the frequency $f = 1/T$ at which the input changes. Voltages V_{dd} and V_{T} are determined by the application and the process. Therefore, the only design parameters that affect P_2 are β and the rise and fall times (τ) of the inverter's input signal. For an inverter with a capacitive load, the transistor β values are determined by the required output rise and fall times. In this case, the short-circuit dissipation only depends on the input signal's rise and fall times, i.e., τ_{r} and τ_{f}, respectively. This is particularly true for buffer circuits which have transistors with large β values.

In the chapter on low-power design (chapter 8), the CMOS power contributions are discussed extensively. However, the design of large buffer circuits is discussed in this section on basic CMOS circuit design.

CMOS buffer design

Large capacitances associated with integrated circuits include those presented by bus lines and 'off-chip' circuits. These capacitances must often be driven at high frequencies. The required 'buffer' driving circuits dissipate a relatively large part of the total power consumed by a chip. Optimising these buffers therefore requires considerably more effort than the approach adopted for CMOS logic. Formula (4.11) shows that short-circuit dissipation is directly proportional to the rise and fall times (τ) of an input signal. The input signals of buffers which drive bus lines connected to large numbers of different sub-circuits on a chip must therefore have particularly short rise and fall times.

Suppose the signal on a bus line with capacitance C_{L} must follow a signal at the output node A of a logic gate which is capable of charging and discharging a capacitance C_0 in τ ns. An inverter chain such as illustrated in figure 4.21 can be used as a buffer circuit between node A and the bus line.

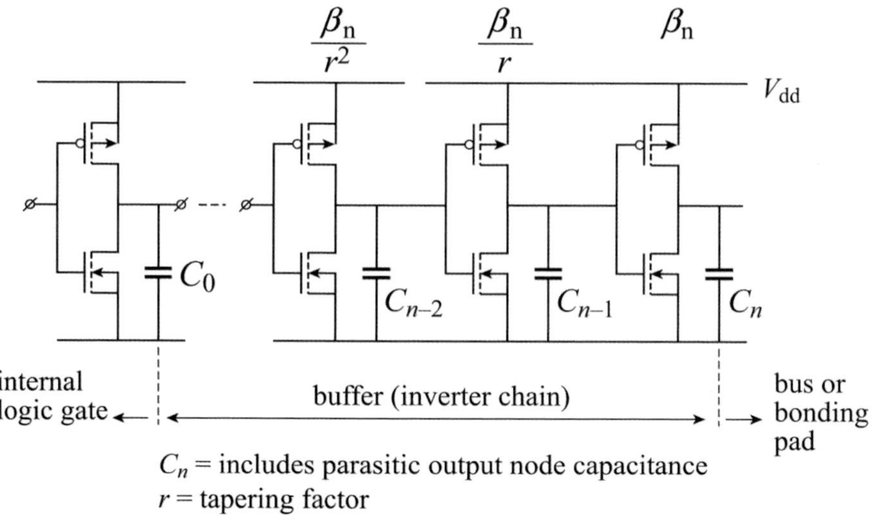

Figure 4.21: *A buffer circuit comprising an inverter chain*

From formula (4.11), it is clear that the rise and fall times on each input of the inverters in the above chain should be short. Moreover, it has been shown in literature [12] that minimum dissipation can be achieved when the rise and fall times on each of these inputs are equal to the rise and fall times at the buffer output. The inverter chain must therefore be designed so that the rise and fall times on the inputs of each of its component inverters are τ ns. According to literature [4], a minimum propagation delay time for the buffer is obtained when the *tapering factor* r between the βs of successive inverters is e, the base of natural logarithm. In terms of dissipation and silicon area, however, this will not lead to an optimum design. Design optimisation for minimum dissipation and silicon area requires a different approach [12].

When a signal is produced by a logic gate and must be buffered to drive a relatively large capacitive load $C_L = C_N = 10\,\mathrm{pF}$, then the design of this buffer is not defined by the power consumption only. Also area and switching-current peaks (di/dt) are important parameters that have to be seriously dealt with.

The size of the final driver stage n, is defined by the timing specifications of (dis)charging the output load C_n. The driving strength of the $n-1$ stage is defined by the required tapering factor. In conventional CMOS technologies, with around $1\,\mu m$ channel lengths, the optimum tapering factor was close to ten, in order to achieve identical rise and fall times on the input and output of the individual driver stages. A larger tapering factor would lead to smaller pre-driver stages and so to more delay and larger short-circuit power consumption. A smaller tapering factor would lead to larger pre-driver stages, less delay, but to increased switching (di/dt) noise. When porting the driver to the next technology node, assuming a scaling factor $s \approx 0.7$ and maintaining the same W/L ratio, the transistor sizes W and L of the final driver stage N also scale with an average factor of 0.7 ($= s$). Because the gate-oxide thickness scales with the same factor, the total fan-in capacitance of the final stage becomes 1.4 times smaller. In order to achieve the same rise and fall times on the input of the final stage, this allows an increase of the tapering factor with the same amount. In other words: the tapering factor r is dependant on the technology node. Figure 4.22 shows six different drivers (inverter chains) with different tapering factors designed in a 1.2 V 65 nm CMOS technology.

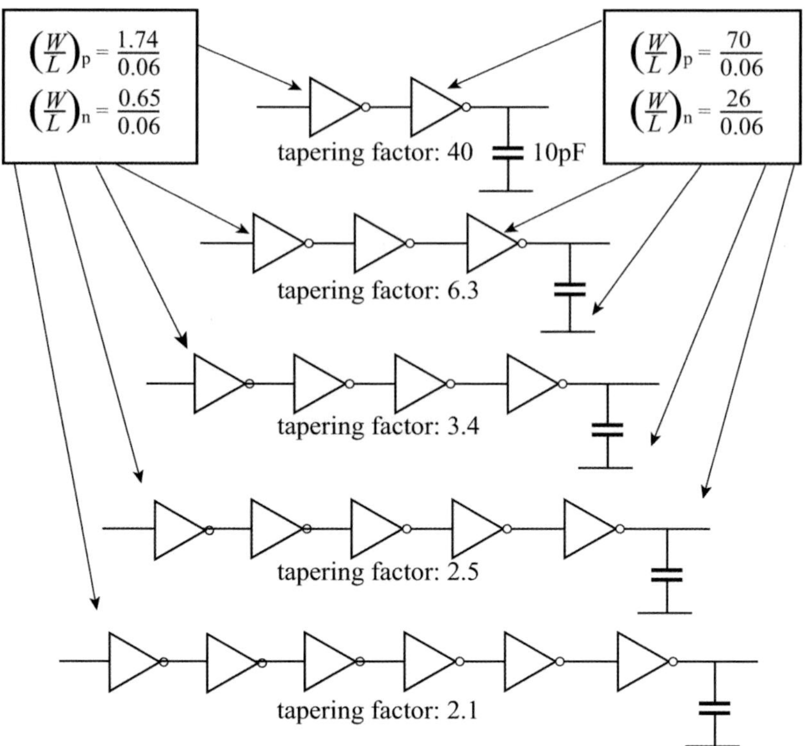

Figure 4.22: *Inverter chains with different tapering factors, all driving the same load*

The input driving stage, which mimics the logic gate drive strength, and the output driving stage, driving a 10 pF load, are the same for all six drivers. The circuit simulations are done for a clock frequency of 50 MHz. Table 4.1 shows the major characteristics of these six drivers.

Table 4.1: Characteristics of various 10 pF inverter chains with different tapering factors

Number of inverters	2	3	4	5	6	unit
Tapering factor	40	6.3	3.4	2.5	2.1	
Relative area	1	1.15	1.35	1.61	1.85	
Total power	738	742	746	750	754	μW
Dynamic power	721	722	723	723	723	μW
Short-circuit power	12	12	11.5	11.5	11.5	μW
Max. di/dt (rel.)	1	5	10	13	14	
Output delay	896	657	590	580	575	ps

Because supply noise (see chapter 9), which has a linear relation with the di/dt, is a real issue in nanometer CMOS ICs, it is a dominant factor in choosing the right driver (tapering factor). The first column refers to the first driver of figure 4.22. It shows 36% more delay than the second inverter chain. This second inverter chain, however, only shows 10% more delay than number three, etc.

The short-circuit power consumption in this table only represents the short-circuit power consumed by the final inverter stage, because the pre-drivers short-circuit power is negligible. The table shows that the short-circuit power consumption is only a fraction of the dynamic power consumption. Generally, if a tapering factor equal to or larger than ten is chosen, then, in nanometer CMOS ICs, the short-circuit power maybe completely neglected in the power discussions. Also the di/dt value is relatively low for this tapering factor. These considerations hold for on-chip clock drivers and bus drivers, as they usually consume more power than an average logic gate.

Because many output drivers still use 3.3 V, 2.5 V or 1.8 V supply voltages, these definitely will show different optimum tapering factors, which reduce with increasing output voltages. An example of a 1.8 V output driver circuit is given in figure 4.22. In such output drivers also a tapering factor of close to ten or larger would be the best choice: it guarantees a relatively short delay, a small short-circuit power consumption and a relatively low switching noise (di/dt).

Noise margins

The maximum amplitude of a noise signal that can be superimposed on all nodes of a long inverter chain without causing the output logic level to change is called *noise margin*. Figure 4.23 shows the transfer characteristic of a CMOS inverter for three different gain factor ratios. The noise margins for both high and low levels are very large because of the almost rectangular shape of these transfer characteristics. For the symmetrical inverter, with $\beta_n=\beta_p$ and $V_{T_n}=-V_{T_p}$, the noise margins are equal for both levels. Of course, not every inverter is symmetrical. In such cases, the noise margin is different for the two levels. However, the difference is only significant for highly asymmetrical inverters.

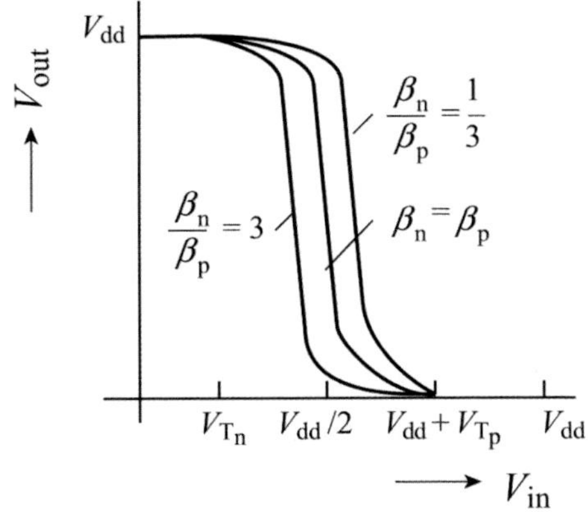

Figure 4.23: *CMOS inverter transfer characteristics for different aspect ratios*

Generally, the operation of CMOS logic circuits is very robust. Even when the supply voltage is reduced to below the lowest of the threshold voltages (V_{T_n} and V_{T_p}), it will still deliver a correct logical output. However, this so-called *subthreshold logic* will then operate at relatively low frequencies.

4.4 Digital CMOS circuits

4.4.1 Introduction

CMOS circuits can be implemented in static or dynamic versions. The choice is mainly determined by the type of circuit and its application.

Two important factors which influence this choice are chip area and power dissipation. The differences between these factors for the two types of implementation are treated in this section.

4.4.2 Static CMOS circuits

A logic function in static CMOS must be implemented in both nMOS and pMOS transistors. An nMOS version only requires implementation in nMOS transistors. A single load transistor is then used to charge the output. This load transistor also conducts when the output is 'low'. A current therefore flows from supply to ground and causes DC dissipation while the output of an nMOS logic gate is 'low'.

In a CMOS logic gate, a current only flows between supply and ground during output transitions. Figure 4.24 shows some static CMOS logic gates. Back-bias connections for both the nMOS and the pMOS transistors are indicated in the inverter in figure 4.24(a). The respective back-bias voltages, V_{sb} and V_{ws}, are both 0 V. The back-bias connections are no longer shown in figures 4.24(b), 4.24c and all subsequent figures. Unless otherwise stated, the substrate voltages are assumed to be V_{ss} for the nMOS transistors and V_{dd} for the pMOS transistors. Figures 4.24(b) and 4.24(c) show nMOS and pMOS transistors, respectively, connected in series. The sources of some of these transistors are not connected to V_{ss} or V_{dd}. The back-bias effect has a considerable influence on nMOS and pMOS transistors whose sources are not connected to V_{ss} and V_{dd}, respectively. This is particularly true when the source is loaded.

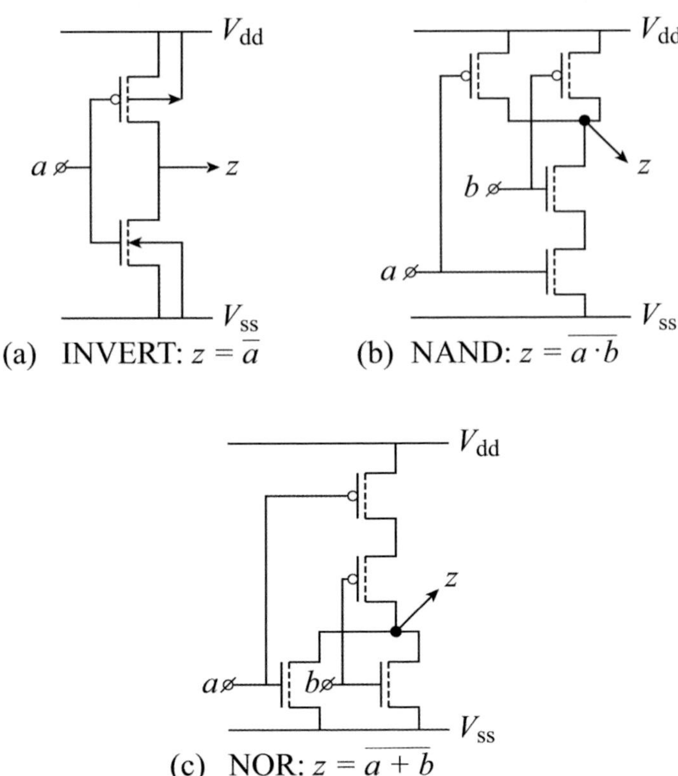

(a)   INVERT: $z = \overline{a}$ (b)   NAND: $z = \overline{a \cdot b}$

(c)   NOR: $z = \overline{a + b}$

Figure 4.24: *Examples of static CMOS logic gates*

In general, a series connection of transistors in the nMOS section of a CMOS logic gate will reflect a parallel connection of transistors in the pMOS section and vice versa. This is illustrated in figure 4.25, which shows an example of a static CMOS implementation of a complex logic function and its equivalent logic gate diagram.

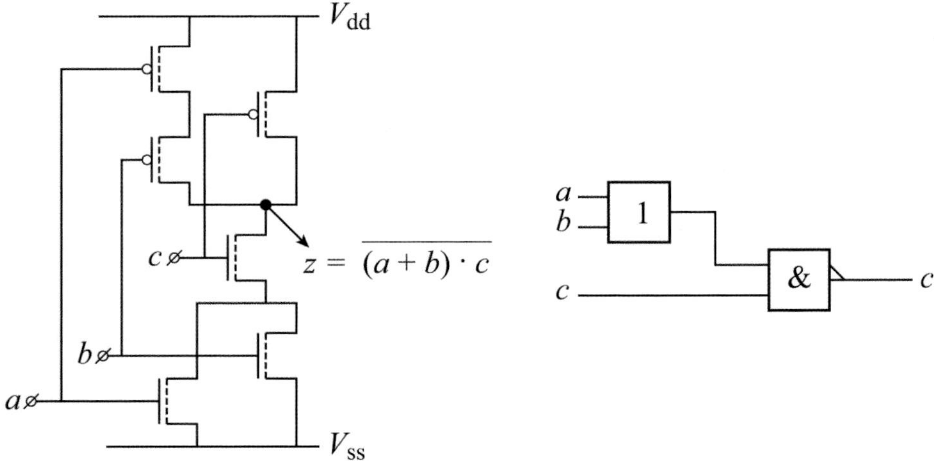

Figure 4.25: *An example of a more complex static CMOS logic gate*

The performance of a pMOS transistor is less than that of an nMOS transistor. The number of pMOS transistors in series in a CMOS logic gate should therefore be minimised. If this number becomes very large then, only in exceptional cases, can a pseudo-nMOS implementation be used.

Figure 4.26 is an example of a pseudo-nMOS implementation of the CMOS equivalent in figure 4.25. The pseudo-nMOS version is identical to its nMOS counterpart except that the nMOS load element is replaced by a pMOS transistor with its gate connected to V_{ss}. Both nMOS and pseudo-nMOS logic gates have the advantage of the same low input capacitance. The output rise time of a pseudo-nMOS logic gate is determined by only one pMOS transistor and should therefore be short. A disadvantage of such a gate is the static power dissipation when the output is 'low'. The output low level and noise margins are determined by the ratio of the widths of the nMOS and pMOS transistors. Pseudo-nMOS logic is therefore also a form of ratioed logic, as discussed in section 4.2.2. This type of logic is only very sparingly used, and certainly not in today's low-power designs.

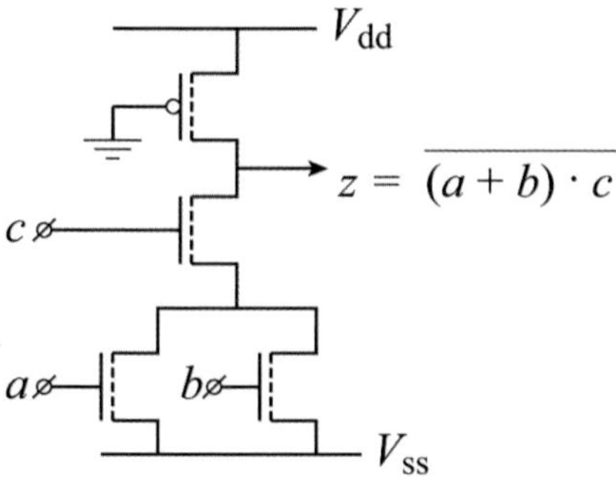

Figure 4.26: *A pseudo-nMOS logic gate*

The CMOS transmission gate (pass transistor)

Figure 4.27 shows a *transmission gate* comprising a complementary pair of transistors. This is an important component in both static and dynamic circuits. It is used to control the transfer of logic levels from one node to another when its control signals are activated. A single nMOS enhancement transistor can also be used to implement a transmission gate. Such an implementation has only one control signal but is disadvantaged by threshold loss. The threshold voltage of the transistor may be relatively high because of the body effect and the maximum high output level equals a threshold voltage below the control voltage. For this reason, the CMOS implementation is preferred.

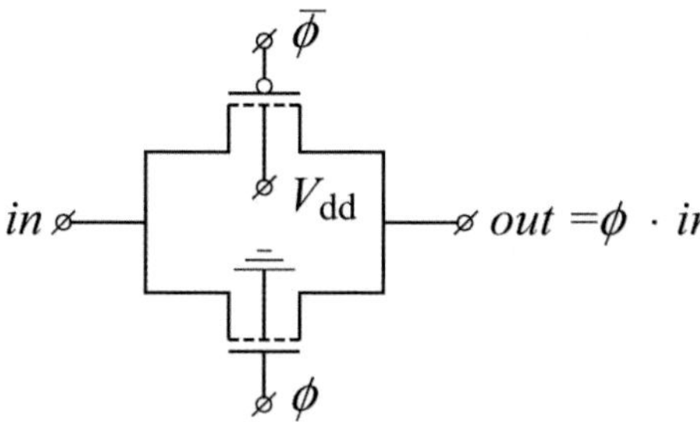

Figure 4.27: *CMOS transmission gate*

If the gate of the nMOS transistor in the CMOS transmission gate is controlled by a signal ϕ, the gate of the pMOS transistor must be controlled by the complementary signal $\overline{\phi}$. When the input voltage is $0\,\mathrm{V}$ and ϕ is 'high', the output will be discharged to $0\,\mathrm{V}$ through the nMOS transistor. The complementary behaviour of the pMOS transistor ensures that the output voltage equals V_{dd} when the input voltage is at V_{dd} level and $\overline{\phi}$ is 'low'.

Figure 4.28 shows the contributions of both MOS transistors to the charge and discharge characteristics of a CMOS transmission gate. The pMOS and nMOS transistors prevent threshold loss on the output 'low' and 'high' levels, respectively.

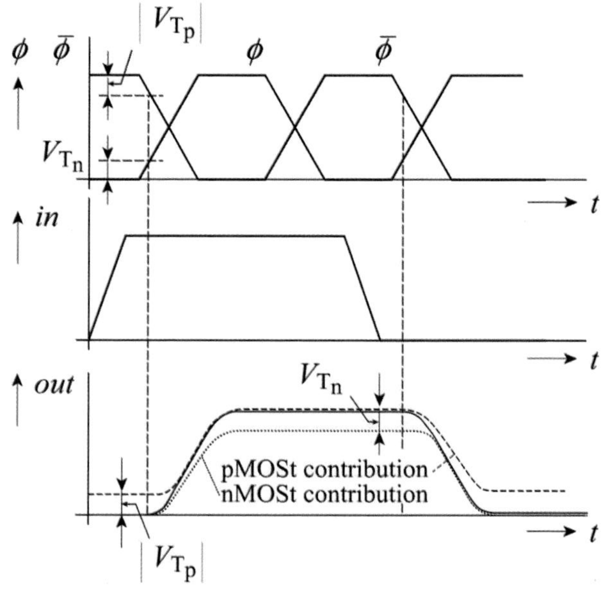

Figure 4.28: *CMOS transmission gate behaviour and the individual contributions of the nMOS and pMOS transistors to the charge and discharge characteristics*

Pass-transistor logic

In static CMOS circuits, transmission gates are used in latches, flip-flops, 'pass-transistor logic' and in static random-access memories. Examples of pass-transistor logic are exclusive OR (EXOR) logic gates and multiplexers. Figure 4.29 shows pass-transistor logic implementations of an EXOR gate. The nMOS transmission gate implementation in figure 4.29(a) is disadvantaged by high threshold loss resulting from body effect. The complementary implementation in figure 4.29(b) yields shorter gate delays at the expense of larger chip area. When connecting the outputs of these gates to a latch circuit (e.g., two cross-coupled pMOS loads), a static CMOS logic family is created (figure 8.17). The threshold voltage loss over the nMOS pass gates is compensated by the level restoring capability of the latch.

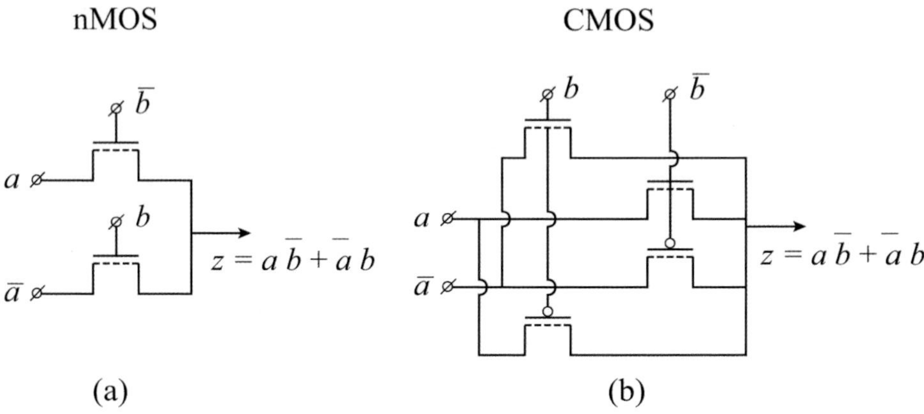

Figure 4.29: *Pass-transistor logic implementations of an EXOR logic gate with (a) nMOS pass transistors (b) CMOS pass-transistor gates*

A general disadvantage of pass-transistor logic as presented in figure 4.29 is the series resistance between the inputs a and $\bar{a}$ and the output z. The charging and discharging of a load at the output through the pass transistor causes additional delay. Other disadvantages include the need for complementary control signals. The potentials of pass-transistor logic challenge the creativity of the designers. Several alternatives have been published. These are discussed in detail in the low-power chapter 8, together with their advantages and disadvantages.

Finally, circuit designs implemented with pass-transistor logic must be simulated to prevent unexpected performance degradation or even erroneous behaviour caused by effects such as charge sharing (section 4.4.4). With decreasing voltages in current and future processes, the performance of pass-transistor logic tends to drop with respect to standard static CMOS logic. Therefore, the importance and existence of pass-transistor logic is expected to decrease in the coming years. The forms of CMOS logic discussed above can be used in both asynchronous circuits and synchronous, or 'clocked', circuits. The latter type of circuits are the subject of the next section.

4.4.3   Clocked static CMOS circuits

Signals which flow through different paths in a complex logic circuit will
ripple through the circuit asynchronously if no measures are taken. It is
then impossible to know which signal can be expected at a given node
and time. Controlling the data flow inside a circuit therefore requires
synchronisation of the signals. Usually, this is done by splitting all
the different paths into sub-paths with a uniform delay. The chosen
delay is the worst case delay of the longest data ripple. In synchronous
static CMOS circuits, the sub-paths are separated by means of 'latches'
and/or 'flip-flops' which are controlled by means of periodic clock signals.
Dynamic circuits may also use latches and flip-flops. Alternatively, data
flow in dynamic circuits may be controlled by including the *clock signals*
in every logic gate.

Static latches and flip-flops

Latches and flip-flops are used for temporary storage of signals. Figure 4.30 shows an example of a static *CMOS latch* and an extra transmission gate. The transmission gate on the left-hand side is an integral part of the latch, which also comprises two cross-coupled inverters.
Complementary logic values can be written into this latch via the transmission gates when the clock signal is high, i.e., when $\phi = 1$ and $\overline{\phi} = 0$.
Feedback in the latch ensures that these values are held when $\phi = 0$ and
$\overline{\phi} = 1$. This basic principle is used in static full-CMOS memory cells
and flip-flops.

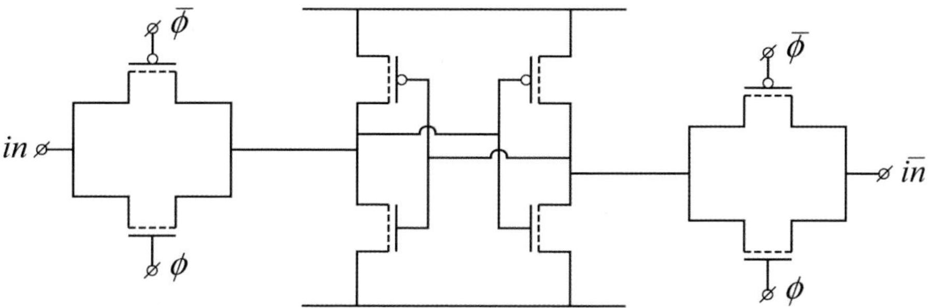

Figure 4.30: *CMOS static latch*

A flip-flop can temporarily store data and is controlled by one or more clock signals. The maximum clock frequency of a clocked static CMOS circuit is determined by the worst case delay path between two flip-flops. This path has the longest propagation delay as a result of a combination of logic gates and/or long signal tracks with large capacitances. There are several implementations of static CMOS flip-flops. The discussions below are limited to different forms of D-type flip-flops.

A *D-type flip-flop* can be built by connecting two latches in series, as shown in figure 4.31. The latches in this example use nMOS transmission gates. When the clock ϕ_1 goes 'high', data at the D input is latched into the 'master' latch of the flip-flop while the 'slave' latch maintains the previous input data. The D-input has to compete with the latch's feedback inverter via the nMOS transmission gate. The W/L aspect ratios of the transistors in the feedback inverter are therefore very small. The threshold voltage loss of the nMOS transmission gate produces a 'poor' high level at the input of the large inverter.

The aspect ratio, as expressed in equation (4.6), used for the large inverter must ensure that its output is 'low' when the poor high level is present at its input. The high level is then regenerated by the small feedback inverter. Static dissipation therefore does not occur. In practice, the aspect ratio of the large inverter must be close to 1. This ensures that the inverter's switching point is lower than half the supply voltage.

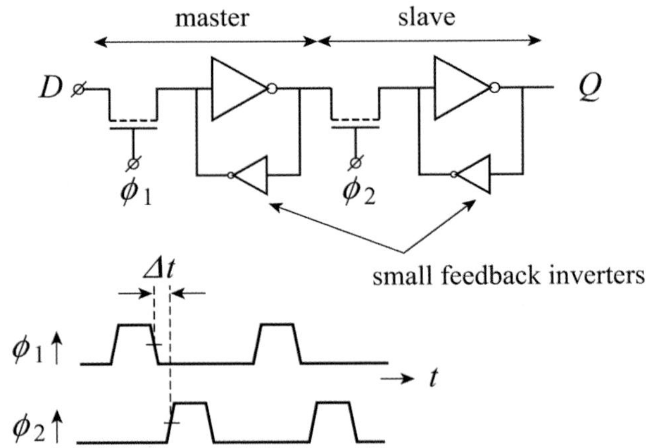

Figure 4.31: *(a) D-type flip-flop with nMOS transmission gates and (b) its 2-phase non-overlapping clock signals*

The flip-flop in figure 4.31 is therefore not very robust in its operation.

Its operation is much more reliable, when it is implemented with complementary transmission gates. In this case, however, the nMOS and pMOS transistors in the first transmission gate are controlled by ϕ and $\overline{\phi}$, respectively. The nMOS and pMOS transistors in the second transmission gate are controlled by $\overline{\phi}$ and ϕ, respectively.

Another implementation of the D-type flip-flop is shown in figure 4.32. The additional transmission gates in the feedback loops of each latch interrupt these loops when data is being written into the latch. This reduces the driving requirements of the input circuit and the master, which makes it easier to change the state of the flip-flop.

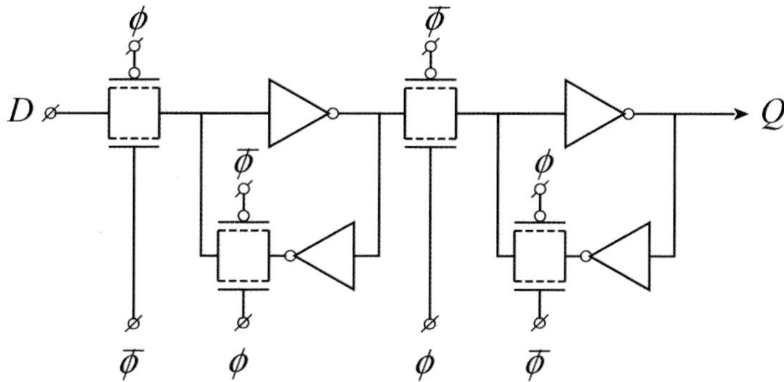

Figure 4.32: *Another implementation of a D-type flip-flop with complementary transmission gates*

Two clocks must be routed in chips with flip-flops which require complementary clocks, such as ϕ_1 and ϕ_2 in figure 4.31 or ϕ and $\overline{\phi}$ in figure 4.32. If the routing area is critical, a single clock flip-flop must be used. Such a flip-flop must then include an inverter to locally generate the inverse of the routed clock. However, there is then an increased risk of 'transparency'. This occurs when the *'clock skew'* causes a flip-flop's transmission gate to simultaneously conduct for a short period of time. This causes the flip-flop to be briefly transparent and data can *'race'* directly from the input to the output. This effect occurs when the flip-flop's complementary clocks arrive via different delay paths. If the clock ϕ_1 in figure 4.31, for instance, is delayed by more than a time period Δt with respect to clock ϕ_2, the flip-flop would be briefly transparent.

Clocks ϕ_1 and ϕ_2 in figure 4.31 are *non-overlapping*, i.e., ϕ_1 is 'low' before ϕ_2 goes 'high' and vice versa. The use of non-overlapping clocks is a good means of preventing transparency in flip-flops.

A discussion of the many more types and variants of static D-type flip-flops is beyond the scope of this book. However, the D-type flip-flop presented in figure 4.33 is particularly interesting. This flip-flop is primarily implemented with NAND logic gates. It requires only a single clock and is very robust. Unfortunately, it consists of 15 nMOS and 15 pMOS transistors and therefore requires considerably more chip area than the 10-transistor flip-flop in figure 4.31. A 'high-density gate array' layout of the flip-flop in figure 4.33 is shown in figure 7.34.

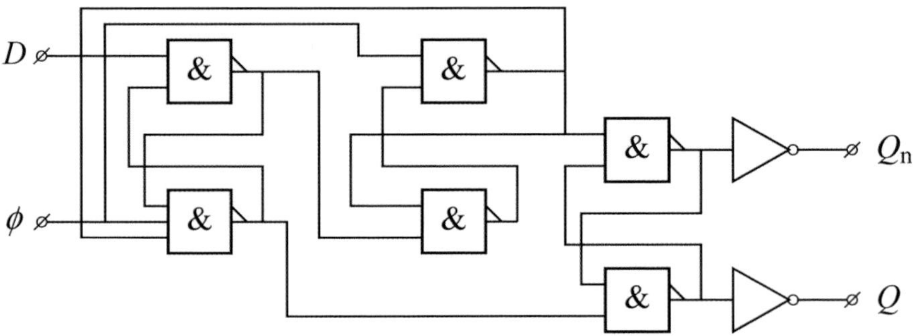

Figure 4.33: *A D-type flip-flop comprising NAND logic gates*

4.4.4 Dynamic CMOS circuits

The main advantage associated with dynamic CMOS circuits is the small chip area that they require. The explanation lies in the fact that logic functions are only implemented in nMOS transistors. Only one pMOS transistor is used per logic gate to charge its output node. Dynamic CMOS circuits are therefore *'nMOS-mostly'* and can occupy significantly less chip area than their static CMOS equivalents. This is particularly true for complex gates.

Figure 4.34 shows a *dynamic CMOS* implementation of a NOR gate. A dynamic CMOS gate of this type requires four different clocks for proper operation, i.e., ϕ_1, $\overline{\phi_1}$, ϕ_2 and $\overline{\phi_2}$. Inputs a and b must be generated by a gate in which ϕ_1 and ϕ_2 are interchanged. The output may also only serve as an input for a gate with ϕ_1 and ϕ_2 interchanged. The operation of the NOR gate is described as follows:

- Node Z is precharged to V_{dd} when clock ϕ_1 is 'low'.

- When ϕ_1 goes 'high', Z will be discharged if either a or b is 'high'.

- Clock ϕ_2 is then 'low' and the transfer gate passes the value on Z to the input of another logic gate.

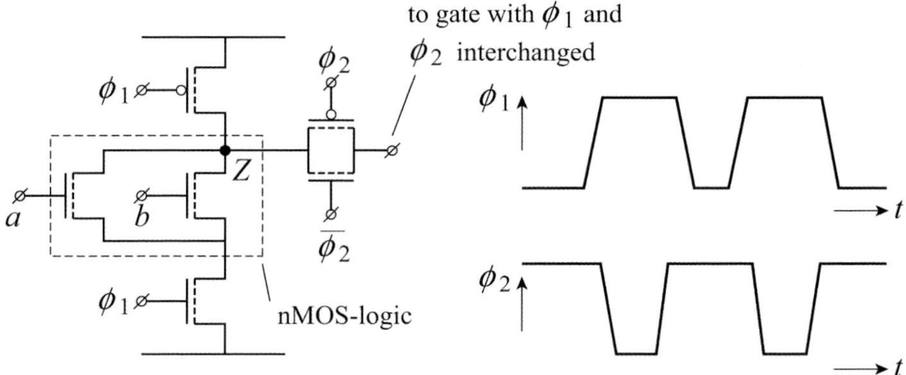

Figure 4.34: *A dynamic CMOS implementation of $Z = \overline{a + b}$*

There is a wide variety of dynamic CMOS logic implementation forms. These include the race-free, pipelined CMOS logic from the Catholic University of Leuven and Bell Labs' DOMINO-CMOS. In contrast to the form of dynamic CMOS shown in figure 4.34, all logic gates in a DOMINO-CMOS circuit are simultaneously precharged during the same part of the clock period. The logic gates sample their inputs when the precharge period ends. In keeping with the domino principle, however, each logic gate can only switch state after its preceding gate has switched. Figure 4.35 shows an example of a DOMINO-CMOS logic gate. The output Y of the dynamic gate is precharged when the clock ϕ is 'low'. The output Z of the static inverter is then 'low'. In fact, the inverter output nodes of all logic gates are 'low' during precharge. These outputs can therefore either stay 'low' or switch to 'high' when ϕ is 'high'. Clearly, each node can only make one transition during this sample period. A node stays in its new state until the next precharge period begins. The data must obviously be given enough time to ripple through the worst case delay path during a sample period. The sample period will therefore be much longer than the precharge period. An important disadvantage of DOMINO-CMOS logic is that all gates are non-inverting. Circuit adaptations are therefore required to implement logic functions with inverse inputs, e.g., an EXOR gate.

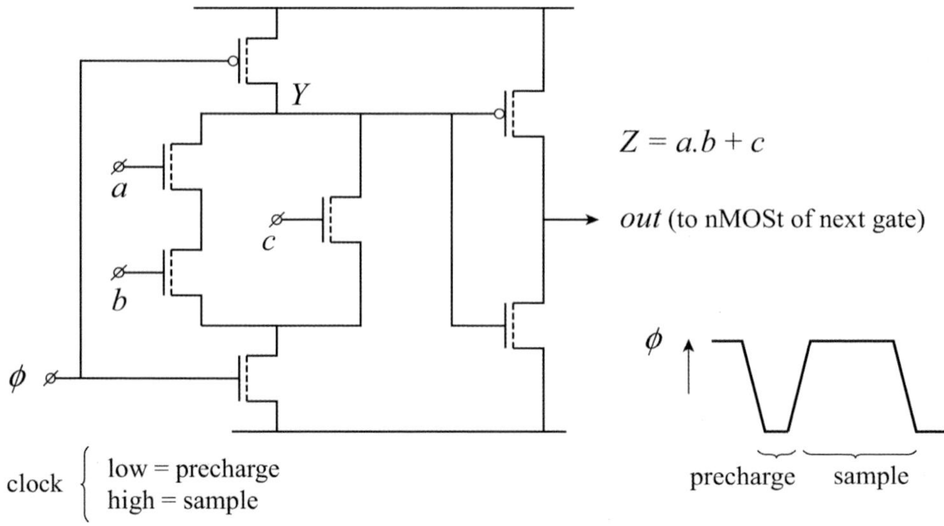

Figure 4.35: *An example of a DOMINO-CMOS logic gate*

Another disadvantage is the need to buffer each logic gate with an inverter; this requires extra silicon area. Today, DOMINO-CMOS logic is often used in high-performance processors. Particularly the most delay-critical circuits, like multipliers and adders are implemented in some style of DOMINO-CMOS [11]. Such high-performance processors require logic with large drive capability. This is particular an advantage of DOMINO logic, because only the inverter stage in a logic gate needs to be upscaled instead of every transistor within the logic function. With respect to power dissipation, several remarks on dynamic circuits are made in chapter 8.

Dynamic CMOS latches, shift registers and flip-flops

There are many variations of dynamic CMOS shift registers. However, most of them (like their static CMOS counterparts) basically consist of inverters and transfer gates. A *shift register* is in fact a series connection of flip-flops. Dynamic versions of latches and flip-flops therefore also exist. A dynamic *flip-flop* is also referred to as a dynamic shift register cell because it dynamically shifts data from its input to its output during a single clock cycle.

A minimum clock frequency is required to maintain information in circuits that use dynamic storage elements. This minimum frequency is usually several hundred Hertz, and is determined by the subthreshold

leakage current and the leakage current of the reverse-biased diffusion
to substrate pn-junctions in both nMOS and pMOS transistors. There
are many different types of dynamic CMOS storage elements.

By deleting the feedback inverters in figure 4.31, we get the dynamic
D-type flip-flop shown in figure 4.36. Of course, this flip-flop comprises
two dynamic latches.

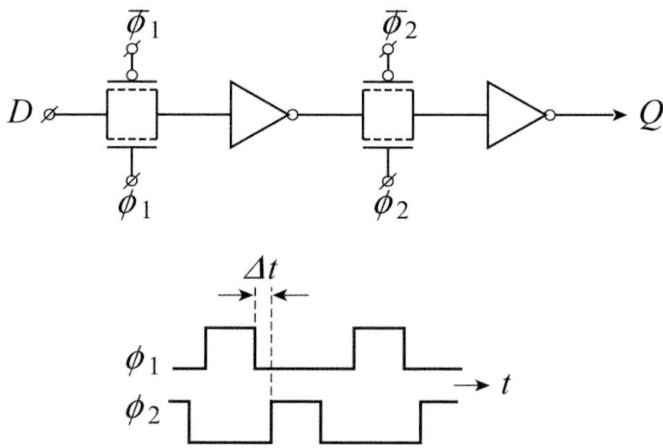

Figure 4.36: *Dynamic D-type flip-flop with non-overlapping clock signals*

The input data D in the above flip-flop is dynamically stored on the
input capacitance of the first inverter when ϕ_1 is 'high'. When ϕ_2 is
'high', the output level of the first inverter is dynamically stored on the
input capacitance of the second inverter. The *non-overlapping clocks* are
intended to prevent the latch from becoming transparent and allowing
data to race through the cell during a clock transition. Just as in the
static flip-flop, however, this flip-flop will become transparent if the clock
skew exceeds Δt. A shift register operates incorrectly when transparency
occurs in its flip-flops.

Figure 4.37 presents another type of dynamic CMOS shift register
cell. An advantage of this implementation is the reduced layout area
resulting from the absence of complementary transfer gates. The clocks
in the first section could also be switched and used in the second section.
The resulting risk of transparency requires considerable attention.

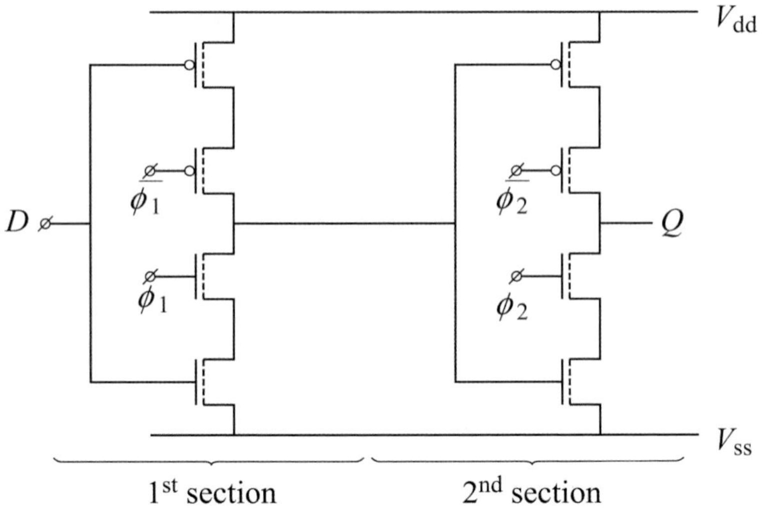

Figure 4.37: *Another dynamic CMOS shift register cell*

Critical phenomena in dynamic circuits

The operation of dynamic MOS circuits relies on the parasitic capacitances that store the logic levels. During a certain period of the clock cycle, several nodes in a dynamic circuit become floating, which makes them very susceptible to such effects as charge sharing and cross-talk.

- **Charge sharing**

  A typical example of *charge sharing* is shown in figure 4.38.

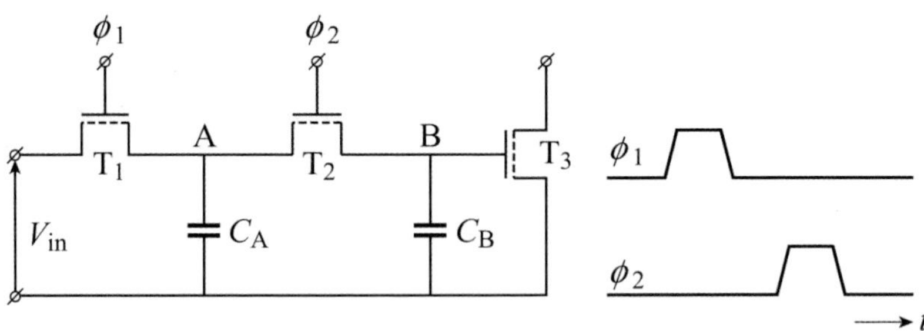

Figure 4.38: *An example of charge sharing*

The high levels of clocks ϕ_1 and ϕ_2 are assumed to cause no threshold loss in transistors T_1 and T_2. When ϕ_1 goes 'high', C_A is charged to the voltage V_{in} and remains at this level when ϕ_1 goes low again. During the period when ϕ_2 is 'high', the charge on C_A is shared between C_A and C_B. The voltages at nodes A and B are then described by:

$$V_A = V_B = \frac{C_A}{C_A + C_B} \cdot V_{in} \qquad (4.12)$$

As long as $C_B \ll C_A$, then $V_A \approx V_{in}$. However, if C_B is relatively large, then a 'high' level will be significantly degraded when charge is shared between C_A and C_B. Charge sharing circuits must therefore be used with caution and, if possible, should be avoided.

- **Cross-talk**
  Figure 4.39 shows a schematic of a situation in which *cross-talk* can occur. A capacitance C exists between node A and a signal track B which crosses it. When ϕ_1 goes from '1' to '0', capacitance C_A is supposed to act as temporary storage for the logic signal that was at A when ϕ_1 was '1'. However, node A has a very high impedance when ϕ_1 is '0', and a voltage change ΔV_B on the signal track B results in the following voltage change at node A:

$$\Delta V_A = \frac{C}{C_A + C} \cdot \Delta V_B$$

The value of the 'cross-over' capacitance C is proportional to the area of the overlap between node A and track B. A large value for C can lead to a disturbance of the logic levels at node A. The area and the number of potentially dangerous crossings must therefore be kept to a minimum during the layout phase of dynamic circuits. Each dynamic node in the finished layout must be checked to ensure that cross-talk noise remains within acceptable margins.

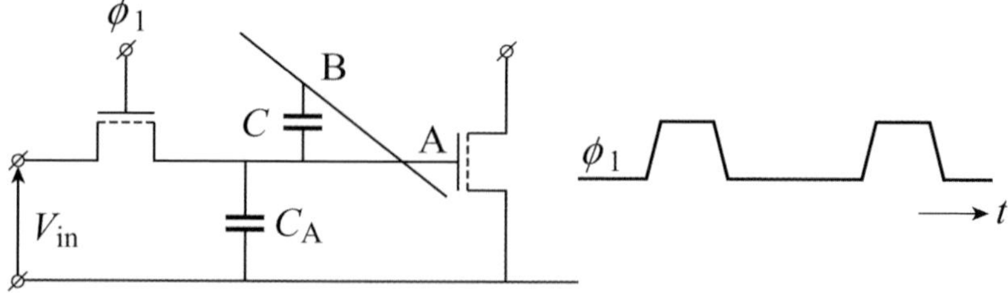

Figure 4.39: *A potential cross-talk situation*

The properties of dynamic MOS circuits can be summarised as follows:

- dynamic MOS circuits have less fan-in capacitance and consume less chip area than static equivalents.

- phenomena such as charge sharing and cross-talk make the electrical design and layout of dynamic nMOS circuits considerably more difficult than for static circuits.

Full CMOS (static CMOS) circuits are currently clearly ahead of dynamic CMOS circuits in the VLSI race. Significant numbers of CMOS ICs, however, still use dynamic CMOS circuits for the implementation of special functions, particularly for high-performance applications.

4.4.5   Other types of CMOS circuits

The most important characteristics of different CMOS circuits have been presented. These include the small chip area associated with dynamic implementations of logic gates, the low power dissipation associated with static implementations, large logic swings and large noise margins, etc. The advantages and disadvantages associated with an implementation choice can therefore be weighed up. Power dissipation, for instance, can be sacrificed for speed, or speed can be achieved when lower noise margins are accepted.

In the past, several articles have appeared on specialised forms of CMOS, including Cascode Voltage Swing Logic (CVSL) [8]. A CVSL logic gate is obtained by replacing the pMOS transistors in a conventional static CMOS logic circuit by nMOS transistors, which require inverse input signals. The reduction in chip area (at the expense of speed) is particularly noticeable when complex logic gates are implemented in static or dynamic CVSL. A modified form of CVSL called Differential Split Level (DSL) Logic uses a reduced logic swing. It therefore operates about 2 to 3 times faster but dissipates more power than CVSL. These type of logic families were more used in conventional CMOS technologies with higher supply voltages.

Some advice which may simplify the task of selecting the right logic implementation is given in the next section.

4.4.6 Choosing a CMOS implementation

An important decision at the start of a new CMOS design is the choice of logic implementation. The choice of a static or dynamic form is determined by a number of factors. The most dominant ones are power dissipation, speed, chip area and noise immunity. These factors are examined below.

Power dissipation

As previously shown, static CMOS circuits do not dissipate power when the circuit is stable. Except for the subthreshold leakage power, power is only dissipated in gates that change state. In clocked static CMOS circuits, most power dissipation occurs during and immediately after clock transitions. In clocked dynamic CMOS, however, each gate output is precharged every clock cycle.

Consider the dynamic inverter as an example. If the input remains 'high' during successive clock periods, then the output should be 'low'. However, the output is precharged during every clock period. This repeated charging and discharging of the output leads to high power consumption. A static CMOS inverter in the same situation would not change state and would therefore consume no power. Circuits for low-power or battery-operated applications and many memory circuits are therefore implemented in static CMOS. Chapter 8 presents extensive discussions on low-power issues.

Speed and area

Dynamic CMOS logic circuits are generally faster than their static CMOS counterparts. The nMOS-mostly nature of dynamic CMOS logic means that pMOS transistors are largely reserved for precharge and/or transfer functions while logic functions are only implemented in nMOS transistors. The input capacitance of a dynamic logic gate is therefore lower than a static equivalent. In addition, complex logic gates implemented in static CMOS may contain many pMOS transistors in series in the 'pull-up' path. A dynamic CMOS implementation offers increased speed and a smaller area because it uses only one pMOS transistor as an active pull-up.

Noise immunity

In a static CMOS logic circuit, there is always a conduction path between a logic gate's output and ground or the supply. Therefore, no logic gate output nodes are floating. Noise-induced voltage deviations on their logic levels are automatically compensated by current flows which restore levels. Dynamic circuits suffer from charge sharing and cross-talk effects, as already mentioned. There is also always a minimum clock frequency required because of the leakage of charge from floating nodes. As a result, static circuits are more robust. For this reason, most semi-custom design libraries are implemented in static CMOS.

4.4.7   Clocking strategies

Advantages and disadvantages of several implementations of single-phase and multi-phase *clocking strategies* have been described in the previous discussions of static and dynamic CMOS circuits. Single-phase circuits are the most efficient in terms of routing area. However, they may require more transistors than multi-phase alternatives. Today's flip-flops include two inverters to generate the intended ϕ and $\bar{\phi}$. The many transistors required for a NAND gate implementation of a flip-flop should also be remembered. In addition, the timing behaviour of *single-phase* circuits is critical and requires many circuit simulations to ensure equivalent functionality for best and worst cases, i.e., when delays are shortest and longest, respectively. *2-phase* circuits that use non-overlapping clocks have less critical timing behaviour.

Clock skew is always present in clocked circuits. Chapter 9 describes clocking strategies and alternatives, and also extensively discusses potential timing problems involved in designs with relatively large clock skew(s).

4.5 CMOS input and output (I/O) circuits

The electrical 'interfaces' between a CMOS IC and its external environment must ensure that data is received and transmitted correctly. These input and output interfaces must be able to withstand dangers that they may be reasonably expected to encounter. CMOS input and output circuits and the associated protection circuits are discussed below.

4.5.1 CMOS input circuits

MOS ICs often have to communicate with several other types of logic, such as ECL and TTL. A *TTL-compatible* input buffer must interpret an input voltage below 0.8 V as 'low' while voltages above 2 V must be interpreted as 'high'. The switching point of a TTL-compatible CMOS inverter must therefore be about 1.5 V. However, the switching point of a symmetric CMOS inverter (i.e., an inverter with equal transconductances for the nMOS an pMOS transistors) is half the supply voltage. The effects of asymmetry on the switching point of an inverter are shown in the transfer characteristic in figure 4.23. This figure clearly illustrates that a TTL-compatible CMOS inverter must be asymmetric.

Figure 4.40 shows a TTL-CMOS input buffer with the approximated transistor aspect ratios. The first inverter converts the TTL input signal to a CMOS level. Today's I/O voltages support 2.5 V or 1.8 V with threshold voltages close to 0.5 V. For this reason the nMOS input transistor is in series with an nMOS diode so that is off when the input is the maximum TTL low-level of 0.8 V.

An input buffer is usually located quite a distance from the logic gates that it drives. The required routing then forms a considerable load capacitance. A clock signal's input buffer is even more heavily loaded. The size of the load capacitance determines the required widths of the nMOS and pMOS transistors in an input buffer's second inverter. To achieve equal rise and fall times, the ratio of these widths must be approximately as shown.

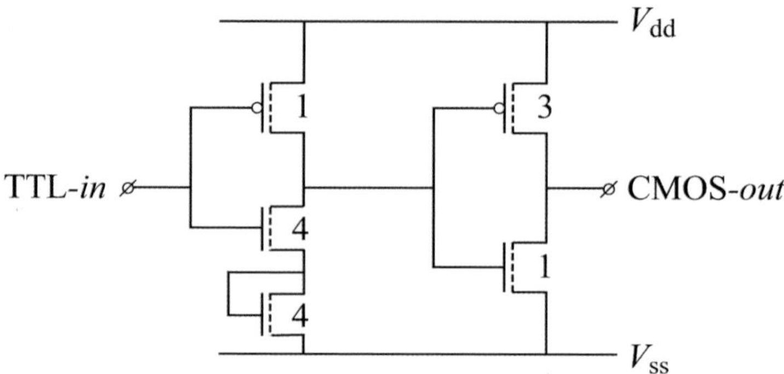

Figure 4.40: *TTL-CMOS input buffer*

The widths and lengths of manufactured transistors may vary independently as a result of processing variations. The effects of these variations are particularly evident for smaller dimensions. Minimum allowed dimensions should therefore not be used to achieve the required accuracy for the switching point of about 1.5 V for the first inverter in figure 4.40. In a 2.5 V CMOS input buffer, for instance, with a minimum channel length of 0.25 μm and minimum channel width of 0.3 μm, the first inverter does not need the additional nMOS diode and could be dimensioned as follows:

$$\left(\frac{W}{L}\right)_{\mathrm{p}} = \frac{0.5}{0.5}\,\mu\mathrm{m} \quad \text{and} \quad \left(\frac{W}{L}\right)_{\mathrm{n}} = \frac{1}{0.5}\,\mu\mathrm{m}$$

Not using the minimum transistor sizes makes them less sensitive to process variations.

4.5.2 CMOS output buffers (drivers)

There are many different output buffer designs. They usually contain a tapered chain of inverters, as discussed in section 4.3.2. Transistor sizes in the output buffer are determined by the specifications of the output load and the clock frequency. Output load capacitances usually range from 10 to 30 pF, and I/O clock frequencies vary between 100 MHz and 1 GHz.

Several problems arise when many outputs switch simultaneously at a high frequency. The resulting peak currents through metal tracks may exceed the allowed maxima. These currents also cause large *voltage peaks* across the intrinsic inductances in the bond wires between a chip's package and its bond pads. The accumulation of peak currents in power and ground lines leads to relatively large noise signals on the chip. These problems (which are also discussed in chapter 9) must be taken into account when designing output buffers.

The very large transistors required in output drivers could result in unacceptably large *short-circuit currents* between supply and ground if the charge and discharge transistors were allowed to conduct simultaneously. Figure 4.41 shows an example of a short-circuit free output buffer. This tri-state buffer is combined with an output flip-flop and can drive a 10 pF load at 250 MHz. Signals 1, 2 and 3 represent the input data, the clock and the tri-state control, respectively. The logic circuits II and III control the gates of the nMOS and pMOS output driver transistors, respectively. These circuits ensure that the driver transistors never conduct simultaneously. The pre-driver logic gates must be designed such that they fulfill the tapering factor requirements as described in section 4.3.2. This output driver is just one example of the many existing driver types. Many alternatives are available through publications and through the internet, each with its own schematic which is targeted at the specific application area.

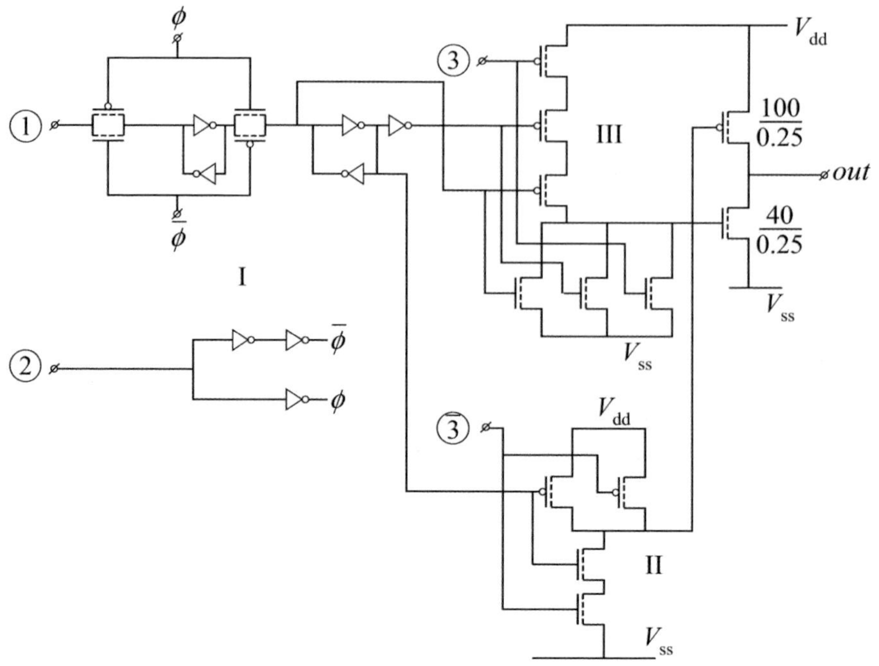

Figure 4.41: *Short-circuit free tri-state CMOS output buffer*

4.6 The layout process

4.6.1 Introduction

In this section, we present a simple set of basic design rules for a *CMOS process* containing a single polysilicon and a single metal layer. These layout *design rules* represent a virtual 50 nm state-of-the-art CMOS process. Although such a process usually incorporates about six to ten metal layers, only one metal layer will be used in this layout design process. This is because many of the libraries only use the first metal layer for the local interconnections inside each library cell. After a description of each individual mask, the creation of a stick diagram and the layout process are demonstrated with an example. Finally, a process cross-section shows the real silicon implementation.

4.6.2 Layout design rules

The process *masks* of the chosen technology are listed below in the order of the process sequence. Many of these masks are described in section 3.9.

ACTIVE (layout colour: green)

> This mask defines the *active areas* inside which the transistors will be created. Outside the active areas, thick oxide will be formed with STI (or LOCOS in the past). The width of an ACTIVE pattern determines the transistor channel width.

NWELL (layout colour: yellow)

> This mask defines the areas where the pMOS transistors will be located. The n-well actually serves as a substrate for the pMOS transistors. As the CMOS process offers complementary transistors, the creation of a p-type substrate (p-well) for nMOS transistors is also required. This is usually automatically generated from the NWELL mask: a p-well will be created everywhere where no n-well pattern is defined. This p-well mask need not be drawn in the stick diagram and layout.

POLY (layout colour: red)

> This mask defines the polysilicon pattern. A transistor channel is formed where POLY crosses an ACTIVE region. On top of thin gate oxide, polysilicon acts as a MOS transistor gate. Outside the active areas, polysilicon is used as a local interconnection only over small distances inside the library cells. The minimum width of the polysilicon determines the transistor channel length.

NPLUS (layout colour: orange)

> The sources and drains of nMOS transistors need n^+ implants. The NPLUS mask defines the areas in which n^+ is implanted. During the n^+ implantation, the STI (thick oxide regions) and the polysilicon gate act as barriers, e.g., we get self-aligned n^+ regions (sources and drains) everywhere within ACTIVE which is surrounded by NPLUS and not covered by POLY.

PPLUS (layout colour: purple)

> Complementary to the NPLUS mask, sources and drains of the pMOS transistor are p-type doped by means of the PPLUS mask.

CONTACT (layout colour: black)
> This mask defines contact holes in the dielectric layer below the first metal layer (METAL). Through these contact holes, the metal layer can contact polysilicon (POLY) and source or drain regions (ACTIVE).

METAL (layout colour: blue)
> This defines the pattern in the first metal layer, which can be aluminium and tungsten in the 180 nm CMOS node and above, and copper in the 120 nm CMOS node and beyond. A track in this layer can be used for both short and long interconnections because its sheet resistance is relatively low.

Note:

Only a one-metal-layer process is used here. Current CMOS technologies use six to ten metal layers. However, for educational purposes a layout with one metal layer gives a very good insight into the layout process. More metal layers only means: "more of the same".

Design rules for a virtual 50 nm CMOS process:

The following set of design rules of a virtual 50 nanometer CMOS process will be used in an example of a layout and in several exercises at the end of this chapter. Figure 4.42 serves as an illustration of each of the design rules.

Design rules for a virtual 50 nm CMOS process

ACTIVE

a.	Track width	70
b.	Track spacing	100

NWELL

c.	Track width	400
d.	Track spacing	400
e.	Extension NWELL over ACTIVE	100

POLY

f.	Track width	50
g.	Track spacing	100
h.	Extension POLY over ACTIVE (gate extension)	100
i.	Extension ACTIVE over POLY (source/drain width)	100
j.	Spacing between ACTIVE and POLY	50

NPLUS

k.	Track width	200
l.	Track spacing	200
m.	Extension NPLUS over ACTIVE (n^+ ACTIVE)	100
m1.	Spacing between n^+ ACTIVE and POLY	120
n.	Spacing between n^+ ACTIVE and NWELL	100

PPLUS

o.	Track width	200
p.	Track spacing	200
q.	Extension PPLUS over ACTIVE (p^+ ACTIVE)	100
q1.	Spacing between p^+ ACTIVE and POLY	120

CONTACT

r.	Minimum and maximum dimensions	70×70
s.	Spacing between contacts	90
t.	Extension ACTIVE over CONTACT	20
u.	Extension POLY over CONTACT	20
v.	Extension METAL over CONTACT	20
w.	Spacing CONTACT and POLY gate	50
x.	CONTACT on gate regions not allowed	!

METAL

y.	Track width	80
z.	Spacing between tracks	80

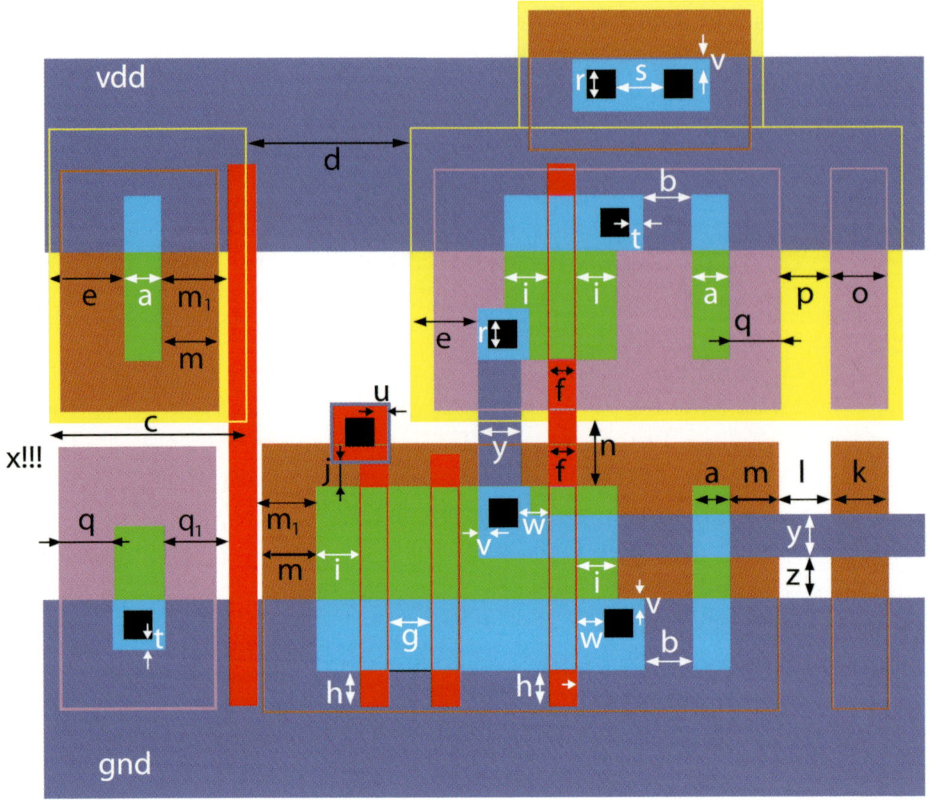

Figure 4.42: *Illustration of each of the design rules of the previous page*

The minimum width and spacing in a certain mask pattern is defined by the different processing steps involved. For instance, the ACTIVE is defined by the STI formation process, while a METAL pattern is the result of deposition and etching techniques. Minimum overlaps or separations between patterns in different masks are defined by alignment tolerances with respect to a common reference location and by the different processing steps involved. The minimum width of the POLY mask pattern determines the channel length of the transistors and is usually referred to in the process notation, e.g., a 50 nm CMOS process means that the minimum POLY width is close to 50 nm. The previous set of design rules are reflecting a virtual 50 nm CMOS process. For educational purposes, these rules have been simplified and rounded. For example, rules t, u and v assume an extension of 20 nm in all directions. However, in reality this could also be 10 nm in one direction and 30 nm in the other.

Usually, when a complex layout has to be developed, a stick diagram is first drawn to explore the different possibilities of layout interconnections. The use of a stick diagram is discussed first.

4.6.3 Stick diagram

A *stick diagram* is used as an intermediate representation between circuit diagram and layout. This topological representation of the circuit is drawn in colours which correspond to those used in the layout. Only the connections of the different mask patterns are depicted, without paying attention to the sizes. The EXNOR circuit of figure 4.43 serves as an example for the development of a stick diagram. This EXNOR circuit represents the Boolean function: $Z = \overline{(a+b)\overline{ab}} = a\,b + \bar{a}\,\bar{b}$

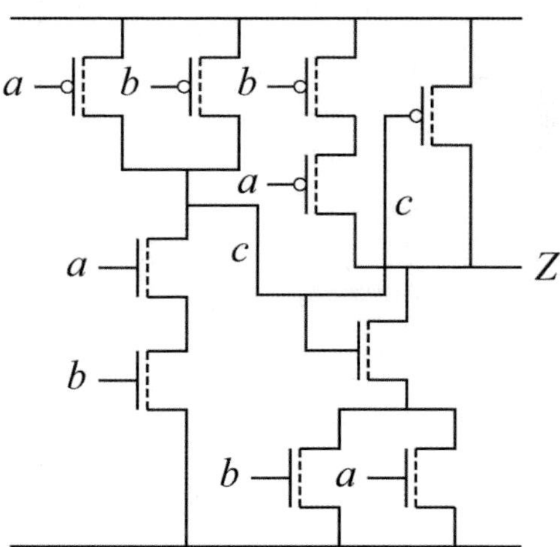

Figure 4.43: *Circuit diagram of a CMOS EXNOR logic gate*

Figure 4.44 illustrates the procedure for the generation of the stick diagram for the EXNOR logic gate.

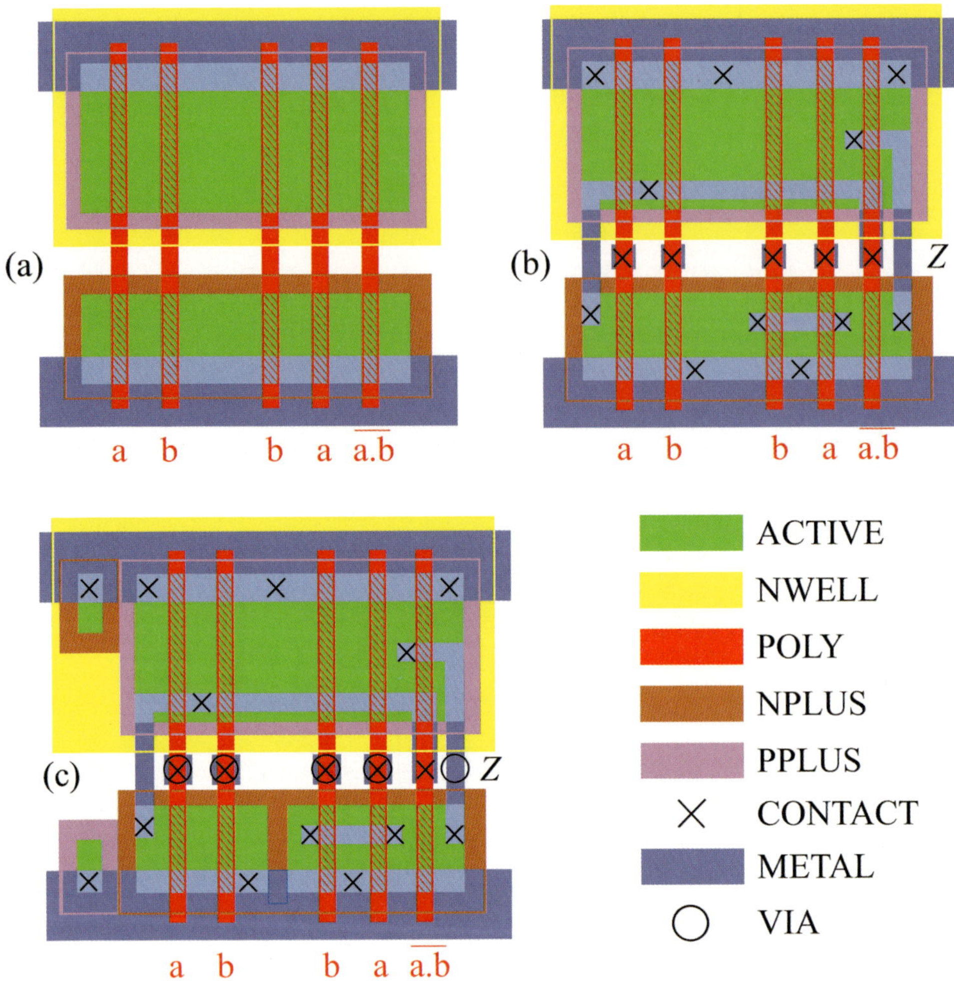

Figure 4.44: *Various steps in the design of a stick diagram*

The creation of this topological view is divided into three phases, represented by (a), (b) and (c) in the figure. These phases are explained as follows:

(a) Two horizontal parallel thin oxide (ACTIVE) regions are drawn. The lower ACTIVE region is usually reserved for nMOS transistors while the upper region is for the pMOS transistors. The envisaged CMOS process uses NPLUS and PPLUS masks to define the n^+ and p^+ diffusion regions of the source/drain areas of the nMOS and pMOS transistors, respectively. An NPLUS boundary is therefore drawn

246

around the lower ACTIVE region in the stick diagram while the upper region is surrounded by a PPLUS boundary. The n-well is indicated by the NWELL area, which overlaps ACTIVE areas surrounded by PPLUS. It is not required to draw the PWELL mask, because it is the inverse of the NWELL mask; everything outside the NWELL area becomes PWELL. Parallel polysilicon (POLY) gates are drawn vertically across both ACTIVE regions. Metal (METAL) supply and ground lines are drawn horizontally over the PPLUS and NPLUS regions, respectively.

(b) Additional METAL and POLY lines indicate transistor connections according to the function to be implemented. The source/drain diffusion areas of neighbouring transistors are merged and black crosses represent contacts. These transistor connections are implemented from left to right. The two nMOS transistors on the left of the stick diagram, for example, correspond to the nMOS transistors of the NAND gate on the left of the circuit diagram in figure 4.43. The drains of two pMOS transistors and one nMOS transistor are connected with METAL to form the NAND gate output. This connection is represented by a metal interconnection of n^+ and p^+ diffusion areas. A direct diffusion connection between an n^+ and p^+ area is not possible as it would form a diode. Connections between n^+ and p^+ areas therefore always occur via metal. The NAND gate output is connected to the gate of the most right nMOS and pMOS transistors.

(c) The third nMOS source/drain area from the left in figure 4.44(b), is connected to ground and to another node. This is clearly not according to the required functionality and such diffusion areas are therefore split into separate diffusion areas in figure 4.44(c). Finally, we have to enable connections to the inputs and the outputs. Because first metal is already used for supply lines and internal cell connections, we have to enable connections to the second metal layer. This is done through adding vias to the input and output terminals of the cell. These vias are represented by the black circles. On top of these vias, small second-metal areas must be positioned to form the real terminals. However, for educational purposes, this is not drawn in this stick diagram, as it would make the figure less clear.

No back-bias voltage is used in the chosen process. The p-type substrate is therefore connected to ground and the n-well is connected to the supply. These substrate and n-well connections are indicated at the left side in the figure.

There should be enough connections from PWELL to ground and from NWELL to V_{dd} to keep latch-up sensitivity to a low level. (latch-up is discussed in section 9.5.5) These contacts reduce the values of R_1 and R_2, respectively, in figure 9.31. In current advanced CMOS libraries, these PWELL and NWELL contacts are included in a separate standard cell, which can be placed according to the needs of the specific design, e.g. 30-40 μm apart. This subject is further addressed in the layout discussion below.

4.6.4 Example of the layout procedure

The following example shows the complete layout process from a basic Boolean function, through Boolean optimisation, circuit diagram and stick diagram to a layout. Consider the following Boolean function:

$$Z = \bar{a}\,\bar{b}\,\bar{c} + \bar{a}\,c\,\bar{d} + \bar{a}\,c\,\bar{d} + \bar{a}\,b\,c\,\bar{d}$$

To optimise this function for implementation in CMOS, an inverse Boolean expression in the format $Z = \bar{f}$ must always be found, because every single CMOS logic gate implements an inverted expression:

$$
\begin{aligned}
Z &= \bar{a}\,\bar{b}\,\bar{c} + \bar{a}\,c\,\bar{d} + \bar{a}\,c\,\bar{d} + \bar{a}\,b\,c\,\bar{d} \\
&= \bar{a}\,(\bar{b}\,\bar{c} + c\,\bar{d} + c\,\bar{d} + \bar{b}\,c\,\bar{d}) \\
&= \bar{a}\,(\bar{b}\,\bar{c} + (\bar{c} + c + \bar{b}\,c)\,\bar{d}) \\
&= \bar{a}\,(\bar{b}\,\bar{c} + \bar{d}) \\
&= \overline{\overline{\bar{a}\,(\bar{b}\,\bar{c} + \bar{d})}} = \overline{a + \overline{(\bar{b}\,\bar{c} + \bar{d})}} = \overline{a + (b + c)d}
\end{aligned}
$$

Therefore, the optimised function for implementation as a single CMOS logic gate is: $Z = \overline{a + (b + c)d}$. The circuit diagram for this logic function is shown in figure 4.45.

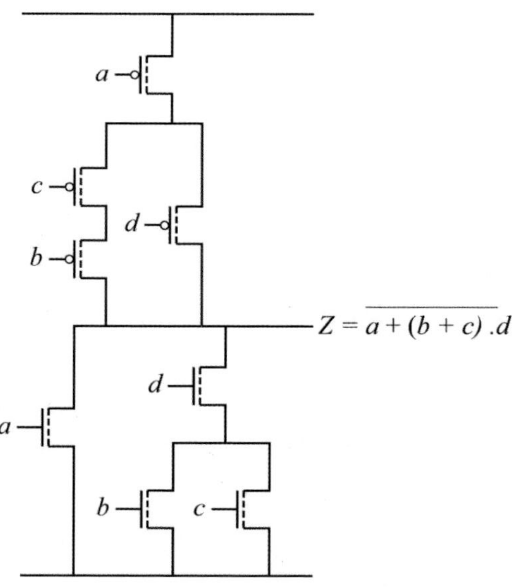

Figure 4.45: *Circuit diagram implementing* $Z = \overline{a + (b + c) \cdot d}$

The corresponding CMOS stick diagram and layout can be found in figure 4.46(a) and figure 4.46(b) respectively. Figure 4.46(c) shows a cross-section through the line D-D' in the layout.

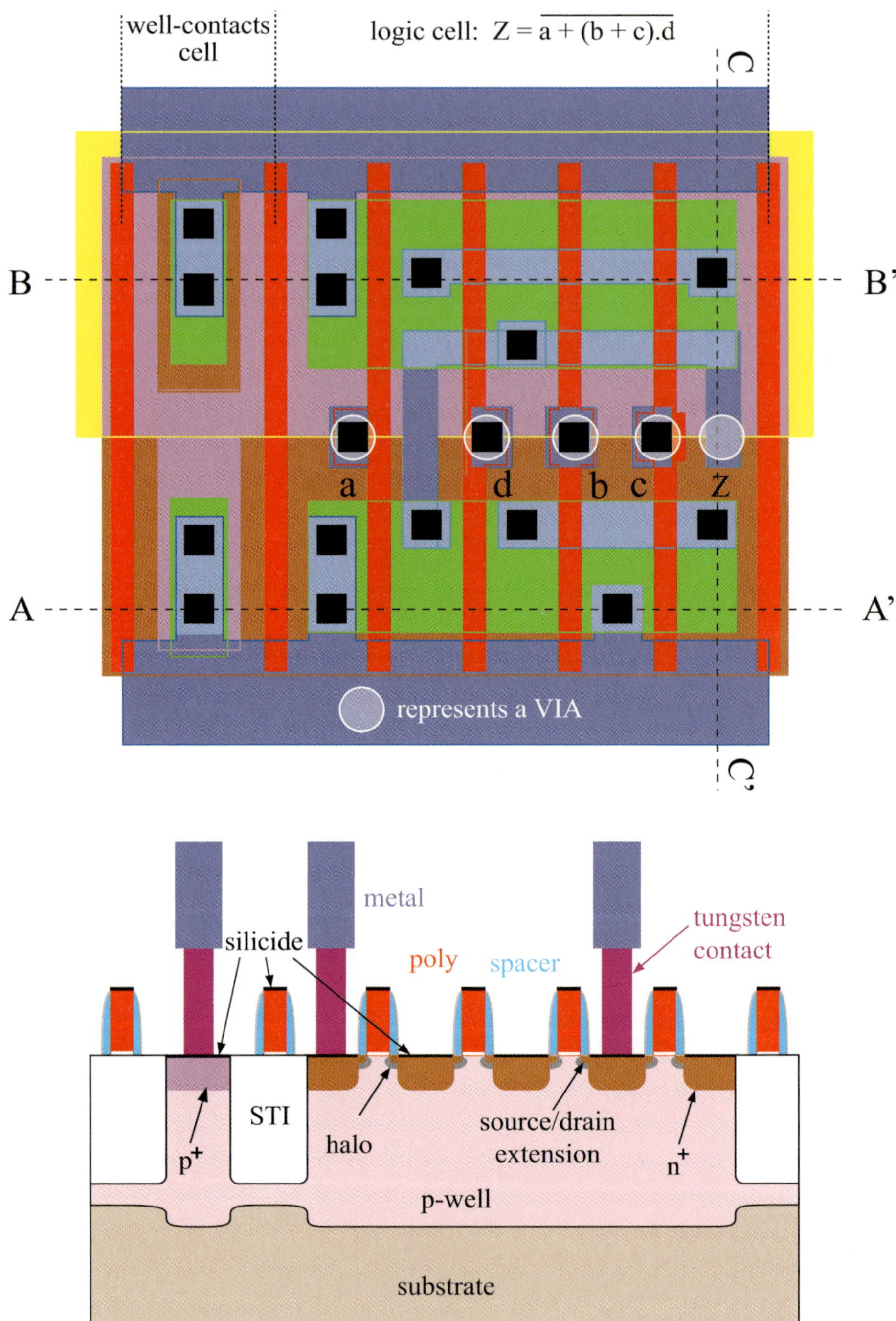

Figure 4.46: *Layout (top) and cross-section (bottom) of the sample logic gate along the line A-A'*

The layout contains one substrate (p-well) and one n-well contact. The use of extra n-well and p-well contacts reduces latch-up sensitivity but may lead to an increased layout area. In a $0.25\,\mu$m $2.5\,$V CMOS technologies and above, a practical compromise was to place at least one substrate and n-well contact per five nMOS and pMOS transistors, respectively. Because the supply voltage of current advanced CMOS technologies is close to $1\,$V, the probability of latch-up has reduced significantly and requires much less PWELL and NWELL contacts. Current libraries in $45\,$nm technologies may include special standard cells which only contain these well contacts. These cells must then be placed at least every 30 to $40\,\mu$m. This rule of thumb applies to logic circuits. The large transistors in driver and I/O circuits which usually operate at higher supply voltages ($3.3\,$V, $2.5\,$V and $1.8\,$V), require considerably more substrate and n-well contacts.

The n-wells in a CMOS circuit layout are usually connected to the supply voltage. Generally, different neighbouring n-wells (which are connected to the same voltage) should be extended to form one large well.

The output node of a static CMOS logic gate is formed by an interconnection of n^+ and p^+ diffusion areas. The p^+ diffusion area is usually the larger. The parasitic capacitance of such an output node is therefore larger than its nMOS counterpart. In addition, the width of a pMOS transistor is usually larger than an nMOS transistor width.

As a result of silicided p^+ diffusion regions, the series resistance of sources and drains are low and usually only one contact is sufficient per connected node. These resistances are only several ohms per square in CMOS technologies with silicided source and drain regions. Minimum source and drain areas can then be used to keep parasitic capacitances small.

The process cross-section in figure 4.46(c) is made along the line A-A'. The cross-section includes n^+ source, drain and gate areas, STI oxide isolation areas, a p-well area, a p-well contact and a source contact. A detailed study of the relationship between the cross-section and the layout should enable the reader to draw a cross-section at a line anywhere in the layout.

Circuit density and performance are often improved by using several polysilicon layers (memories) and seven to ten metal layers (VLSI). The area reduction must compensate for the costs associated with the additional masks and processing steps. However, with the ever-increasing current density, more and more metal layers are required to distribute the power properly across the chip.

In CMOS technologies beyond 100 nm, all nMOS and pMOS transistors are only allowed to have straight (no L-shape or Z-shape) poly silicon gates. In 65 nm CMOS and beyond, the regularity of the transistor patterns in increasing to support *litho-friendly design*. In these technologies an increasing fixed-pitch approach is adopted to ease the lithographic process and to improve reproduction and yield.

4.6.5 Guidelines for layout design

Designing a correct layout involves more than just a translation of the circuit diagram into a layout that meets the relevant design rules. Attention must be paid to several key issues:

- Minimise layout area.
  A minimum layout area will especially reduce the overall silicon costs with the development of a new library that is to be used for the design of numerous chips. Moreover, when ICs become smaller, they generally show a higher performance, consume less power and are cheaper.

- Pay attention to parasitic elements.
  Each design, whether a library cell or a large logic block, must be optimised with respect to parasitic capacitances (source and drain junctions, metal interconnects) and resistances (mainly of long interconnections). This is necessary to achieve better performance and again reduces the power consumption.

- Pay attention to parasitic effects.
  Effects such as cross-talk, charge sharing and voltage drop across supply lines particularly greatly reduce the performance as well as the signal integrity. Such effects are extensively discussed in chapter 9.

Table 4.2 shows some typical values of the capacitances and resistances of different components and materials used in a virtual 50 nm CMOS technology with a gate oxide thickness $t_{\mathrm{ox}} = 12\,\text{Å}$ (1.2 nm).

Table 4.2: Parasitic capacitances and resistance values in a virtual 50 nm process with $t_{\mathrm{ox}} = 1.2\,\mathrm{nm}$)

Material	Capacitances	Resistances
Polysilicon (POLY)	gate cap: $17\,\mathrm{fF}/\mu\mathrm{m}^2$* edge cap: $0.22\,\mathrm{fF}/\mu\mathrm{m}$ *	poly 300-400 $\Omega/\square$ polycide 8 $\Omega/\square$
Copper (Cu) (METAL)	average track cap: $0.19\,\mathrm{fF}/\mu\mathrm{m}$	$\frac{1.7 \cdot 10^{-8}}{H}\ \Omega/\square$
Source/Drain implants (ACTIVE)	track cap: $1\,\mathrm{fF}/\mu\mathrm{m}^2$ thick oxide edge cap: $0.3\,\mathrm{fF}/\mu\mathrm{m}$ cap to POLY edge: $0.3\ \mathrm{fF}/\mu\mathrm{m}$	$\mathrm{n}^+ \approx 100\text{-}250\,\Omega/\square$ $\mathrm{p}^+ \approx 150\text{-}350\,\Omega/\square$ silicided $\mathrm{n}^+ \approx 8\,\Omega/\square$ silicided $\mathrm{p}^+ \approx 8\,\Omega/\square$

Note: * on thin oxide

It is clear that polysilicon and $\mathrm{n}^+/\mathrm{p}^+$ junctions can only be used for very short connections inside library cells as a result of the relatively high sheet resistance values.

Especially nanometer CMOS processes include six to ten layers of metal. In many cases, the upper metal layer has a greater thickness, a larger minimum feature size and a larger spacing. Therefore, this upper level must be used for a structured and proper overall chip power supply network.

The above discussions on CMOS layout implementation conclude this chapter. More information on the design of CMOS circuits and layouts can be found in the reference list.

4.7 Conclusions

CMOS has become the major technology for the manufacture of VLSI circuits, and now accounts for about 90% of the total IC market. The main advantage of CMOS is its low power dissipation. This is an important requirement in current VLSI circuits, which may contain hundreds of millions to more than a billion of transistors.

Static CMOS circuits are characterised by high input and parasitic capacitances and relatively large logic gate structures. The silicon area occupied by a static CMOS logic circuit is about twice that of an nMOS counterpart. Dynamic CMOS circuits are nMOS-mostly and are therefore generally smaller than their CMOS counterparts. The use of a static rather than a dynamic implementation must therefore be justified by a sufficient reduction in power dissipation. Generally, static CMOS shows the lowest τD product and is thus the most power efficient implementation for VLSI. Moreover, its robustness is very important in current nanometer ICs as these show increasing noise, caused by cross-talk and supply voltage drops. Low-power issues and maintaining signal integrity at a sufficiently high level are the subjects of chapter 8 and 9, respectively.

Basic technologies for the manufacture of MOS devices are explained in chapter 3. Various nMOS circuit principles are introduced. This chapter emphasises the most important differences between CMOS and nMOS circuits. These differences are evident in the areas of technology, electrical design and layout design. A structured CMOS layout design style is presented in this chapter while using a limited set of representative design rules. The combination of the CMOS and nMOS circuit design and layout principles discussed in this chapter should afford the reader sufficient insight into the basic operation of different CMOS circuits.

4.8   References

CMOS physics and technology (see also chapter 3)

[1] Richard C. Jaeger,
   'Introduction to Microelectronic Fabrication',
   Modular Series on Solid-State Devices, Volume 5, 1988

[2] Y. Sakai, et al.,
   'Advanced Hi-Cmos Device Technology',
   IEEE IEDM, pp. 534-537, Washington DC, 1981

[3] S.M. Sze,
   'Modern Semiconductor Device Physics',
   John Wiley & Sons, 1997

[3a] S. Wolf and R.N. Tauber,
   'Silicon processing for the VLSI Era',
   Volume 1, Process Technology, Lattice Press, 1986

[3b] S.M. Sze,
   'Modern Semiconductor Device Physics',
   John Wiley & Sons, 1997

CMOS design principles (general)

[4] C. Mead, L. Conway,
   'Introduction to VLSI Systems',
   Addison-Wesley, 1980

[5] N. Weste, K. Eshraghian,
   'Principles of CMOS VLSI Design, a Systems Perspective',
   Addison-Wesley, 1993

[6] L.A. Glasser, D.W. Dobberpuhl,
   'The Design and Analysis of VLSI circuits',
   Addison-Wesley, 1985

[7] M. Annaratone,
   'Digital CMOS circuit Design',
   Kluwer Academic Publishers, 1986

[8] L.G. Heller, et al.,
   'Cascode Voltage Switch Logic',
   IEEE Digest of technical papers of the ISSCC, 1984

[9] Jan M. Rabaey,
    'Digital Integrated Circuits: A Design Perspective',
    Prentice Hall, 1995

[10] Kerry Bernstein, et al.
    'HIGH SPEED CMOS DESIGN STYLES',
    Kluwer Academic Publishers, 1999

[11] International Solid-State Circuits Conference
    Digest of Technical papers, February 2000,
    pp. 90-11, pp. 176-177, pp. 412-413, pp. 422-423

Power dissipation in CMOS

[12] H.J.M. Veendrick,
    'Short-Circuit Dissipation of Static CMOS Circuitry and its Impact
    on the Design of Buffer Circuits',
    IEEE Journal of Solid State Circuits, Vol. SC-19,
    No. 4, August 1984, pp. 468-473

For further reading

[13] 'IEEE Journal of Solid-State Circuits'

[14] 'ISSCC and ESSCIRC conferences,
    VLSI and ISLPED symposia,
    digests of technical papers'

4.9 Exercises

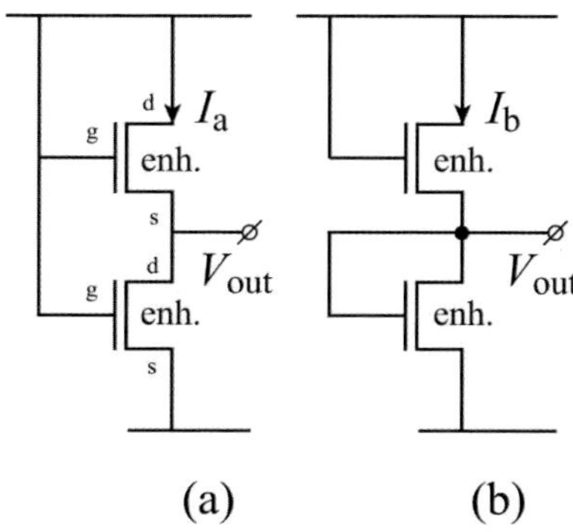

1. The following values apply for the above circuits:
   $V_{dd} = 1\,V$
   $K = 0\,V^{1/2}$
   $|V_x| = 0.25\,V$
   All transistors are of the same size; assume ideal transistors (no leakage currents).

   a) What can be said about V_{out} in circuits a and b: $V_{out} <>= V_{dd}/2$? Explain.

   b) Which of the currents I_a and I_b is larger and why?

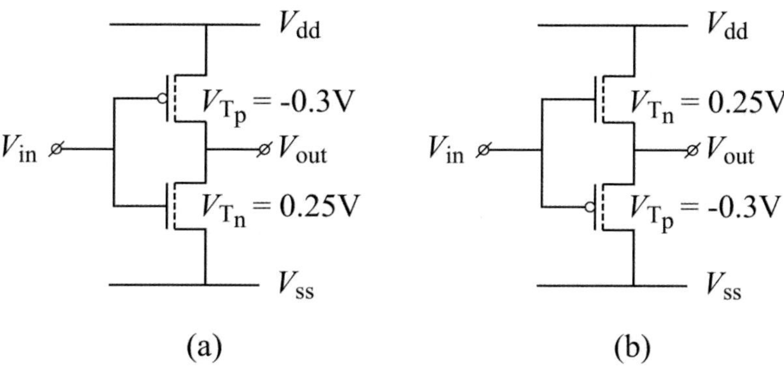

2. a) If $V_{dd} = 0.4\,V$ in the above figure, explain what would happen at

the output of circuit (a) when V_{in} switches from 0 V to V_{dd} and back. Draw this in the inverter characteristic: $V_{out}=f(V_{in})$.

b) Repeat a) for $V_{dd}=1$ V.

c) If $V_{dd}=1$ V in circuit (b) and V_{in} switches from 0 V to V_{dd} and back, draw $V_{in}=f(t)$ and $V_{out}=f(t)$ in the same diagram (assume $K=0\,\mathrm{V}^{1/2}$).

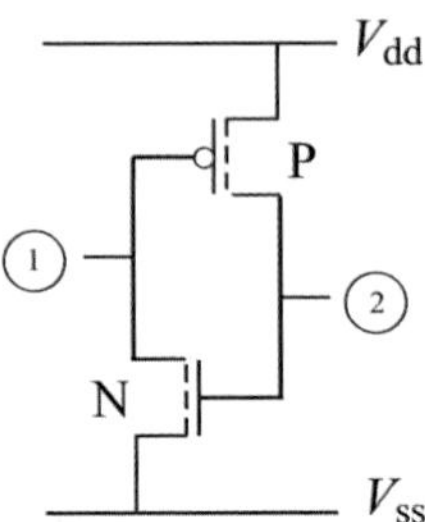

3. The above circuit is called a tie-off cell.

   a) Describe its behaviour during power up.?

   b) What will the voltages at nodes 1 and 2 be after the power-up situation has been stabilised?

   c) What are the major differences between this circuit and a CMOS inverter?

   d) Assume that the transistor sizes in this circuit are identical to the transistor sizes of a CMOS inverter. Assume that the input of the inverter is connected to V_{dd}. Which of the circuits, tie-off cell or inverter, would represent more capacitance between the V_{dd} and V_{ss} terminals and why?

4. a) Explain in no more than ten lines the cause of short-circuit dissipation.

   b) What is the reason that the short-circuit power consumption in large drivers (bus- and output drivers) in modern CMOS processes ($L < 90\,\mathrm{nm}$) has become negligible.

5. Draw a process cross-section along the line indicated by B-B' in the layout in figure 4.46.

6. The following function must be implemented in a CMOS logic circuit: $z = c(\bar{a}\, b + a\, \bar{b})$

   a) Draw a circuit diagram of a static CMOS implementation of the required logic circuit. The required inverse signals must also be generated in this circuit.

   b) Adopt the approach presented in this chapter and draw the CMOS stick diagram and layout of the logic circuit
   Assume $(\frac{W}{L})_\mathrm{n} = \frac{200\,\mathrm{nm}}{50\,\mathrm{nm}}$ and $(\frac{W}{L})_\mathrm{p} = \frac{300\,\mathrm{nm}}{50\,\mathrm{nm}}$
   (scale: $5\,\mathrm{nm} \equiv 1\mathrm{mm}$).

7. Consider the following logic function: $z = c + a\, b + \bar{a}\, \bar{b}$

   a) Rewrite this function such that it is optimised for implementation in MOS.

   b) Draw a circuit diagram of a static CMOS implementation.

   c) Adopt the approach presented in this chapter and draw the CMOS stick diagram and layout of the logic circuit
   Assume $(\frac{W}{L})_\mathrm{n} = \frac{200\,\mathrm{nm}}{50\,\mathrm{nm}}$ and $(\frac{W}{L})_\mathrm{p} = \frac{300\,\mathrm{nm}}{50\,\mathrm{nm}}$
   (scale: $5\,\mathrm{nm} \equiv 1\mathrm{mm}$).

8. A static CMOS inverter has been implemented in a $45\,\mathrm{nm}$ CMOS technology for $1\,\mathrm{V}$ operation with $V_{\mathrm{T_n}} = |V_{\mathrm{T_p}}| = 0.25\,\mathrm{V}$.

   a) For which of the logic gates, NOR or NAND, would you limit the number of inputs and why?

   b) Explain what will happen if you would run this inverter at only $0.2\,\mathrm{V}$?

   c) Which of the library cells would you consider to be the most critical one in reducing the supply voltage to subthreshold voltage levels?

9. The following values are given for the parameters in the adjacent circuit: $V_{X_n} = 0.25\,\text{V}$
$V_{X_p} = -0.3\,\text{V}$
$K_p = K_n = 0\,\text{V}^{1/2}$
$V_{bb} = -1\,\text{V}$

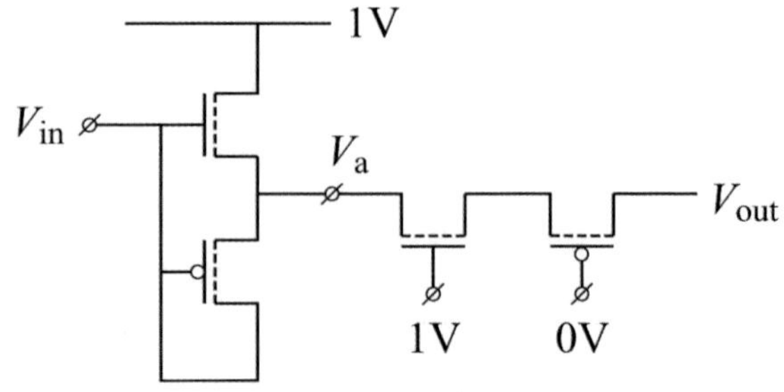

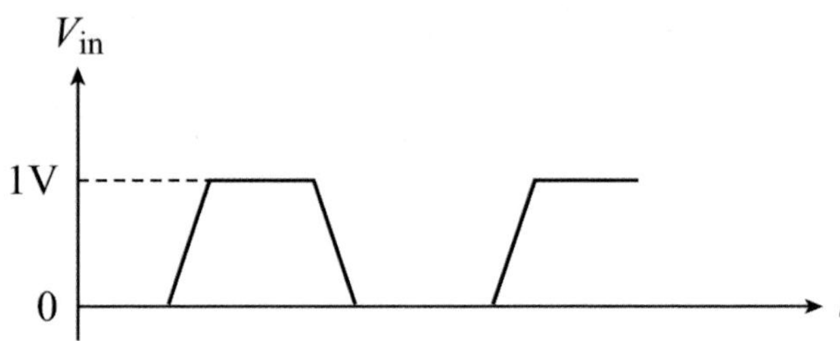

Explain what happens to voltages V_a and V_{out} when V_{in} switches as shown. Draw V_{in}, V_a and V_{out} in one diagram.

Chapter 5

Special circuits, devices and technologies

5.1 Introduction

This chapter discusses a number of special circuits, devices and technologies. These circuits and devices can be used in digital, analogue and mixed analogue/digital applications. They are realised in various MOS technologies or their derivatives, which include the BICMOS technologies discussed in this chapter.

The chapter begins with an explanation of circuits that operate as image sensors. We distinguish *charge-coupled devices (CCDs)* as well as *CMOS image sensors*. Their ability to capture images finds its usage in all kinds of cameras. Their operation is based upon the conversion of light into electrons.

The second category of special devices covered in this chapter are MOS transistors capable of delivering high power. These power MOS field-effect transistors, or *power MOSFETs*, are feasible as a result of improvements in technology, which enable the manufacture of transistors capable of withstanding high voltages as well as large current densities. Power MOSFETs obviously operate according to the same field-effect principle as ordinary MOS transistors. This principle is discussed in chapter 1.

Finally, devices based on mixed bipolar and CMOS technologies are discussed. They were/are particularly used to enhance the performance of both digital and mixed analogue/digital circuits.

5.2 CCD and CMOS image sensors

5.2.1 Introduction

Charged-coupled devices (CCDs) are almost exclusively used as image sensors. They basically operate by transferring charge from below one transistor gate to below another one in a 'channel'. CCD implementations include *surface-channel* (SCCD) and *buried-channel* (BCCD) devices. Also, for analogue applications, there must be a relationship between the size of the packet and the signal which it represents. The packet size must therefore be maintained during transfer. An alternative to CCD imaging is CMOS imaging, which is currently used in large volumes in cheap consumer and communication products, such as digital cameras and mobile phones.

5.2.2 Basic CCD operation

CCD shift registers can be realised with 2-phase, 3-phase and other multi-phase clock systems. The operation of a CCD is explained below with the aid of the 2-phase SCCD structure shown in figure 5.1. A diagram of the 2-phase clocks ϕ_1 and ϕ_2 is also shown in this figure. The gates indicated by bold lines are polysilicon 'storage gates', under which charge is stored. The remaining gates are 'transfer gates' created in a second polysilicon or metal layer. They lie on a thicker oxide than the storage gates and therefore have a much higher threshold voltage ($V_T \approx 1\,\text{V}$). These transfer gates serve as a barrier between the storage gates. Operation of the 2-phase SCCD is explained on the basis of the surface potential distributions under the gates.

Suppose the first and third storage gates contain a full and an empty charge packet, representing the logic levels '1' and '0', respectively.

The charge packet corresponding to the first storage gate is then full of electrons. This is represented by a full 'charge bucket' under the gate in figure 5.1. The charge bucket corresponding to the third storage gate, however, is almost empty, i.e., it is practically devoid of electrons. At time point 1, both ϕ_1 and ϕ_2 are 'low' and the storage gates are separated from each other. At time point 2, ϕ_1 has switched from a low to a high level and the charge is transferred from the ϕ_2 storage gates to the ϕ_1 storage gates. At time point 3, both ϕ_1 and ϕ_2 are 'low' again and the charge is now stored under the ϕ_1 storage gates. The description of the shift behaviour at time points 4 and 5 is obtained by replacing ϕ_1

by ϕ_2 in the above descriptions for time points 1 and 2, respectively.

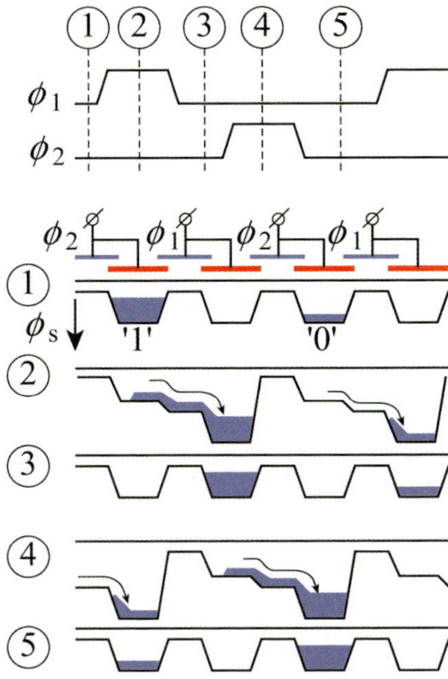

Figure 5.1: *The shift operation in a basic 2-phase SCCD*

A comparison of the time points 1 and 5 in figure 5.1 shows that the charge has been transferred from the first to the third bucket in one complete clock period. In fact, the charge is transferred from one CCD 'cell' to another during a single clock period. Each cell clearly requires two storage elements which each comprise a bucket, a *transfer gate* and a *storage gate*. The two storage elements in a CCD cell are analogous to the master and slave latches in a *D-type flip-flop*. Clearly, the implementation of a 2-phase CCD register comprising 1 Mb, for example, requires 2 million storage elements on a chip. In practice, a better ratio between the number of storage elements and the number of register cells is obtained by using another type of clocking strategy.

The discussion of charge transfer in figure 5.1 is based on the assumption that one bucket was full with electrons and another one was empty. The operation of an SCCD clearly relies on the filling of these buckets. Figure 5.2(a) shows a simplified SCCD comprising some sensor cells and an output section.

In an image sensor photons reach the silicon surface through a lens.

The silicon then converts the photons into electrons locally. A complete image is then captured in an array, which is read out by shifting (transferring) its contents to the CCD array output. The *charge transfer* in an SCCD occurs right at the silicon surface under the gates.

Unfortunately, the surface is inhomogeneous and therefore plagued by *surface states*. These surface states have a certain energy and can trap electrons which have higher energy. During charge transfer, the associated change in surface potential profile causes the surface states to release the trapped electrons. If this occurs before the transfer is complete, then the released electrons will simply rejoin the rest of the electrons in the packet and *'transfer efficiency'* is maintained. However, if an electron is released from a surface state when the transfer is complete, then it cannot rejoin its charge packet. This reduces transfer efficiency. The surface states continue to release the trapped electrons until a new charge packet arrives. The new packet will not be degraded by surface states that are still full when the packet arrives. The empty surface states will, however, be filled by the new packet and the process will repeat itself.

Clearly, transfer efficiency depends on the number of surface states. In previous generations of CCDs, transfer efficiency was increased by using a small charge to represent a '0'. This *'fat zero'* ensures that surface states remain filled. Transfer efficiency is also reduced by incomplete transfer of charge packets at high clock frequencies.

Leakage current accounts for another problem related to CCDs and, of course, to other dynamic memories as well. This *'dark current'* is caused by thermal generation of minority carriers and slowly fills the buckets of a CCD. The result is a *'maximum storage time'*, during which the data in a CCD will remain correct. In addition, dark current causes a fixed noise pattern on the data that is read from a CCD.

Both transfer efficiency and dark current largely determine the operating limits of a CCD. These factors therefore require considerable attention during CCD design.

The above section clearly indicates that surface states form an important limiting factor for the performance of SCCDs. These surface states are unavoidable. Therefore, the only way to improve performance is to realise a CCD in which storage and transfer of charge occurs in a channel which is 'buried' a short distance below the silicon surface. A buried n-channel can be realised by creating a thin n-type layer on top of a p-type substrate. Compare the SCCD and BCCD structures in figure 5.2(a) and (b) respectively.

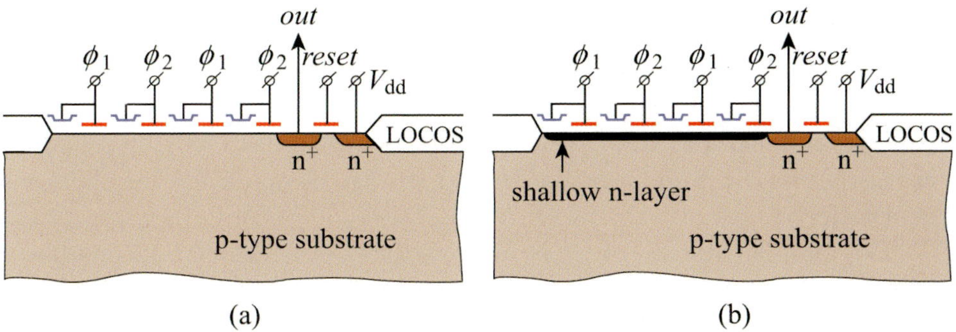

Figure 5.2: *(a) Basic structure of an SCCD and (b) a BCCD*

The operation of an SCCD is closely related to the characteristics of a MOS capacitor with a thick depletion layer. However, the operation of a BCCD is related to the characteristics of a MOS capacitor with a fully depleted layer. Therefore, the first requirement for the successful operation of a BCCD is that the thin n-type layer is fully depleted of electrons. This is achieved by using clock and control signals with an offset voltage. This voltage produces a potential maximum a short distance below the silicon surface. Electrons (representing data bits) injected into the device are stored at this potential maximum. The depleted n-type layer prevents the charge carriers from reaching the surface states and a high transfer efficiency is therefore achieved. The operation of a BCCD is otherwise identical to that of an SCCD.

Buried-channel CCDs were developed for two important reasons. The first is their *immunity* to surfaces states. The second is the increased *operating frequency* which they afford compared to surface-channel CCDs. The increase is caused by the fact that charge is transferred at a speed determined by the bulk mobility instead of the surface mobility. The maximum clock frequency of a BCCD is therefore twice that of an SCCD of equivalent dimensions. However, the definition of the buried channel in a BCCD requires an extra mask. BCCDs are also subject to many problems when their dimensions are reduced. In addition, it is inherently difficult to control the charge in a BCCD because it is stored at a distance from the gate which is longer than for an SCCD. Currently, all image sensor CCDs are implemented as BCCDs. Because of the large number of pixels, a lot of transfers are required. The immunity to surface states then outweighs the disadvantages of BCCDs.

The charge-coupled device principle can be used in both analogue and digital applications. As stated, the bulk part of the applications is in image sensors. Professional cameras now use sensors with over 100 Megapixels. In video camera applications, conventional CCD image ICs consisted of separate sensor and memory parts. Currently the chip only contains a sensor (figure 5.3), which captures the image when the shutter is open and it temporarily acts as a storage device when the shutter is closed.

The main advantage of CCD image sensors over the CMOS imaging devices (discussed in the next section), is the outstanding image quality of the CCDs. Their fabrication technology is optimized with one main goal: imaging performance. For that reason CCDs are still very popular in established markets such as digital still photography, camcorders, but also in high-end markets such as broadcast, astronomy, etc.

Figure 5.3: *Example of a 11 million pixel CCD image sensor ($\approx 9cm^2$)* *(Source: Dalsa BV)*

5.2.3   CMOS image sensors

MOS image sensors already exist since the late 1960s. Due to problems with noise, sensitivity, scalability and speed, CCD sensors became much more popular. In the early 1990s however, CMOS image sensors regained their popularity. The efforts were driven by low-cost, single-chip imaging systems solutions. Today the developments in, and applications of CMOS imaging have intensified so much that complete sessions at the major IC conferences, like IEDM and ISSCC, are devoted to them [1].

Another driving factor for an increased activity in CMOS image sensors is the continuous improvement in CMOS technology. Scaling of the sensor pixel size is limited by both optical physics and costs [2] and occurs at a lower pace than the scaling of the CMOS feature size, see figure 5.4(a). This allows to combine the CMOS image sensor with image processing on a single chip at relatively lower costs.

The ability to capture low-light images depends on the efficiency to convert incoming photons into electrons, which subsequently discharge the pixel capacitor. We distinguish between both passive and active pixels. An *Active Pixel Sensor (APS)* includes an active amplifier in every pixel. Figure 5.4 shows three different pixels. When the pass transistor in figure 5.4(b) is accessed, the photodiode is connected to a bit line. Its charge is converted into a voltage by the readout circuit (amplifier) located at the bottom of a bit line. Due to the small pass gate, this single transistor pixel allows the smallest pixel size and consequently, the highest *fill factor* (ratio of sensor area to total area of sensor plus support electronics). The performance of a pixel was improved by adding active amplifier circuitry to the cell, see figure 5.4(c), resulting in average fill factors between 20% and 30%. The photogate APS in figure 5.4(d), integrates charge under the gate. Its readout architecture looks simular as in CCDs [2].

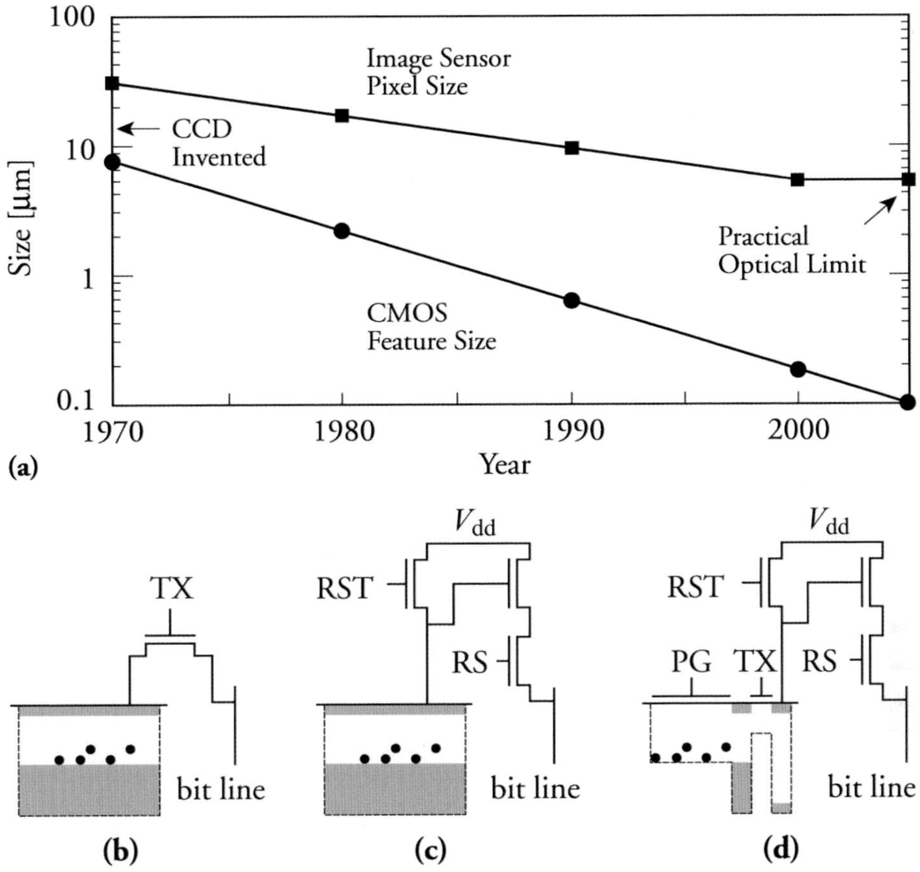

Figure 5.4: *(a) Scaling of MOS pixel and feature size, (b) passive pixel, (c) photodiode active pixel sensor (APS) and (d) photogate APS pixel. (Source: [2])*

The low power consumption, the limited number of external connections and supplies, high level of integration, and low production costs allow CMOS sensors to be used in a variety of applications: multimedia, low cost cameras [3], PC camera, security and machine monitoring, video phone, fax, etcetera. It is expected that further development of CMOS image sensor technology will create completely new imaging markets that were not accessible in the past with CCDs. Moreover, application of CMOS imaging technology is no longer restricted to consumer applications. Examples are the automotive industry and the ambient intelligence applications. More sophisticated and high-resolution imaging applications will become available as CCD and CMOS imagers continue to improve [1]. However, the development of very powerful signal processing chips enable CMOS imaging systems with high image quality. Today's image processing cores can correct for single pixel defects and defect columns without any noticeable effect for the customer.

5.3 Power MOSFET transistors

5.3.1 Introduction

The invention of the bipolar junction transistor in 1947 provided the foundation for modern integrated circuits and power electronic circuits. The first power devices based on semiconductor technology were demonstrated by Hall in 1952. He used germanium stacked junctions to achieve a continuous forward current of 35 A and a punch-through voltage of 200V. Since about 1955, silicon has been preferred for power devices. By 1960, such junctions allowed the implementation of 500 V rectifiers. Currently, silicon rectifiers are available with continuous current ratings of 5000 A and reverse voltages of 6000 V. The application of MOS technology in *power transistors* has been a major focus point for the industry since the late seventies.

The prospects of high speed and high input impedance in many low-voltage applications are particularly attractive. *Double-diffused MOS transistors* were originally introduced during the mid-seventies. The *DMOS transistor* allowed increased performance without reducing the source-drain distance, whilst excessive electric fields were avoided. Originally, the introduction of DMOS power FETs was seen as a major threat to the bipolar power transistor. However, their advantages only render *power MOSFETs* suitable for a limited part of the power electronics application area.

Improvements in technology and yield have resulted in better performance for MOS power transistors. Power MOSFETs can be implemented as discrete devices or can be integrated with other devices on a single chip. Usually the integrated power MOS devices deal with lower voltages and less power consumption than the discrete ones. Breakdown voltages over 1000 V are now possible with discrete devices. The *breakdown voltage* V_{B} of a power MOSFET is related to its typical resistance $(R_{\mathrm{on}} \cdot Area)$. Typical corresponding values might be $(R_{\mathrm{on}} \cdot Area) = 0.1\,\Omega\text{·mm}^2$ at $V_{\mathrm{B}} = 100\,\text{V}$ for a discrete power n-type MOSFET and $(R_{\mathrm{on}} \cdot Area) = 0.15\,\Omega\text{·mm}^2$ at $V_{\mathrm{B}} = 100\,\text{V}$ for an integrated nMOS de-

vice. In practice, power dissipation is limited by the maximum power
rating of the power MOSFET's package. Figures between 100 W and
350 W have been realised for packaged discrete power MOSFETs. Discrete power MOSFETs with die sizes of 200 mm^2 have been reported
in the literature. Large-area low-voltage devices are designed for use as
synchronous rectifiers, replacing diodes in power supplies (e.g., in PCs
and laptops). When the current levels of power devices exceed about
1 A at operating voltages in excess of 150 V, monolithic integration of
the power devices with the rest of the circuit is no longer cost effective.

5.3.2   Technology and operation

All high-voltage devices use a so-called drain extension (as discussed in
chapter 3), which is used as drift zone to distribute the voltage across.
The resistivity of this drift zone strongly depends on the current and
gate voltage.   Most discrete power MOSFETs use vertical drain extensions, while integrated power MOSFETS usually apply lateral drain
extensions.

This section will first focus on the technology aspects of discrete
power devices and then continue with an example technology for the
integration of power MOSFETs with analog and digital circuits onto
one chip.

The high-voltage *vertical double-diffused MOS (VDMOS)* transistor
shown in figure 5.5 is an example of a discrete power MOSFET.

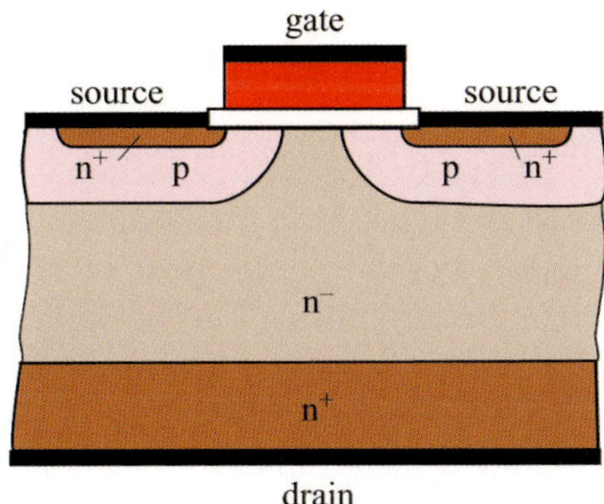

Figure 5.5: *Cross-section of a VDMOS discrete power transistor*

The threshold voltage of the above VDMOST is generally 2 to 3 Volts.

When the gate voltage is increased from $0\,\mathrm{V}$ to about $12\,\mathrm{V}$, the p-well area near the surface is inverted (see section 1.3). A channel then exists between the n^+ source and the n^- epitaxial layer. The charge carriers will flow *vertically* from the source to the drain when a high voltage is applied to the latter. The drain voltage can vary from $50\,\mathrm{V}$ to $1000\,\mathrm{V}$, depending on the application. VDMOS transistors are usually n-type rather than p-type because of their higher channel mobility. Because of the scaling of the gate oxide thickness, devices with a gate voltage below $5\,\mathrm{V}$ and a threshold voltage below $1\,\mathrm{V}$ have become available as well.

A power MOSFET is a regular array of n transistors connected in parallel, with an equivalent resistance (RDS_on) equal to $1/n$ of the individual transistor's RDS_on. So, a larger die results in a lower on-resistance, a larger parasitic capacitance and so in a reduced switching speed. Therefore there is a continuous drive to minimize the size of the individual transistor in order to reduce the on-resistance while keeping the capacitances constant. The use of the trench technology in power MOSFETs has resulted in significant improvements of on-resistance for low voltage devices. Instead of being planar, the gate structure is now built in a trench, which may reduce the transistor area by about 50%, depending on the operating voltage range.

Figure 5.6 shows a schematic and a SEM photograph of a cross-section of a trench MOSFET [4] for applications with a voltage range up to $100\,\mathrm{V}$. For these applications, transistors are used in which the gate is incorporated in a trench. The way to reduce the on-state resistance of these devices is by increasing the cell density. This is done by using innovative integration technologies. Transistor densities as high as 640 million cells per square inch are available in recent power MOSFET trench technologies.

Power MOSFETs, integrated with a mixture of analog and digital circuits, are often realised with a kind of *Bipolar-CMOS-DMOS (BCD)* process. This technology combines the best of the three worlds and enables products that are used in a variety of applications, e.g., mobile phones, motor drivers, automotive bus transceivers and LED drivers. They can handle drain voltages up to $150\,\mathrm{V}$, while they use a gate oxide thickness of around $12\,\mathrm{nm}$, which allows gate voltages close to $5\,\mathrm{V}$.

Some high-end automotive applications need optimum isolation to limit mutual interference of the different circuits (analog, digital and high voltage) and are fabricated on an SOI substrate [4].

Other applications may use the same devices and operate with the same operating voltages, but do not require the full isolation. They are then fabricated on bulk wafers, which are cheaper.

Most of these BCD technologies are currently using the $0.35\,\mu$m node for volume production. Over the last couple of years vendors are porting their products more towards the 180 nm and 140 nm nodes. Because high voltages don't scale for the category of products that use voltages between 100 V to 150 V, it is not expected that these products need further scaling to beyond 100 nm, unless they are integrated with large amounts of mainstream CMOS logic.

The 20 V to 50 V category BCD products, such as power management units, e.g., to drive the LEDs in a mobile phone, are made on bulk wafers. The power MOSFETs in these devices can carry 20V to 50V on the drain, but also only about 5 V on the gate. These products run in high volumes in the $0.25\,\mu$m node, but there are currently also example products that run in the 130 nm node [5].

The trend in this category is to port the products to the 65 nm node, while maintaining the high voltages on the drain, but with a reduced gate voltage of 2.5 V, which is equal to the commonly used I/O voltage standard.

The drive for smaller form factors and reduced system costs will force the power device products toward low-cost *high-voltage CMOS (HV-CMOS)* technologies, which begin to offer functionality comparable to BCD technologies but with much less process complexity. 40 V LCD driver products are already produced in high volumes in a HV-CMOS process. It is expected that other applications, such as printer head drivers and bus transceivers, will soon follow [6].

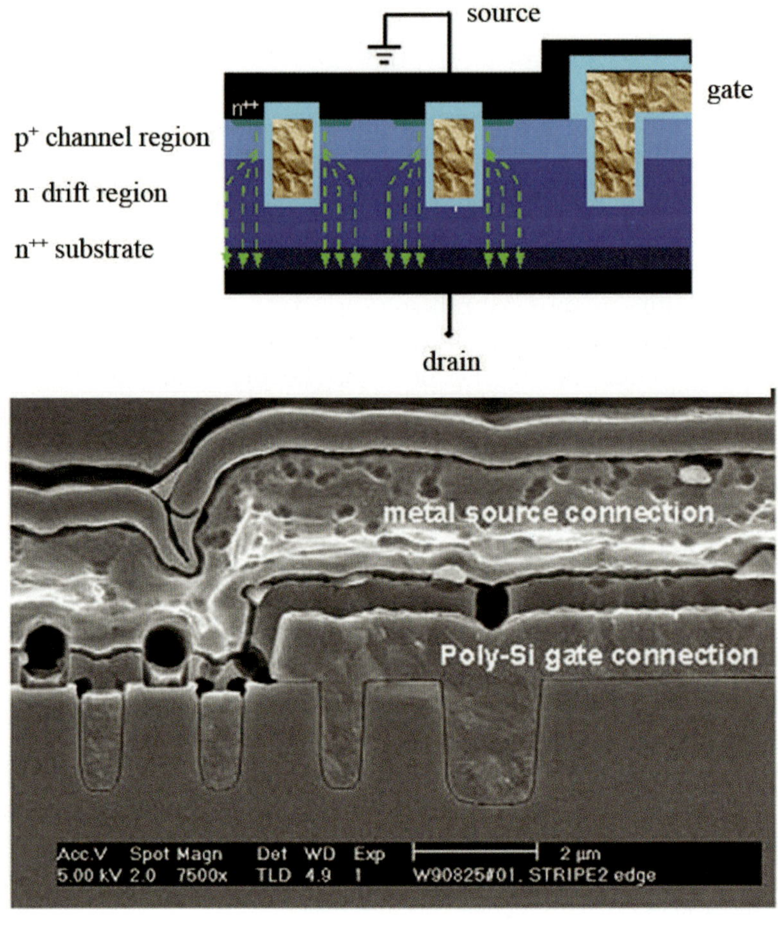

Figure 5.6: *Schematic cross-section, showing the electron flow in the on-state (positive voltage on the gate), and a SEM photograph of a trench MOSFET (Source: NXP Semiconductors)*

5.3.3   Applications

Power MOSFETs have a relatively long history as discrete power switches in fluorescent lamp ballasts, switch-mode power supplies, and automotive switching applications. In electrical shavers, they are used both in the form of discrete devices and as part of larger integrated circuits, e.g., automatic supply voltage adaptors and battery chargers. Their high current capability makes power MOSFETs suitable for use in driver circuits, e.g., for stepper motors. On the other hand, plasma display drivers, for example, only require relatively small driving currents.

Power MOSFETs are easily integrated in bipolar and BICMOS circuits because they do not inject minority carriers. The combination of low-voltage bipolar transistors and high-voltage lateral DMOS transistors of both n and p types facilitates production of analogue high-voltage circuits [7]. Examples include video output amplifiers [8] and [9].

In the world of today, electronic devices are more and more getting mobile. Obvious examples are phones, laptops, PDA's, etc. The absence of direct connection to the mains, as well as the ever increasing demands on environmental friendliness put strong demands on the power consumption of the circuitry in these devices. Large operating times and low energy consumption when charging the battery both require the use of efficient power converters. Power MOSFETs are also key components in these converters. Improving these components implies the search for the best trade-off between the off-state breakdown voltage and the power losses. For a given breakdown voltage, which is determined by the application, these power losses should be as low as possible, in order to make the power converters as efficient as possible.

From the above we can conclude that there are several important factors in the development of a power MOSFET technology: device architecture, device density, current capability, on-resistance, break-down voltage, etc. The order of priority, however, depends on the application area.

5.4 BICMOS digital circuits

5.4.1 Introduction

Since the mid-eighties, a growing interest in BICMOS technologis has resulted in a lot of commercially available ICs. The *BICMOS technology* facilitates a combination of both bipolar and CMOS devices on a single IC and enables the simultaneous exploitation of the advantages of both device types.

The penalty of more complex processing restricted the use of BICMOS technologies to fairly specialised applications. It is estimated that a BICMOS wafer after full processing will cost 20% to 30% more than a CMOS wafer. In several applications, this price increase will be offset by the performance enhancement. Performance characteristics of BICMOS devices and their technology are explained below. Future expectations and market trends are also discussed.

5.4.2 BICMOS technology

There are several ways of obtaining a BICMOS process. It could, for instance, be based on an existing bipolar process or a completely new BICMOS process could be developed. The conventional approach, however, was to start from a CMOS process. An associated advantage was that existing CMOS design and CAD tools could then be used for BICMOS designs. A BICMOS process based on an n-well CMOS process is considered here. This is a logical choice because of the considerable similarities between this BICMOS process and the n-well CMOS process discussed in chapter 3.

The development of the BICMOS process from an n-well CMOS process is explained with the aid of the cross-sections in figure 5.7. The source and drain implants are typically less than a tenth of a micron deep. The depth of the n-well is less than a micron. The realisation of an npn transistor requires an additional p-type implant in the n-well. This implant forms the base of the npn transistor and is shown in figure 5.7.

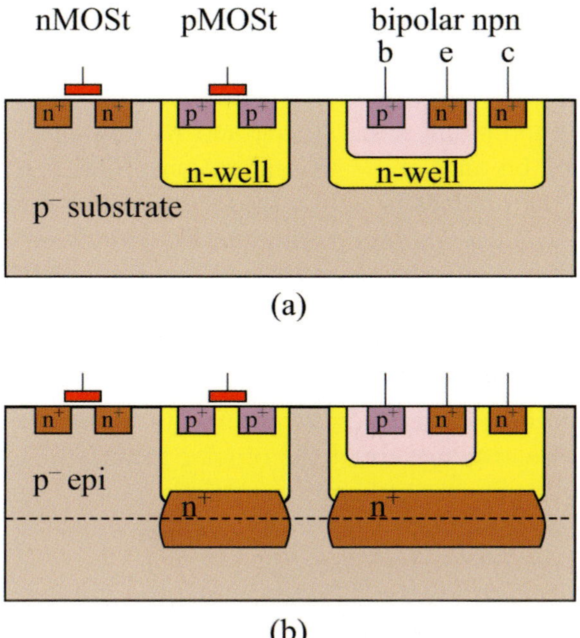

Figure 5.7: *The development of a BICMOS process from an n-well CMOS process*

The npn transistor exhibits a relatively high collector series resistance. This is also the base resistance of the pnp transistor in the *parasitic thyristor*, as discussed in chapter 9 (latch-up). This 'simple' structure is therefore rather susceptible to *latch-up*.

The above disadvantages are largely overcome when the structure shown in figure 5.7b is used. In the associated process, n^+ implants are created in the p-type substrate prior to the growth of a p^- epitaxial layer. The resulting 'buried layer' n^+ areas subsequently become part of the n-wells. The npn transistor obtained in this process is basically isolated and latch-up via the parasitic pnp transistor is largely prevented by the n^+ buried layer. The creation of the buried collector areas and the base implant requires two more masks than in a standard n-well CMOS process.

In the late eighties and early nineties, BICMOS was also used for digital circuit categories that needed to operate beyond the performance limits of CMOS. The temporary increase in market volumes around 1996 was caused by falling MOS memory prices (of DRAMs in particular) and to the growing high-end market for microprocessors (high demand for

BICMOS-based Pentium^{TM} chips), ASICs and SRAMs. At voltages below 2.5 V, the performance gap between bipolar and CMOS gradually narrowed, in favor of full-CMOS products. Over the years, as a result of intensive R&D, more and more analog and RF functions have been realised in CMOS, because of cost considerations. This had reduced BICMOS usage in the nineties to only mixed-signal and RF-circuits. Growing demands for multi-Gb data communication and wide-bandwidth radio communication systems caused renewed interest in BICMOS technology. This has put stringent requirements on the technology for providing sufficient performance at affordable cost. A combination of SiGe heterojunction bipolar transistors (HBT) and CMOS was the most promising technology solution to meet these requirements. It needed full optimisation of the device architecture of the bipolar transistor: a shallow-base, a dedicated Ge profile in combination with low-power concepts and low-cost solutions [10]. Figure 5.8 shows an example cross section of a SiGe HBT for excellent RF performance [11,12].

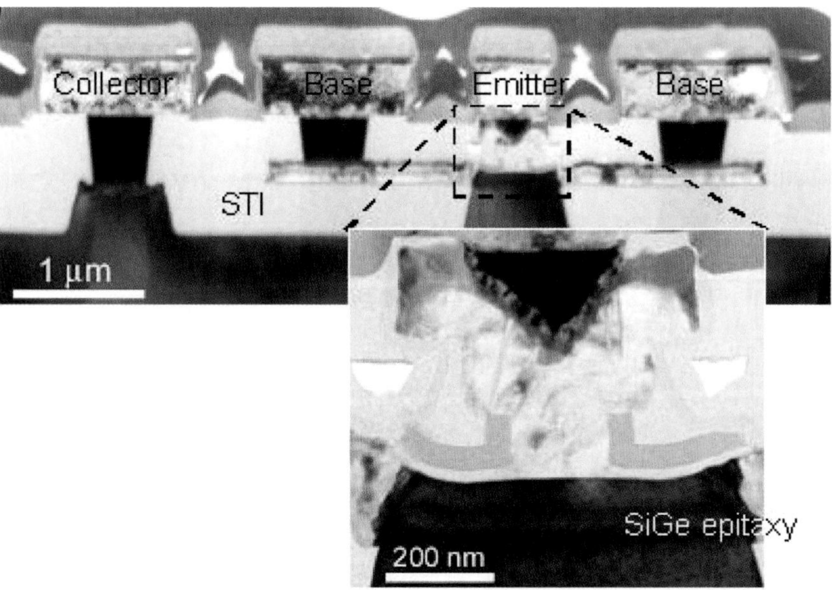

Figure 5.8: *Transmission Electron Microscope (TEM) picture of a SiGe HBT device cross section, showing the Emitter, Base and Collector metal connections and the shallow trench isolation (STI). The inset shows a magnification of the actual intrinsic device, with an epitaxially grown SiGe base layer as the engine for excellent RF performance. (Photo: NXP Semiconductors)*

The Ge profile plays a dominant role in the characteristics of the SiGe
HBT device. It is obvious that a SiGe BICMOS technology would allow
to combine the performance and density requirements for the above-
mentioned communication systems [13,14].

Some vendors offer a BICMOS process which has built-in options
that allows a more optimal integration of passives (resistors, capacitors
and inductors) for RF system-in package products [15].

5.4.3   BICMOS characteristics

Its higher *gain factor* and lower *noise* generally renders bipolar technology more suitable than CMOS for analogue applications. However,
CMOS is more attractive for digital control, storage and signal processing circuits because of its low quiescent power, reasonable performance
and high packing density. The mixture of the two technologies offers
unique possibilities in both analogue, digital and mixed analogue/digital
applications.

BICMOS was first introduced in digital I/O circuits, where it provided increased output driving capability. It was subsequently applied
in the peripheral circuits of SRAMs to shorten the access times. These
circuits included sense amplifiers, word line and bit line drivers.

Low-voltage bipolar transistors and high-voltage lateral DMOS transistors, incorporating both n-type and p-type channels, are combined
in some BICMOS processes. These processes allow the integration of
truly analogue high-voltage circuits, such as the video output amplifiers
mentioned in section 5.3.3.

The previously-mentioned applications of BICMOS technologies illustrate their potential benefits. However, in addition to the increase in
costs compared to an average CMOS technology, there are other draw-
backs associated with BICMOS. For instance, the CMOS digital parts
of a BICMOS chip may generate considerable transient noise on the
supply and ground lines. This 'bounce' is discussed in chapter 9. Considerable efforts are required to prevent it from entering analogue parts
of the chip. Moreover, the reduced density of BICMOS logic limits its
usage to critical functions on a VLSI chip. This reduces the potential
performance advantage. The commercial use of BICMOS technology
for digital ICs is therefore only justified when the additional costs are
compensated by increased performance.

5.4.4 BICMOS circuit performance

BICMOS logic gates usually employ CMOS transistors to perform the logic function and bipolar transistors to drive the output loads. The two typical BICMOS implementations of a *NAND gate* shown in figure 5.9 illustrate this two-stage structure.

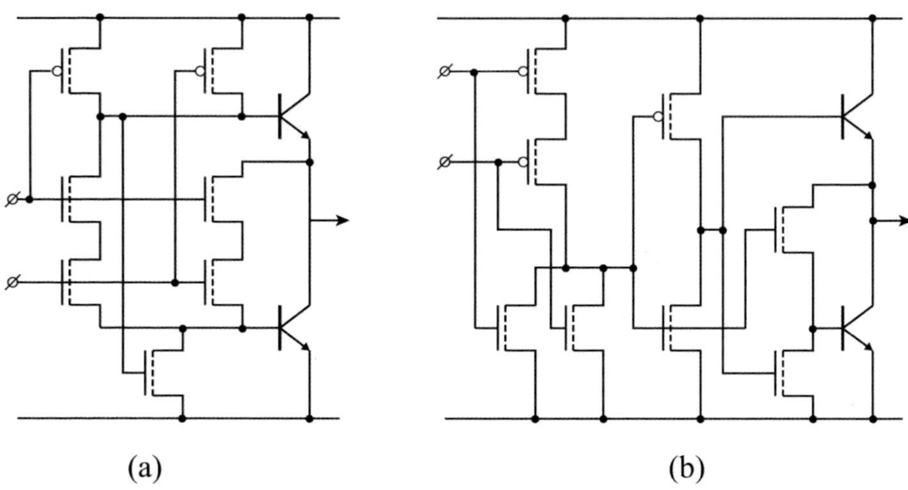

(a)   (b)

Figure 5.9: *Typical BICMOS implementations of a NAND gate*

The two-stage structure of a BICMOS logic gate leads to a larger propagation delay for an unloaded BICMOS gate than for its CMOS counterpart. The performance advantage of a BICMOS implementation over a CMOS implementation therefore only applies in the case of gates with larger fan-outs. Figure 5.10 shows a frequently published comparison of the propagation delay as a function of fan-out for typical CMOS and BICMOS NAND gates. The comparison was made for nMOS and pMOS transistor widths of $4\,\mu$m and $7\,\mu$m, respectively, in a process with a $0.35\,\mu$m gate length. The cross-over point lies between a fan-out of two and three. For higher fan-outs, the performance of a BICMOS circuit is better.

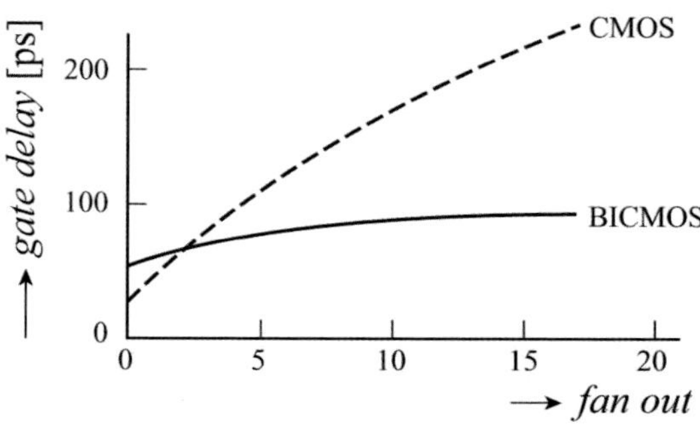

Figure 5.10: *Gate delays of typical CMOS and BICMOS NAND gates*

Figure 5.10 suggests that BICMOS is preferable to CMOS even for relatively low fan-outs. For large capacitive loads, the figure shows that the propagation delay can be reduced by a factor of 2.5 when BICMOS is used. However, the presented comparison does not account for the extra area required by the driver stage in the BICMOS implementation. A more representative comparison is obtained when the CMOS logic gate is also equipped with a CMOS output driver. The resulting comparison is shown in figure 5.11 for BICMOS and CMOS NAND gates implemented as NOR gates followed by bipolar and CMOS drivers, respectively. Such a comparison shows a dramatic reduction in speed advantage and reveals that BICMOS only affords a small performance improvement for gates with a high fan-out. In practice, this means that implementation of logic gates in BICMOS is not cost effective for low to medium speed applications. Its usage in VLSI circuits and *Application-Specific ICs (ASICs)* is therefore limited to circuits that have to drive large capacitances, e.g., driver and I/O circuits. BICMOS is also used in ICs that have to operate beyond the performance limits of CMOS.

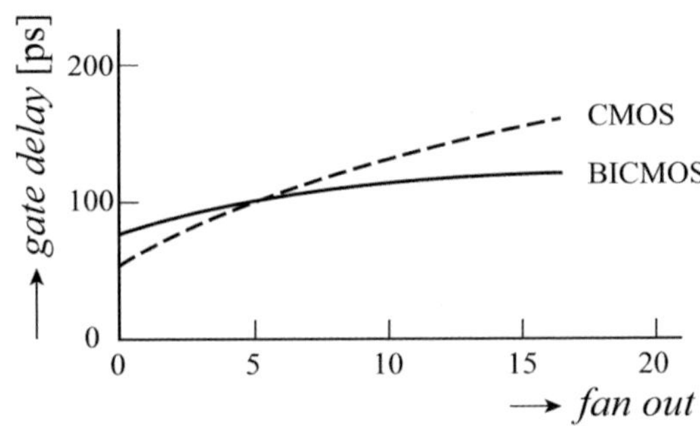

Figure 5.11: *Propagation delays of CMOS and BICMOS NAND gates implemented as NOR gates with CMOS and bipolar drivers, respectively*

Supply voltage dependence, temperature dependence and process parameter dependence are also important factors that must be included in a comparison of the performance of BICMOS and CMOS circuits. These factors are compared below.

CMOS current levels are quadratically reduced when the supply voltage is reduced. This results in a reduction of the speed of both CMOS and BICMOS circuits. Bipolar circuits, however, are also hampered by inefficient operation at lower supply voltages. Manufacturers of BICMOS ICs will therefore face a dilemma when supply voltage standards below 1.8 V become more accepted as minimum feature size decreases to below 180 nm. Innovative design may reduce this dilemma.

The influence of temperature on the performance of CMOS and BICMOS circuits is closely related to the different origins of transistor currents. In bipolar transistors, the current is caused by diffusion. This current is less affected by temperature than the MOS transistor drift current discussed in section 2.3. As a consequence, the switching speed of BICMOS is less dependent on temperature than that of CMOS.

It has been empirically found that variations in CMOS parameters caused by processing spread have a greater influence on circuit performance than variations in bipolar process parameters.

Finally, it should be noted that a BICMOS driver implementation shows a reasonable power dissipation advantage over a CMOS driver.

It is clear that the application of BICMOS technology is not trivial. This explains its limited application in semiconductor products.

5.4.5 Future expectations and market trends

From a performance point of view, the future for BICMOS technologies originally looked promising. However, a fair comparison of BICMOS and CMOS circuit performance reveals that the advantages afforded by BICMOS are really only significant in mixed analog/RF/digital circuits.

The major issue with BICMOS, as compared to a combination of separated bipolar and CMOS ICs in one package, is costs. But also such aspects as performance, power consumption and integration density are essential to determine the overall system benefits. Intensive research on performance improvements has led to highly complex HBT device architectures with f_T and f_{max} values of 300 GHz and 350 GHz, respectively [14]. As long as BICMOS enables the integration of different functions onto a single die in a cost-effective way, it will remain as an effective platform for those systems that require performance beyond the limits of CMOS.

5.5 Conclusions

A number of devices and technologies that can be used in both purely digital as well as mixed analogue/digital ICs are discussed in this chapter. Because this is the only link between the presented topics, no general conclusions are presented here. The reader is therefore referred to the application sections associated with the CCD and MOS power transistor topics and the section on future expectations and market trends associated with the BICMOS topic.

5.6 References

[1] 'Image Sensor' Session at the ISSCC conferences:
ISSCC Digest of Technical Papers, 2000 and onwards

[2] Eric R. Fossum,
'CMOS Image Sensors: Electronic Camera-On-A-Chip',
IEEE Transactions on Electron Devices, Vol. 44, October 1997

[3] Kwang-Bo Cho, et al.
'A 1/2.5 inch 8.1 Mpixel CMOS Image Sensor for Digital Cameras',
ISSCC Digest of Technical Papers, 2007, pp 508-509

[4] F. Udrea, et al.
'SOI-based devices and technologies for high voltage ICs',
BCTM, 2007, pp 74-79

[5] H. Yeates, et al.
'Single chip for mobile phone has low component count',
Electronics weekly, October 26th, 2005

[6] H. Gensinger,
'High-Voltage CMOS Technologies for Robust System-on-Chip Design',
HVCMOS_FSA Forum, June 2006

[7] A. Ludikhuize, 'A versatile 250/300V IC process for Analog and Switching Applications',
IEEE Trans. on Electron Devices, Vol. ED-33, pp 2008-2015, December 1986

[8] P. Blanken, P. van der Zee,
'An integrated 8MHz video output amplifier',
IEEE Trans. on Consumer Electronics, Vol. CE-31, pp 109, 1985

[9] P. Blanken, J. Verdaasdonk,
'An integrated 150 V_{pp}, 12kV/μs class AB CRT-driving amplifier',
ISSCC, Digest of Technical Papers, 1989, New York

[10] K. Washio, et al.,
'SiGe HBT and BiCMOS Technologies',
IEDM, Digest of technical papers, session 5.1.3., 2003

[11] J. Donkers, et al.,
'Vertical Profile Optimisation of a Self-Aligned SiGeC HBT Process with an n-Cap Emitter',
IEEE/BCTM, 2003

[12] J. Donkers, et al.,
'Metal Emitter SiGe:C HBTs',
IEDM, Digest of technical papers, 2004

[13] L.J. Choi, et al.,
'A Novel Isolation Scheme featuring Cavities in the Collector for a High-Speed $0.13\,\mu$m SiGe:C BiCMOS Technology',
SiRF, 2007

[14] M. Khater, et al.,
'SiGe HBT technology with $f_{max}/f_T = 350/300\,$GHz and gate delay below $3.3\,$ps',
IEDM, Digest of technical papers, 2004, pp 247-250

[15] P. Deixler, et al.,
'QUBiC4plus: a cost-effective BiCMOS manufacturing technology with elite passive enhancements optimized for 'silicon-based' RF-system-in-package environment',
Bipolar/BiCMOS Circuits and Technology Meeting, 2005, pp 272 - 275

5.7 Exercises

1. A dynamic shift register can be implemented as discussed in the chapter on CMOS circuits. It can also be implemented as a charge-coupled device (CCD). What are the main differences between the former implementations and the CCD implementation? State advantages and disadvantages to support your answer.

2. Assume that the transfer of a logic '1' through an SCCD is represented by a full charge packet. Explain what happens if the temperature increases when a series of data bits consisting of a hundred '1's, one '0' and again a hundred '1's, i.e., 111...1111011111...111, is transferred through the device.

3. Explain the main differences between a low-voltage MOS transistor which operates at $1.2\,\mathrm{V}$ and a power MOSFET.

4. Explain why BICMOS circuits exhibit a longer propagation delay than their CMOS counterparts for small capacitive loads and a shorter propagation delay for large capacitive loads.

5. Explain why BICMOS circuit performance relatively reduces with technology scaling, compared to CMOS circuit performance.

Chapter 6

Memories

6.1 Introduction

Memories are circuits designed for the storage of digital values. In a computer system, memories are used in a large variety of storage applications, depending on memory capacity, cost and speed. Figure 6.1 shows the use of memory storage at different hierarchy levels of a computer system.

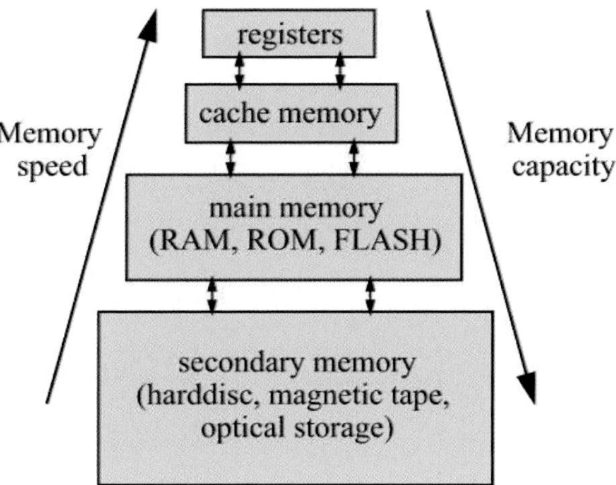

Figure 6.1: *Memory hierarchy in a computer system*

The high-level memories directly communicate with the computer's processor. They must deal with its high data communication bandwidth and therefore need high performance, but they are expensive. As we move down the hierarchy, both the memory capacity and the access time increase, resulting in a reduction of the cost per bit. A cache memory acts as an intermediate storage between the CPU and the main memory and stores the most-frequently and/or most-recently used data and instructions for fast access.

A memory may constitute a single IC or be part of a larger IC. These types are referred to as *stand-alone* and *embedded* memories, respectively.

The digital values in a memory are each stored in a '*cell*'. The cells are arranged in a *matrix* or *array*, which affords an optimum layout.

Memories that lose their data when power is removed are referred to as *volatile*. Memories that retain their data in the absence of power are called *non-volatile* memories. The '*data retention time*' of a memory is the period for which it keeps its data when the supply voltage is removed. A finer division of memories yields the following four types:

- Serial memory;

- Content-addressable memory (CAM);

- Random-access memory (RAM);

- Read-only memory (ROM).

Figure 6.2 presents an overview of the various implementation possibilities for memories. This figure also shows the respective market shares in 2005. The increased market share gained by the DRAMs is mainly the result of the rise of new high-speed architectures, which make them particularly suited for the growing high memory bandwidth applications such as games, video and graphics applications, and printers, etc.

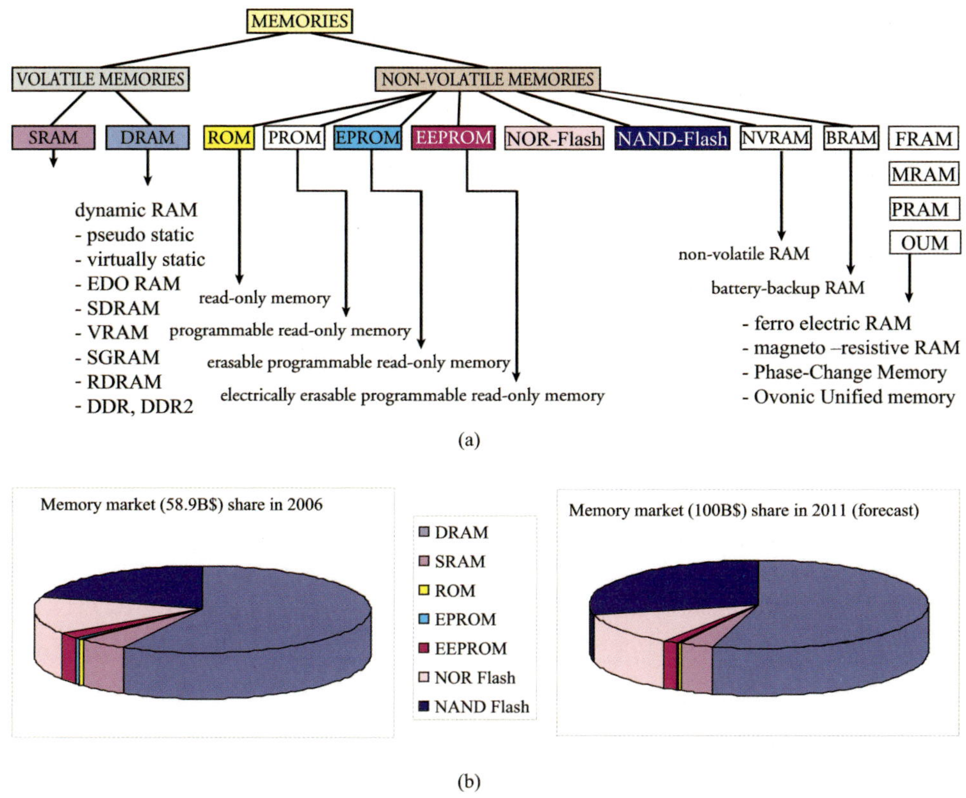

Figure 6.2: *(a) Overview of different types of memories (b) Relative memory market shares in 2006 and expected market shares 2011 (Source: WSTS, IC Insights)*

Volatile memories include *'static'* and *'dynamic'* RAMs. Electrical feedback in the memory cell of a *static RAM (SRAM)* ensures that voltage levels are maintained and data is retained as long as the power supply remains. The data in a *dynamic RAM (DRAM)* memory cell is stored as a charge on a capacitor. Gradual leakage necessitates periodic refreshing of the stored charge. A dynamic RAM that internally refreshes its own data automatically is called a *pseudo-static* or *virtually static* RAM.

The cells in serial memories form one or more shift registers, which can each store a 1-bit data stream. The 'first in, first out' *(FIFO)* operation of shift registers ensures that data enters and leaves a serial memory in the same sequence. Examples of their use include delay lines in video applications.

The cells in a content-addressable memory may contain an SRAM cell plus additional comparison circuitry, because an access is not based on offering a dedicated address, but on the comparison of input data bits with stored data bits. When a match occurs, the corresponding output data bits are returned.

The cells in a RAM or ROM array must have individual unique 'addresses'. Alternatively, they may be connected in parallel groups. In this case, each group or 'word' has a specific address. The capacity of a RAM or ROM that is divided into words is specified by the number of words and the number of bits per word. Examples are $1\,Gb{\times}4$, $512\,Mb{\times}8$ and $256\,Mb{\times}16$. These three specifications all refer to a $4\,Gb$ memory, which can store over 32,000 newspaper pages or 9 hours of MP3 music. The quantification of bits in this chapter is according to international conventions: 1 Gb equals 1 gigabit (1 Mb=1 megabit, etc.), 1 GB equals 1 gigabyte (1 MB=1 megabyte, etc.). The word *byte* is a short notation for *by eight* and so it contains 8 bits.

The data in a ROM can only be read, whereas the data in a RAM can be written and read. The sequence in which data is read from a ROM or RAM is unrestricted. Therefore, *access* is in fact *random* in both cases. The term RAM, however, is generally only used to refer to memories that allow reasonably high frequency read and write operations at random locations.

A RAM requires both data and address inputs and data outputs. Figure 6.3 is a general schematic representation of an addressable memory.

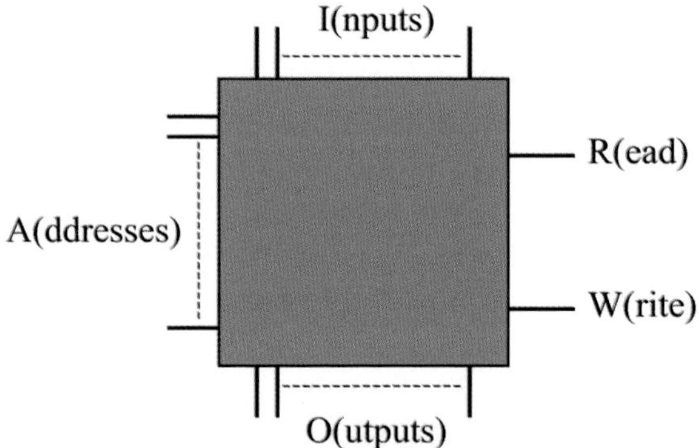

Figure 6.3: *General representation of a memory*

The memory shown is obviously a RAM. The read (R) and write (W) inputs are often combined in one single input which controls the mode of operation. A ROM requires no data inputs but does require address inputs and data outputs. The schematic of a ROM is therefore obtained if the data (I) and W inputs in figure 6.3 are removed. The schematic of a serial memory is obtained if the address inputs are removed. Flash memories have either random access (NOR flash) or serial access (NAND flash) (see section 6.5.4).

The 'access time' of a memory is the time interval between the initial rising clock edge in a read operation and the moment at which the data is available at the output terminals. The 'cycle time' of a memory is the minimum possible time between two successive accesses. The cycle time of an SRAM may be greater than, smaller than or equal to its access time, while the typical cycle time of a DRAM is about twice the access time. This is because the accessed cells in a DRAM must be refreshed after each read and write operation. Although access times are often used for comparison of the different memories available from different manufacturers, cycle time comparison would be better for benchmarking purposes. There are many techniques that improve the access and cycle times of DRAMs. These are discussed in section 6.4.4.

The various types of memories are discussed in this chapter. Their operation and properties are explained and possible applications are given. A brief discussion of the structure of a simple 4 kb SRAM provides considerable insight into memory operation.

6.2   Serial memories

Serial memories are usually encountered in the form of static or dynamic shift registers. Modern *video memories* are an important exception. These memories are serial by nature and random access is therefore often not required. However, they are implemented as DRAMs, in which the cells are serially accessed. Many of these memories include a buffer memory, such as a *FIFO* (first-in first-out) or *LIFO* (last-in first-out) to change the sequence of the data bits. Serial memories are used in video and graphics applications. Such a memory is sometimes called a *video RAM* or *VRAM* (see section 6.4.4).

Small serial memories may be implemented using the CMOS shift register cells presented in chapter 4. The discussions on shift registers in chapter 4 makes further elaboration on serial memories unnecessary.

6.3 Content-addressable memories (CAM)

In a *content-addressable memory (CAM)*, also called *associative memory*, each cell has its own comparison circuit to detect a direct match between the search bit and the stored bit. Instead of supplying a memory address, in a CAM, an input word is supplied and a search operation is executed through the whole memory within one clock cycle. The response to a search is a list of one or more addresses, depending on a single or multiple match of the stored data in the memory. Some CAM memories directly return the full corresponding data contents on those addresses. A more detailed description of a CAM is beyond the scope of this book. A rather complete tutorial on CAM can be found in [1].

6.4 Random-access memories (RAM)

6.4.1 Introduction

Random-access memories can be subdivided into the two following classes:

- Static RAM (SRAM);

- Dynamic RAM (DRAM).

These two types of RAM are discussed separately below. The basic operation of a RAM is explained with the aid of a 4 kb SRAM. A subsequent discussion of the major differences between SRAMs and DRAMs illustrates the considerable difference in their operation.

6.4.2 Static RAMs (SRAM)

A true static memory is characterised by the time between a change on its address inputs and the presence of valid bits at its data outputs. Dynamic memories often require a considerably more complex pulse pattern with very stringent timing requirements.

SRAM block diagram

For most stand-alone SRAMs, every possible combination of address inputs can be decoded. A memory with n address inputs therefore contains 2^n addresses. An SRAM with twelve address inputs, for example, therefore has at least 4096 memory words. Figure 6.4 shows the block diagram of such a 4 kb SRAM. This example represents a so-called by-1 memory (4k by-1), meaning that at each address selection only one cell (one bit) is accessed. This has been done for educational purposes. In most memories a complete word, which may contain 4, 8, 16, 32, or even 64 bits, is stored at one address.

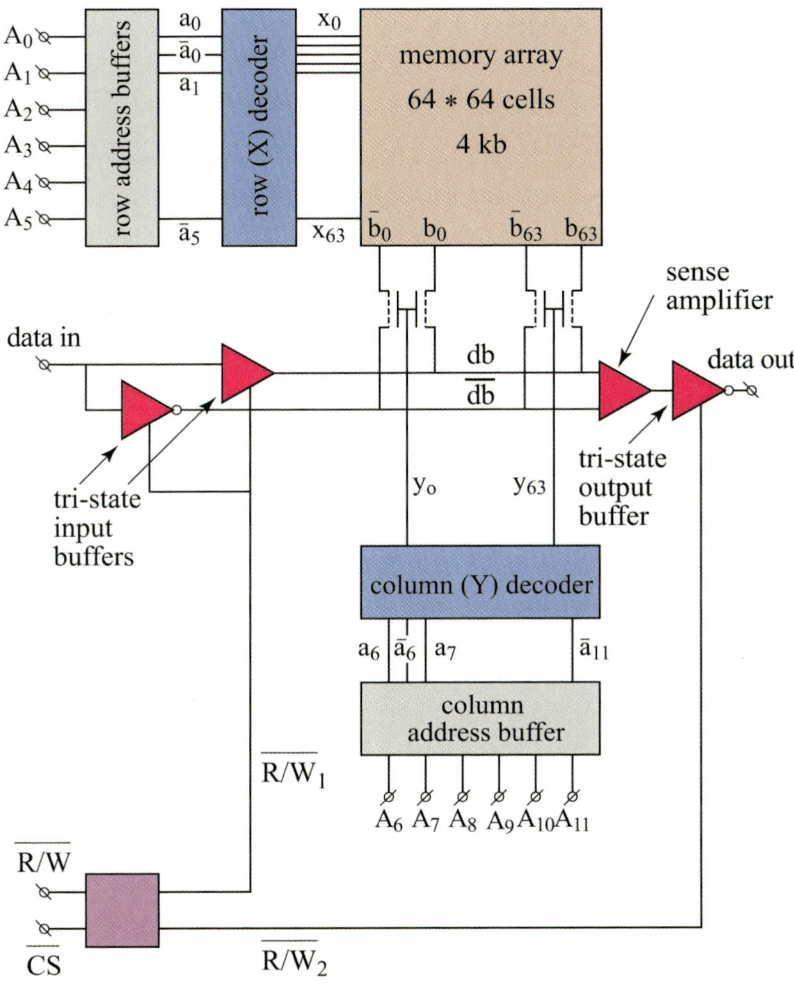

Figure 6.4: *Block diagram of a 4 kb SRAM*

Its 4096 memory cells are organised in an array of 64 rows and 64 columns. Each row and column can therefore be addressed by 6 address inputs. In addition to an array of memory cells, an SRAM also requires control logic circuits. These circuits will now be described.

- A *row decoder* selects the 'word line' x_i of the row in which the addressed word (or cell) is located. The row decoder is also known as an *x-decoder*.

- A *column decoder* selects the 'bit line select' line y_j of the column in which the addressed word (or cell) is located. The column decoder is also known as a *y-decoder*. The addressed cell is located at the point of intersection of the selected row and column and is referred to as cell x_i,y_j. The y_j signal selects the *bit lines* b_j and $\overline{b_j}$ of the addressed cell.

- *Address buffers* connected to the address inputs drive the row and column decoders. The output lines of the row and column address buffers traverse the length and width, respectively, of the array. They therefore form large capacitive loads for the address buffers.

- The tri-state *data input buffers* drive data buses db and $\overline{db}$ when the memory is being written. These buffers drive the large capacitive load of the data bus line and the selected bit line. They must also be capable of forcing the memory cell into another logic state. Current memories have separated data bus drivers and bit line drivers.

- A *sense amplifier* detects the contents of the selected cell via the complementary bit lines b_j and $\overline{b_j}$ and data bus lines db and $\overline{db}$. The detection must occur as fast as possible so that the access time is reduced to a minimum. The sensitivity of the sense amplifier may be as low as 70 to 100 mV. Current sensing and differential voltage sensing are alternative techniques for optimized memory performance. Because of the reducing voltage headroom, current sensing becomes less popular, because it consists of more analogue circuitry. The 4 kb SRAM in this example only includes one sense amplifier. Preferably one sense amplifier per column should be used, but it does no longer fit in the memory cell pitch. Therefore, today's memories may include one sense amplifier for every four columns combined with a multiplexer circuit for selection.

- The tri-state *data output buffer* transfers the data from the sense amplifier to the SRAM output when the memory is being read. Because memories are often used in a memory bank, it must be possible to enable one single memory, while others are disconnected from the bus. This requires a high-ohmic third state of the output (tri-state).

The SRAM control signals

The control signals required in an SRAM are described below. For the sake of simplicity, the commonly-used *output enable (OE)* signal is omitted.

- The *write enable* $(\overline{WE})$ signal determines whether data is written to the selected cell or read from it. During writing, the bit line signals are derived from the input. Depending on the data to be stored, only one of the two bit lines is pulled to ground, while the other is kept at V_{dd} level by the precharge transistors. During reading, the cell pulls one of the bit lines low and the bit line signals are then transferred to the output.

- The *chip select* $(\overline{CS})$ signal facilitates selection of a single SRAM when many are combined to form a large memory system. Such a system consists of one or more *memory banks*. The memories in such a system may be connected to common address and data buses. Although more than one memory (or even a complete bank) can be selected at the same time, only one at a time can put data on the data bus. The $\overline{CS}$ signal of the relevant memory is activated by decoder logic in the memory bank. This logic produces 'high' logic levels on the $\overline{CS}$ inputs of the remaining memories. Their output buffers are therefore placed in the high-impedance mode and the memories are isolated from the data bus.

Normal memory architectures are by 2, by 4, by 8, etc., meaning that 2, 4, respectively 8 memory arrays can be accessed simultaneously. Figure 6.5 shows the physical representation of a x8 RAM architecture.

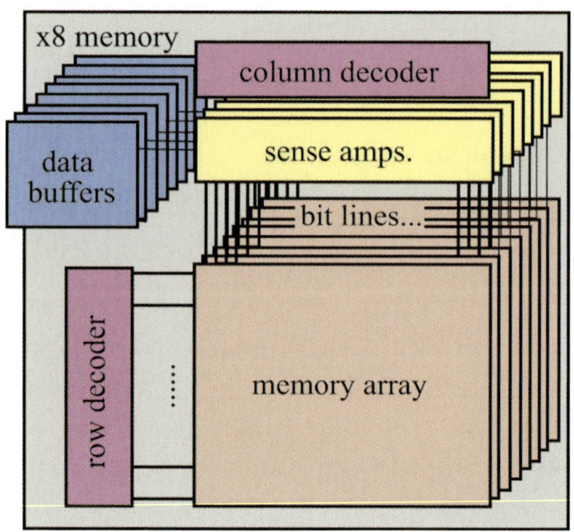

Figure 6.5: *Physical representation of a x8 RAM architecture*

The more parallel accesses a memory allows, the higher the communication bandwidth to interfacing CPUs.

The SRAM read operation

The read operation performed in an SRAM is explained with the aid of an example in which data is read from the cell x_{35}, y_{40}, see figure 6.4. The steps required to achieve this are as follows:

- The word line x_{35} is activated by placing the decimal value 35 on address inputs A_5 to A_0: $A_5 A_4 A_3 A_2 A_1 A_0 = 100011$.

- The bit line select signal y_{40} is activated by placing the decimal value 40 on the address inputs A_{11} to A_6:
  $A_{11} A_{10} A_9 A_8 A_7 A_6 = 101000$.

- The $\overline{CS}$ signal is driven 'low' to select the memory.

- The $\overline{WE}$ signal is driven 'high' so that the information in the selected cell can be read via the selected bit lines, the sense amplifier and output buffer. The logic '1' on the $\overline{WE}$ signal activates the output buffer and places the tri-state input buffers in the high-impedance state. At the beginning of each read cycle, all bit lines b_i and $\overline{b_i}$ are precharged through clocked transistors (not drawn in figure 6.4) to the high level (other precharge levels, such as half-V_{dd}

or low (V_{ss}) levels are also used). If the value '0' is subsequently read from the selected cell, then bit line $\overline{b_{40}}$ remains 'high' while bit line b_{40} discharges slightly via the cell. The bit line voltage levels are transferred to the respective $\overline{db}$ and db data buses. The sense amplifier quickly translates the resulting voltage difference to a logic '0', which is then transferred to the output via the buffer. A similar explanation applies when the value '1' is read from the selected cell, but then b_{40} remains high and $\overline{b_{40}}$ will discharge.

The SRAM write operation

The write operation performed in an SRAM is explained with the aid of an example in which data is written to the cell x_{17},y_{15}, see figure 6.4. The steps required to achieve this are as follows:

- The word line x_{17} is activated by placing the decimal value 17 on the address inputs A_5 to A_0: $A_5A_4A_3A_2A_1A_0 = 010001$.

- The bit line select signal y_{15} is activated by placing the decimal value 15 on the address inputs A_{11} to A_6: $A_{11}A_{10}A_9A_8A_7A_6 = 001111$.

- The $\overline{CS}$ signal is driven 'low' to select the memory.

- The $\overline{WE}$ signal is driven 'low' so that the information on the data input can be written to the selected cell via the data input buffers and the selected bit lines. The value on the db data bus is then equal to the value on the data input while the $\overline{db}$ data bus has its inverse value. The logic '0' on the $\overline{WE}$ signal activates the input buffers and places the tri-state output buffer in the high-impedance state.

SRAMs are designed in a variety of synchronous and asynchronous architectures and speeds. An asynchronous SRAM is activated when an address change is detected. As a result, a clock signal is generated and stored data is accessed. However, this type of SRAM is limited in its speed. Therefore, the fastest SRAMs are generally synchronous. Controlled by one or more clocks, synchronous SRAMs show reduced access and cycle times, boosting their clock frequencies to the same level as those of the high-performance RISC processors and PCs. Improved performance can be achieved when several words are selected simultaneously by a single address. In *burst mode* operation, the address is incremented by an on-chip counter and the parallel read words are serialised to form a large sequence of high-speed data bits. Several burst addressing sequences can be supported, including those used in Pentium^{TM} and PowerPC^{TM} processors.

Static RAM cells

Access time is an important RAM specification and is mainly determined by the signal propagation time from the memory cell to the output. A satisfactory *access time* requires an optimum design of the memory cell, selection circuits, bit lines, sense amplifiers and output buffers. Possible *memory cell implementations* for SRAMs are discussed in detail below.

1. *Six-transistor/full-CMOS SRAM cell*

   Figure 6.6 shows a memory cell consisting of six transistors T_1 to T_6. Transistors T_1 to T_4 comprise two cross-coupled inverters which function as a latch. Pass transistors T_5 and T_6 provide access to the latch. During a write operation the write data is transferred to the bit lines, the word line goes 'high' and the data on the bit lines is transferred to the latch through pass transistors T_5 and T_6.

   During a read operation, first both bit lines are precharged to V_{dd}, by switching signal ϕ only shortly to zero. Then the word line goes 'high' and the contents of the cell cause a slight discharge on one of the precharged bit lines. The discharge takes place through the relevant pass transistor, T_5 or T_6, and inverter nMOS transistor, T_1 or T_3.

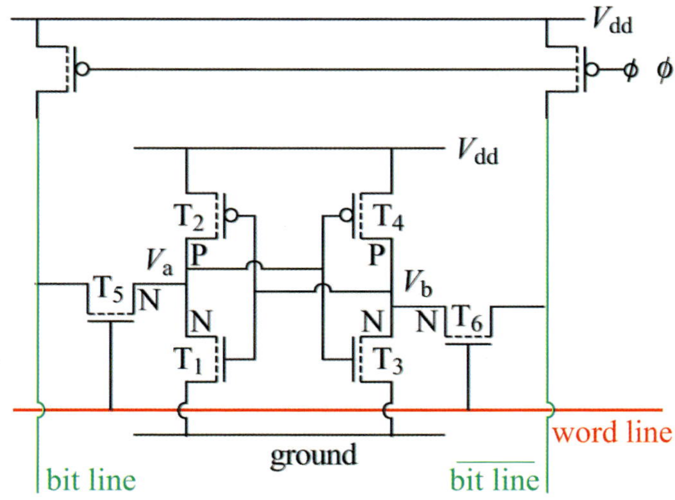

Figure 6.6: *Six-transistor static RAM cell*

A small voltage difference of about 70-100 mV between the two bit lines is sufficient for an SRAM sense amplifier to determine the logic level in the memory cell. This logic level is then transferred to the output pin via the output buffer.

The small subthreshold and gate leakage currents are the only currents that flow in the six-transistor cell when it is simply retaining data. Memories containing full-CMOS cells are therefore suitable for low-power applications. However, the relatively large distance required between nMOS and pMOS transistors requires quite a large chip area for this memory cell.

2. *Four-transistor/R-load SRAM cell*

   Figure 6.7 shows a memory cell consisting of four transistors. This cell contains two cross-coupled inverters with resistive loads. These types of inverters are discussed in section 4.2 and they lead to continuous static power dissipation in the memory cell. This dissipation is kept as low as possible by forming the resistors in an extra high-ohmic polysilicon layer. Typical values are 10 GΩ or more. This polysilicon layer necessitates a more complex manufacturing process than for the full-CMOS cell. An advantage of the four-transistor cell, however, is its reduced cell area, because the resistors are implemented in a second polysilicon layer and folded over the transistors. These memories are hardly or no longer used today, mainly because of their reduced operating margins, but also because of their relatively large power consumption, in both active and standby modes.

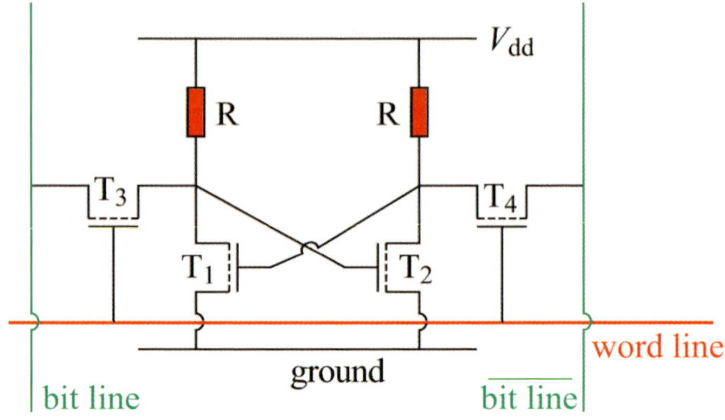

Figure 6.7: *Four-transistor static RAM cell*

3. *Four-transistor loadless SRAM cell*
   The introduction of a loadless four-transistor cell [2,3] allows a 35% area reduction using the same design rule. Comparing figure 6.7, in the loadless cell, the resistors R and the V_{dd} connection are completely removed and transistors T_3 and T_4 are replaced by pMOS transistors. This allows the cell nodes to store full-swing signals after writing. In the standby mode, bit lines are precharged to V_{dd} and the data is maintained in the cell when the leakage current of the pMOS transistors is more than an order of magnitude larger than that of the nMOS transistors. However, because its operation partly depends on the value of the leakage currents, it will be very difficult to realise large memories with it, since leakage currents are not very well controlled. This is particularly due to the large intra-die variations in V_T in current CMOS technologies.

The word lines in both the six-transistor and four-transistor memory cells are implemented in a stack of polysilicon and metal. The considerable parasitic capacitance and resistance of long word lines causes the cells furthest from a row decoder in an SRAM to exhibit a greater

RC-delay than those closest to the decoder. This situation is often redressed by dividing the arrays of large memories into several smaller sections with separate row decoders between them. The resulting word lines have lower parasitic capacitance and resistance and their *RC*-delays are at least a factor four lower than for a single array. The silicides mentioned in chapter 3 are also used to reduce resistance of polysilicon word lines.

Previous discussions made it clear that the 6-transistor SRAM cell is most commonly used in stand-alone and embedded SRAMs. Due to the continuous scaling of threshold and supply voltages, the noise margins of SRAM memories *(SNM = static-noise margin; WM = write margin)* have reduced dramatically, basically due to the increase in transistor parameter spread, combined with a reduction of the supply voltages. Particularly the threshold-voltage variation in the SRAM narrow-width transistors due to random doping fluctuations has a great influence on the SRAM robustness of operation. Due to the strict area requirements, an SRAM cell may easily loose its state when variations in transistor sizes (W and L), in supply voltage (V_{dd}), in threshold voltage (V_T), in temperature and STI stress, to name a few, become too large. The SNM of an SRAM cell describes how well it can cope with these variations. Figure 6.8 shows a diagram representing the stability of the SRAM cell of figure 6.6.

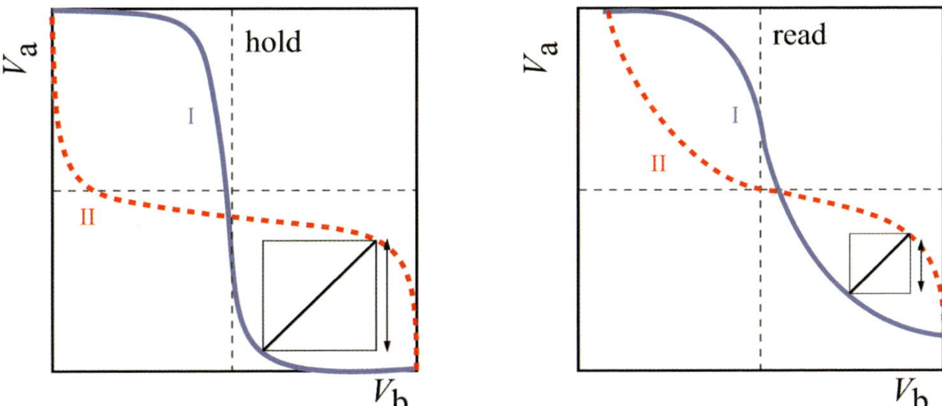

Figure 6.8: *Stability diagram of the SRAM cell of figure 6.6 and its SNM during hold and read mode*

The curves I in the diagram shows the response of V_a on the stimulus of V_b, while the response of V_b on the stimulus of V_a is represented by curves II. For each mode of memory operation (hold or read mode), the SNM is represented by the largest square that can be drawn between the related curves I and II. A large size of the square represents a large SNM. Traditional worst-case values for SNM were in the order of 120 mV-200 mV. Because of the reducing supply voltages and increasing process parameter spread, current SRAMs show SNMs of only a few tens of millivolts.

Usually a design parameter is targeted at a certain mean value, with 3-σ margins ($\approx 0.1\%$ of the parts fail) at each side of the mean value (figure 6.9), assuming a normal distribution of the probability density function of a parameter with a mean value μ and a standard deviation σ.

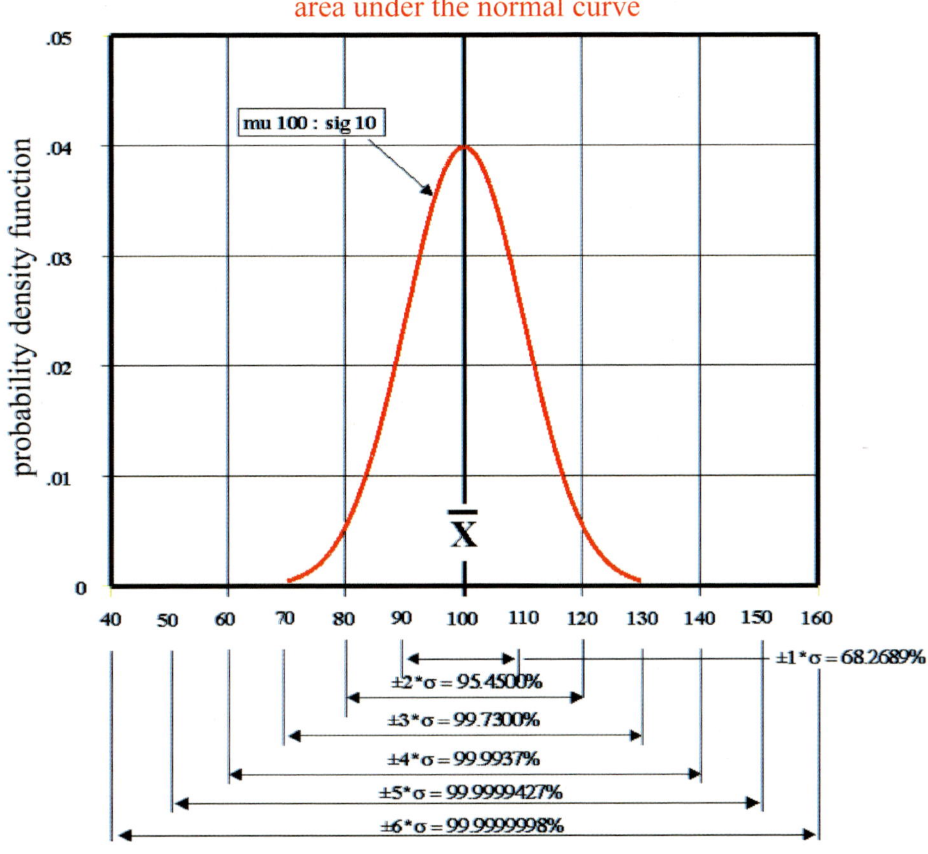

Figure 6.9: *Probability density function versus the spread σ around the mean value μ of a certain parameter*

By including a 3-σ margin at both sides of the mean value, the probability of a failure per cell is $\approx 0.1\%$ at each side. For the early memories, a 3-σ parameter spread was considered sufficient to design the relatively small SRAM memories with several kilobits capacity. *Worst-case (WC)* and *best-case (BC) corner simulations* do not cover all possible combinations already for a long time, due to the increasing occurrence of transistor mismatch, both in value as in number. In other words, due to mismatch, not all memory cell transistors are in the same point in the same corner at the same time, like in the *WC* and *BC simulations*. For most of the embedded memories with capacities of up to 10 Mb a 6-σ parameter spread is taken into account, due to the reduced voltage margins and increased number of memory cells. According to figure 6.10, which shows this *parametric yield* loss as a function of the read or write margin, the yield of that memory is close to 99% (equivalent to 1% yield loss). This yield loss is derived from the probability density function (figure 6.9), which shows that in 0.0000002% of the trials the 6-σ margins would be exceeded. In many cases we only need to take one side of this probability density function, e.g., when a parameter exceeds the +6-σ value (at the right side of the diagram), leading to a failure, then in case the same parameter would exceed the -6-σ value (at the left side of the diagram), it would usually not lead to a failure.

Sometimes even a 6.4-σ spread is taken for this size of memory to achieve 99.9% yield without redundancy (see section 6.4.7), which means that only 1 in 10 billion cells fails. A 7-σ spread, which is also under discussion for 45 nm SRAM design, refers to 99.99% yield in a 100 Mb memory without redundancy.

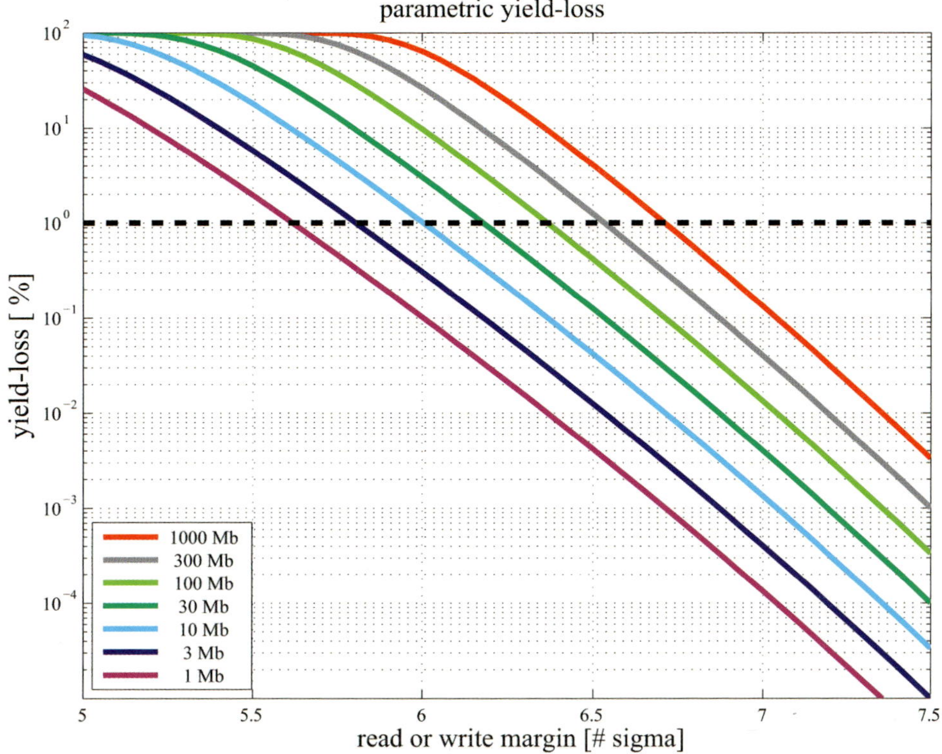

Figure 6.10: *Parametric yield loss as a function of the read or write margin*

The previously mentioned yield numbers can be achieved without redundancy. The diagram in figure 6.11 shows the number of redundant cells that are required for various memory capacities to achieve 90% yield. It shows that we can exchange redundancy with read or write margin in the design. The decision, which solution will be supported, is most commonly based on the amount of area overhead.

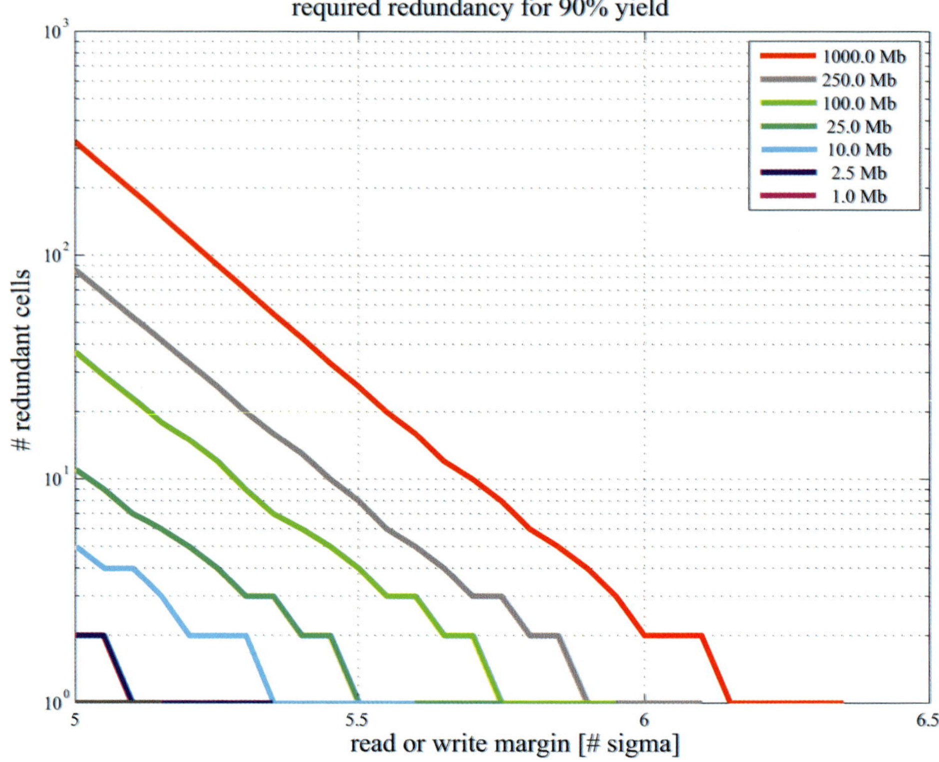

Figure 6.11: *Assuming a yield of 90%, then this figure shows the amount of redundancy required for different memory capacities*

In 90 nm CMOS technologies and below it is very hard to run the memories at voltages below 1 V. This is a combined result of the increasing number of bits and the different and almost conflicting transistor sizing requirements from the read and the write operation. Another problem is the increasing leakage currents, particularly when many memory cells share the same bit line.

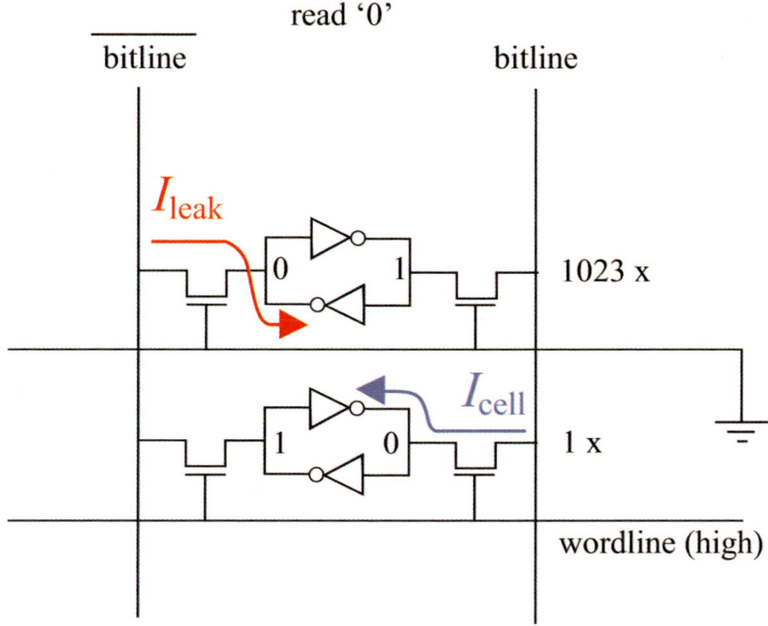

Figure 6.12: *Representation of a disturbed read operation*

In the example in figure 6.12, we assume that a bit line is connected to 1 k memory cells and that the lower cell is to be read. If the total leakage current of 1023 cells (I_{leak}) is about equal to or higher than the cell read current (I_{cell}), the sense amplifier may read the wrong state, because also $\overline{\text{bitline}}$ is discharged.

SRAM margins still get worse and do not allow the cell area to be scaled each next technology generation with a factor of two anymore. As a consequence, the memory density increase will no longer double each new generation, but is expected to increase by only 50%. Common stand-alone SRAM cell areas are between 80-100F^2/bit, compared to 6F^2/bit for a stacked DRAM cell (see section on DRAMs) and only 2F^2/bit for a multi-level NAND-flash cell. For *embedded SRAM (eS-RAM)* memories it holds that it will face the same problems as the stand-alone SRAMs, but only one or a couple of technology generations later. Massive simulations are required to completely validate SRAM designs. One solution to maintain future SRAM operation robustness is to use larger transistor sizes in the cell, since $\sigma \propto 1/\sqrt{WL}$. Alternatives are to use a more complex cell (7-10 transistors/bit) [4] or to use a more complex technology, such as the 3-D approach by Samsung [5], where the pMOS load and nMOS pass transistor are stacked on top of the planar pull down. Because this boosts the cost of embedded memories, more and more emphasis is put on embedded alternatives, such as embedded DRAM (*eDRAM*) with 1T, 3T, or 4T cells. Some vendors even use embedded ZRAM (section 6.4.4) on their microprocessors.

6.4.3 Dynamic RAMs (DRAM)

The basic block diagram of a DRAM is quite similar to that of an SRAM. The main difference between an SRAM and a DRAM is the way in which information is stored in the respective memory cells. All stand-alone DRAMs consist of n-type cells because of the high-performance requirements. DRAMs may use back-bias voltages to have a better control on the threshold voltage to limit leakage for improving refresh characteristics, and to reduce junction capacitances. When DRAMs are embedded in a logic chip, p-type cells were often chosen, because the n-well in which the DRAM is located can then be separately connected to an additional positive back-bias to achieve the previous advantages. In triple-well technologies this is no longer necessary, because each individual n- and pwell are electrically isolated.

Figure 6.13 shows the basic circuit diagram and a water model of a *single-transistor DRAM cell*, which is also called a *1 T-cell*. A logic '1' is written into the 1 T-cell by placing a high level on the bit line while the word line is active. The capacitor in the cell is then charged to a high level. This is also applicable with reverse polarities for p-type cells. The data in a cell is thus determined by the presence or absence of a charge on its capacitor. Parasitic junction leakage and transistor subthreshold leakage cause this charge to leak away, just like the water in the pond evaporates as time progresses. The information in the cell must therefore be frequently refreshed.

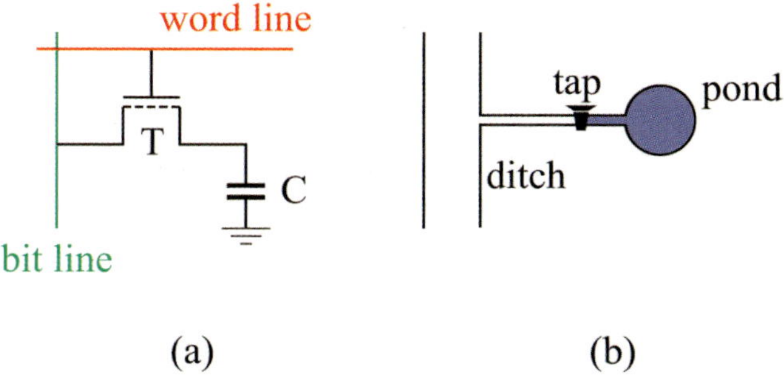

Figure 6.13: *(a) Circuit diagram of a DRAM (b) Water model of a DRAM cell.*

In addition to leakage, the information in a DRAM memory cell is also destroyed when it is read. This so-called *destructive read-out (DRO)* is caused by the cell capacitance being much smaller than the bit line capacitance. The cell contents must therefore be restored immediately after each read operation. For this reason, each bit line is equipped with a *refresh amplifier*, which consists of a sense amplifier and some restore circuits. This sense amplifier detects the bit line level and writes its amplified value back into the cell. The operation is called a *row refresh* because it is done simultaneously for all cells that are addressed by an active word line.

In practice, the *refresh operation* for each cell must be performed every 2 to 256 ms, depending on the cell size and the technology. In many large memories, today, the static refresh period is 64 ms. A 1 Gb DRAM, e.g., a 1 Gb DDR SDRAM, where two internal rows are refreshed in parallel, 16,368 row addresses require a periodic refresh interval time of (64 ms/16,368)/2= 7.81 μs [6]. During the refresh cycle, the internal control keeps track of which rows have been refreshed and which have not.

130 nm DRAMs apply a negative bias (e.g. -1.5 V, generated on chip) to reduce subthreshold leakage in standby mode. However, this additional field between gate and drain increases the GIDL leakage. 90 nm DRAM process generations and beyond therefore use a special *recessed-channel array transistor (RCAT)* as access transistor in the DRAM cell to reduce GIDL. This transistor is therefore only used in the memory array [7]. It has a much longer channel length (in the vertical direction, so that the lateral cell sizes can still scale) to reduce the subthreshold leakage current. Gate leakage is reduced by creating a stack of different dielectrics, fabricated by using atomic layer deposition (see chapter 3)

resulting in an overall high-ϵ equivalent dielectric in this transistor. Both leakage reduction techniques are driven by the need to enhance the capacitors' data retaining properties and minimize the refresh frequency in order to reduce the power consumption in both active and standby mode. A DRAM can therefore not be accessed for a certain percentage of the time. This percentage is typically between one and five percent and is one of the reasons why DRAMs are more difficult to use than SRAMs. The requirements for the DRAM leakage currents are much tighter than for logic circuits. The total sum of all leakage components may not be more than about $1\,\text{fA}$ per cell [8].

The read operation in a DRAM requires a reasonable signal level on the bit line. For a long time, smaller cell sizes came hand in hand with smaller storage capacitance values. However, this value determines the sensing voltage, the data retention times, the sensing speed and the soft-error rate. Particularly the increasing sensitivity to soft errors (chapter 9) has put a lower limit to this capacitance value. Independent of the memory generation ($512\,\text{Mb}$, $1\,\text{Gb}$, $4\,\text{Gb}$ or more), a value between $25\,\text{fF}$ and $35\,\text{fF}$ is currently generally applied. Due to the continuous scaling of the parasitic node capacitances in an SRAM, these memories have become more susceptible to soft errors than DRAMs.

The use of planar (C)MOS processes for the implementation of DRAM memories was limited to capacities of up to $1\,\text{Mb}$. A typical example of the *planar DRAM cell* used in these DRAMs is shown in figure 6.14.

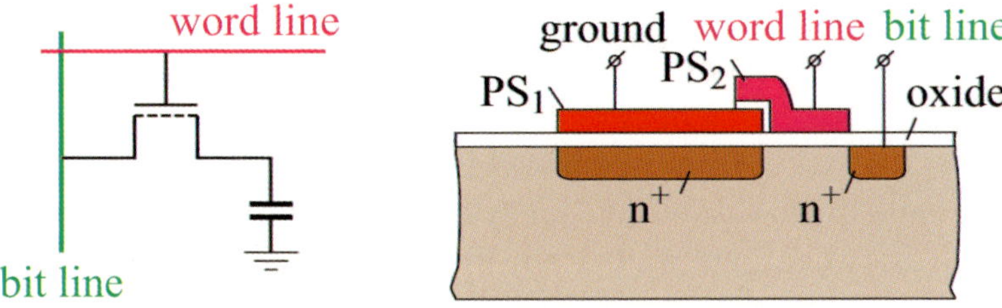

Figure 6.14: *The planar DRAM cell*

An unacceptably small capacitance renders planar cells unsuitable for current DRAMs. Three-dimensional cells which afford increased storage capacitance in a reduced planar surface area are therefore used for large

DRAMs. These include the *stacked capacitance cell (STC)* and the *trench capacitance cell* shown in figure 6.15.

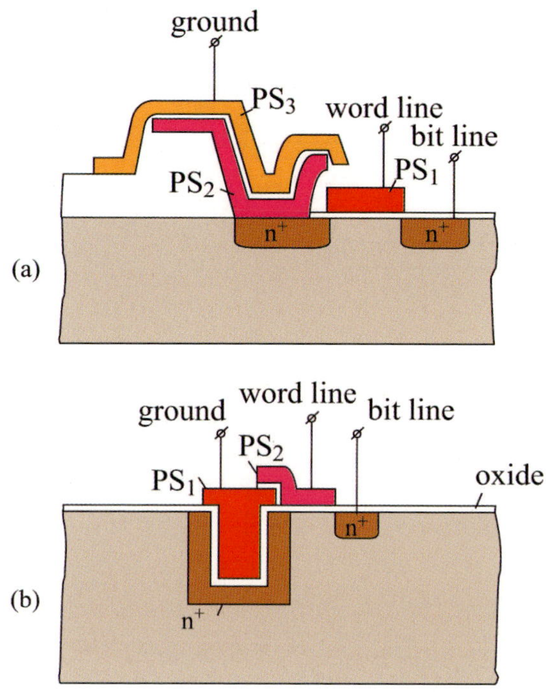

Figure 6.15: *(a) Stacked capacitance and (b) trench capacitance three-dimensional DRAM cells*

These manufacturing processes are much more complex. In a stacked-capacitor technology, the capacitor is commonly fabricated after the transistors with less thermal processing (figure 6.17.a). The dielectric can be sandwiched between various electrode materials that allow the use of high-ϵ materials. Until $0.15\,\mu$m DRAMs, an *oxide-nitride-oxide (ONO)* dielectric has been used, while Ta_2O_5 (Tantalum Pentoxide, with $\epsilon_r \approx 25$) has been used in the $0.12\,\mu$m generation. These high-ϵ dielectrics cannot be used as normal transistor gate oxide that needs to contact polysilicon. Increase of the capacitance value is achieved by reducing the dielectric thickness. The minimum thickness, however, is typically limited by the above mentioned leakage current of $1\,$fA per cell. The ONO sandwiches achieve a thickness of about $3.5\,$nm. Another way to increase the capacitance in a stacked-capacitor cell is to fabricate an uneven surface on the bottom electrode of the capacitor (figure 6.16), by using a so-called *hemispherical grain (HSG)* process flow.

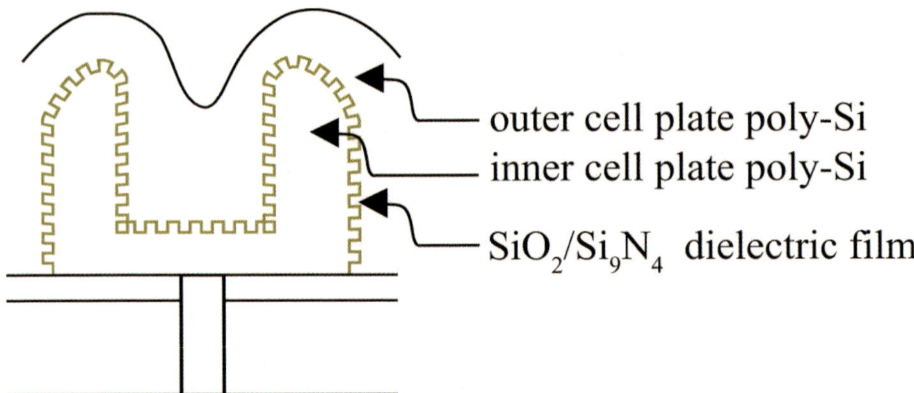

Figure 6.16: *Cross section of a cylindrical stacked-capacitor cell using a hemispherical process flow*

This increases the surface area as well as the total capacitance value. Compared to a cell with an even surface, the HSG cell shows a capacitance increase of more than a factor of two.

In a trench-capacitor technology (figure 6.17.b) the capacitor is fabricated before the selection transistor.

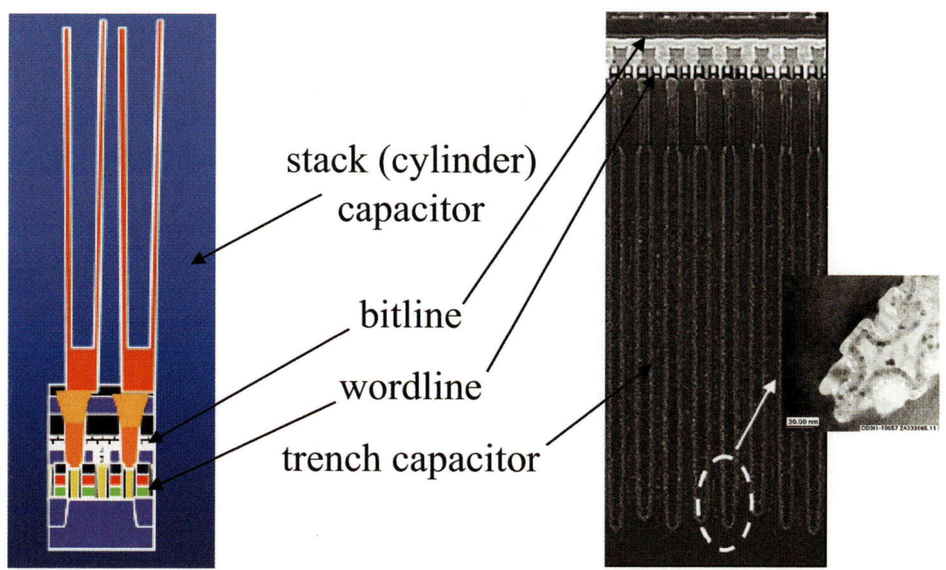

Figure 6.17: *Example of a stacked capacitor cell architecture (a) and a trench-capacitor cell in 70 nm CMOS (b) (drawing + photo: Qimonda)*

As the transistor fabrication needs a high-temperature step, (future) high-ϵ capacitor dielectrics need to survive this harsh treatment, which is an important materials selection criterion. The above mentioned Ta_2O_5 is suitable for stacked capacitor applications but cannot be used in trench cells, since these then need to survive the temperature cycles needed for transistor fabrication. Embedded DRAMs therefore use trench-capacitor cells, because then the memory cells are already fabricated, before the normal CMOS process flow starts, including the necessary temperature cycles.

Some companies use Al_2O_3 *silicon-insulator-silicon (SIS)* trench capacitors [9]. *BST* ($Ba_{1-x}Sr_xTiO_3$) materials show dielectric constants, which are even higher, i.e., about 250-600. However, the processes for producing BST are not yet fully developed.

Comparing stacked and trench capacitor DRAM cell trends leads to the following observations:

- Both types of memories use a high-ϵ dielectrics which could not have been fabricated without the so-called *atomic layer deposition (ALD)* step (see chapter 3).

- The smaller capacitance ($\approx 30\,\mathrm{fF}$) value in a stacked capacitor requires a larger voltage across the memory array to store an equal amount of charge compared to a trench-capacitor cell ($\approx 35\,\text{fF}$), which also leads to a larger power consumption.

- Due to the high aspect ratio (= height/width ratio $\approx$ 60-80) of the storage capacitor for both trench- and stacked-capacitor cells, their series resistance can be a bottleneck in further scaling. This is due to the fact that the planar trench-cell area scales quadratically with the feature size, while its depth almost remains constant.

- While trench capacitors are buried beneath the silicon surface, stacked capacitors create significant topographies across these designs and put stringent demands on both lithography and mechanical stability.

- One of the most important life-time reliability tests is the *burn-in test* (see chapter 9), to identify devices which are subject to infant mortality or excessive parameter drift. During a standard burn-in test the device is operated for 160 hours (or 240 hours) at an elevated temperature, usually 125°C. This test activates very substantial wafer stress mechanisms, which dramatically affect the dielectric layer homogeneity in the deep-trench capacitor cells of a DRAM. Particularly in nanometer CMOS processes (65 nm and beyond) the dielectric layer may be damaged, causing unacceptably high cell leakage currents. This can be a reason that future stand-alone DRAMs may all use a stacked capacitor cell.

- Trench-capacitor storage nodes are more sensitive to soft errors and to substrate noise.

- Stacked capacitor DRAMs are built from twin cells, meaning that two cells share the same bit line contacts. Due to a different topology, trench capacitor cells show etching problems which do not allow to share the bit line contact between two neighbouring cells. Trench-capacitor areas ($8F^2$/bit) are therefore usually larger in area than stacked-capacitor cells ($6F^2$/bit), with F being the half pitch (in 65 nm CMOS process: F=65 nm). Beyond 50 nm only stacked-capacitor memories will be manufactured. Because of the use of *vertical pillar transistors (VPT)* (drain stacked on top of source, separated by the channel length; see figure 6.18) a single memory cell area is expected to reduce to $4F^2$/bit.

- Due to the high-aspect ratio (60 or more) of the trench capacitor, the trench edge requires an etching angle of approximately 89 degrees in order to still have sufficient width in the bottom of the trench. Beyond 90 nm technologies it becomes much more difficult to fulfill this requirement. A trench-capacitor DRAM wafer is full of holes, causing a lot of highly leaking capacitors (yield loss) after burn-in.

- From the above it is clear that the scaling of trench-capacitor cells incorporates much more problems than those of stacked-capacitor cells and we see that some original trench-DRAM vendors gradually move toward the stacked DRAM camp. These stacked DRAMs can basically be fabricated with either of the two different memory-cell architectures, dependent on the processing sequence of the capacitor and the bit line: *capacitor-over-bitline (COB)* and *capacitor-under-bitline (CUB)* architecture. Because of the so-called *overlay problem* (alignment problem of the contact hole connecting the bit line, along two storage capacitors down to the common source of the access transistors of the twin cells) in the CUB architecture, all stacked DRAMs now use the COB cell architecture. Figure 6.18 shows the trend in (stacked) COB DRAM architectures.

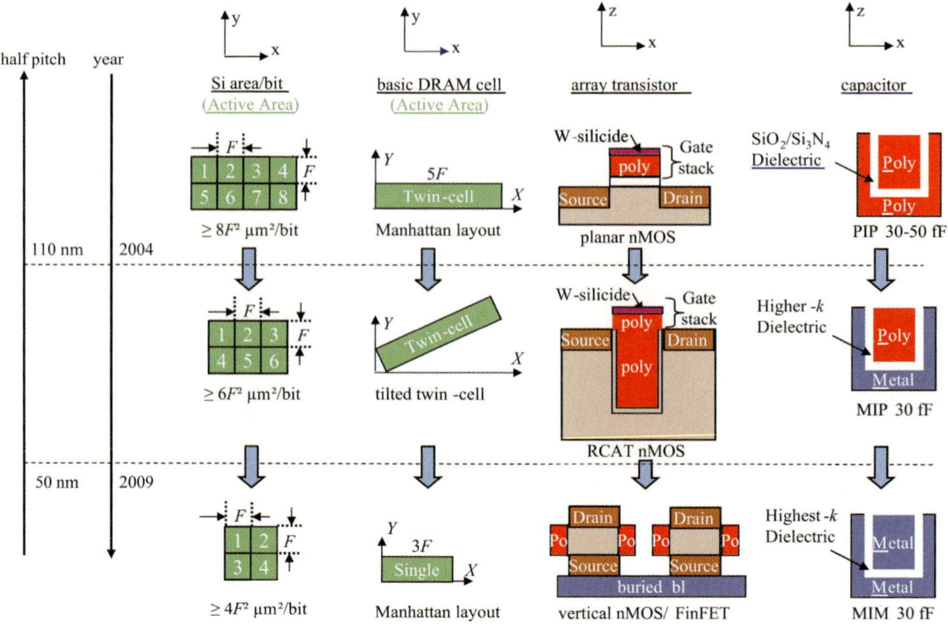

Figure 6.18: *(Stacked) COB DRAM architecture trends*

The figure clearly shows the trend in cell area: from $8F^2$ to $4F^2$, the trend in transistor architecture: from planar through RCAT to vertical pillar transistor or Finfet and the trend in storage capacitor architecture: from PIP (poly-insulator-poly), through MIP (metal-insulator-metal) to MIM (metal-insulator-metal). The target is to keep the storage capacitance per cell almost constant and roughly equal to 30 fF.

In the past, DRAM memory density increased with a factor of four every new technology node. A factor of two (s^{-2}) could be achieved from conventional technology scaling, assuming a size-scaling factor of $s \approx 0.7$. A factor of 1.4 was achieved by shrinking the lateral cell area with more than s^2. Another factor of 1.4 in density was achieved by increasing the chip size with this factor. Today, for reasons of process complexity and yield, the chip sizes can no longer be increased. Since the third dimension (stacked or trench cells) is fully exploited today, also the lateral cell area cannot be scaled more than with s^2. Therefore, today, current DRAM capacity only doubles every new technology node.

Despite associated processing and operational disadvantages, the D-RAM has achieved a dominant market position. This is mainly because of the relatively low area per bit, which is generally 15 to 20 times smaller than those of SRAMs. This leads to cost advantages of a factor of five to ten.

General remarks on DRAM architectures

There are important differences between the basic DRAM and SRAM operation. Both SRAMs and DRAMs have similar secondary and sometimes even tertiary amplifiers in the I/O path.

The access time of a DRAM was approximately two to four times longer than that of an SRAM. This is mainly because most SRAMs were designed for speed, while DRAM designers concentrate on cost reduction.

DRAMs are generally produced in high volumes. Minimising the pin count of DRAMs by row and column address multiplexing makes DRAM operation slower but cheaper as a result of the smaller chip size. Because of the optimisation of DRAM cells for a small area, the higher DRAM processing costs can be regained by the larger number of dies on the wafer. Moreover, DRAM technologies only use two metal layers up to the 65 nm node. In 45 nm technologies three metal layers may be used. This low number of metal layers is possible due to the very simple and regular structure of bit lines, word lines and supply rails.

In addition to minimising cell area, other techniques are also used to reduce the total area of memories. One such technique reduces the number of bond pads on stand-alone DRAMs by multiplexing the row and column addresses through the same bond pads.

Stand-alone SRAMs use separate bonding pads for the row and column addresses to achieve fast access times. The access time of a stand-alone SRAM is therefore considerably shorter than that of an equivalent stand-alone DRAM. This is illustrated in figure 6.19(a). The *RAS* and *CAS* signals represent the *row-address signal* and *column-address signal*, respectively. This figure compares the access times of a stand-alone SRAM and a stand-alone DRAM, which uses row and column address multiplexing. The access time of the SRAM is only determined by the time interval t_1 whereas the total access time of the DRAM is determined by the sum of several set-up, hold and delay times. The improved DRAM access time in figure 6.19(b) is achieved by omitting the column address latches and implementing a *static column access*.

The data rate of a RAM is determined by the cycle time. This has already been defined as the minimum possible time between two successive accesses to a memory. The cycle time of an SRAM can be equal to its access time. In a DRAM, however, the cycle time is the sum of the access time, the precharge time of the bit lines and the refresh time. This holds for full random access. In page mode (or EDO), precharge and refresh times do not add to the (page mode) cycle time. Therefore, page mode cycle times are about two to three times shorter than full random-access cycle times.

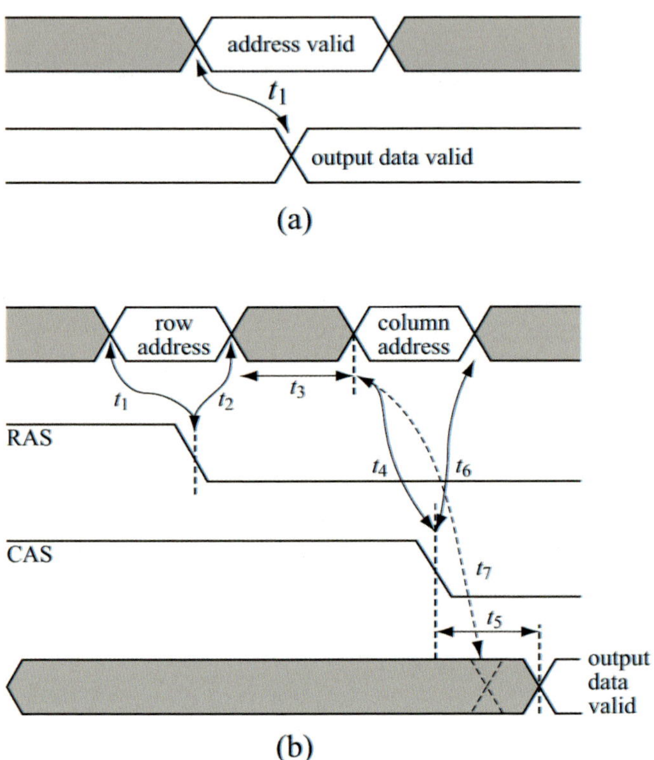

Figure 6.19: *Access times of (a) an SRAM: access time=t_1 and (b) a DRAM: access time=$t_1 + t_2 + t_3 + t_4 + t_5$ or improved access time=$t_1 + t_2 + t_3 + t_7$*

One of the biggest problems over the last decades is the fact that microprocessors showed much larger speed improvements than the DRAMs (see figure 6.20). This gap is the primary drive for DRAM manufacturers to dramatically increase random access and data rates and offer Gb-bandwidth interfaces.

6.4.4 High-performance DRAMs

The increased overall system performance required the DRAM performance to increase at the same pace. Several solutions have been developed to improve DRAM performance during reading. These relatively new generation DRAMs include *Fast Page Mode (FPM)*, *Extended Data Out (EDO)* Mode, burst data using *synchronous DRAMs (SDRAM)* and *Rambus DRAM (RDRAM)*.

All four approaches are based on the ability to access complete pages without requiring the start of a new memory cycle. A *page*, which represents all the memory cells that share a common row address, can have a length of as many as several kilobit. The drawback of page mode is the segmentation of the data, increasing the chance that the required data will not be on the accessed page. Particularly graphics applications benefit from page mode access.

Another advantage of page mode architectures is their reduced power consumption, because there are no sense and refresh currents during page mode access. Most DRAMs are asynchronous; these include conventional DRAMs, FPM and EDO RAMs. A memory operation is then initiated on the arrival of input signals.

The differences between a synchronous and an asynchronous DRAM involve more than just the presence or absence of a clock signal. With SDRAMs, for instance, a precharge cycle is independent of a *RAS*, to allow multiple accesses on the same row. Internally, a refresh cycle is identical to a read cycle. No column addresses are needed during refresh, since no output data is required. FPM DRAM, EDO DRAM, SDRAM and RDRAM are all based on the same core memory. Therefore, their internal timing diagrams look very similar. The differences are mainly determined by how they communicate with the outside world. These differences include the speed at which address and control signals can propagate through the DRAM and the speed at which data propagates from the DRAM to the memory controller [10]. In the following, a brief overview of the different high-speed DRAM architectures is presented.

Fast Page Mode DRAM

An *FPM DRAM* offers faster access to data located within the same row because the row command doesn't need to be repeated. This means that only one *RAS* signal needs to be given, followed by four *CAS* signals, because the four words all come from the same row. The column address set-up starts as soon as the column address is valid, so that the column address can be latched at the falling edge of CAS. This is different from conventional page modes in which a column address access was initiated by the falling edge of the CAS signal. It was, therefore, required to wait with the column address set-up until the falling edge of CAS. In this way, a reduced page cycle can be achieved in comparison to conventional page mode DRAMs.

Extended Data Out DRAM

The *EDO DRAM* architecture looks very similar to the FPM DRAM. However, it contains an additional register that holds the output data. This allows the start of the next cycle before the previous one is finished. The possibility to "overlap" output data with input data of a next cycle results in a 30% speed improvement over comparable page mode DRAMs. Most EDO DRAMs contain a single bank architecture and must therefore process memory operations serially. A memory operation cannot start before the previous one is completed.

Synchronous DRAMs

When the transfer of address, data and control signals to a DRAM is synchronised by the system clock, such a DRAM is called a synchronous DRAM. Both *SDRAMs* and *RDRAMs* have synchronous architectures and interfaces. Different synchronous DRAM architectures are presented here.

1. SDRAM architectures

   In an SDRAM, in addition to a given external starting address, the next column addresses during a burst are generated by an on-chip counter, while an asynchronous DRAM requires the memory controller to generate a new column address for each access. SDRAMs and RDRAMs are generally built with multiple memory banks (two, four...). Each bank is a memory of its own [11], allowing individual and parallel operation for maximum performance. SDRAM architectures use burst features to accommodate fast external transfer at increasing burst rates. Synchronous DRAMs (SDRAM, SGRAM and RDRAM) use the system clock as their clock input. Therefore, they are targeted to meet the speed requirements of commonly-used PC systems. A trend in increasing the memory's bandwidth is the use of *Double Data Rate (DDR)*

I/Os, which are already available since 1998. Figure 6.20 shows
how the memory interface speed tries to keep pace with the increasing CPU speed [12]. The *memory controller* plays a dominant role
in the interface between memory and CPU. It is therefore required
that these controllers also exhibit sufficient performance.

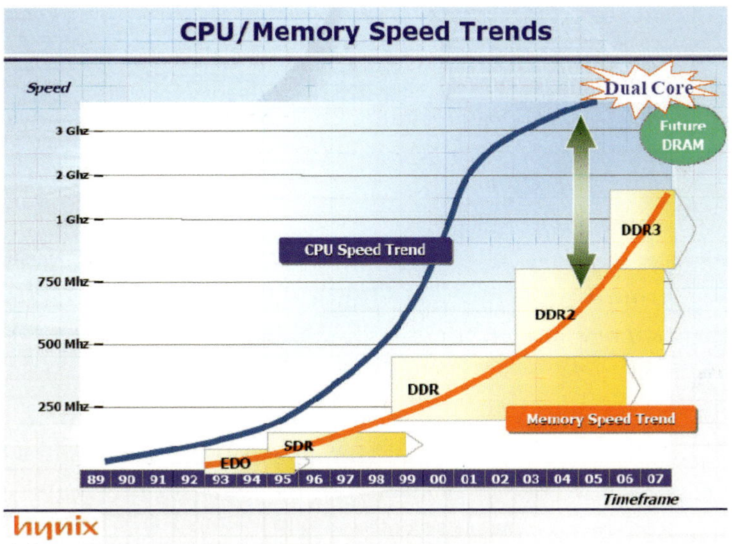

Figure 6.20: *Memory versus CPU speed trends (Source: Hynix)*

In the DDR mode, both the falling and rising edges of the clock are
used to double the data throughput, with data sample frequencies
from 200-400 MHz at 2.5 V. New versions of the DDR concept drive
the data rates rapidly further, but also at reduced voltages: DDR2
with 400-800 MHz bus rates at 1.8 V and DDR3 with bus rates to
above 1 GHz (800-1600 MHz) at 1.5 V. Although the bandwidth of
DDR3 is doubled compared to DDR2, the transition from DDR2
to DDR3 did not require a major speed change of the DRAM core
(the memory cell array). To achieve this double bandwidth, DDR3
uses a prefetch of eight words instead of a four-word prefetch for
DDR2. This means that for every single read or write operation, a
total of eight words are accessed in parallel in the DRAM core to
enable the high data rate at the interface. Table 6.1 summarises

some DDR memory speed parameters.

Table 6.1: Summary of DDR memory speed parameters

Type	V_{dd} [V]	Clk freq. [MHz]	Data rate [Mb/s/pin]
SDRAM	3.3	100	100
DDR-1	2.5	100-200	200-400
DDR-2	1.8	200-400	400-800
DDR-3	1.5	400-800	800-1600
DDR-4 (expected)	1.2 ?	800-1600	1600-3200

Even a new fast graphics 512 Mb GDDR4 memory chip, using 80 nm CMOS, is offered in a 32-bit data bus configuration. The memory has an ultra-high-speed data processing rate of 2.4 Gb/s [13].

Many SDRAMs can also operate in a random-access mode, in which they show similar timing as FPM or EDO DRAMs. SDRAMs may have 64-bit or even 128-bit wide I/O formats. Besides commodity DRAM applications, this allows them to also serve in applications with extremely high memory bandwidths. For this purpose, an SDRAM architecture includes: burst feature, more than one memory bank for parallel memory operation and a clocked or synchronous interface. Particularly graphics applications (which are characterised by high-speed and wide I/O buses) require extremely high bandwidths. Video RAMs (VRAMs) and Synchronous Graphics RAMs (SGRAMs) are specially designed for graphics applications.

2. Video RAM architectures

   As the pixels on a computer terminal or a TV are refreshed serially, the first *Video RAMs (VRAM)* provided continuous streams of serial data for refreshing the video screen. It acts as a buffer between the (video) processor and the display. Most VRAMs were a dual-port version of a DRAM meaning that the display is reading its image from the video RAM, while the processor is writing a new image into it. The standard DRAM had to be extended with a small serial access memory and a serial I/O port to support the storage of video pictures [14]. However, all VRAMs still have the original standard random-access DRAM port also available. During a serial read, the DRAM array is accessible via the DRAM port for a separate read or write operation. Special features, such as block write and flash write, etc. are supported by additional circuits. However, the rapid rise of special SDRAM architectures, such as SGRAMs became so cheap and dense, that they made the original VRAMs obsolete.

3. SGRAM architectures

   Synchronous Graphics RAM (SGRAM) architectures are similar to SDRAMs but optimised for graphics applications. They contain similar additional hardware, such as registers and mask registers to support block write and write-per-bit functions. This results in faster and more efficient read and write operations. These features are supported by special registers and control pins. Colour registers are mainly used to store the colour data associated with large areas of a single colour, such as a filled polygon [14]. The data in these colour registers can be written in consecutive column locations during block-write operation. Write-per-bit allows the masking of certain inputs during write operations; it determines which memory locations are written. Most SGRAMs are specially designed for use in video cards in PCs.

   They lag by about a factor two in memory capacity behind commodity DRAMs. 32 to 256-bit wide SGRAMs are being developed. A major difference with a VRAM is the additional synchronous interface of the SGRAM. Current SGRAMs have I/O data rates of several GBps (gigabytes per second). DDR interfaces can push the SGRAM's graphics peak bandwidth even further. The popularity of SGRAMs has increased such that it is currently used in many graphics systems. Another DRAM version, called the *Rambus*^{TM} *DRAM (RDRAM)*, is gaining popularity as well, particularly in graphics applications.

4. RDRAM architectures

   The *RDRAM* (particularly the *Direct RDRAM*) provides high bandwidth for fast data transfer between the memory and the programming parts in a system. The Rambus^{TM} interface is licensed to many DRAM manufacturers and, at certain engineering fees, they can get customised interfaces to their existing products.

Because of the high bus clock rates (600 to 800 MHz) and the use of DDR, RDRAMs claim extremely high bandwidths, competing with that of SDRAMs and require fewer data lines than the wide-word DRAM. The Direct RDRAM has only little overhead on a standard DRAM architecture and offers several modes from power-down (only self-refresh) to selective powered-down memory blocks [15].

An alternative to the Direct RDRAM is the *XDR RDRAM (extreme data rate RDRAM* which offers extremely high bandwidth and low latency. It can offer several times the bandwidth of a DDR3 memory.

There are several other memory types with high to extremely high bandwidths. This offers system designers a wide choice in creating the optimum solution for their particular application.

Currently, DRAMs have passed the gigabit level, with production versions available of 1 Gb and 4 Gb, and with 8 Gb and 16 Gb versions in development. As the application area increases, the hunger for increased densities and higher speeds will drive the complexity and performance of SDRAMs and DRAMs to incredibly high levels. Even a *Deca-Data Rate SDRAM* with an I/O error detection scheme for high-end servers and network applications has already been developed [16]. In many applications, there is also a pressure on the DRAM standby power consumption. Another example of a DRAM shows an extended data-retention sleep mode, with longer refresh cycles to reduce standby power consumption. This, of course, led to more bit failures, which, on their turn, were then corrected by using *error-correction code (ECC)* combined with the conventional redundancy [17]. These last two examples show in which direction memory vendors are thinking when optimizing their memories for high speed or low power.

The move of high-speed microprocessors from bulk CMOS to an SOI technology, has also initiated the exploration of SOI for memories. This has resulted in the presentation of the *Zero Capacitor DRAM*, which is a capacitor-less, single transistor DRAM, also called *Z-RAM*. In the operation of MOS transistors in an SOI technology, the floating-body effect (section 3.2.4), was seen as a parasitic phenomenon. In the Z-RAM, the floating-body charge is even enhanced and used to store ones and zeros.

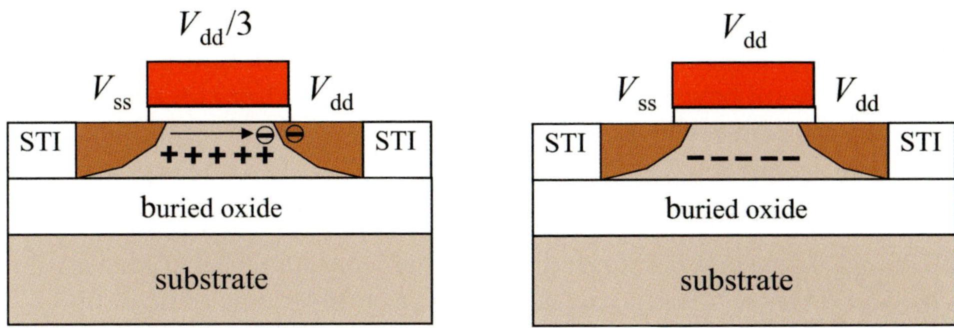

Figure 6.21: *Storage of a logic 1 and logic 0 in a Z-RAM cell*

If the body of an nMOS device is positively charged, its threshold voltage is reduced and the device carries a "large" current, representing a logic "1". A negatively charged body has the opposite effect and represents a logic "0". Because it uses no specific capacitor device, its scalability potentials are expected to be better than that of SRAM and DRAM [18].

6.4.5 Single- and dual port memories

Most modern memories are *single-port memories*. These memories can not perform a read and write operation at the same time, because these operations share the same port. Several applications, e.g., communications, video and graphics processing, etc., will benefit from an architecture in which both operations cane be performed simultaneously. A *dual-port memory* (or *multi-port memory*) supports writing on one memory position through one port while reading from another position through the other port, which increases the communication bandwidth. As discussed before, video (D)RAMs are often available as dual-port memory. Also many SRAMs are available as dual-port memory. Compared to the single-port SRAM cell of figure 6.6, in a dual-port cell, all transistors and connections are doubled, except for the two feed-back inverters. This leads to an area increase of approximately 70 to 80%. Moreover, because of the increased cell capacitances, an individual access takes more time, leading to a bandwidth improvement of much less than a factor of two.

The increased complexity and layout area of a multi-port memory, compared to a single-port memory, comes at a cost. The relatively small production volumes drive the costs of these memories even higher. In summary: compared to the use of two single-port SRAMs, for certain applications, a dual-port SRAM may only offer minor advantages such that the choice between applying a dual-port SRAM or two single port SRAMs becomes difficult.

6.4.6 Error sensitivity

The logic value in a RAM cell may change as a result of radiation caused by *α-particles* and *cosmic particles*. The α-particles may come from impurities in the metal layer (e.g., aluminium), from the package, or from other materials in the close vicinity of the chip. The particle radiation generates a relatively large number of electron-hole pairs, which may randomly change the data in memory cells. This random loss of stored information occurs in both DRAM and SRAM cells. SRAMs are particularly prone to the resulting *'soft errors'*, which become more influential as densities increase and stored charges decrease. DRAMs based on CMOS technology have reduced susceptibility to α-particles, because the storage capacitance per cell remains constant.

Memories can also be covered with a *polymide layer* to protect them against external α-particle radiation. This reduces soft-error rates by a factor of 1000 or more. This does not apply to the cosmic particles, which can even pass through half a meter of concrete. This is one of the reasons why the cell charge is not decreased every new DRAM generation. Chapter 9 presents more details on soft errors.

6.4.7 Redundancy

Stand-alone memories are sold in very high volumes and must therefore be very cheap to produce. Methods to achieve a low price include yield-improvement techniques which may, for example, result in a yield in excess of 70 % for areas greater than $100\,\mathrm{mm}^2$. However, many stand-alone memories have one or more cells that do not function properly. For this reason, most stand-alone memories include several redundant memory rows and/or columns which can be used to replace defective cells. The faulty cells are detected by means of memory testers and a laser beam is used to isolate their corresponding rows or columns. This so-called *laser-fusing* technique is also used to exploit spare rows and columns and re-address the faulty ones to the redundant (spare) ones. Currently, the *poly fuse* has replaced the traditional laser fuse, because it does not require the special laser equipment. Poly fuses are smaller and can be programmed by a tester by providing normal signals and voltages. Redundancy techniques may be used to improve the yield by a factor of as much as 20 to 50 during the initial development phase of new memory process generations. During memory production ramp up, the memory may include more redundancy (e.g. between 2 to 6% additional bits) than during high-volume production in mature processes (e.g., less than 5$_0 additional bits).

6.5 Non-volatile memories

6.5.1 Introduction

Since their introduction in the early 70's, non-volatile memories have become key components in many electronic systems. Until the explosion in the growth of flash memories, the market was relatively small and mainly driven by mobile applications. Today, the flash memory market has become the second largest after the DRAM market.

As discussed in section 6.1, a non-volatile memory keeps its stored data when the memory is disconnected from the supply. Non-volatile memories include ROM, PROM, EPROM, EEPROM, flash (E)EPROM, FRAM, MRAM and PCM. In the following paragraphs, their basic operation is discussed in some detail, including their fundamental properties.

6.5.2 Read-Only Memories (ROM)

A *ROM*, also known as *mask ROM* or *mask-programmable ROM*, is in fact a random-access memory which is written during the manufacturing process. The information is therefore lasting and non-volatile. It can be read but it can never be altered. With the exception of the write facility, the architecture of a ROM is similar to that of a RAM. Subsequent discussions are therefore restricted to the different techniques for writing the information during the manufacturing process. The ROM memory cells required by each technique are examined separately.

Different processing layers could, in principle, be used to store information in a ROM. Two choices, made for educational purposes, are the diffusion and contact layers. ROM cells and structures based on the corresponding ACTIVE and CONTACT masks are discussed below.

ROM cell with the information in the ACTIVE mask

Figure 6.22 shows the structure of a ROM which is programmed by means of the ACTIVE mask, see section 4.6. The ROM cell is enclosed by a dashed line in the figure. An example of the layout of such a cell is given in figure 6.23.

All bit lines in the ROM in figure 6.22 are precharged when ϕ is 'low'. The V_{SS_1} line is switched to ground when ϕ goes 'high'. The cell enclosed by a dashed line is read when the corresponding word line WL_3 goes 'high'. Bit line bl_2 will then be discharged if ACTIVE is present in the cell. Otherwise, bl_2 will remain charged. The information in the ROM is therefore stored in the ACTIVE mask, corresponding to the presence or absence of a memory transistor at the selected cell position.

Figure 6.24 shows a photograph of a ROM array based on the cell of figure 6.23.

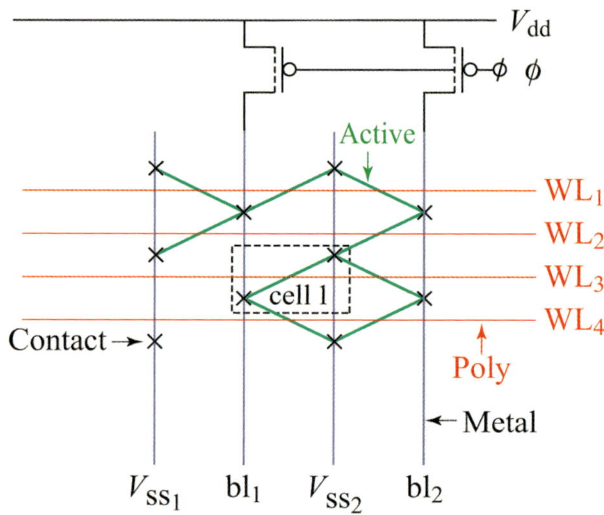

Figure 6.22: *ROM with information in the ACTIVE mask*

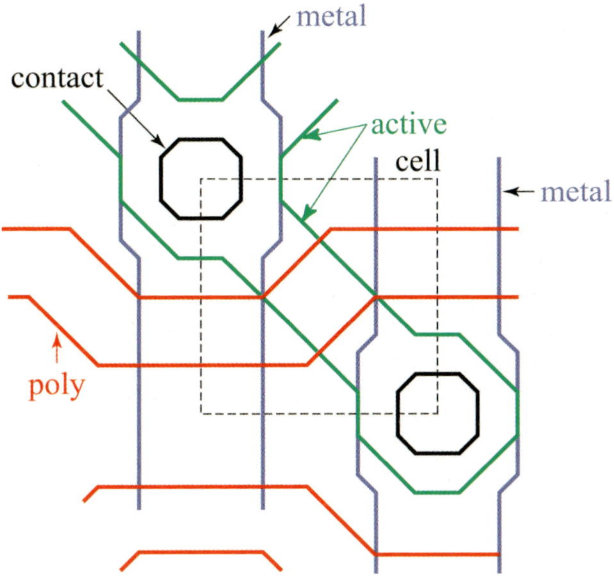

Figure 6.23: *Layout of an ACTIVE-mask programmed ROM memory cell*

Figure 6.24: *Photograph of an array of ROM cells (Source: NXP Semi-conductors)*

ROM cell with the information in the CONTACT mask

Figure 6.25 shows the structure of a ROM which is programmed by means of the CONTACT mask. All bit lines in this ROM are precharged through the pMOS transistor when ϕ is 'low'. A word line is activated when ϕ goes 'high'. The bit lines of cells connected to the selected word line and containing a CONTACT hole are then discharged. The CONTACT hole in the cell locally connects the aluminium (METAL) bit line to the drain of a transistor, which has its source connected to a grounded diffusion (ACTIVE) track. The series resistance of the ACTIVE tracks is reduced by means of an extra aluminium ground line which is implemented every 8 to 10 bit lines.

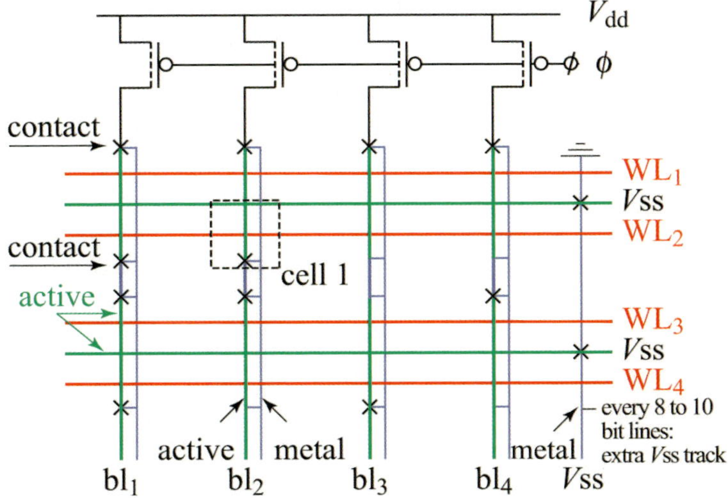

Figure 6.25: *ROM with information in the CONTACT mask*

Comparison of the ACTIVE-mask and CONTACT-mask programmed ROM cells

A fair comparison of the chip area of the ACTIVE-mask and CONTACT-mask programmed ROM memory cells requires the inclusion of a suitable fraction for the area of the extra ground line in the latter cell. This gives the following values for a 65 nm CMOS process:

ROM cell in figure 6.22 : $0.06\,\mu\mathrm{m}^2$ $\leftrightarrow$ ROM cell in figure 6.25 : $0.1\,\mu\mathrm{m}^2$

Although the second cell is the larger of the two, it has the advantage that its information is programmed in the CONTACT mask. This mask is used in one of the last steps in the manufacturing process. Therefore, ROMs which store information in the CONTACT or one of the METAL or VIA masks can be largely prefabricated. Now, only a small number of manufacturing steps are required to realise a ROM with specific contents. In contrast, the ACTIVE mask is usually one of the first in the manufacturing process. The *turn-around time* between order and delivery is therefore much shorter for a ROM with its information in the CONTACT or METAL or VIA masks than for a ROM with information in the ACTIVE mask. Therefore, in multi-metal layer processes, the programming is increasingly done in one of the last mask layers, most commonly a via mask.

There are some other types of ROMs as well. In a *serial ROM*, a NAND type of structure is used to discharge the bit line. In such a ROM, a V_T-implant is used for program storage (enhancement or depletion type of memory transistor). The series connection of the cells allows a much smaller number of contacts. This results in a small area, but also in a relatively low speed of operation.

In certain applications, the contents of a ROM can be directly copied into a part of a RAM, from which it can be accessed much faster. This (part of a) RAM is then called *shadow RAM*. The BIOS-code in a PC was usually stored in a ROM, however, in most PCs it was directly copied into a shadow RAM during booting.

In general, the maximum ROM memory capacity is lacking behind the flash memory (four to eight times) and in many of its original applications ROM is replaced by flash memory, which have experienced an incredibly high growth of both capacity and market volume. A ROM is used in high-volume applications, where it is absolutely certain that the contents need not to be changed.

6.5.3   Programmable Read-Only Memories

Introduction

The three different types of programmable Read-Only Memory are PROM, EPROM and EEPROM. Respectively, these ROMs are *programmable, electrically-programmable* and *electrically-erasable programmable*. They are programmed by users rather than during manufacturing. Although they are programmed by users, these memories are still called read-only memories because the number of programming/erasing cycles is rather limited in normal usage.

PROMs (Programmable Read-Only Memories)

A *PROM* is a one-time programmable read-only memory which can be programmed only once by the user. Each cell contains a fuse link which is electrically blown when the PROM is programmed. Traditional PROMs were usually manufactured in a bipolar technology. The fuses were then implemented in a nickel-chromium (NiCr) alloy. The resulting cell is relatively large and is about four times the size of a ROM cell.

Today, poly-fuse cells are used in standard CMOS processes, which can also be electrically blown. These fuses are then silicided, so that larger programming currents are generated at the same programming voltages. These cells are smaller than the traditional NiCr cells.

Currently, PROMs move towards the 3-D dimension, where four or eight layers of memory arrays are stacked on top of each other. Memory cells are located between two successive layers of metal and positioned at the crossroads of the metal wires in each of these layers, which run in perpendicular directions. Each cell consists of a diode in series with an anti fuse. By applying a large electrical field across the anti fuse (by selecting the corresponding metal tracks in two successive layers, between which the cell is located), its physical state changes, causing a dramatic reduction of its resistance. The cells that are not programmed, maintain their high-resistive state. Since the cells are only fabricated between layers above the silicon, the silicon can be used to implement all selection, programming and reading circuitry. Each cell is only $4F^2$ in size, but because of the four or eight layers, their effective area is only $1F^2$ or $0.5F^2$ respectively. These 3-D *one-time-programmable (OTP) memories* exhibit a relatively large *area efficiency (AE)*, which may be larger than 85%. In the 45 nm node, this memory may show a bit capacity as high as 64 Gb. A disadvantage of using a fuse or anti fuse for programming is that the memory array cannot be tested.

The wish for rewritability in many applications has increased the demand for erasable architectures. These are discussed in the following sections.

EPROMs

Figure 6.26(a) shows a schematic representation of an *EPROM* memory cell.

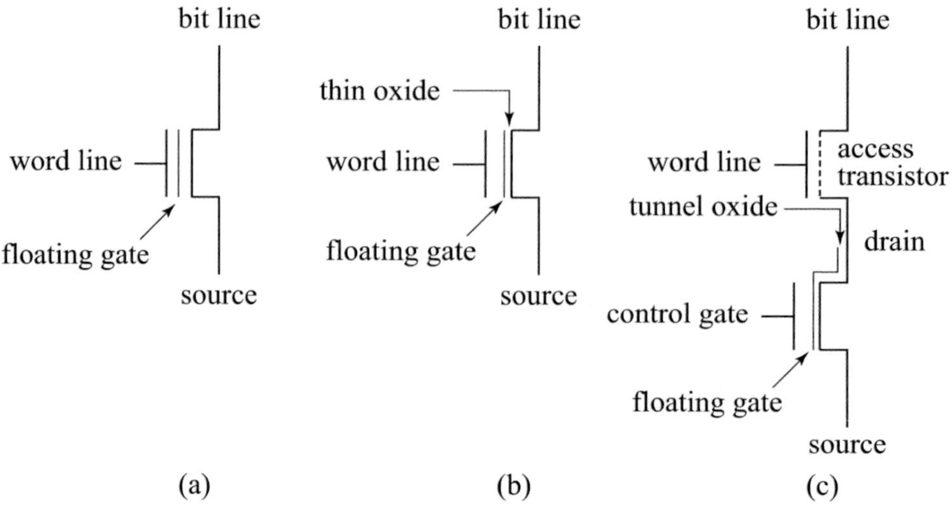

Figure 6.26: *Schematic representation of (a) an EPROM cell, (b) a flash (E)EPROM cell and (c) a full-featured EEPROM cell*

The data in this cell, as in an EEPROM cell, is represented by the presence or absence of charge on the 'floating gate' of the memory transistor. The floating gate is charged by means of a large electric field between the transistor's source and drain. This accelerates electrons in the channel to very high energy levels. Some of the resulting 'hot electrons' (see chapter 9) penetrate through the gate oxide to the floating gate. This type of programming is called *channel hot electron injection (CHEI)*. Sufficient charge is collected on the floating gate when high drain-to-source voltages of over $3.2\,\text{V}$ (in a $65\,\text{nm}$ process) and gate-source voltages of about 8 to $9\,\text{V}$ are applied. This causes currents of the order of $0.3\,\text{mA}$ in the cell. The number of programming/erasing cycles in an EPROM is limited (10.000 to 100.000 cycles). Currently, the higher voltages are often generated on-chip by means of charge pumps. Alternatively, an EPROM can be removed from the system and programmed in a special PROM programmer. It then uses a second power supply of around 8 to $9\,\text{V}$, depending on the technology node.

EPROMs are erased by exposing the cells to ultraviolet (UV) light. This is done through the transparent (quartz) windows in EPROM packages. In many applications, EPROMs are only programmed once. They are therefore also available as *one-time-programmable (OTP)* devices in cheap standard plastic DIL packages with no transparent windows. As a result of its complex reprogramming operation (non field-programmable

UV erase), the use of EPROMs, today, is very limited.

6.5.4 EEPROMs and flash memories

Floating-gate PROM structures, which allow electrical erasing and programming, were developed at the end of the seventies. The two resulting categories are electrically-erasable PROM (EEPROM) and flash memories.

EEPROM

Unlike with EPROM and flash memory, EEPROM data can be changed on a bit-by-bit basis. This is also called a *full-featured EEPROM* or *double EPROM*, whose basic cell architecture is shown in figure 6.26(c). Because of the separate access transistor in the cell, EEPROMs feature relatively low bit densities compared to EPROM and flash memories. This transistor allows selective erasure of cells. Erasure is often done per byte.

Figure 6.27 shows a cross-section of the storage transistor of a full-featured EEPROM cell.

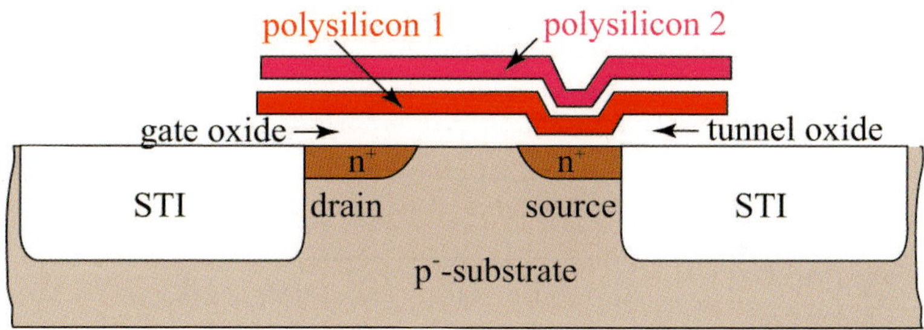

Figure 6.27: *Example of floating-gate EEPROM cell*

Data storage and erasure are achieved by moving electrons through a small thin oxide tunnel region between the floating gate and drain. This is done by applying a high electric field of about 10 MV/cm across the tunnel oxide, which induces so-called *Fowler-Nordheim (FN) tunnelling*. The cell is programmed by applying a high voltage to the drain with respect to the top gate, which causes the electrons to flow back to the drain. The cell is erased when a voltage of about 12 to 15 V is applied between the gate and drain (substrate or source, depending on the technology). Now, electrons tunnel through the thin oxide and produce a negative charge on the floating gate. This increases the threshold voltage of the memory transistor. Therefore, the memory transistor in an erased cell has a high threshold voltage. The small currents involved in the tunnelling mechanism used in full-featured EEPROMs facilitate on-chip generation of the 12 to 15 V for programming and erasing the memory.

An important characteristic of a full-featured EEPROM is the variation in memory transistor threshold voltage associated with successive programming/erasing cycles. Eventually, the threshold-voltage difference between a programmed and an erased cell becomes too small due to charge trapping in the oxide. This imposes a limit on the number of times that a cell can be erased and programmed. The plot of the threshold-voltage variation is called the *endurance characteristic*, see figure 6.28 for an example. The threshold difference enables a few hundred thousands to more than a million programming/erasing cycles for the individual cells.

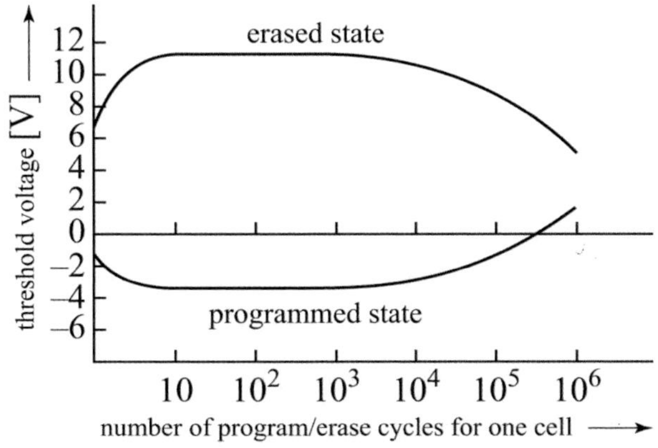

Figure 6.28: *Endurance characteristic of a full-featured EEPROM cell*

The *data retention time* of all EEPROMs is more than ten years. The various applications [19] of EEPROMs: conventional consumer applications, universal remote controls, cordless telephones, garage door openers, cameras, automotive, home audio and video and smart cards. Attention is also focused on cellular telephones and pagers. Innovative features have been added to EEPROMs, such as Block Lock which allows users to combine alterable data with secured data.

Although EEPROM technology offers more flexibility, it is facing increased competition from flash memory, which allows much higher densities, as a result of the absence of the bit-by-bit change feature of an EEPROM, see figure 6.26.

Flash memories

A *flash memory* is in fact an EPROM or EEPROM in which the complete memory or complete sectors (blocks) of memory can be erased simultaneously, in one flash. Today, the most important flash-memory categories are: *NAND-flash* and *NOR-flash*.

All flash memories are arranged into blocks. The typical block capacity is 128 kB for a NOR-flash and 8 kB for a NAND-flash, respectively. Erasure cannot be done individually, but is done by complete blocks in "one flash". The lifetime of a flash chip is determined by the maximum number of erase cycles per block, which is typically specified around 100,000 cycles. It is therefore crucial that the erase cycles are evenly distributed over the blocks. Today's flash memories include control circuitry which distributes the number of program/erase cycles evenly over the total memory.

Each cell in a NOR-flash (Figure 6.29) contains a bit line contact, which makes it relatively large, but gives it a relatively fast random access. A cell in a NAND-flash is part of a serial chain. It is therefore small (cell area is only $4F^2$/bit, compared to $6F^2$/bit for a DRAM with stacked-capacitor cell and $80\text{-}100F^2$/bit for a 6-T SRAM cell), has a slow random access (typically $20\,\mu s$) but allows fast sequential access (typically $20\,ns$). Programming a NOR-flash is done by using *channel hot-electron injection (CHEI)* , for example by connecting the source to GND, the drain to $+5\,V$ and the top gate to $+12\,V$, respectively, while FN-tunnelling is used for its erasure, by connecting the gate to GND and the source to $+12\,V$. In a NAND-flash FN-tunnelling is used for both programming and erasure. Programming (typically $200\,\mu s$ for a 2112B page) can be done by connecting the gate to $+20\,V$ and the substrate to GND and erasure (typically $2\,ms$ for a 128 kB block) can be done by reversing these voltages.

45 nm NAND-flashes use 16 to 32 transistors in series (figure 6.29). The number of cells on a word line is typically between 2048 and 16348. This is also called a *page*. During a read operation, first the bit lines (figure 6.29) are precharged to V_{dd}. Then the selection transistors (sel) are turned on and all unselected word lines are set to such a high positive voltage that all cells (programmed or not) function as pass tran-

sistors. If there is only one bit stored per cell, the word line of the selected cells is set to GND, assuming that the erased cells have a negative V_T (normally-on) and will conduct and discharge the corresponding bit line. The programmed cells, with a positive V_T will not conduct and cannot discharge the bit line.

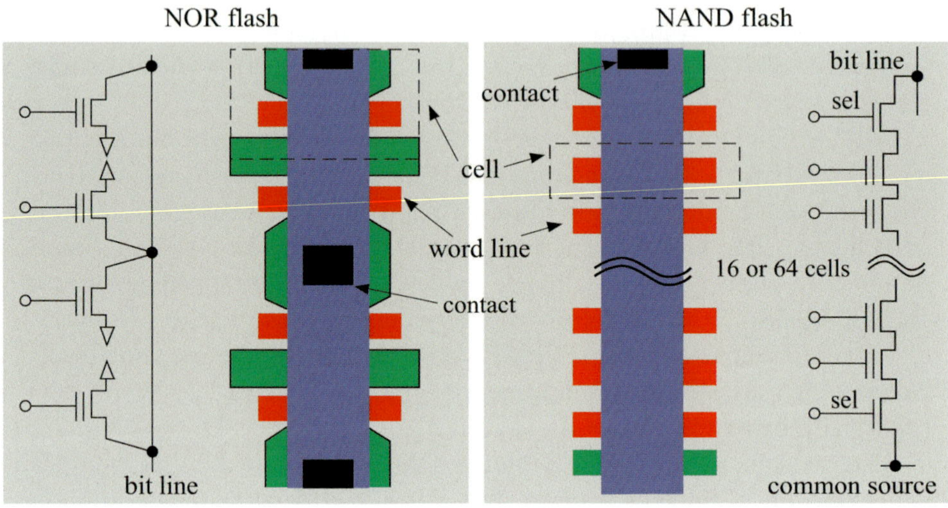

Figure 6.29: *Basic difference between NOR and NAND-flash architecture (Source: Samsung)*

Figure 6.30 shows an example architecture of a 2 Gb NAND-flash. Data is loaded from a page in the array into the data register. Like with SRAM and DRAM, also NAND-flash architectures are being optimized for enhanced throughput. Some NAND-flash architectures therefore have an additional *cache register*. In this case a copy of the data register is added in between the memory array and the drawn data register. This top data register is then called the cache register. This allows loading the next sequential access from the array into the data register, while reading the previously accessed data from the cache register. This increases the data throughput by about one third.

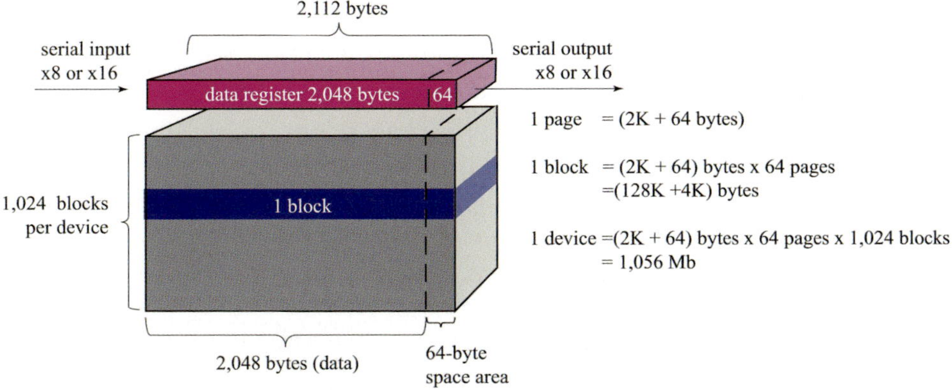

Figure 6.30: *Example architecture of a 2 Gb NAND-flash memory (Source: Micron Technology, Inc.)*

Because of its random access, NOR-flash has traditionally been used to store relatively small amounts of executable code in applications such as mobile phones and organisers.

Because of its serial architecture, the NAND flash has a longer read access. However, the need for low-cost high-density memory drove the NAND-flash into the newer mobile devices with increasing performance, such as mobile phones, MP3 players, cameras and streaming media. Today's feature phones combine all of these applications in one device, which are commonly supported by NAND-flash. NAND-flash is also replacing NOR-flash in code-storage applications as well. NAND-flash architecture looks similar to hard-disk drive, in that it is also sector-based and may also have bad blocks. It therefore requires error-correcting code to guarantee correct data retention. The capacity of NAND flash chips, today, is a result of the combined advances in memory capacity increase of the individual dies and the increase of the number of dies that can be packaged into one single MCM module. Several NAND dies are currently packaged into an MCM, creating total chip capacity of 16 Gb, or more, particularly suited for memory hungry consumer electronics devices [20]. Also the first samples of 16 Gb and 32 Gb single chip NAND flashes have been announced [21].

In all (E)EPROM and flash memories the minimum thickness (most commonly 7-8 nm) of the dielectric layers above and below the floating gate is determined by the accumulated dielectric leakage charge over the specified data-retention time (usually > 10 years). This has limited the scaling of the vertical dimensions and voltages in these memories, which

341

also has a negative impact on the lateral scaling. The relatively large signal-to-noise ratio in these types of memories allows to store more levels in one cell to further reduce the cost per bit. In such a *Multi-Level Cell (MLC)* different amounts of electron charge on the floating gate may represent one of four possible combinations of two bits. During a read cycle, the control gate is set to a high level and the current through the cell is inversely proportional to the charge on the floating gate. Current sensing requires three differential sense amplifiers, which each compare the cell current with that from one of three reference cells. The outputs of these sense amplifiers directly represent the stored cell data. Multilevel storage has been known for quite some time. However, reduced noise margins and increased design complexities created a lack of commercial interest. The first *multilevel-storage* memory has been delivered since 1991. In a serial-EEPROM technology, analogue data samples were directly stored at a resolution of 256 levels in each cell, without the need for conversion of the analogue sample to binary words. The first commercial multi-level flash memory products were announced at the end of 1996. Another, more recent example of a multi-level NAND-flash memory is the 8 Gb [22], fabricated in 65 nm CMOS. Four-bit memory cells are also in development. In a multi-level memory, the distance between adjacent threshold voltage charge distributions on the floating gate is becoming very small and may lead to a decrease in reliability with respect to read operation and data retention. Therefore, a multilevel flash memory may allow only about ten thousand program/erase cycles per physical sector, while a single-level flash memory is capable of a hundred thousand of these cycles. The use of on-chip error correction coding (ECC) may alleviate these problems.

The flash memory is penetrating many markets which were previously dominated by magnetic and optical discs, ROMs, EPROMs an EEPROMs. Being able to continuously increase the density of flash memories would speed this process up even more. Next to using more electrical levels in the memory cell to increase the density of flash memories, also multiple layers of stacked memory cells are introduced. Figure 6.31 and 6.32 show a drawn cross section and a SEM photograph of 3-D stacked strings of NAND flash cells, respectively [23]. These additional layers only require a limited amount of mask and processing steps and only support the fabrication of the cells. Selection of these cells is also performed by the selection circuits located in the bulk silicon wafer.

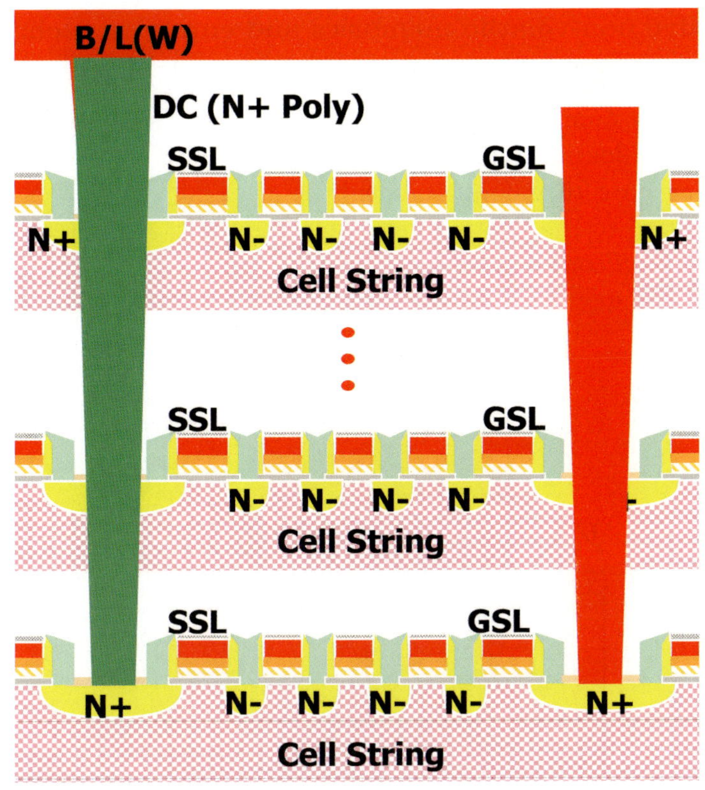

Figure 6.31: *Drawn cross section of 3-D stacked strings of NAND flash cells. (Source: Samsung [23])*

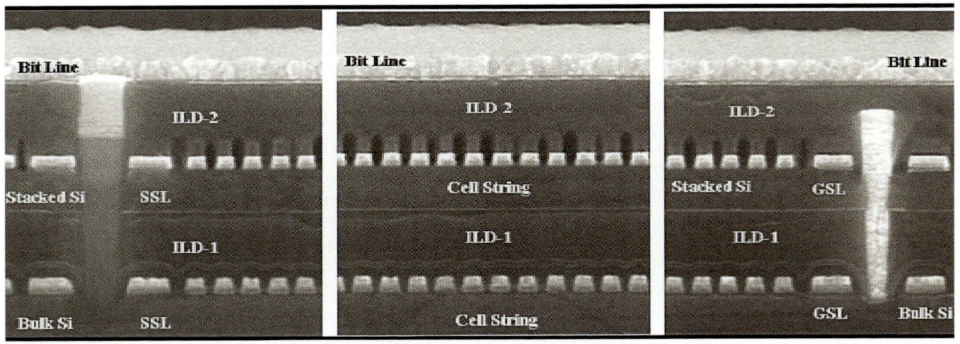

Figure 6.32: *SEM photograph of 3-D stacked strings of NAND flash cells. The second active layer is similar to the perfect single crystal layer used in SOI fabrication. (Source: Samsung [23])*

Alternative non-volatile memories and emerging technologies

One of the problems related to the scaling of floating-gate devices is the relatively large minimum thickness of the dielectric isolation layers above and below the floating gate. This is required to limit charge leakage from the floating gate to guarantee a sufficiently long data-retention time. An alternative to the floating-gate device is the *SONOS (silicon-oxide-nitride-oxide-silicon)* device.

Figure 6.33 shows a cross sectional view of a SONOS memory cell.

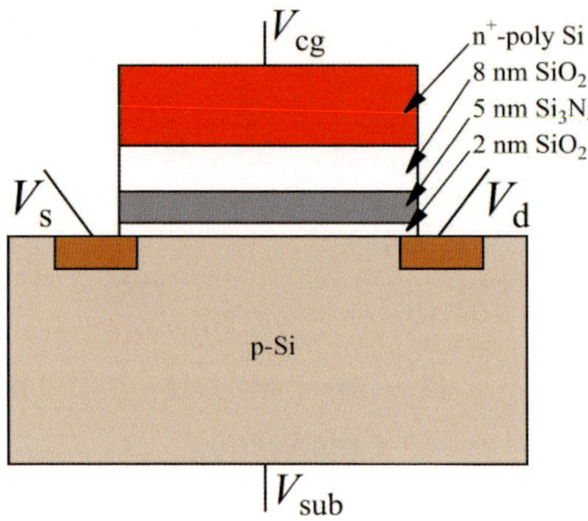

Figure 6.33: *Cross section of a SONOS memory cell*

Basically, the cell is a conventional nMOS transistor, in which an oxide-nitride-oxide stack is embedded between the polysilicon gate and the channel. This stack is built from a gate dielectric consisting of a thermal oxide layer of approximately 2 nm thickness, a silicon nitride layer of about 5 nm and a second oxide layer with a thickness between 5 and 10 nm. Programming of the cell is done as follows. When the gate is connected to a sufficiently large positive voltage, electrons will tunnel from the substrate through the ultra-thin oxide layer to the nitride layer. Because silicon nitride has the intrinsic property to trap electrons, it will trap this negative charge, which causes the threshold voltage of the transistor to increase. Likewise, during erasure, the threshold voltage can be decreased with a negative voltage on the gate, by which holes are injected through the same tunnel oxide.

This nitride layer acts about the same as a floating gate, but shows some different properties. Compared to conventional floating-gate devices, a SONOS device offers a dramatic improvement of the *radiation hardness*. This is due to the fact that the charge is stored in discrete isolated deep traps in the nitride dielectric layer, which is therefore not a conductive layer like the polysilicon floating gate. A high density of traps in the nitride layer, which is somewhat dependent on the nitride film thickness and growth technique, allows sufficient charge storage and memory operation of SONOS devices.

This isolated-charge storage effect makes SONOS devices highly immune to tunnel oxide defects, while even a single defect in the tunnel oxide of conventional floating-gate devices can completely destroy the stored data. This offers a strong reliability advantage, particularly after many program and erase cycles and improves the data retention time. This property can also be exchanged with thinner oxide layers.

A large dose of cosmic charge particles, originated from solar or galactic radiation, may completely discharge floating-gate devices and limit their radiation hardness. In SONOS devices, such a high-dose radiation will only discharge the nitride locally, making them particularly suited for high robustness and reliability demanding applications, e.g., medical, aviation, and military. An example of a data-storage memory using an ONO gate dielectric can be found in [24], which presents a 4 bit per cell 1 Gb *NROM (nitride ROM)*.

Another trend in the flash memories is that their density increase is also achieved by fabricating different memory array layers on top of each other. These so-called stacked-memory cell arrays only contain memory cells, which are controlled and read through the peripheral circuits of the first memory array layer (in the wafer). An additional memory array layer requires therefore only three more masks in the processing (see section 3.3.1; 3-D stacking).

6.5.5 Non-volatile RAM (NVRAM)

A *non-volatile RAM* combines SRAM and EEPROM technologies. This kind of memory is sometimes called a *shadow RAM*. Read and write actions can be performed at the speed of an SRAM during normal operation. However, the RAM contents are automatically copied to the EEPROM part when an on-chip circuit detects a dip in power. This operation is reversed when power returns. An NVRAM therefore combines the retention time of an EEPROM with the high performance of an SRAM.

6.5.6 BRAM (battery RAM)

A *BRAM* comprises an SRAM and a battery which provides sufficient power to retain the data when the memory is not accessed, i.e., when the memory is in the *standby mode*. The battery is used when power is absent. An SRAM is chosen because of its low standby power consumption. The battery is included in the BRAM package and the data retention time is close to 10 years.

6.5.7 FRAM, MRAM, PRAM (PCM) and RRAM

A lot of research effort is devoted to develop the Holy Grail: a universal memory that could someday replace SRAM, DRAM and flash. Several alternatives are currently in development and/or small-volume production.

Ferroelectric RAM technology has "been available" for quite some time. The basic concepts of *FRAM* operation have been known since the fifties. However, with the focus on the costs and the quality of silicon memories, progress in FRAM technology is at a much lower pace than that in SRAM and DRAM technology.

The first FRAM realised on silicon was unveiled in 1988 [25]. It contained 256 bits, which were built up from a six-transistor, two-capacitor array per cell. Compared to DRAM technology, this FRAM consumed a lot of area. Using a two-transistor, two-capacitor cell from 1992, current densities up to 8 Mb are being commercialised, with many different standard interfaces, in nanometer technologies with one-transistor, one capacitor per bit. This basic cell looks very similar to a basic DRAM memory cell, see figure 6.34.

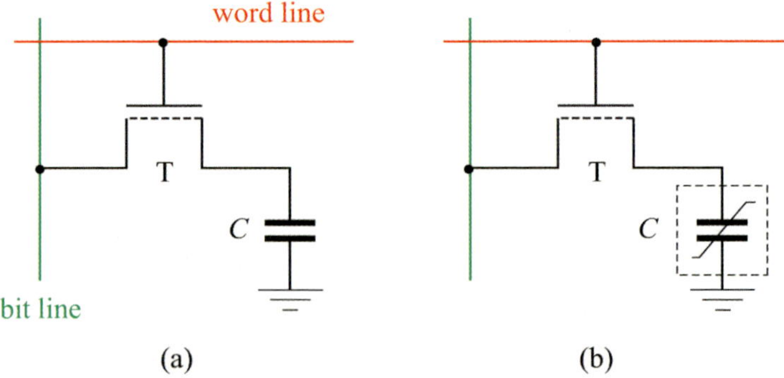

Figure 6.34: *(a) Basic DRAM cell, (b) basic 1T, 1C FRAM cell*

The operation of a DRAM cell is discussed in section 6.4.3. The operation of an FRAM cell is based on the polarisation state of its ferroelectric capacitor. The dielectric material used in this capacitor belongs to a certain class of dipole materials, which are called Perovskite crystals. By applying an electric field across this dielectric, these crystals polarise. This polarised state is maintained after the electric field is eliminated. The dielectric is depolarised when an electric field of the opposite direction is applied.

During a read operation, an electric field is applied across the capacitor. Similar to a DRAM, the current is detected by a sense amplifier. When the dipoles switch state, the sense current is higher. Again similar to a DRAM, the data in a FRAM cell is destroyed during reading (*Destructive Read-Out (DRO)*). The cell contents must therefore be rewritten (refreshed) immediately after each read operation. A complete read cycle includes a precharge period, a read operation and a rewrite operation. Because of higher dielectric constants, an FRAM's cell charge density is higher than that of DRAM cells, allowing smaller cell sizes.

Advances in FRAM technology have resulted in trench capacitor and stacked capacitor architectures, analogous to DRAM technology evolution. Currently, several manufacturers are offering or developing FRAMs [26], which reached a complexity of 4 Mb in 2007. Basically, an FRAM operation depends on voltages rather than currents. This makes FRAMS particularly suited for low power applications. FRAMs are therefore considered as the ideal memory for emerging low-power applications, such as smart cards and RF identification [27]. Potential applications include digital cellular phones and Personal Digital Assistants (PDAs) and automotive applications. Compared to EEPROM and flash memories, the number of read/write operations (endurance cycle) for FRAMs is several orders of magnitude higher (up to 10^{10} to 10^{12}), however, several wearout/fatigue problems of the ferro-electric material have not really been solved yet.

Next to FRAM technology, there are a few other alternative memory technologies in development. *Magneto-resistive RAM (MRAM)* was one of the emerging memory technologies. An MRAM acts as the magnetic counterpart of an FRAM. An MRAM cell consists of a selection transistor and a *magnetic tunnel junction (MTJ)* stack for data storage. This stack is built from a sandwich of two ferro-magnetic layers separated by a thin dielectric barrier layer. One of the magnetic layers has a fixed polarisation direction, while the polarisation direction of the other one can be controlled by the direction of the current in the bit line. The MTJ stack resistance is increased in the case of anti-parallel magnetisation orientation. During a read operation, a current tunnels from one ferro-magnetic layer to the other through the dielectric layer and the resistance state is detected.

This state is compared with the electrical resistance of a reference cell, which is always in the low resistance state.

The MTJ cell can be integrated above the selection transistor to achieve a small cell size and a cost-effective memory solution. It allows a virtually unlimited number of fast read and write cycles, comparable to DRAM and SRAM. The first prototype 16 Mb MRAM was already announced in 2004 [27]. Volume production of the first commercial 4 Mb stand-alone MRAM has been announced in 2007. An MRAM requires a relatively high current (several milli-amps) to program one bit. This high program current, combined with the reducing quality (robustness and reliability) of its magneto-resistive operation, form severe roadblocks for scaling MRAM devices beyond the 65 nm node.

Another interesting non-volatile RAM alternative is the so-called *Phase-Change Memory (PCM)*, also known as *PRAM* and *Ovonic Unified Memory (OUM)*. Its basic operation uses a unique property of poly-crystalline *chalcogenide alloy*. This so-called phase-change property is also used for recording and erasing in optical media (re-writable CD and DVD). In these media the required heat for programming is generated through exposure to an intense laser beam.

Figure 6.35 shows a cross section of a basic PRAM storage cell.

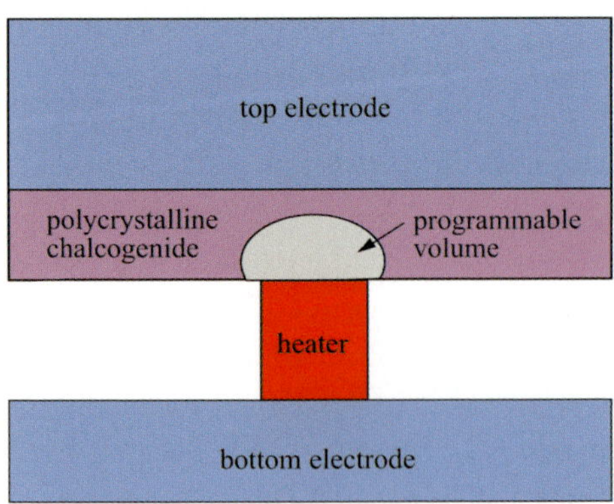

Figure 6.35: *Basic cross section of a Phase-Change Memory*

Under the influence of heat generated by Joule heating, the polycrystalline state can be changed into an amorphous state and back. Each of these states corresponds to a different value of electrical resistivity representing a logic one and logic zero. When heated above melting point ($\approx 650°C$), the chalcogenide alloy totally loses its crystalline structure. In the cell in figure 6.35 only a small programmable volume of the material is locally heated to above its melting point. This heat can be supplied by a current pulse through the heater. When rapidly cooled, the chalcogenide material is locked into its amorphous high-impedance state. By heating the material above its crystallisation but below its melting temperature the cell switches back to its low-impedance crystalline state. The difference in impedance between the two states is between one and two orders of magnitude. During a read operation, a voltage is applied to the cell and the read current is measured against the read current of an identical reference cell with a fixed logic state. Similar to all memories that use a reference cell during reading, this cell must be placed well within the memory array to reduce array edge effects, or must be surrounded by dummy cells.

Today's PRAM complexity is still far behind that of DRAM and SRAM, but due to the extending application areas with limited power budgets, particularly in hand-held devices, there is more pressure to develop a real non-volatile RAM. Volume production of the first commercial PRAM was announced for the second half of 2007. Examples of PRAM designs can be found in [28] and [29]. PRAMs are currently seen as the most promising successor of NOR flash.

Many other non-volatile techniques are currently in basic R&D phase. There are two that look promising and which I only want to mention here. The first one is the *resistive RAM (RRAM)*, a memory, whose operation is also based on resistance change. It consists of a metal/perovskite-oxide/metal sandwich structure, in which a reversible resistance switching behavior can be triggered by the application of short voltage pulses [30, 31]. Finally, the second one is the *conductive bridging memory (CBRAM)*, in which the cell consists of a thin electrolyte layer, sandwiched between two electrodes. The logic state of a cell is defined by deposition or removal of metal, e.g., Ag, within this electrolyte layer. The mechanism is promising because of its scalability, its multi-level capability and low-power potentials. A 2 Mb demonstrator has recently been published [32].

6.6 Embedded memories

The integration of complete systems-on-a-chip (SoC) include the combination of logic circuits (logic cores) with memories (memory cores). There are several reasons to put memories and logic on the same chip. In many cases this is (and will be) done to:
- offer higher bandwidth
- reduce pincount
- reduce system size
- offer a more reliable system
- reduce system power
Also the low cost of interconnect at chip level may be a good reason to embed memories or other cores. The diagram [33] in figure 6.36 shows the relative cost of interconnect as a function of the distance from the center of the chip. It clearly shows that the chip level interconnect is by far the cheapest one.

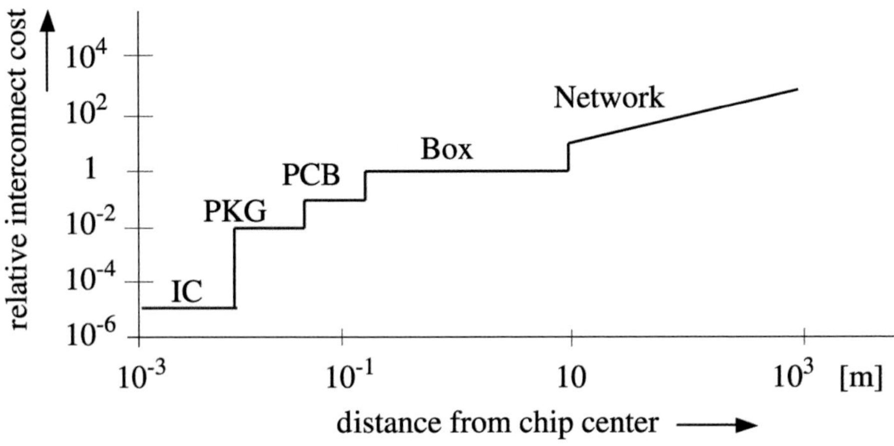

Figure 6.36: *Relative cost of interconnect*

Another reason to embed memories is to fill the design productivity gap. Figure 6.37 shows this gap with respect to the growth in IC complexity according to the ITRS roadmap [34]. The solid line represents the number of logic transistors per chip. The dotted line shows the design productivity. Many of the transistors made available by the technology, but unused by the design, may be used to increase the amount of embedded memory.

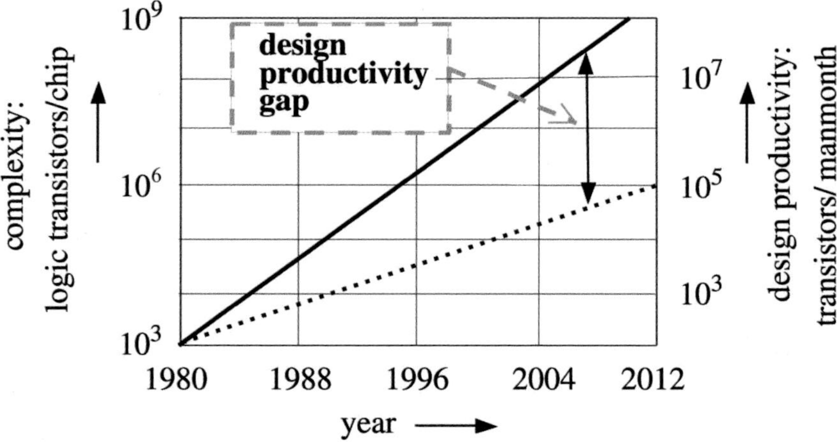

Figure 6.37: *The design productivity gap with respect to the growth in IC complexity*

Basically, there are three different approaches to implement a SOC.

The first one is to embed memories in a logic chip, integrated in a logic-based process (*embedded memory*). Basically all common types of memories can be embedded: SRAM, DRAM, ROM, E(E)PROM and flash memories. Except for the SRAM, they all require several additional masks and processing steps to efficiently embed these memories. Traditionally, the advantage of stand-alone SRAMs was twofold: they offered a higher speed than DRAMs and consumed less power in standby mode. Both advantages, however, are losing ground. The speed of the SRAMs has become so high, that their usage is more and more hampered by the limited bandwidth of the interface between ICs. As already discussed in section 6.4.2, SRAM leakage has reached so high levels, that, in combination with increasing variability, a read operation can cause a parasitic write, which may flip the cell and results in different and almost conflicting transistor sizing requirements for the read and the write operation. The increasing leakage is not only a limiting factor in achieving low-standby power levels, it also contributes to a reduced static noise margin of an SRAM cell. This has resulted in a decrease of interest and production of high-density stand-alone SRAMs. This is also reflected by the papers in the SRAM sessions at the ISSCC conference over the last couple of years [35]. The focus is no longer on or towards stand-alone Gb SRAMs, but more on high-performance high-density caches in high-performance CPUs and alternative SRAM cells with improved robustness (SNM).

Compared to stand-alone memories, embedded memories show a much lower bit density. For embedded SRAMs this difference may be in the order of a factor of two. For embedded DRAMs this factor can be much more (from three to eight), depending on the amount of extra masks and processing steps that are added to the standard CMOS logic process.

Embedded SRAM and/or DRAM can be found in practically every application. Also the usage of embedded flash spans a wide range of applications: micro controllers, industrial, office automation, networking, consumer, smart cards and RFID tags. Today, the increasing requirements in number of features and bandwidth has led to a growth of both the capacity and number of distributed embedded memory instances on a single chip. Complex microprocessors may incorporate level-one, level-two and sometimes even level-three caches and contain several hundreds of millions to even more than a billion transistors, of which most ($>$ 80%) are located in these embedded memories.

Certain applications demand even higher memory capacities and require the highest possible density of the memory blocks. Therefore, the second approach is to embed logic (processors etc.) in a memory (mostly DRAM) process (*embedded logic*). A DRAM in a logic-based process will not be as compact as in a DRAM-based process, because this process has been optimised for it. Analogous to this, logic in a DRAM-based process will not be as compact as in a logic-based process, partly because DRAM processes use fewer metal layers than current logic processes. Next to a higher memory density, embedded logic potentially offers much higher memory bandwidth due to the fact that bus width can be much larger ($>$ 1024b) since these are not pin limited. There are several examples of embedded logic processes: flash-based embedded logic process [36] and DRAM-based embedded logic process [37].

However, the emerging graphics market requires very high speed DRAMs (see section 6.4.4) at limited power consumption, which drives the need for merged DRAM + logic processes (*Merged Memory Logic (MML)*). This is the third approach: to achieve the required logic density, an increased number of metal layers is added to a DRAM-based process. The decision to start from a DRAM with *embedded logic*, or from a logic process with embedded DRAM depends largely on the required memory capacity, the complexity of the logic part, the yield and the possible integration of IP cores.

Testing is a problem that arises with the merging of huge memory blocks with logic on a single chip. In a large-complexity, stand-alone memory, true memory performance can be measured because of the accessibility of the memory through the I/O pads. When such complex memories are embedded, direct accessibility through the pads is obviously less, because this is often done by multiplexing to I/O pads. BIST techniques are required to minimise testing costs and wafer handling (see chapter 10).

6.7 Classification of the various memories

Table 6.3 provides an overview of the different types of memories with respect to some important parameters that characterise them. The numbers in table 6.3 are orders of magnitudes and may vary between different memory vendors. The characteristic values of these parameters render each type of memory suitable for application areas. These areas are summarised in table 6.2.

Table 6.2: Application areas for the various memory types

Memory type	Application areas
SRAM	Super-fast systems, low-power systems, cache memories in PCs (hard disks, DVD R/W caches), workstations, telecommunication, multimedia computers, networking applications, mobile phones, supercomputers, mainframes, servers, embedded memories
DRAM	Medium to high speed, main memory in computer systems, desktop, server, low-cost systems, networking, large volumes, PC, hard disk drives, graphics boards, printer applications, PDAs, camcorders, embedded memories, embedded logic
FRAM	Low-power, non-volatile applications, smart cards, RF Identification, replacement of non-volatile RAM and potentially high-density SRAM
ROM	large volumes, video games, character generators, laser printer fonts, dictionary data in word processors, sound source data in electronic musical instruments embedded memories, e-book
EPROM	CD-ROM drives, modems, code storage, embedded memories
EEPROM	Military applications, flight controllers, consumer applications, portable consumer pagers, modems, cellular and cordless telephones, disk drives, printers, air bags, anti-lock braking systems, car radios, smart card, set-top boxes, embedded memories
FLASH	Portable systems, communication systems, code storage, digital TV, set-top boxes, memory PC cards, BIOS storage, digital cameras, PDAs, ATA controllers, flash cards, palm tops, battery powered applications, mobile phones embedded memories, MP3 players
NVRAM BRAM	Systems where power dips are not allowed, medical systems, space crafts, etc, which require fast read and write access

6.8 Conclusions

The MOS memory market turnover currently represents about one third of the total IC market turnover. This indicates the importance of their use in various applications. Most applications have different requirements on parameters such as memory capacity, power dissipation, access time, retention time and reprogrammability, etc. Modern integrated circuit technology facilitates the manufacture of a wide range of memories that are each optimised for one or more application domains. The continuous drive for larger memory performance and capacity is leading to ever-increasing bit densities and the limits are not yet in sight. The DRAM and flash markets show the largest volumes and, not surprisingly, the highest demand for new technologies (figure 6.2.b).

This has resulted in the presentation of the first 4 and 8 Gb versions at the 2005 and 2006 ISSCC conferences. By the year 2015, it is expected that the cost per bit of non-volatile semiconductor memories has reached the level of magnetic and mechanic harddisks and may even replace them by that time.

This chapter presents the basic operating principles of the most popular range of stand-alone memory types. Their characteristic parameters are compared in table 6.3 and their application areas are summarised in table 6.2.

Table 6.3: Characteristics of different memory types

DEVICE		SRAM	DRAM	ROM	PROM	NOR-flash	NAND-flash	FRAM
physical cell size		$150\text{-}200F^2$	$4\text{-}8F^2$	$4F^2$	$4F^2$	$8\text{-}10F^2$	$4\text{-}5F^2$	$15F^2$
max. number of programming cycles		∞	∞	1	1	$10^4\text{-}10^5$	$10^3\text{-}10^4$	$10^{10}\text{-}10^{12}$
programming time (write)		5-40 ns	20-100 ns	-	5-80 ms	5-10 μs	100-300 μs	80-120 ns
access time (read)		5-20 ns	10-70 ns	5-20 ns	5-20 ms	random: 80-150ns serial: 80-120 ns	random: 10-20μs serial: 5-50ns	80-120 ns
retention time	no power supply	0	0	∞	∞	> 10 years	> 10 years	> 10 years
	power supply	∞	2-20ms					

6.9 References

Information about memories is usually confidential and is often proprietary. Many of the relatively few books available on the subject are therefore outdated. This reference list therefore only contains a few published books and the titles of interesting journals and digests on relevant conferences. In this edition it is extended with many references on state-of-the-art material from conferences, publications, and internet sites.

[1] K.Pagiamtzis, et al.,
'Content-addressable memory (CAM) circuits and architectures: A tutorial and survey',
IEEE Journal of Solid-State Circuits, Vol. 41, No. 3, pp. 712-727, March 2006

[2] K. Noda, et al.
'A $1.9\,\mu\text{m}^2$ Loadless CMOS Four Transistor SRAM Cell in a $0.18\,\mu\text{m}$ Logic Technology',
IEDM Digest of Technical Papers, December 1998, pp 643-646

[3] K. Takeda, et al.
'A 16 Mb 400 MHz loadless CMOS 4-Transistor SRAM Macro',
ISSCC Digest of Technical Papers, February 2000

[4] L. Chang, et al.,
'Stable SRAM Cell Design for the 32 nm Node and Beyond',
2005 Symposium on VLSI Technology, Digest of Technical Papers, pp. 128-129

[5] S.M Jung, et al.,
'Highly Area Efficient and Cost Effective Double Stacked S (Stacked Single-crystal Si) peripheral CMOS SSTFT and SRAM Cell Technology for 512 Mb SRAM',
IEDM 2004, Digest of Technical Papers, pp. 265-268

[6] 'Designing for 1 GB DDR SDRAM',
Micron Technology, Technical Note, 2003

[7] Changhyun Cho, et al.,
'A 6F^2 DRAM Technology in 60 nm era for Gigabit Densities',

2005 Symposium on VLSI Technology, Digest of Technical Papers, pp. 36-37

[8] J.A.Mandelman, et al.,
'Challenges for future directions for the scaling of DRAM',
IBM J. Res. & Dev. Vol. 46, No.2/3, March/May 2002

[9] H.Seidl, et al.,
A fully integrated Al_2O_3 trench capacitor DRAM for sub-100 nm technology',
IEDM, 2002

[10] C. Hampel,
'High-speed DRAMs keep pace with high-speed systems',
EDN, February 3, 1997, pp 141-148

[11] C. Green,
'Analyzing and implementing SDRAM and SGRAM controllers',
EDN, February 2, 1998, pp 155-166

[12] Reza Faramarzi,
'High Speed Trends In Memory Market',
Keynote adress, Jedex conference, Oct. 25-26, 2006, Shanghai
http://www.jedexchina.org/program.htm

[13] 'Samsung Develops Ultra-fast Graphics Memory: A More Advanced GDDR4 at Higher Density',
Press Release (Feb 14, 2006 / SEC)

[14] *www.chips.ibm.com/products/memory.*
'Understanding Video (VRAM) and SGRAM Operation'

[15] D. Bursky,
'Graphics-Optimized DRAMs deliver Top-Notch Performance',
Electronic design, March 23, 1998, pp 89-100

[16] Kyu-hyoun Kim, et al.,
'An 8 Gb/s/pin 9.6 ns Row-Cycle 288 Mb Deca-Data Rate SDRAM with an I/O error-detection Scheme',
ISSCC Digest of Technical papers, Feb. 2006, pp.154-155

[17] Takeshi Nagai,
'CA 65 nm Low-Power Embedded DRAM with Extended Data-Retention

Sleep Mode',
ISSCC Digest of Technical papers, Feb. 2006, pp.164-165

[18] P.Fazan,
'Z-RAM zero capacitor Embedded Memory Technology addresses
dual requirements of die size and scalability',
http://clients.concept-web.ch/is/en/technology_white_paper.php#

[19] B. Dipert,
'EEPROM, survival of the fittest',
EDN, January 15, 1998, pp 77-90

[20] Mike Clendenin,
'Samsung wraps up 16 NAND die in multi-chip package',
EETimes, 11-01-2006

[21] Peter Clarke,
'Samsung takes 16-Gbit NAND flash to 50 nm',
EETimes, 03-01-2007

[22] Dae-Seok Byeon, et al.,
'An 8 Gb Multi-Level NAND Flash Memory with 63 nm STI CMOS
Process Technology',
ISSCC Digest of Technical papers, Feb. 2005, pp.46-47

[23] Soon -Moon Jung, et al.,
'Three Dimensionally Stacked NAND Flash Memory Technology Using Stacking Single Crystal Si Layers on ILD and TANOS Structure
for Beyond 30nm Node',
EDM 2006 Digest of technical papers, pp. 37-40

[24] R.Micheloni, et al.,
'b/cell NAND Flash Memory with Embedded 5b BCH ECC for 36MB/s
System Read Throughput',
ISSCC Digest of Technical papers, Feb. 2006, pp.132-133

[25] B. Dipert,
'FRAM: ready to ditch niche?',
EDN, April 10, 1997, pp 93-107

[26] K.Hoya, et al.,
'A 64 Mb Chain FeRAM with Quad-BL Architecture and 200MB/s
Burst Mode',
ISSCC Digest of Technical Papers, Feb. 2006, pp. 134-135

[27] Richard Wilson,
'MRAM Steps to 16Mbit',
Electronics Weekly, June 23, 2004

[28] W.Y. Cho, et al.,
'A 0.18 μm 3.0 V 64 Mb Nonvolatile Phase-Transistion Random Access memory (PRAM)',
IEEE Journal of Solid-State Circuits, Vol. 40, Jan. 2005

[29] S.Kang, et al.,
'A 0.1 μm 1.8 V 256 Mb 66 MHz Synchronous Burst PRAM',
ISSCC Digest of Technical Papers, Feb. 2006, pp.140-141

[30] Xin CHEN, et al.,
'Buffer-Enhanced Electrical-Pulse-Induced-Resistive Memory Effect in Thin Film Perovskites',
Jpn. J. Appl. Phys. Vol. 45 (2006) Part 1, No. 3A, pp. 1602-1606

[31] Peter Clarke,
'Resistive RAM sets chip companies racing',
EETimes, 04-24-2006

[32] H. Hnigschmid, et al.,
'A Non-Volatile 2 Mbit CBRAM Memory Core Featuring Advanced Read and Program Control',
Proc. 2006 Symposiumn on VLSI Circuits, pp. 138-139

[33] J.S. Mayo,
Scientific American, 1981

[34] Semiconductors Industrial Associations,
ITRS roadmap, yearly update, http://www.itrs.net

[35] SRAM sessions,
International Solid States Circuits Conference 2005 and 2006,
ISSCC Digest of Technical Papers, 2005 and 2006

[36] Al Fazio, et al.,
'ETOXTM Flash Memory Technology: Scaling and Integration Challenges',
May 16, 2002,
http://developer.intel.com/technology/itj/2002/volume06issue02/art03_flashmemory/vol6iss2_art03.pdf

[37] Linley Gwennap,
'Day dawns for eDRAM',
EETimes, 04/14/2003,
http://www.eetimes.com/op/showArticle.jhtml?articleID=16500906

Further reading

[37] B. Prince,
'Semiconductor Memories: A Handbook of Design, Manufacture and Application',
John Wiley & Sons, New York, 1996

[38] W.J. McClean,
'Status 1999, A report on the IC industry',
ICE corporation, Scottsdale, Arizona, 1999

[39] B. Prince,
'High Performance Memories',
John Wiley & Sons, New York, 1996

[40] 'IEEE digest of technical papers of the International Solid State Circuit Conference'.
The ISSCC is held every year in February in San Francisco.

[41] IEEE Journal of Solid-State Circuits

[42] IEDM Digest of technical Papers, since 1984.

6.10 Exercises

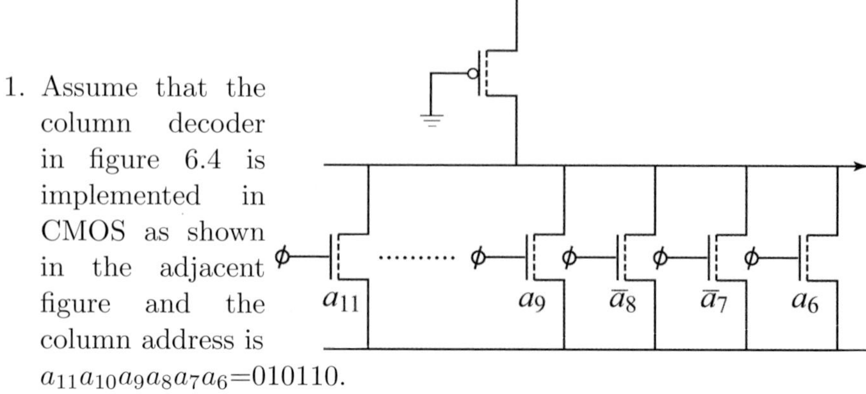

1. Assume that the column decoder in figure 6.4 is implemented in CMOS as shown in the adjacent figure and the column address is $a_{11}a_{10}a_9a_8a_7a_6{=}010110$.

   a) Describe the data flow in figure 6.4 during a read operation when word line x_{20} is also selected.

   b) What is the major disadvantage of such a decoder?

   c) What would be the problem if this decoder were implemented in static CMOS?

2. Describe the major differences between the ROM realisations of figures 6.22 and 6.25. Explain their relative advantages and disadvantages.

3. Why does a stand-alone flash EPROM sometimes require one more power supply than a full-featured EEPROM?

4. Table 6.3 gives a summary of some important memory parameters.

   a) Explain the difference in chip area between a non-volatile RAM and an SRAM.

   b) Explain the difference in access times between an SRAM and a DRAM.

5. The adjacent figure shows a dynamic memory cell which consists of three transistors. This is a so-called *3 T-cell*.

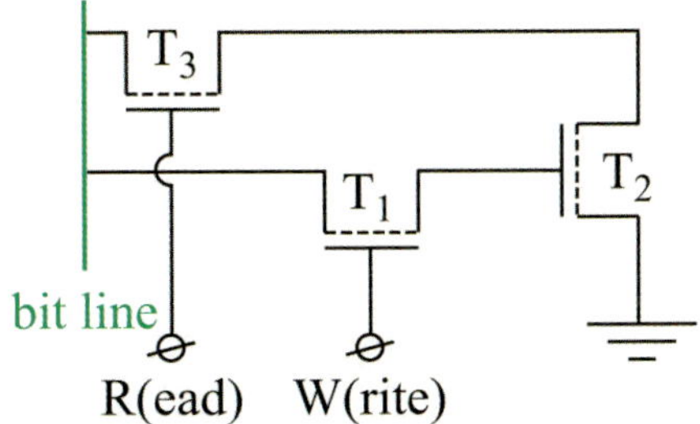

a) Explain the operation of the 3 T-cell.

b) What can be said about the read-out data after one write and read cycle?

c) Comment on the size of the storage nodes in the 3 T-cell and the 1 T-cell?

6. What is a multilevel flash memory?
What is the reason of its existence?
What would be their main problem for future process generations?

7. Explain the difference between an embedded memory and embedded logic.

Chapter 7

Very Large Scale Integration (VLSI) and ASICs

7.1   Introduction

The continuing development of IC technology during the last couple of decades has led to a considerable increase in the number of devices per unit chip area. The resulting feasible IC complexity currently allows the integration of a complete *system on a chip (SOC)*, which may comprise hundreds of millions to a few billion transistors.

Consequently, the design of such chips no longer simply consists of the assembly of a large number of logic gates. This poses a problem at a high level of design: how to manage the design complexity. Besides this, the growing influence of parasitic and scaling effects (see chapters 2, 9, and 11), which may reduce chip performance dramatically, requires a lot of additional design resources to take and implement adequate measures.

Such ICs combine signal processing capacity with microprocessor or microcontroller cores and memories. The dedicated signal processing parts take care of the computing power (workhorse), while the microprocessor or controller serves to control the process and possibly performs some low performance computation as well. The memories may store program code and data samples. The development of such heterogeneous systems on one or more ICs, for instance, may require tens to even hundreds of man-years, depending on their complexity.

A significant amount of total IC turnover is generated in the "low-end market". This market consists of low-complexity ICs and was originally controlled by the large IC vendors. During the eighties and nineties, however, a change took place and the low-end market is now dominated by *Application-Specific Integrated Circuits (ASICs)*. These are ICs which are realised for a single end-user and dedicated to a particular application. ASICs therefore implement customer-specified functions and there are various possibilities for the associated *customisation*. This can be an integral part of an IC's design or production process or it can be accomplished by programming special devices.

ASICs do not include ICs whose functionality is solely determined by IC vendors. Examples of these *"Application-Specific Standard Products" (ASSPs)* include digital-to-analogue (D/A) converters in DVD players. These ASSPs are so-called vendor-driven ICs, of which the vendor wants to sell as many as possible to every customer he can find. ASICs are customer-driven ICs, which are only tailored to the specific requirements of one single customer. Actually, *User-Specific Integrated Circuits (USICs)* would be a more appropriate name for ASICs. The use of USICs would clearly be preferable because it emphasises the fact that the IC function is determined by the customer's specification and not simply by the application area.

The *turn-around time* of an ASIC is the period which elapses between the moment a customer supplies an IC's logic *netlist* description and the moment the vendor supplies the first samples. The turn-around time associated with an ASIC depends on the chosen implementation type. A short turn-around time facilitates rapid prototyping and is important to company marketing strategies. In addition, ASICs are essential for the development of many real-time systems, where designs can only be verified when they are implemented in hardware. There exist many different market segments for which we can distinguish different ASIC products:

- Automotive: networking, infotainment, GPS, tire pressure monitor, body electronics

- Mobile communications: mobile phones (GSM, UMTS), modems, wireless local loop (WLL)

- Medical: patient monitoring, diagnostics, ultrasound

- Display: LCD TV, flat panel, projection TV

- Digital consumer: CD/DVD, MP3, audio, TV, set-top box, encoders/decoders

- Connectivity: WLAN, Blue Tooth, USB, FireWire

- Identification: smart cards and RF-ID tags

- Industrial: robotics, motor/servo control

- Military: image, radar and sonar processing, navigation

Suitable computer aided design (CAD) tools are therefore essential for the realisation of this rapidly expanding group of modern ICs. Growing design complexity combined with shorter product market windows requires the development of an efficient and effective design infrastructure, based on a (application-) domain-specific *SoC design platform*. In this respect, a platform is an integrated design environment, consisting of standard-cell libraries, IPs and application-mapping tools, which is aimed at providing a short and reliable route from high-level specification to correct silicon. The convergence of consumer, computing and communications domains accelerates the introduction of new features on a single chip, requiring a broader range of standards and functions for an increasing market diversity. This makes a design more heterogeneous, with a large variety of domain-specific, general-purpose IP and memory cores. Next to this, there is a dramatic grow in the complexity of embedded software, which may take more than 50% of the total SoC development costs, particularly in multi-processor design.

This puts very high demands on the flexibility and reusability of a platform across a wide range of application derivatives, requiring a large diversity of fast-compiling IPs in combination with efficient verification, debug and analysis tools. Such a platform needs to be scalable and must also allow adding new IP cores without the need for changing the rest of the system.

The design process is discussed on the basis of an ASIC design flow. The various implementation possibilities for digital VLSI and ASICs are discussed and factors that affect a customer's implementation choice are examined. These implementations include: standard-cell, gate-array, field-programmable gate-array (FPGA) and programmable logic devices (PLD). Market trends and technological advances in the major ASIC sectors are also explained.

7.2 Digital ICs

Digital ICs can be subdivided into different categories, as shown in figure 7.1. ASICs can be classified according to the processing or programming techniques used for their realisation. A clear definition of the types and characteristics of available digital ICs and ASICs is a prerequisite for the subsequent discussion of the trends in the various ASIC products. Figure 7.1 presents quite a broad overview of digital ICs but excludes details such as the use of *direct slice writing (DSW)* or masks for IC production. Several terms used in this figure and throughout this chapter are explained on the next page.

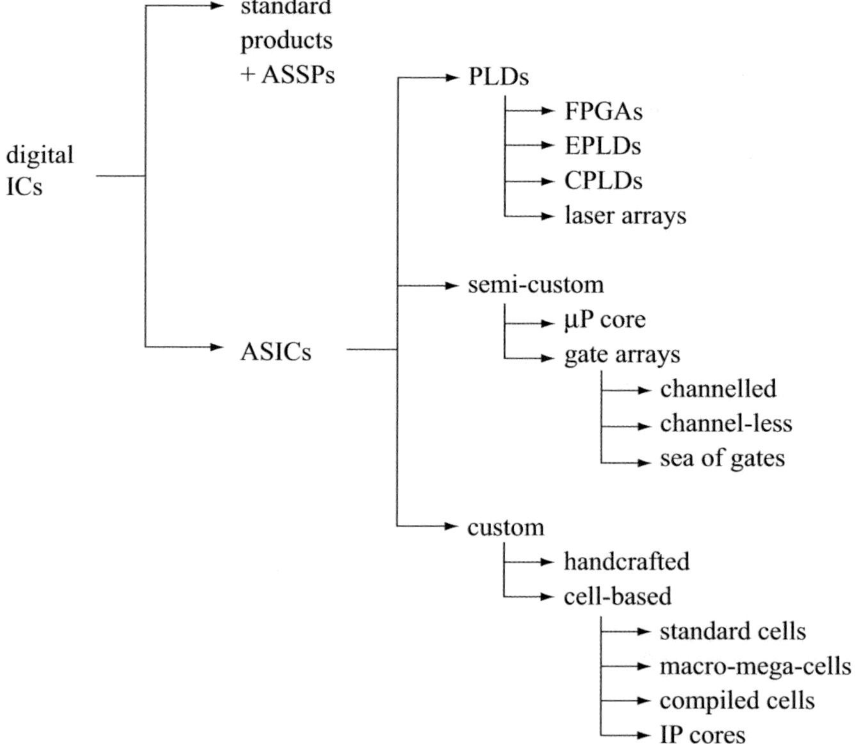

Figure 7.1: *An overview of digital ICs*

<u>**Definitions:**</u>

ASSP: *Application-Specific Standard Products* are ICs that are suitable for only one application but their availability is not restricted to a single customer. Examples include video ICs for teletext decoding and ICs for D/A conversion in DVD players.

Core: Pre-designed industry (or company) standard building block: RAM, ROM, microprocessor (e.g., ARM, MIPS and Sparc), etc.

Custom: A *custom IC* is an IC in which all masks are unique for a customer's application. The term *full-custom IC* is often used to refer to an IC in which many sub-circuits are new handcrafted designs. In this book, full-custom ICs fall under the category of custom ICs. Cell-based custom-IC designs are based on standard cells, *macro cells*, *mega cells* and possibly *compiled cells*. Macro and mega cells, or cores are large library cells like multipliers, RAMs, ROMs and even complete microprocessors and signal processors. Compiled cells are automatically generated by modern software libraries. These cells are used for dedicated applications and are generated as a function of user-supplied parameters.

The customisation of PLD-based ASICs takes place after IC manufacture. Customisation of custom and semi-custom ASICs, however, is an integral part of IC manufacture. The *turn-around time* of ASICs from database ready to first silicon varies enormously and depends on circuit complexity and the customisation technique. This time can range from a few hours for a PLD to between six and twelve weeks for a custom design.

HDL: *Hardware description language.* This language is used for formal description of the behaviour and specification of electronic circuits. It provides the circuit designer to describe (model) a circuit before it is physically implemented. Synthesis tools are able to read this language, extract logic operation, and transfer these into a netlist of logic gates.

IP: *Intellectual Property.* With the complexity of ICs reaching a billion or more transistors, the traditional way of designing can no longer be continued. Therefore, the concept of *Virtual Component* has been introduced by the Virtual Socket Interface Alliance (VSI Alliance: www.vsi.org), which is an international forum trying to standardise reusable cores, concepts, interfaces, test concepts and support, etc. Licensing and royalty issues of IP must also be addressed. This standardisation is a prerequisite to fully exploit the potentials of design reuse. The cores (or IP) can be represented in three forms.

A *soft core* is delivered in the form of synthesizable HDL, and has the advantage of being more flexible and the disadvantage of not being as predictable in terms of performance (timing, area, power). Soft cores typically have increased intellectual property protection risks because RTL source code is required by the integrator.

Firm cores have been optimised in structure and in topology for performance and area through floor planning and placement, possibly using a generic technology library. The level of detail ranges from region placement of RTL sub-blocks, to relatively placed data paths, to parameterised generators, to a fully placed netlist. Often, a combination of these approaches is used to meet the design goals. Protection risk is equivalent to that of soft cores if RTL is included, and is less if it is not included.

Finally, *hard cores* have been optimised for power, size or performance and mapped to a specific technology. Examples include netlists fully placed, routed and optimised for a specific technology library, a custom physical layout or the combination of the two. Hard cores are process- or vendor-specific and generally expressed in the GDSII format. They have the advantage of being much more predictable, but are consequently less flexible and portable because of process dependencies. The ability to legally protect hard cores is much better because of copyright protections and there is no requirement for RTL.

Figure 7.2 is a graphical representation of a design flow view and summarises the high level differences between soft, firm and hard cores.

Due to the convergence of digital communications, consumer and computer, there is an increasing number of real-time signals to be processed: voice, professional audio, video, telephony, data streams, etc. This is usually performed by high-performance analog and digital signal processors.

Today's integrated circuits are complex heterogeneous systems: they consist of many different types of processing, storage, control and interface elements. Many of these elements are available as a kind of (standard) IP. Examples of IP are:

- Microprocessors (CPU): use software to control the rest of the system
    - Intel, SPARC, PowerPC, ARM, MIPS, 80C51, ...
- Digital signal processors (DSP): manipulate audio, video and data streams
    - Omaps, TMS320 and DaVinci (TI), DSP56000 series (Freescale), DSP16000 series (Agere), EPICS and Trimedia and EPICS (NXP), Oak, Teaklite
    - Most DSPs are for wireless products
- (F)PGA-based accelerators: decoders, encoders, error correction, encryption, graphics or other intensive tasks
- Memories
    - Virage, Artisan, embedded memories and caches
    - Memory controllers (Denali): controlling off-chip memories
- Interfaces: external connections
    - USB, FireWire, Ethernet, UART, Bluetooth, keyboard, display or monitor
- Analog
    - A/D, D/A, PLL (e.g., for use in clock generation), oscillator, operational amplifier, differential amplifier, bandgap reference

PLD: *Programmable Logic Devices* are ICs that are customised by blowing on-chip fuses or by programming on-chip memory cells. Most PLDs can be customised by end-users themselves in the field of application, i.e., they are *field-programmable* devices (FPGA). The customisation techniques used are classified as *reversible* and *irreversible*. PLDs include *erasable* and *electrically erasable* types, which are known as *EPLDs* and *EEPLD*, respectively. The former are programmed using EPROM techniques while the EEPROM programming technique is used for the latter devices. These programming techniques are explained in sections 6.5.3 and 6.5.4 respectively. Complex PLDs (CPLDs) are often based on the combination of PAL^{TM} and PLA architectures.

Reuse: Future design efficiency will increasingly depend on the availability of a variety of pre-designed building blocks (IP cores). This *reuse* not only requires easy portability of these cores between different ICs, but also between different companies. Standardisation is one important issue, here (see IP definition). Another important issue concerning reuse is the quality of the (IP) cores. Similar to the Known-Good Die (KGD) principle when using different ICs in an MCM, we face a Known-Good Core (KGC) principle when using different cores in one design. The design robustness of such cores must be so high that their correctness of operation will always be independent of the design in which it is embedded.

RTL: Register transfer level. See section 7.3.4.

Semi-Custom: These are ICs in which one or more but not all masks are unique for a customer's application. Many semi-custom ICs are based on 'off-the-shelf' ICs which have been processed up to the final contact and metal layers. Customisation of these ICs therefore only requires processing of these final contacts and metal layers. This results in short turn-around times. A gate array is an example in this semi-custom category.

Standard product: Standard products, also called *standard commodities*, include microprocessors, memories and standard-logic ICs, e.g., NAND, NOR, QUAD TWO-INPUT NAND. These ICs are produced in large volumes and available from different vendors. Their availability is unrestricted and they can be used in a wide variety of applications. They are often put into a product catalogue.

Usable gates: The number of gates that can actually be interconnected in an average design. This number is always less than the total number of available gates (gate array).

Utilisation factor: The ratio between that part of a logic block area which is actually occupied by functional logic cells and the total block area (gate array and cell-based designs).

	Design flow	Representation	Libraries	Technology	Portabillity
Soft not predictable very flexible	system design RTL design	behavioural RTL	N/A	technology independent	unlimited
Firm flexible predictable	floor planning synthesis placement	RTL & blocks netlist	reference library • footprint • timing model • wiring model	technology generic	library mapping
Hard not flexible very predictable	routing verification	polygon data	process-specific library & design rules • characterised cells • process rules	technology fixed	process mapping

Figure 7.2: *Graphical representation of soft, firm and hard cores (Source: VSIA)*

7.3 Abstraction levels for VLSI

7.3.1 Introduction

Most of today's complex VLSI designs and ASICs are synchronous designs, in which one or more clock signals control the data flow to, on and from the chip. On a chip, the data is synchronised through flip-flops, which are controlled by a clock ϕ (figure 7.3). Flip-flops temporarily store the data and let it go on clock demand. At any time the positions and values of all data samples are known (by simulations).

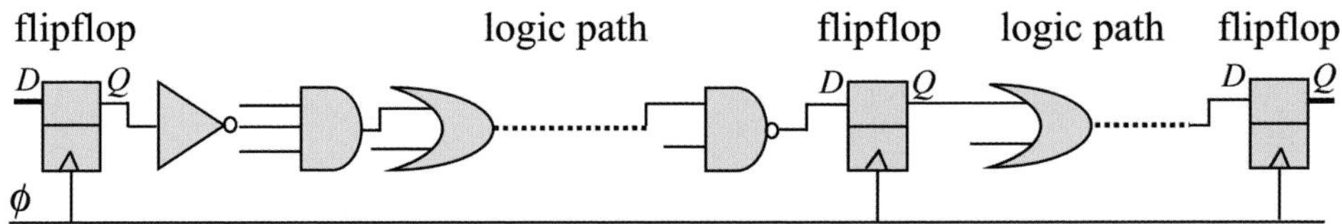

Figure 7.3: *Representation of a logic path in a synchronous design*

The logic gates in between the flip-flops perform the functionality of the logic block from which they are part. So, in a synchronous chip, the signal propagates through the logic path from one flip-flop to the next. The logic path with the longest propagation delay (usually one

373

with many complex gates) is called the *worst-case delay path*. This path determines the maximum allowed clock frequency. Next to many different functional logic blocks, most systems also contain memory, interface and peripheral blocks.

The implementation of a complete system on one or more ICs starts with an abstract system level specification. This specification is then analysed and transformed into a set of algorithms or operations. Next, an optimum architecture that efficiently performs these operations must be chosen.

A signal processor serves as an example. The chosen processor must perform an adaptive FIR filter. As a consequence, this processor must repeatedly fetch numbers from a memory, multiply or add them and then write the result back into the memory. Such a chip may contain several ROM and/or RAM memory units, a multiplier, an adder or accumulator, data and control buses and some other functional modules.

The design of an IC comprises the transformation of a specification into a layout. The layout must be suitable for the derivation of all process steps required for the manufacture of the IC's functional modules and their interconnections. Clearly, the *design path* starts at the top (or system) level and ends at the bottom (or silicon) level. This '*top-down*' process is illustrated in figure 7.4.

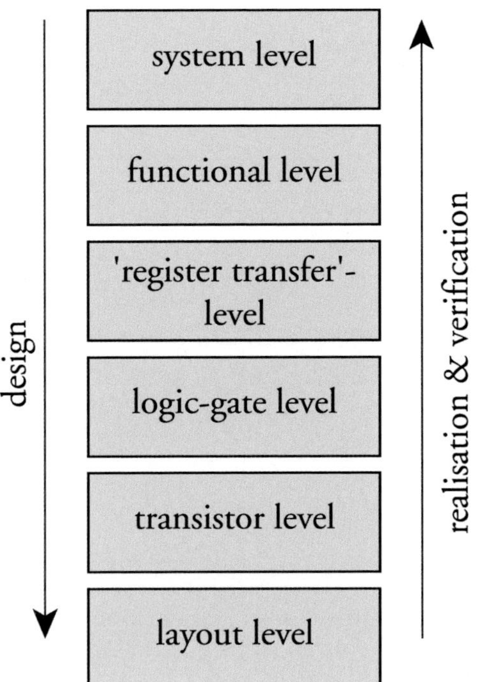

Figure 7.4: *Abstraction levels in the design and implementation/verification paths of VLSI circuits*

The various design phases are accompanied by several different *abstraction levels*, which limit the complexity of the relevant design description. The top-down design path allows one to make decisions across abstraction levels and gives high level feedback on specifications. The '*bottom-up*' path demonstrates the feasibility of the implementation of (critical) blocks. This process begins at the layout level of a single part and finishes with the verification of the entire IC layout. The abstraction levels that are used in the design path are described on the following pages. Table 7.1 shows the design complexity at these levels of abstraction.

Table 7.1: Design complexity at different levels of abstraction

Level	Example	Number of elements	
system	heterogeneous system	10^7-10^9	transistors
functional	signal processor	10^5-10^7	transistors
register	digital potentiometer	10^3-10^5	transistors
logic gate	Library cell (NAND, full adder)	2-30	transistors
transistor	nMOSt, pMOSt	1	transistor
layout	total SoC	10^8-10^{10}	rectangles

7.3.2 System level

A system is defined by the specification of its required behaviour. Such a system could be a multiprocessor system and/or a *heterogeneous system*, consisting of different types of processing elements: microprocessor, DSP, analog, control, peripheral and memory cores. Advanced heterogeneous architectures, today, also include the integration of graphics processing units (GPU) to increase graphics processing speed between one or two orders of magnitude, compared to running it on a CPU. Figure 7.5 shows a heterogeneous system, containing a signal processor, a microprocessor, embedded software, some glue logic (some additional overall control logic), local buses, a global bus, and the clock network. The transformation of a system into one or more ICs is subject to many constraints on timing, power and area, for example.

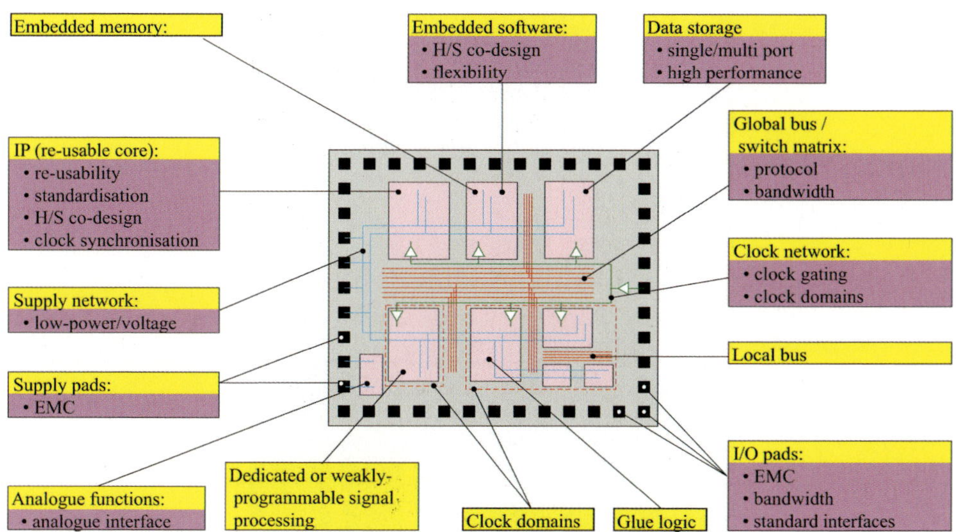

Figure 7.5: *Systems on a chip; an example of a heterogeneous system*

While a heterogeneous system consists of several different types of processing and storage elements, there is today also an increased focus on architectures with multi-processor cores and even architectures built from only a limited number of different cores. In the ultimate case, an architecture can be built from a multiple of identical cores (tiles) to create a *homogeneous sytem*. Figure 7.6 (top) shows a layout of a massively-parallel processor for video scene analysis implemented as a homogeneous design [1], as opposed to the heterogeneous chip (bottom).

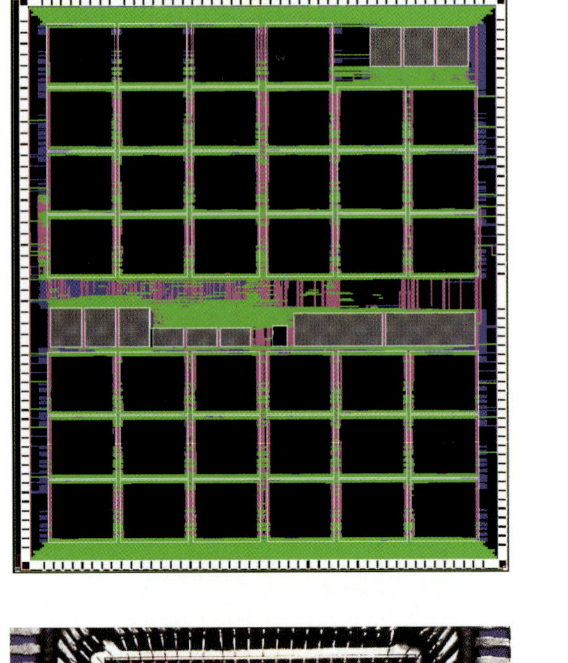

Figure 7.6: *Example of a homogeneous design, consisting of a multiple of identical cores (tiles) and a heterogeneous chip consisting of various different cores (Source: NXP Semiconductors)*

System decisions taken at the highest level have the most impact on the area and performance parameters. Decisions regarding functions that are to be implemented in hardware or software are made at the system level. Filter sections, for example, are frequently programmed in software. A system-level study should also determine the number of chips required for the integration of the chosen hardware. It is generally desirable to sub-divide each chip into several sub-blocks. For this purpose, *data paths* and *control paths* are often distinguished. The former is for data storage and data manipulation, while the latter controls the information flow in the data path, and to and from the outside world. Each block in the data path may possess its own *microcontrol unit*. This usually consists of a decoder which recognises a certain control signal and converts it into a set of instructions.

The block diagram shown in figure 7.7 represents a description of the signal processor of figure 7.5 at the system abstraction level. The double bus structure in this example allows parallel data processing. This is typically used where a very high data throughput is required. For example, data can be loaded into the Arithmetic Logic Unit (ALU) simultaneously from the ROM and the RAM. In this type of architecture, the data path and control path are completely separated. The control path is formed by the program ROM, which may include a program counter, control bus and the individual microcontrol units located in each data path element.

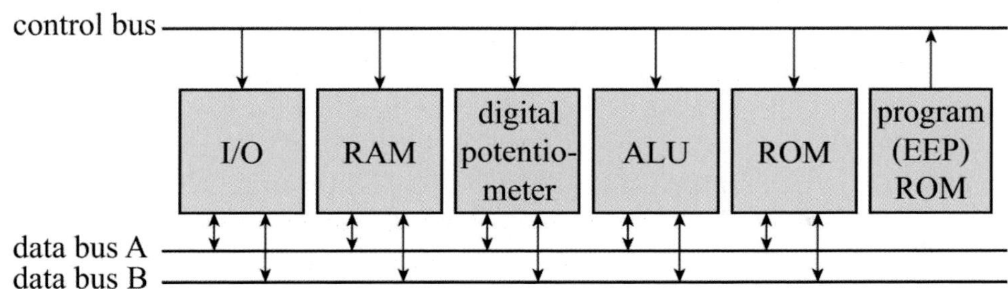

Figure 7.7: *Block diagram of a signal processor*

Other system implementations may not show such a clear separation of data and control paths.

7.3.3 Functional level

A description at this level of abstraction comprises the behaviour of the different processing elements and other cores of the system. In case of the signal processor of figure 7.7, we distinguish: an ALU, a digital potentiometer, a RAM, a ROM, and the I/O element.

RAMs, ROMs and I/O elements are usually not very complex in their behaviour. As a result of the simplicity of their behaviour, however, they are mostly described in the next, lower level of abstraction, the RTL level.

Let us take the digital potentiometer as an example. Also this one, because of its simple architecture, will be described at the lower RTL level.

There are some tools, mainly in development, that allow a description of complex blocks at functional level. The maturity and ease of use of these tools is not yet such that they are common part of current design flows.

The chosen potentiometer, at this hierarchy level, consists of different arithmetic units (adder, mutiplier, subtractor), which are functions as well, so the RTL level and functional level show some overlaps (see also figure 7.14).

7.3.4   RTL level

RTL is an abbreviation for Register-Transfer Language. This notation originates from the fact that most systems can be considered as collections of registers that store binary data, which is operated upon by logic circuits between these registers. The operations can be described in an RTL and may include complex arithmetic manipulations. The *RTL description* is not necessarily related to the final realisation.

To describe a function at this level is a difficult task. A small sentence in the spec, e.g., *performs MPEG4 encoding*, will take many lines of RTL code and its verification is extremely difficult. Logic simulation and/or even emulation may help during the verification process, but can not guarantee full functionality, since it is simply impossible to fully cover all possible cases and situations. Let us return to our digital potentiometer example. The behaviour of this potentiometer can be described as:

$$Z = k \cdot A + (1 - k) \cdot B$$

When $k = 0$, Z will be equal to B and when $k = 1$, Z will be equal to A. The description does not yet give any information about the number of bits in which A, B and k will be realised. This is one thing that must be chosen at this level. The other choice to be made here is what kind of multiplier must perform the required multiplications. There are several alternatives for multiplier implementation, of which some are discussed as examples.

- *Serial-parallel multiplier:* Input Ra input is bit-serial and the Rb input is bit-parallel, see figure 7.8.

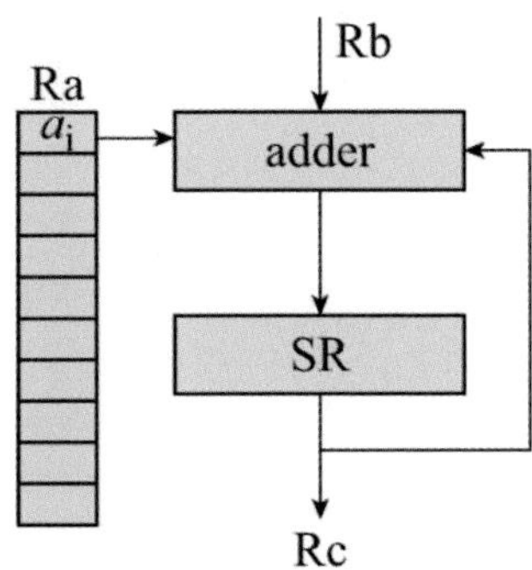

Figure 7.8: *Example of a bit-serial iterative multiplier*

During the execution of a multiplication, the *partial product* is present on the multiplier's parallel output bits (Rc). These are initially zero.

If $a_i{=}1$, for instance, then the Rb bits must be added to the existing partial product and then shifted one position to the left. This is a *'shift-and-add'* operation. When $a_i{=}0$, the Rb bits only have to be shifted one place to the left in a 'shift' operation and a zero LSB added to it.

- *Parallel multiplier:* The bits of both inputs Ra and Rb are supplied and processed simultaneously. This *'bit-parallel'* operation requires a different hardware realisation of the multiplier. Options include the *array or parallel multiplier*, schematically presented in figure 7.9.

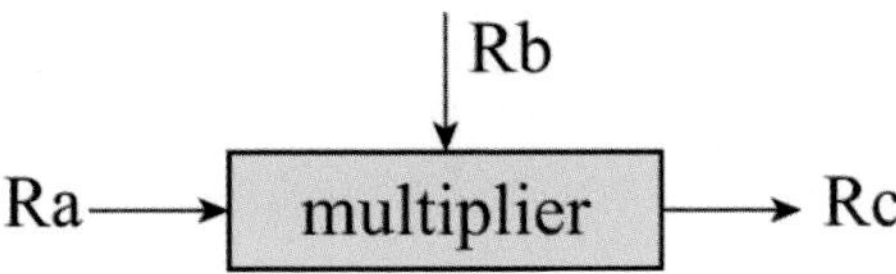

Figure 7.9: *A parallel multiplier*

The array multiplier necessitates the choice of a structure for the addition of the partial products. The possibilities include the following:

- *Wallace tree*: Here, bits with equal weights are added together in a tree-like structure, see figure 7.10. An advantage of the architecture is that the two input signals for each single adder always arrive at the same time, since they have propagated through identical delay paths. This will reduce the number of glitches at the outputs of the individual adder circuits, which may occur when there is too much discrepancy between the arrival times of the input signals.

- *Carry-save array*: Figure 7.11 illustrates the structure of this array, which consists of AND gates that produce all the individual $x_i \cdot y_i$ product bits and an array of full adders which produce the total addition of all product bits.

As an example, at this level, we choose the array multiplier (parallel multiplier) with carry-save array. This would lead to a different behaviour from the serial multiplier, and thus to a different RTL description.

An example of RTL-VHDL description for the potentiometer is given in figure 7.16.

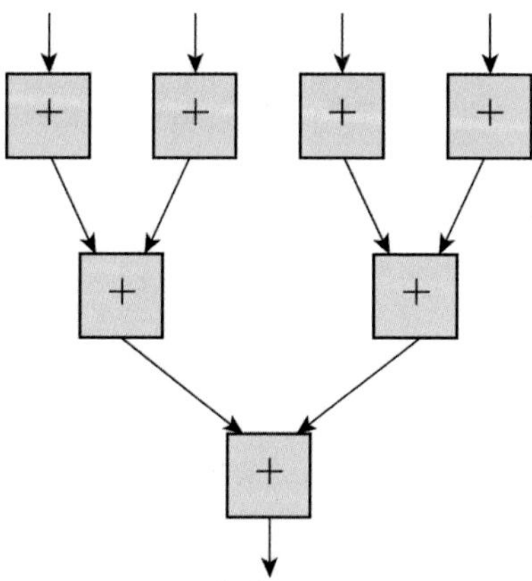

Figure 7.10: *Wallace tree addition*

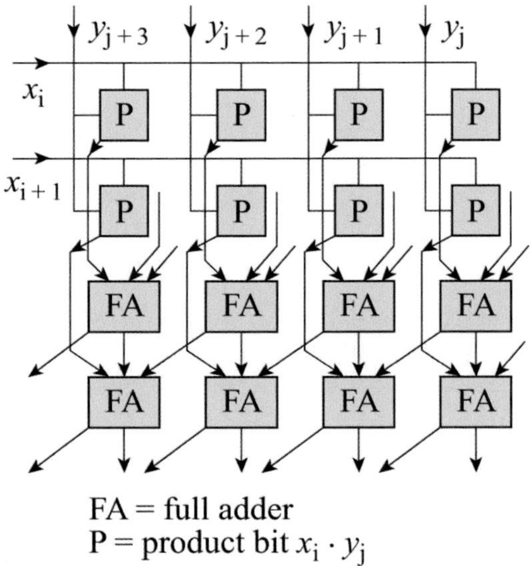

Figure 7.11: *Array multiplier (parallel multiplier) with carry-save array*

7.3.5 Logic-gate level

As stated in section 7.4, the RTL description is often specified through *hardware description languages (HDL)*, such as VHDL and Verilog. It is then mapped onto a library of cells (logic gates). This is done by a logic synthesis tool, which transforms a VHDL code into a netlist (see example in figure 7.23). A *netlist* contains a list of the library cells used and how they are connected to each other. Examples of such library cells (logic gates) are: AND, NAND, flip-flop and full adder, etc. As an example of the decisions that need to be taken at this logic level, we choose the full adder, from which we will build the array multiplier of figure 7.11. A full adder performs the binary addition of three input bits $(x, y$ and $z)$ and produces sum (S) and carry (C) outputs. Boolean functions that describe the operation of a *full adder* include the following:

(a) Generation of S and C directly from x, y and z:

$$
\begin{aligned}
C &= x\,y + x\,z + y\,z \\
S &= x\,\overline{y}\,\overline{z} + \overline{x}\,\overline{y}\,z + \overline{x}\,y\,\overline{z} + x\,y\,z
\end{aligned}
$$

(b) Generation of S from C:

$$
C = x\,y + x\,z + y\,z
$$

383

$$S \;=\; \overline{C}(x+y+z) + x\,y\,z$$

(c) Generation of S and C with exclusive OR gates (EXORs).

The choice of either one of these implementations depends on what is required in terms of speed, area and power. Implementation (b) will contain fewer transistors than (a), but will be slower because the carry must first be generated before the sum can evaluate. The implementation in (c) is just to show another alternative. Suppose our signal processor is used in a consumer video application where area is the most dominant criterion, then, at this hierarchy level, it is obvious that we choose implementation (b) to realise our full adder. A logic-gate implementation is shown in figure 7.12.

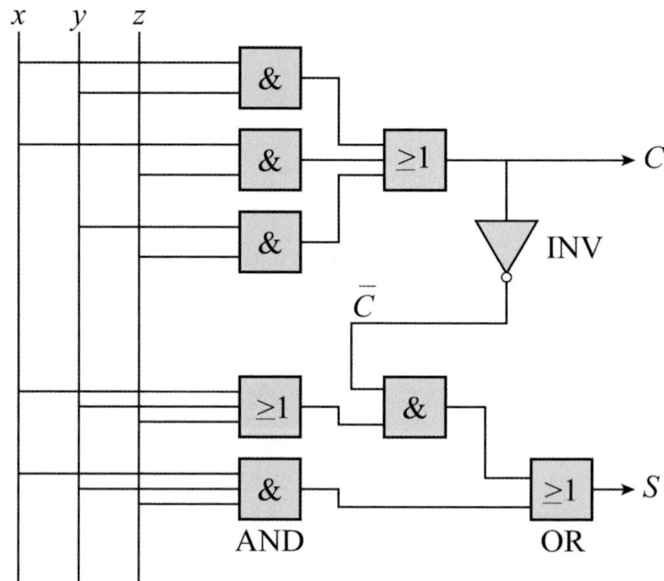

Figure 7.12: *Basic logic-gate implementation of a full adder*

7.3.6   Transistor level

At this level, the chosen full adder must be mapped onto a number of transistors. In some design environments, the logic-gate level is not explicitly present and the higher level code is directly synthesized and mapped onto a 'sea of transistors'. These are discussed in section 7.6.6.

The transistor level description depends on the chosen technology and the chosen logic style, such as dynamic or static CMOS. For the realisation of our full adder, we choose a static CMOS implementation, as shown in figure 7.13.

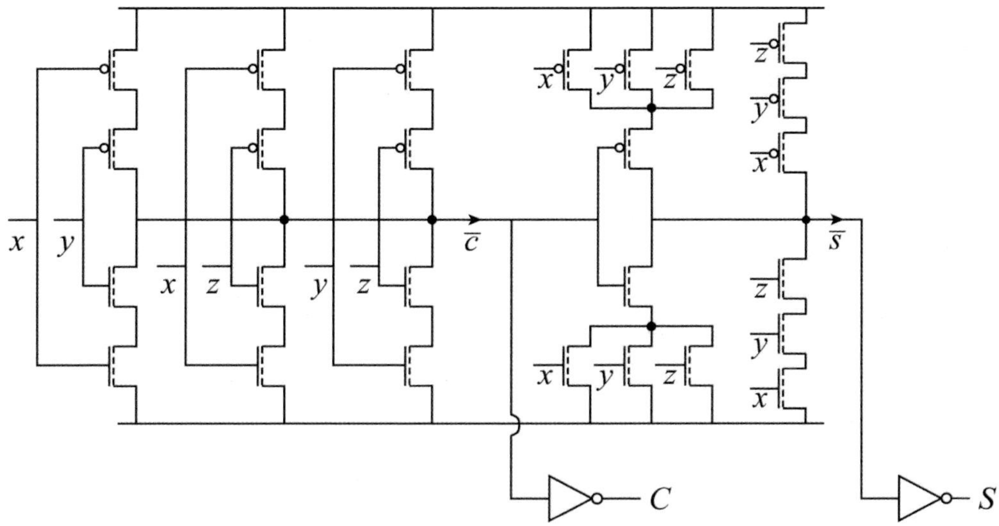

Figure 7.13: *Static CMOS realisation of the chosen full adder cell*

As this full adder consists of a relatively low number of transistors (30), it is efficient, both in terms of area and power dissipation, compared to the one realised with AND, OR and INVERT gates in figure 7.12. Note that both the sum S and carry C circuits are symmetrical with respect to their nMOS and pMOS transistor schematics, because the full adder is one of the few symmetrical logic functions, next to the half adder and the multiplexer.

Thus, the transistor level implementation of the logic gate is determined by either speed, area or power demands, as is actually every IC implementation. In this example we choose the implementation of figure 7.13 for our full-adder.

7.3.7 Layout level

The chosen transistor implementation must be translated to a layout level description at the lowest abstraction level of a design. Most of the time, these layouts are made by specialists, who develop a complete library of different cells in a certain technology. To support high performance, low-power and low-leakage applications, today, a library may consist of 1500 different cells. There may be different cell versions of the same logic function, but with a different drive strength, a different threshold voltage and/or a different gate oxide thickness. However, special requirements on high speed or low power may create the need for custom design, to optimise (part of) the chip for that requirement. In chapter 4, the layout process is explained in detail.

7.3.8 Conclusions

As shown in the signal processor example before, in the top-down design path, decisions have to be made at each level about different possible implementations. In this way, a *decision tree* arises. Figure 7.14 shows an example of a decision tree for the previously discussed signal processor system.

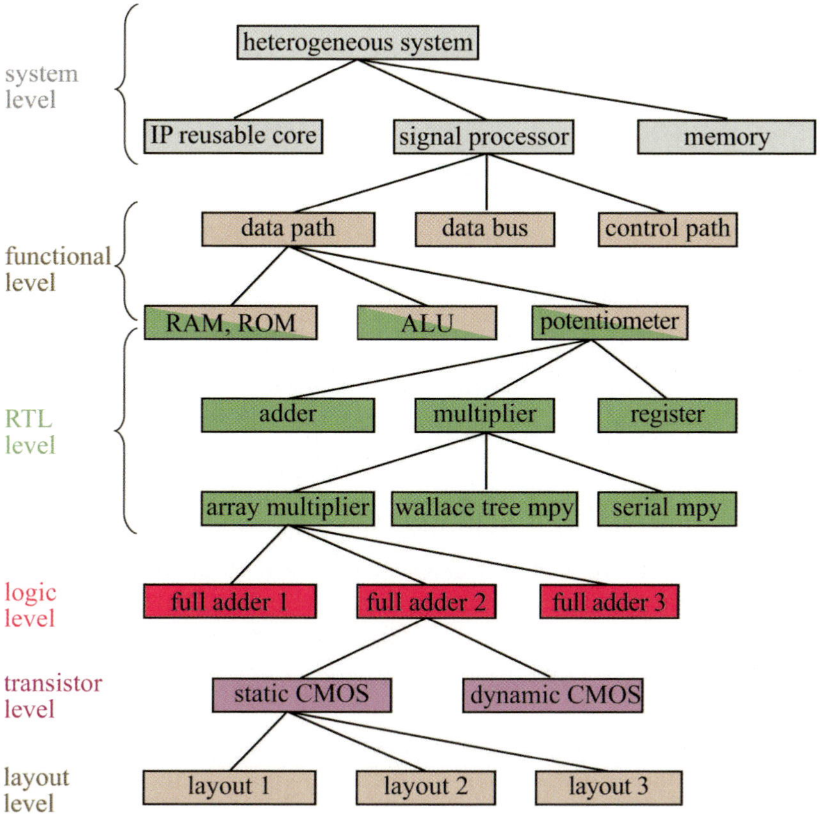

Figure 7.14: *Decision tree for a complex system on a chip*

The decision tree starts at the highest level, i.e., the system level. Every time we move one level down in the tree, we focus on a smaller part of the design, which allows us to add sufficient details to take the right decision at this level and then move to the next level. However, the decisions at each level can be strongly dependent on the possibilities available at a lower or at the lowest level. System designers who wish to achieve efficient area implementations therefore require a reasonable knowledge about the consequences of their decision at implementation level. For instance, the decision to implement a double data bus structure (figure 7.7) requires twice as many interconnections as a single bus implementation. As a result, the implementation of a double bus will take twice the area, but it also doubles the noise contribution since it doubles the level of the simultaneously switching current.

Decision trees and abstraction levels basically reduce the complexity of design tasks to acceptable levels. However, the abstraction levels are also accompanied by verification problems. More levels can clearly increase verification difficulties. Requirements at a certain level of abstraction depend on details at a lower level. Details such as propagation delays, for example, can influence higher level timing behaviour.

For example the final layout implementation of a full adder clearly influences its electrical behaviour. Delay times are also determined by factors such as parasitic wiring capacitances.

The bottom-up implementation and verification process begins at the layout level. Cell layouts are assembled to form modules, and these are combined to form the larger units that are indicated in the *floor plan* of the IC. The floor plan is a product of the top-down and bottom-up design process and is an accurate diagram which shows the relative sizes and positions of the included logic, analog, and memory cores. Cores that are identified as *critical* during the design path are usually implemented first. These are cores which are expected to present problems for power dissipation, area or operating frequency. Verification of their layouts reveals whether they are adequate or whether an alternative must be sought. This may have far-reaching consequences for the chosen architecture.

The inter-dependence of various abstraction levels and implementations clearly prevents a purely top-down design followed by purely bottom-up implementation and verification. In practice, the design process generally consists of iterations between the top-down and bottom-up paths.

Abstraction level descriptions which contain sufficient information about lower-level implementations can limit the need for iterations in the design path and prevent wasted design effort. The maximum operating frequency, for example, of a module is determined by the longest delay path between two flip-flops. This *worst-case delay path* can be determined from suitable abstraction level descriptions and used to rapidly determine architecture feasibility. As an example, the multiplier in the previously-discussed signal processor is assumed to contain the worst-case delay path.

The dimensions of logic cells in a layout library, for example, could be used to generate floor plan information such as interconnection lengths. These lengths, combined with specified delays for the library cells (e.g., full adder, multiplexer, etc.) allow accurate prediction of performance. The worst-case delay path can eventually be extracted from the final multiplier layout and simulated to verify that performance specifications are met.

The aim of modern IC-design environments is to minimise the number of iterations required in the design, implementation and verification paths. This should ensure the efficient integration of systems on silicon.

7.4 Digital VLSI design

7.4.1 Introduction

The need for CAD tools in the design and verification paths grows with increasing chip complexity. The different abstraction levels, as discussed in the previous subsection, were created to be able to manage the design complexity at each level.

7.4.2 The design trajectory and flow

The continuous growth in the number of transistors on a chip is a drive for a greater integration of synthesis and system level design. The increasing complexity of the system level behaviour, combined with an increasing dominance of physical effects of devices (e.g., variability), supply lines (e.g., voltage drop and supply noise), and interconnections (e.g., propagation delay and cross-talk), is a drive for a greater integration of synthesis and physical design.

Figure 7.5 shows a *heterogeneous system on a chip (SOC)*. First, the entire design must be described in a complete specification. For several existing ICs, such a specification consists of several hundreds of textual pages. This design specification must be translated into a high-level behavioural description, which must be executable and/or emulatable.

In many cases, software simulation is too slow and inaccurate to completely verify current complex ICs. Also, the interaction with other system components is not modelled. Logic *emulation* is a way to let designers look before they really act. Emulation allows the creation of a hardware model of a chip. Here, proprietary emulation software is used, which is able to map a design on reprogrammable logic, and which mimics the functional behaviour of the chip. Emulation is usually done in an early stage of the design process and allows more effective *hardware/software co-design*. The validation/verification problem has also led to the introduction of hybrid simulator tools [2], which claim to speed up simulation by 10 to 100 times for a full-chip or multi-chip system. Once the high-level behavioural description is verified by simulation or emulation, all subsequent levels of design description must be verified against this top-level description. Figure 7.15 shows a general representation of a design flow.

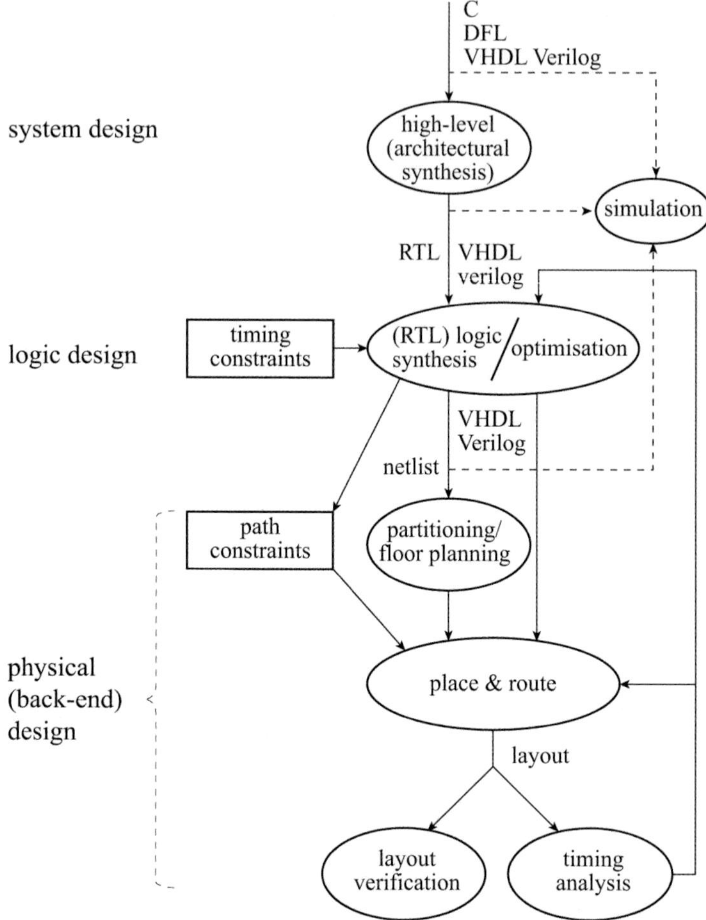

Figure 7.15: *General representation of a design flow*

Synthesis tools automatically translate a description at a higher hierarchy level into a lower level one. These tools are available at several levels of abstraction. High-level synthesis transforms a behavioural description into a sequence of possible parallel operations which must be performed on an IC. Such a behavioural description usually expresses functionality in a high-level computer programming language such as algorithmic SystemC, behavioural VHDL, C^{++}, etc. The derivation of ordering operations in time is called *scheduling*.

The *allocation* (or *mapping*) process selects the required data-path components. These high-level components include complete signal processor and microprocessor cores, as well as co-processors, ALUs, RAMs and I/O blocks, etc. However, high-level synthesis at system level is still in the R&D phase and its use is restricted to specific application areas, such as the design of digital signal processor ICs. For telecom and audio processor ICs, there are tools which are different from those that are created and used for the development of video signal processors. *Behavioural synthesis* tools generate RTL hardware descriptions in *VHDL* or *Verilog* from the system specification. The *RTL code* of a logic block describes its functionality in detail, in fact, it describes the behaviour of every bit in that block at every clock cycle.

Current and future *systems on silicon* (figure 7.5) are, and will be, designed by using a wide variety of pre-designed building blocks. This design *reuse* requires that these *Intellectual Property (IP)* parts, such as microcontrollers and microprocessors, can be easily ported from one chip design to another. Such a reuse must be supported by tools. Design reuse will be fuelled by the sharing of cores among companies. In many cases, a Reduced Instruction Set Computer (RISC) microprocessor core (ARM, MIPS, Sparc) is used. If we include the application (program) in an on-chip ROM or other type of memory, this is called *embedded software*.

Synthesis tools must play a key role in integrating such pre-designed building blocks with synthesised glue logic onto one single chip. The most-used type of synthesis is from the RTL level to a netlist of standard cells. Each *system on a chip* can be considered to consist of many registers which store binary data. Data is operated upon by logic circuits between these registers. The operations can be described in a *Register-Transfer Language (RTL)*. Before the VHDL code (or Verilog) is synthesised at this level, the code must be verified by simulation.

At higher functional levels, software (VHDL) simulators are often sufficiently fast. However, in many cases, RTL level simulation is a bottle-neck in the design flow. Besides an increase in the complexity of ICs, longer frame times (as in MPEG video and DAB) must also be simulated. Such simulations may run for several days, resulting in too long iteration times and allowing only limited functional validation of an RTL design.

A *hardware accelerator*, with accompanying software, is a VHDL simulator platform in which the hardware is often realised with reconfigurable logic, e.g with field-programmable gate arrays (FPGAs) or with a large multiprocessor system, which is connected to the network or a host system. Gate level descriptions as well as memory modules can be downloaded into a hardware accelerator. However, most non-gate level parts (RTL and test bench) are kept in software. The accelerator hardware speeds up the execution of certain processes (i.e., gates and memory) and the corresponding events. In fact, the accelerator is an integral part of the simulator and uses the same type of interface. Generally, the raw performance of a hardware accelerator is less than with *emulation*.

When the RTL description is simulated and proven to be correct, RTL synthesis is used to transform the code (mostly VHDL or Verilog) into an optimised netlist. Actually, the described function or operation at RTL level is *mapped* onto a library of (standard) cells. Synthesis at this level is more mature than high-level synthesis and is widely used. The synthesis of the functional blocks and the composition of the complete IC is the work of the physical or back-end designer. Next to the logic synthesis, *back-end design* tasks also include the place and route of the logic cells in the generated netlist, and the floor planning, which assigns the individual logic blocks, memories and I/O pins to regions in the chip. It also includes tasks that maintain signal integrity (crosstalk, supply noise, voltage drop, etc.), variability (parameter spread, transistor matching, etc.), reliability (electromigration, antenna rules, etc.) and *design for manufacturability (DfM)* (via doubling, metal widening or spreading, dummy metals, etc.). This back-end design is no longer a straightforward process, but it requires many iterations to cover all of the above design objectives simultaneously. This shows that the back-end design has become a very complex task, which needs to be supported by appropriate tools, smoothly integrated in the design flow.

Finally the *design verification* is also a growing part of both the front-end and back-end design trajectory. CAD tools are also used for the validation in the IC-design verification path. Simulation is the most commonly used *design-verification* method. *Behavioural simulation* is usually done on an IP block basis at a high abstraction level (algorithm/architecture). It runs quickly because it only includes the details of the behaviour and not of the implementation. *Logic simulation* is performed at RTL or netlist level and relates to the digital (or Boolean) behaviour in terms of logic 1's and 0's. *Circuit simulation* is the transistor level simulation of the behaviour of a schematic or extracted layout. It usually includes all device and circuit parasitics and results in a very accurate and detailed analog behaviour of the circuit. Due to the rapid increase in the IC's complexity, it is impossible to completely simulate a system on a chip and verify that it will operate correctly under all conditions. Moreover, it is very difficult to envision and simulate all potential event candidates that may lead to problems. Achieving 100% verification coverage would require huge time-consuming simulations with an unlimited number of input stimuli combinations.

Luckily, there are other verification methods that complement the simulation. *Formal verification* is a mathematical method to verify whether an implementation is a correct model for the specification. It is based on reasoning and not on simulation. This verification may include the comparison of design descriptions at different levels of abstraction. Examples of this so-called *equivalence checking* are the comparison between behavioural description and RTL description, which checks whether the synthesis output is still equivalent to the source description, and the comparison between the RTL description and the synthesized netlist to prove equal functional behaviour. It does not prove that the design will work.

Timing verification is done at a lower hierarchy level. During a *static-timing analysis (STA)* each logic gate is represented by its worst-case propagation delay. Then, the worst-case path delay is simply the sum of the worst-case delays of the individual gates in that path. Due to the increasing process-induced parameter spread in devices and interconnect structures, these worst-case numbers are often so high that this type of static timing analysis leads to design overkill, to less performance than in the previous technology node, or to incorrect critical paths. This has led to the introduction of a *statistical static timing analysis (SSTA)* tool, which tries to find the probability density function of the signal arrival times at each internal node and primary output. This type of analysis is considered necessary, particularly for complex high-performance ICs. However, probability density functions are difficult to compute and the method needs to be simplified to make it a standard component of the verification process.

As a result of the growing number of transistors on one chip and with the inclusion of analogue circuits or even sensors on the same chip, verification and analysis have become serious bottle-necks in achieving a reasonable design turn-around time. Extensive verification is required at each level in the design flow and, as discussed before, there is a strong need for cross-verification between the different levels. Verification often consumes 20 to 50 percent of the total design time. With increasing clock speed and performance, packaging can be a limiting factor in the overall system performance. Direct attachment of chip-on-board and flip-chip techniques continue to expand to support system performance improvements. Verification tools are therefore needed across the chip boundaries and must also include the total interconnect paths between chips.

7.4.3 Example of synthesis from VHDL description to layout

This paragraph discusses the design steps of the digital potentiometer (see section 7.3.4), starting at the RTL description level (in VHDL) and ending in a standard cell layout. Figure 7.16 shows the RTL-VHDL description of this potentiometer.

```vhdl
LIBRARY IEEE;

USE IEEE.std_logic_1164.ALL;
USE IEEE.std_logic_arith.ALL;
USE IEEE.std_logic_unsigned.ALL;

ENTITY potmeter IS

    GENERIC (par_width: natural := 4;
             operand_width : natural := 12);

    PORT (A, B: IN std_logic_vector(operand_width-1 DOWNTO 0);
          K: IN std_logic_vector(par_width-1 DOWNTO 0);
          Z: OUT std_logic_vector(par_width+operand_width-1 DOWNTO 0));

END potmeter;

ARCHITECTURE behaviour OF potmeter IS

BEGIN

    PROCESS (A, B, K)

        CONSTANT K_max: integer := 2**par_width-1;
        VARIABLE K_int: integer;

    BEGIN

        K_int := conv_integer(K);
        Z <= K*A + conv_std_logic_vector(K_max-K_int, par_width) * B;

    END PROCESS;

END behaviour;
```

Figure 7.16: *RTL-VHDL description of potentiometer*

Figure 7.17(a) shows a high abstraction level symbol of this potentiometer, while a behavioural level representation is shown in figure 7.17(b).

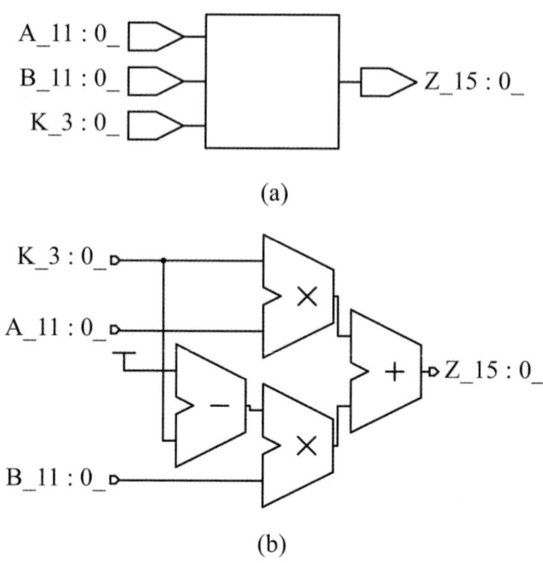

Figure 7.17: *(a) Abstraction level symbol and (b) behavioural level representation of the potentiometer*

After synthesis, without constraints, our potentiometer looks as shown in figure 7.18

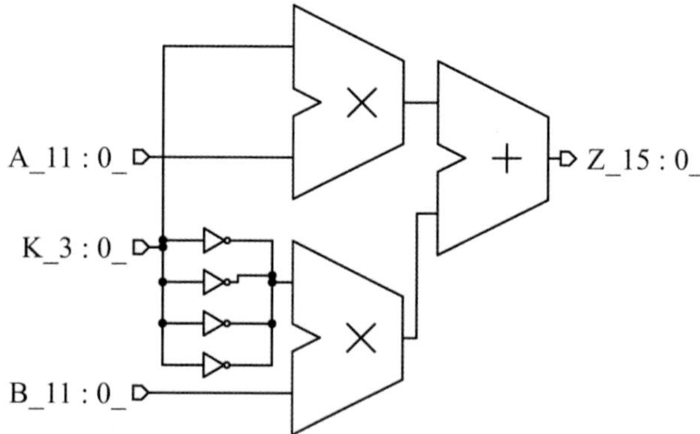

Figure 7.18: *Potentiometer schematic after synthesis with no constraints*

Figure 7.19 shows the multiplier and adder symbolic views after synthesis.

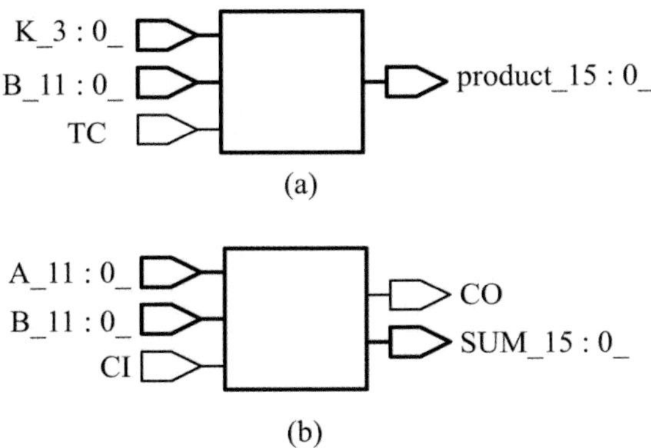

Figure 7.19: *Multiplier and adder symbolic views*

Figure 7.20 shows the schematics of the adder, after synthesis with no constraints.

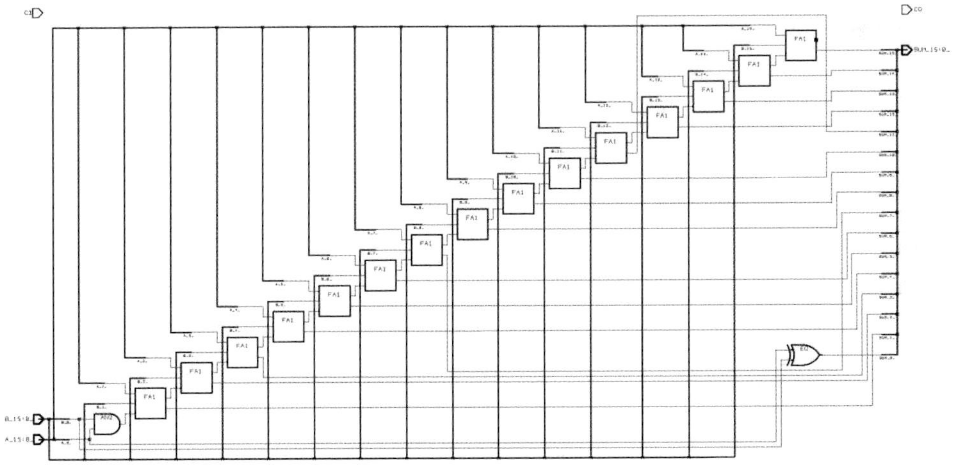

Figure 7.20: *Adder schematics after synthesis with no constraints*

Figure 7.21 shows the schematics of the adder, after synthesis with a timing constraint for the worst-case delay path.

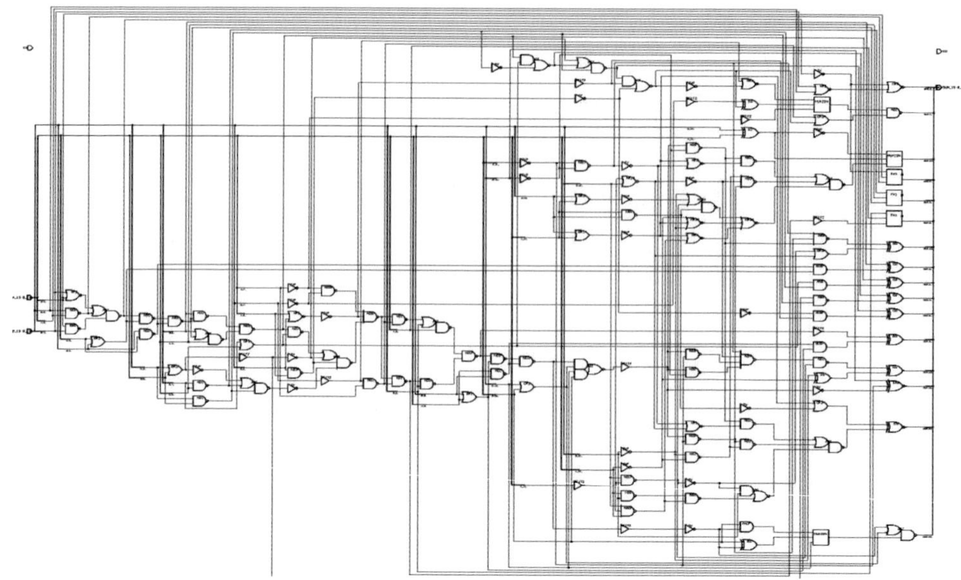

Figure 7.21: *Adder schematics after timing-constraint synthesis*

The additional hardware in figure 7.21 compared to that of figure 7.20 is used to speed up the carry ripple by means of carry look-ahead techniques. Figure 7.22 shows the relation between the delay and the area. The figure clearly shows that reducing the delay by timing constrainted synthesis can only be achieved with relatively much additional hardware (area).

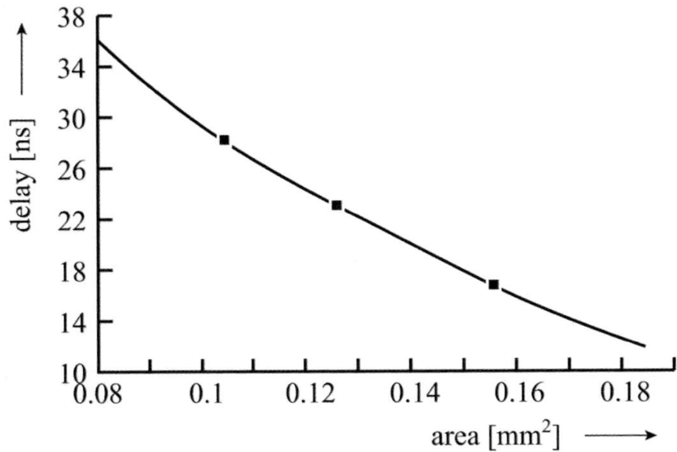

Figure 7.22: *Relation between maximum delay and the amount of hardware (area)*

Figure 7.23 shows a part of the *netlist* of library cells onto which the potentiometer function has been mapped. The figure shows the different library cells and the nodes to which their inputs and outputs are connected.

```
module potmeter_DW01_add_14_1 ( A, B, CI, SUM, CO );
input   [13:0] A;
input   [13:0] B;
output [13:0] SUM;
input   CI;
output CO;
    wire n52, n53, n54, n55, n56, n57, n58, n59, n60, n61, n62, n63, n64, n65,
        n66, n67, n68, n69, n70, n71, n72, n73, n74, n75, n76, n77, n78, n79,
        n80, n81, n82, n83, n84, n85, n86, n87, n88, n89, n90, n91, n92, n93,
        n94, n95, n96, n97, n98, n99, n100, n101, n102, n103, n104, n105;
    BF1T1 U5 ( .Z(SUM[2]), .A(A[2]) );
    BF1T1 U6 ( .Z(SUM[0]), .A(A[0]) );
    BF1T1 U7 ( .Z(SUM[1]), .A(A[1]) );
    AO6 U8 ( .Z(n52), .A(n53), .B(n54), .C(n55) );
    AO6 U9 ( .Z(SUM[3]), .A(n56), .B(n57), .C(n58) );
    AO32 U10 ( .Z(n59), .A(n60), .B(n61), .C(n62), .D(n63) );
    AO32 U11 ( .Z(n64), .A(n59), .B(n65), .C(n54), .D(n55) );
    NR2 U12 ( .Z(n66), .A(n67), .B(n68) );
    AO6 U13 ( .Z(n69), .A(A[7]), .B(B[7]), .C(n70) );
    NR2 U14 ( .Z(n71), .A(n65), .B(n63) );
    NR2 U15 ( .Z(n72), .A(n73), .B(n74) );
    AN2 U16 ( .Z(n75), .A(n76), .B(n77) );
    EO U17 ( .Z(SUM[9]), .A(n66), .B(n78) );
    EO U18 ( .Z(SUM[8]), .A(n79), .B(n80) );
    EO U19 ( .Z(SUM[6]), .A(n81), .B(n82) );
    EO U20 ( .Z(SUM[5]), .A(n71), .B(n83) );
    MUX21N U21 ( .Z(SUM[13]), .A(B[13]), .B(n84), .S(n85) );
    IV U69 ( .Z(n84), .A(B[13]) );
    IV U70 ( .Z(n105), .A(A[10]) );
    IV U71 ( .Z(n96), .A(B[7]) );
    IV U72 ( .Z(n79), .A(n95) );
endmodule
```

Figure 7.23: *Potentiometer netlist after synthesis with 14 ns timing constraints*

After the use of place and route tools, a standard cell design of the potentiometer is created, see figure 7.24 for the result. This netlist and layout are the result of the chosen description of the potentiometer's functionality according to:

$$Z = k \cdot A + (1 - k) \cdot B$$

This implementation requires two adders and two multipliers. However, an obvious optimisation of the same function may lead to a more efficient implementation. The following description

$$Z = k \cdot (A - B) + B$$

requires only two adders and one multiplier. This example shows that the decision taken at one hierarchy level can have severe consequences for the efficiency of the final silicon realisation in terms of area, speed and power consumption.

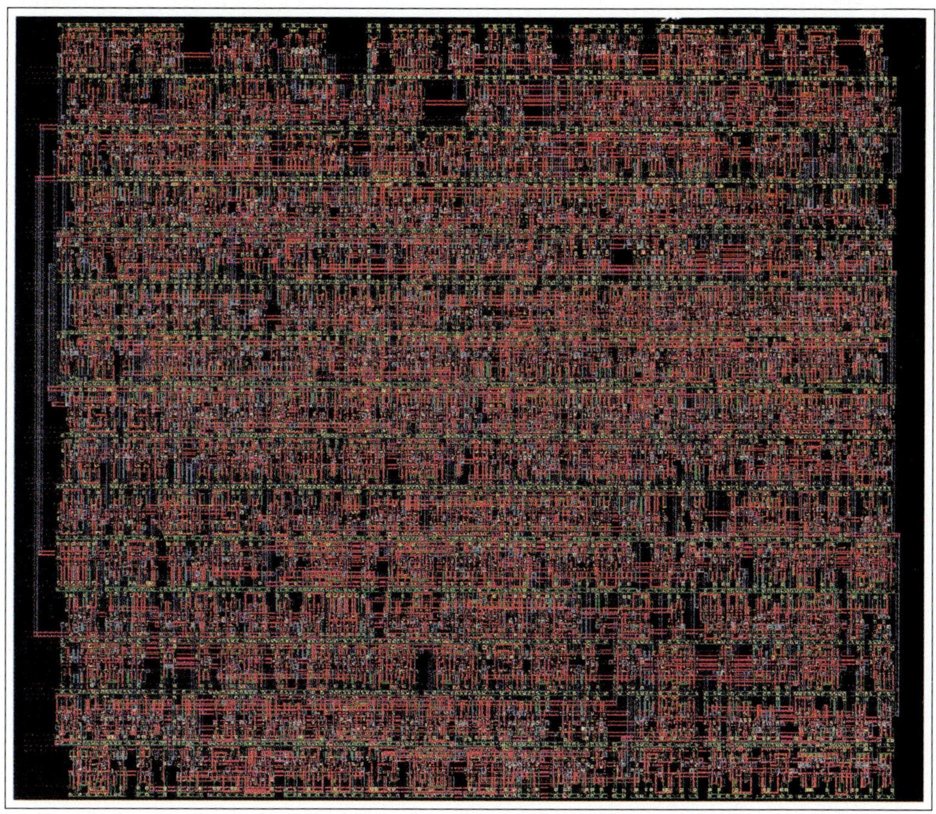

Figure 7.24: *Standard cell implementation of potentiometer*

Although the synthesis process uses tools which automatically generate a next level of description, this process is controlled by the designer. An excellent design is the result of the combination of an excellent tool and a designer with excellent skills in both control of the tools and knowledge of IC design.

7.5   The use of ASICs

The growth in the ASIC business is primarily the result of the increasing number of application areas and of the general increase in the use of ICs. *ASICs* often provide the only solution to problems attributed to speed and/or space requirements. Another incentive for the use of ASICs is the degree of concealment which they afford. This concealment poses extra difficulties to competitors interested in design duplication.

ASICs make it reasonably easy to add new functionality to an existing system without an extensive system redesign. In addition, the increased integration of system parts associated with the use of ASICs has the following advantages:

- Reduced physical size of the system

- Reduced system maintenance costs

- Reduced manufacturing costs

- Improved system reliability

- Increased system functionality

- Reduced power consumption.

The advantages afforded by ASICs can have a positive influence on the functionality/price ratio of products and have led to the replacement of standard ICs in many application areas. However, there are also disadvantages associated with the use of ASICs. These include the following:

- The costs of realising an ASIC are quite substantial and less predictable than those associated with standard ICs.

- Unlike standard products, ASICs are not readily available from a diverse number of suppliers. Inaccurate specifications or errors in the design process may cause delays in ASIC turn-around time and result in additional *non-recurring engineering (NRE)* costs. These are costs incurred prior to production. Typical NRE costs include the cost of:

- Training and use of design facilities
- Support during simulation
- Placement and routing tools
- Mask manufacturing (where applicable)
- Test development
- The delivery of samples.

Furthermore, standard products are always well characterised and meet guaranteed quality levels. Moreover, small adjustments to a system comprising standard products can be implemented quickly and cheaply.

The advantages and disadvantages associated with the use of ASICs depend on the application area and on the required ASIC type and quantities. Improved design methods and production techniques combined with better relationships between ASIC customers and manufacturers will have a considerable influence on the transition from the use of standard products to ASICs.

An ASIC solution in the above discussions does not necessarily imply a single chip or *system-on-a-chip (SoC)* solution, but it might also refer to a *system-in-a-package (SiP)* solution. For a discussion on SoC versus SiP system solutions, the reader is kindly requested to read the appropriate subsection in chapter 10.

7.6 Silicon realisation of VLSI and ASICs

7.6.1 Introduction

In addition to the need for computer programs for the synthesis and verification of complex ICs, CAD tools are also required for the automatic or semi-automatic generation of layouts. The development of Intel's Pentium and Xeon processors, for example, took several thousands of man-years. The same holds for the IBM PowerPC. Figure 7.25 shows a photograph of the Intel Xeon processor. This Tulsa chip in the Xeon family combines two processor cores with 1 MB L2 cache and 16 MB L3 cache per core, resulting in a chip with 1.3 billion transistors. It runs at a maximum clock frequency of 3.4 GHz, while consuming 150 W. In fact, the increased use of CAD tools in recent years has very often merely facilitated the integration of increasingly complex systems without contributing to a reduction in design time. This situation is only acceptable for very complex high-performance ICs such as a new generation of microprocessors. Less complex ICs, such as ASICs, require fast and effective design and layout tools. Clearly, the need for a fast design and layout process increases as the lifetimes of new ICs become shorter. The lifetime of a new generation of ICs for DVD players, for instance, is about close to one year. This means that the design process may take only a couple of months. Each layout design must be preceded by a thorough floor plan study. This must ensure that the envisaged layout will not prove too large for a single chip implementation in the final design phase. A floor plan study can take considerable time and only leads to a definite floor plan after an iterative trial-and-error process. Layouts of some parts of the chip may be required during the floor plan study. Although we distinguish between the different ASIC categories of custom ICs, semi-custom ICs and PLDs in this book, the differences are rapidly diminishing as a result of the pace at which improvements in IC technologies are realised. PLDs are moving towards gate arrays, gate arrays are moving towards cell-based designs and cell-based designs may use sea-of-gates structures such as embedded arrays to implement the glue logic as well as for mapping of cores onto such arrays. Each category uses the best features of the others.

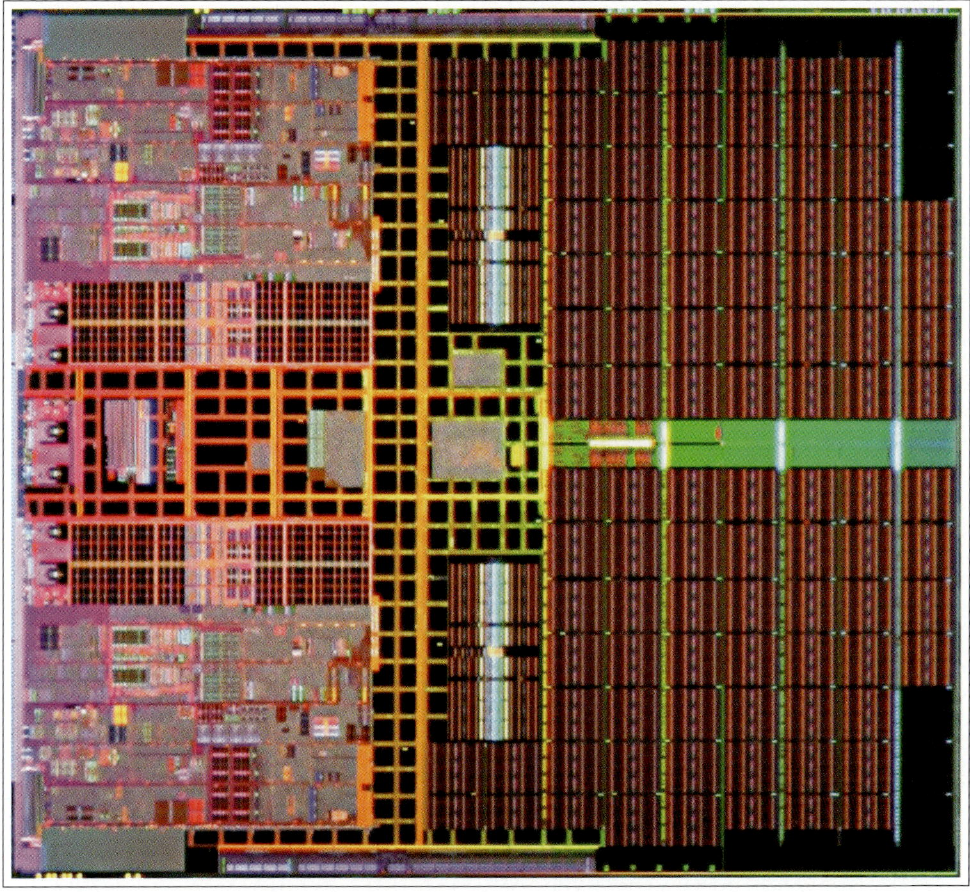

Figure 7.25: *The Intel Pentium4 Xeon^{TM} processor, containing 1.3 billion transistors (Source: Intel)*

The choice of implementation is determined by the required development time, production volume and performance. Table 7.2 summarises the performance of various *layout implementation forms*. This table is only valid in general terms.

Table 7.2: Comparison of performance of different layout implementation forms

Implementation form	Performance	
	speed	area
Handcrafted layout	$+++++$	$+++++$
Bit-slice layout	$-++++$	$-++++$
Cell based design	$--+++$	$--+++$
Structured array ASIC	$---++$	$---++$
(Sea-of-gates) gate array	$----+$	$----+$
PLD (FPGAs and CPLDs)	$----+$	$----+$

The different layout implementation forms are discussed separately in the next subsections.

7.6.2 Handcrafted layout implementation

A *handcrafted layout* is characterised by a manual definition of the logic and wiring. This definition must account for all relevant layout design rules for the envisaged technology. The design rules of modern technologies are far more numerous and complex than those used in the simple initial nMOS process. However, various CAD tools have emerged which ease the task of creating a handcrafted layout. These include interactive computer graphic editors (or *polygon pushers*), compactors and *design-rule-check (DRC)* programs.

An example of a handcrafted layout is illustrated in figure 7.26. Such an implementation yields considerable local optimisation. However, the required intensive design effort is only justified in MSI circuits and limited parts of VLSI circuits. The use of handcrafted layout is generally restricted to the design of basic and analog cells. These may subsequently be used in standard-cell libraries, module generators and bit-slice layouts, etc.

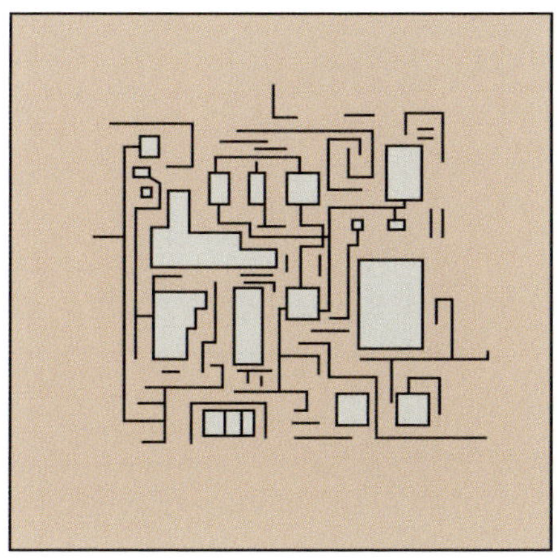

Figure 7.26: *Typical contours of a handcrafted layout*

7.6.3 Bit-slice layout implementation

A *bit-slice layout* is an assembly of parallel single-bit data paths. The implementation of a bit-slice layout of a signal processor, for example, requires the design of a circuit layout for just one bit. This bit slice is subsequently duplicated as many times as required by the word length of the processor. Each bit slice may comprise one or more vertically-arranged cells. The interconnection wires in a bit slice run over the cells with control lines perpendicular to data lines. CAD tools facilitate the efficient assembly of bit-slice layout architectures. The bit-slice design style is characterised by an array-like structure which yields a reasonable packing density. Figure 7.27 illustrates an example of a bit-slice layout architecture. A bit-slice section is also indicated in the chip photograph in figure 7.52.

The AMD Am2901 is an example of a bit-slice architecture. Today this layout style has become less popular, because it requires a lot of manual design effort compared to the availability of a fully synthesizable alternative with the standard-cell approach, discussed in section 7.6.5

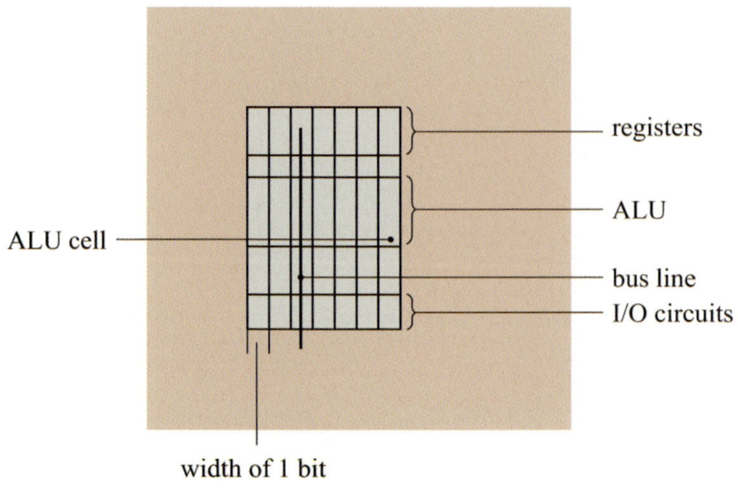

Figure 7.27: *Basic bit-slice layout*

7.6.4 ROM, PAL and PLA layout implementations

In addition to serving as a memory, a ROM can also be used to implement logic functions. An example is shown in figure 7.28.

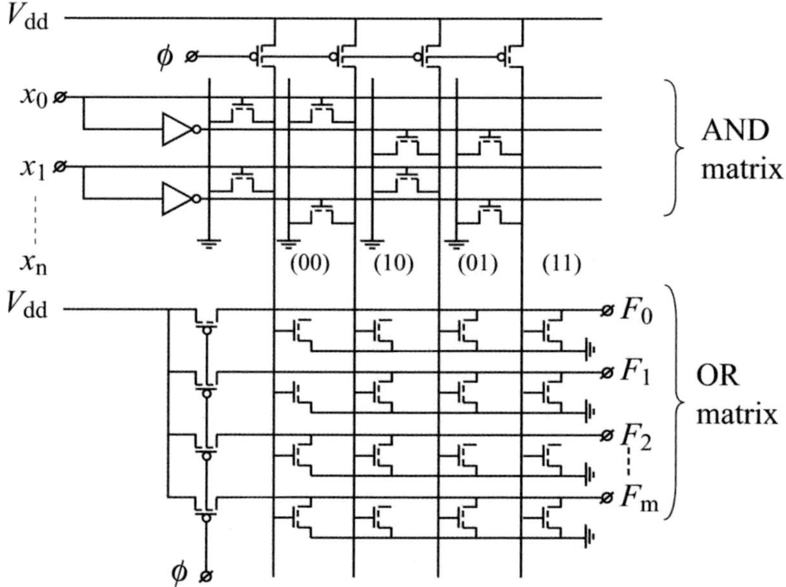

Figure 7.28: *Logic functions realised with a ROM*

Only one vertical line in this ROM will be 'high' for each combination of address inputs $x_n \cdots x_0$. This vertical line drives the gates of $m + 1$ transistors in the OR-matrix. The outputs F_j, that are connected to the drains of these transistors, will be 'low'. If, for example, the address inputs are given by $x_0 x_1 = 10$, then the second column line will be 'high'. A 'low' will then be present on outputs F_1 and F_2. The information stored in the ROM in figure 7.28 is thus determined by the presence or absence of connections between MOS transistor drains and the output lines. In this way, the structure of a ROM can easily be used to realise logic functions. Table 7.3 shows a possible truth table, which could be implemented with the ROM in figure 7.28.

Table 7.3: Example of a truth table implemented with the ROM in figure 8.16

x_n	-	-	-	x_1	x_0	F_m	-	-	-	F_1	F_0
0	-	-	-	0	0	0	-	-	-	1	1
0	-	-	-	0	1	1	-	-	-	0	1
0	-	-	-	1	0	0	-	-	-	0	0
0	-	-	-	1	1	0	-	-	-	0	0

Clearly, the set of logic functions that can be realised in a ROM is merely limited by the number of output and address bits. The regular array structure of a ROM leads to a larger transistor density per unit of chip area than for random logic. A large number of logic functions could, however, require an excessively large ROM while the use of a ROM could prove inefficient for a small number of logic functions. In general, a ROM implementation is usually only cheaper than random logic when large volumes are involved.

Unfortunately, there are no easy systematic design procedures for the implementation of logic functions in ROM. Other disadvantages are as follows:

- Lower operating frequency for the circuit

- The information in a ROM can only be stored during manufacturing

- Increasing the number of input signals by one causes the width of the ROM to double

- A high transistor density does not necessarily imply an efficient use of the transistors.

It is clear from figure 7.28 that the vertical column lines in a ROM represent the *product terms* formed by the address inputs x_i. These product terms comprise all of the logic AND combinations of the address inputs and their inverses. Only the OR-matrix of a ROM can be programmed.

Figure 7.29 illustrates the basic structure of a *programmable logic array (PLA)* . Its structure is similar to that of a ROM and consists of an *AND matrix* and an *OR-matrix*. In a PLA, however, both matrices can be programmed and only the required product terms in the logic functions are implemented. It is therefore more efficient in terms of area than a ROM. Area requirements are usually further reduced by minimising the number of product terms before generating the PLA layout pattern.

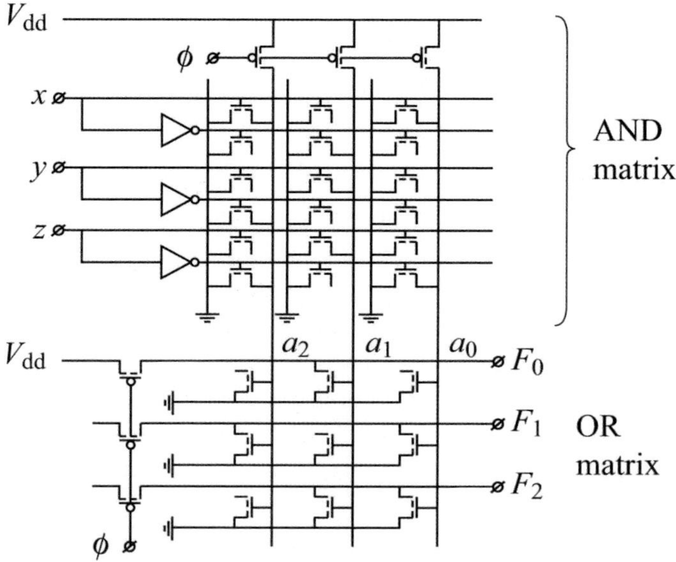

Figure 7.29: *Basic PLA structure*

The logic functions implemented in the PLA in figure 7.29 are determined as follows: a_0 is 'high' when x and $\overline{z}$ are low, i.e., $a_0 = \overline{x}\,\overline{z}$. Similarly, $a_1 = x\,\overline{y}\,\overline{z}$ and $a_2 = \overline{x}\,y\,z$.

The outputs are therefore expressed as follows:

$$F_0 = \overline{a_1} = \overline{x\,\overline{y}\,\overline{z}}$$

$$F_1 = \overline{a_0 + a_2} = \overline{\overline{x}\,\overline{z} + \overline{x}\,y\,z}$$

$$F_2 = \overline{a_0 + a_1} = \overline{\overline{x}\,\overline{z} + x\,\overline{y}\,\overline{z}}$$

A PLA can be used to implement any combinatorial network comprising AND gates and OR gates. In general, the complexity of a PLA is characterised by $(A + C) \times B$, where A is the number of inputs, B is the total number of product terms, i.e., the number of inputs for each OR gate, and C is the number of outputs, i.e., the number of available logic functions.

Sequential networks can also be implemented with PLAs. This, of course, requires the addition of memory elements. A PLA can be a stand-alone chip or an integral part of another chip such as a microprocessor or a signal processor. PLAs are frequently used to realise the logic to decode *microcode instructions* for functional blocks such as memories, multipliers, registers and ALUs. Several available CAD tools enable a fast mapping of logic functions onto PLAs. As a result of the improvements in cell-based designs, ROM and PLA implementations are becoming less and less popular in VLSI designs. Another realisation form is the *Programmable Array Logic (PAL)*. In this concept, only the AND plane is programmable and the OR plane is fixed.

Table 7.4 summarises the programmability of planes (AND, OR) in the ROM, PAL and PLA devices. Programmable techniques include fuses (early and smaller devices), floating gate transistors ((E)EPROM) and flash devices. In some cases, a ROM (PLA) block is still used in a custom design; the programming is done by a mask. These are then called mask-programmable ROMs (PLAs).

Table 7.4: Programmability of AND and OR planes in ROM, PAL or PLA devices

Device	Programmable	
	AND-plane	OR-plane
ROM	no	yes
PAL	yes	no
PLA	yes	yes

7.6.5 Cell-based layout implementation

Figure 7.30 shows a basic layout diagram of a chip realised with *standard cells*.

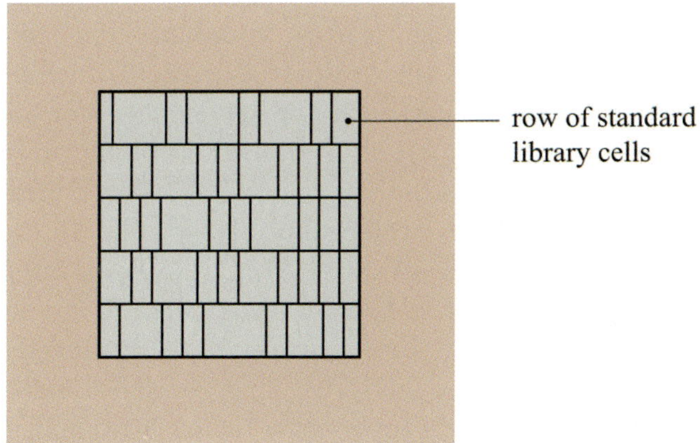

Figure 7.30: *Basic standard-cell layout*

In this design style, an RTL description of the circuit is synthesized and mapped onto a number of standard cells which are available in a library, see section 7.4.2. The resulting netlist normally contains no hierarchy. The standard-cell library usually consists of a large number of different types of logic gates, which are all of equal height (figure 7.31).

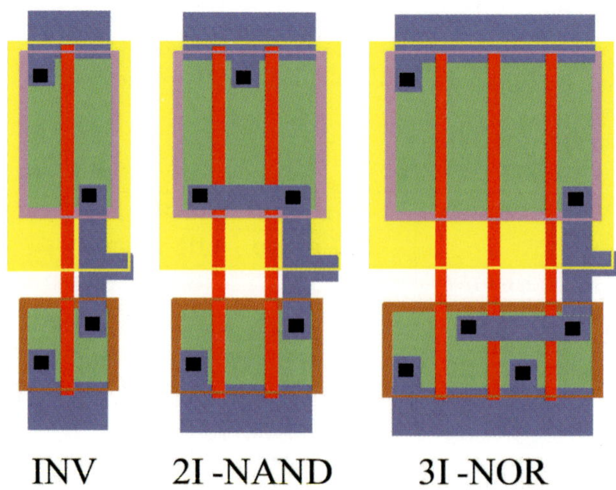

Figure 7.31: *Example of three basic standard cells*

Today's libraries may contain between 500 to 2000 cells, due to a large variety of drive strengths and different threshold voltages (HV_T, SV_T, and LV_T, respectively referring to high, standard, and low-V_T). This enables the synthesis tools to better adapt a design to such performance requirements as high speed, low power or low leakage, for example.

The standard-cell layout method is supported by mature CAD tools for *placement* and *routing* . Routing is done at a fixed grid across the logic gates. The supply lines are specially structured to create a supply network with minimum resistance and is usually an integral part of the standard cell design approach. The clock network is usually generated by a clock-tree synthesis tool, which creates balanced clock trees to reduce intrinsic clock skew and also deals with timing constraints. However, many clock-synthesis tools often balance different clock paths by compensating interconnect RC delay in one path with buffer delays in another, leading to a different path sensitivity to PVT variations. High-speed processors, however, use relatively large clock grids leading to less clock skew and less sensitivity to PVT variations, but at increased power levels. In addition, they require a detailed analysis of all parasitic resistive, capacitive and inductive effects, including the modelling and simulation of the current return paths. Modern standard-cell design environments facilitate the inclusion of larger user-defined cells in the library. These *blocks*, *macros* or *cores* may include multipliers, RAMs, signal processor cores, microprocessor cores, etc.

414

During the late eighties, extra attention was paid to advanced circuit test methods. These include *scan test* and *self-test* techniques, see section 10.2.1. The scan technique uses a sequential chain of intrinsically available flip-flops to allow access to a large number of locations on an IC or on a printed circuit board. The self-test technique requires the addition of dedicated logic to an existing design. This logic generates the stimuli required to test the design and checks the responses. The result is a logic circuit or a memory which is effectively capable of testing itself. Details of IC testing are discussed in chapter 10.

7.6.6 (Mask programmable) gate array layout implementation

Gate arrays are also referred to as *mask-programmable gate arrays*. A conventional *gate array* contained thousands of logic gates, located at fixed positions. The layout could, for example, contain 10,000 3-input NAND gates. The implementation of a desired function on a gate array is called *customisation* and comprises the interconnection of the logic gates. The interconnections were located in dedicated *routing channels*, which were situated between rows of logic gates. In these conventional channelled gate arrays, the routing was often implemented in two metal layers.

This type of gate array is depicted in figure 7.32a. The channels are essential for interconnecting the cells when production processes with one or even two metal layers are involved.

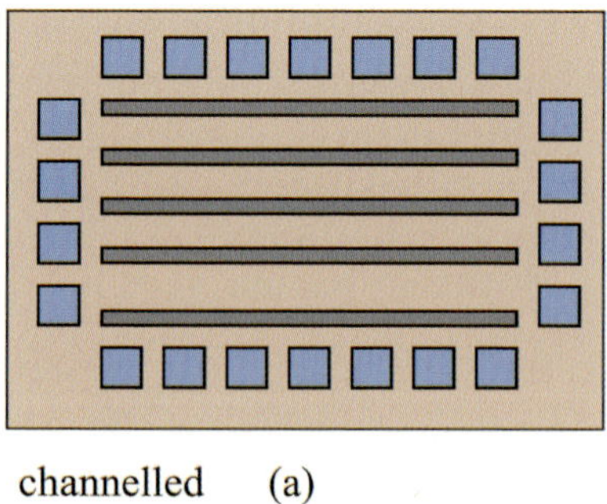

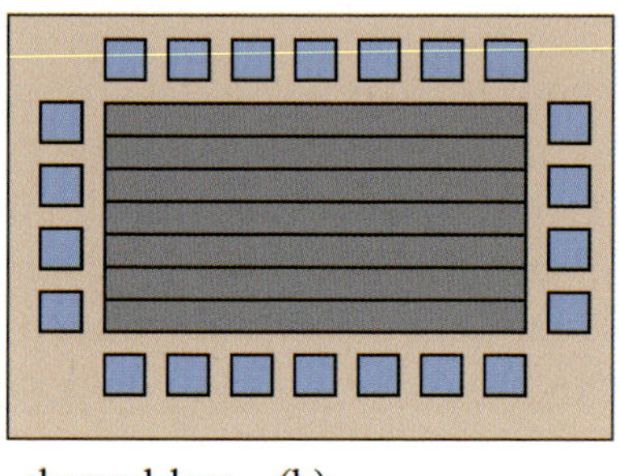

Figure 7.32: *Floor plan for (a) conventional and (b) channel-less gate arrays*

In a conventional gate array, the ratio between the available cell and routing channel areas was fixed. Obviously, the actual ratio between the areas used was dependent on the type of circuit. In practice, the available area is rarely optimally used. This feature is especially important for larger circuits. Furthermore, larger circuits require more complex interconnections and this increases the density in routing channels. The *channel-less gate array* architecture was therefore introduced. Other names encountered in literature for this architecture include: *high-density gate array (HDGA)*, *channel-free gate array*, *sea-of-gates*, *sea-of-transistors* and *gate forest*.

Figure 7.32b shows the floor plan for a channel-less gate array. It consists of an array of transistors or cells. It does not contain any specially reserved routing channels. Modern gate arrarys comprise an array of *master cells*, which consist of between four and ten transistors. In some cases, the master cells are designed to accommodate optimum im-

plementations of static RAMs, ROMs or other special circuits. A given
memory or logic function is implemented by creating suitable contact
and interconnection patterns in three or more metal layers. The master
cells in a gate array can be separated by *field oxide isolation*, which is
created by using the STI technique described in chapter 3. An example
of such a gate array master-cell structure is shown in figure 7.33, which
also shows an example of a gate array floor plan.

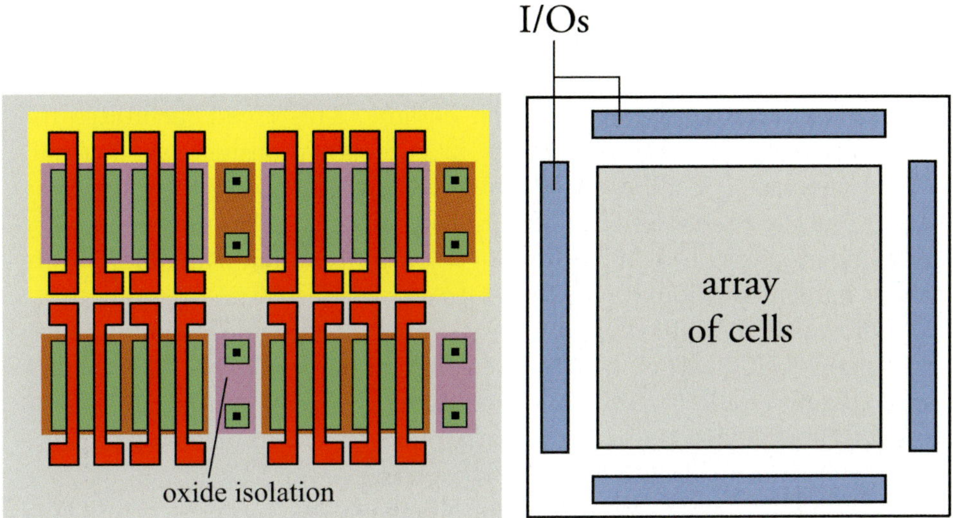

Figure 7.33: *An example of a gate array master-cell structure and floor
plan*

Figure 7.34 shows a section of a sea-of-transistors array, which comprises
a row of pMOS and nMOS transistors. The complete array is created by
copying the section several times in the horizontal and vertical directions.
These gate arrays are also often called *continuous arrays* or *uncommitted
arrays*. The rows are not separated by routing channels and the floor
plan is therefore the same as shown in figure 7.32b. These gate arrary
architectures facilitate the implementation of large VLSI circuits on a
single gate array using a large number of metal layers. The logic and
memory functions are again realised through the interconnection and
contact hole patterns.

The various logic gates and memory cells in a sea-of-transistors architecture are separated by using the *gate-isolation technique* illustrated
in figure 7.34.

417

The layout in the figure is a D-type flip-flop, based on the logic diagram shown. The gate-isolation technique uses pMOS and nMOS isolation transistors, which are permanently switched off by connecting them to supply and ground, respectively. This technique obviously requires both an nMOS and a pMOS isolation transistor between neighbouring logic gates [3].

The NRE costs of these devices depend on circuit complexity and are in the order of 100 k\$-1 M\$. Small transistors placed in parallel with larger transistors facilitate the integration of logic cells with RAMs, ROMs and PLAs in some of these HDGA architectures [4].

The design methods used for gate arrays are becoming increasingly similar to those used for cell-based design. This trend facilitates the integration of scan-test techniques in gate array design. As a result of the increasing number of available cells, the software for gate array programming resembles that of cell-based designs. Also, the availability of complete cores that allow reuse (IP) are becoming available to gate array implementation.

Off-the-shelf families of gate arrays are available and include the full transistor manufacture with source and drain implants. Customisation therefore only requires the processing of several contact and metal masks. This facilitates a short *turn-around time* in processing and renders gate arrays suitable for fast prototyping.

Gate array publications include advanced low-power schemes and technologies (SOI). For high speed gate arrays, gate delays (3-input NOR with a fan-out of two) below 50 ps have been reported. The complexity of advanced gate arrays has exceeded several tens of millions of gates. The popularity of these (mask programmable) gate arrays reached a maximum during the nineties. This decade shows a dramatic reduction in new gate array design starts, mainly due to the rapid cost reduction and gate complexity increase of the field-programmable gate arrays. These are the subject of the next paragraph.

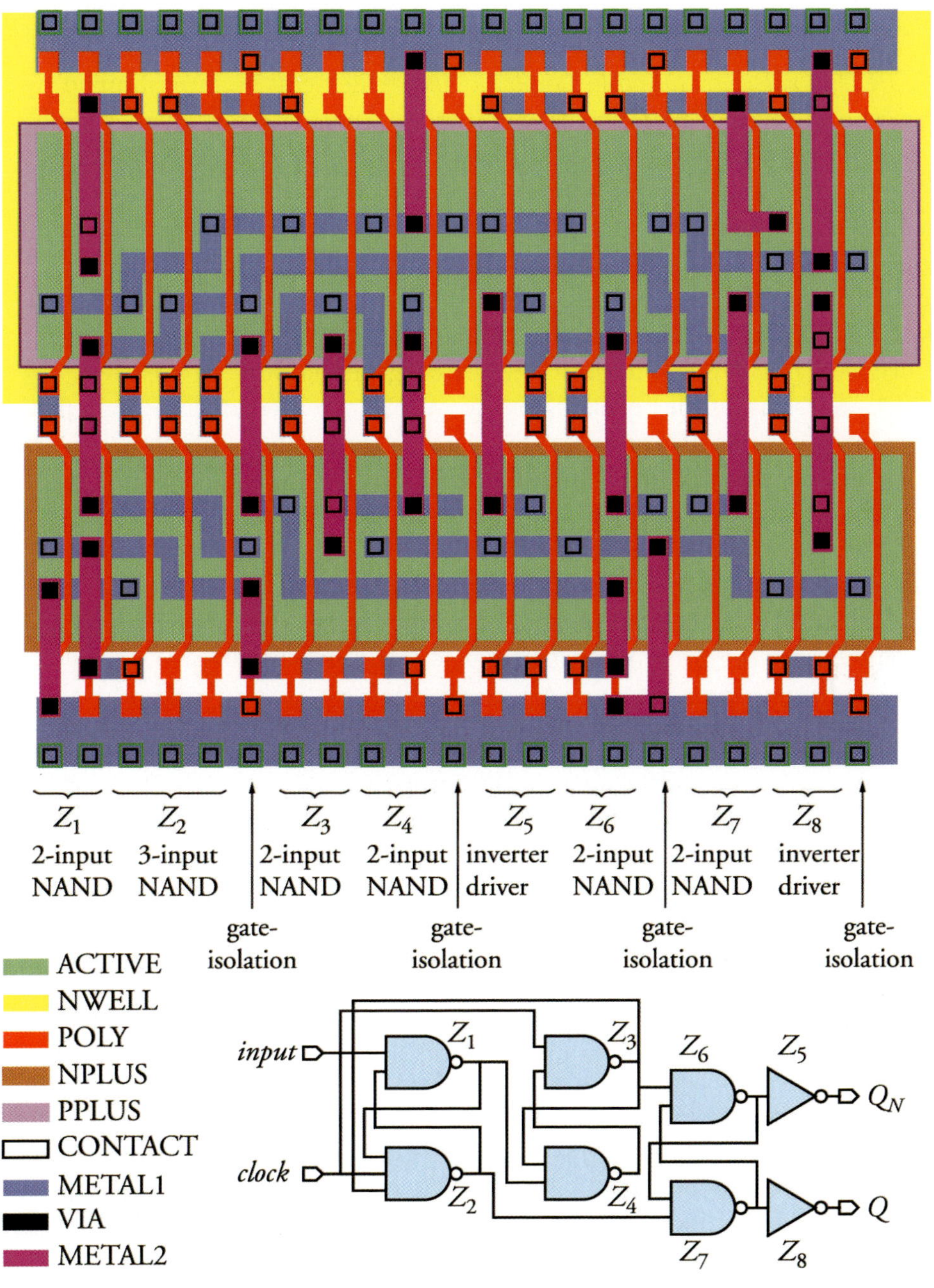

Figure 7.34: Sea-of-transistors array with gate isolation

7.6.7 Programmable Logic Devices (PLDs)

A *PLD* is a *Programmable Logic Device*, which can be programmed by fuses, anti-fuses or memory-based circuits. Another name currently also used for a certain category of these devices is *Field Programmable Device (FPD)*. The first user-programmable device that could implement logic was the programmable read-only memory (PROM), in which address lines serve as logic inputs and data lines as output (see also sections 6.5.3 and 7.6.4). PLD technology has moved from purely bipolar technology, with a simple fuse-blowing mechanism, to complex architectures using antifuse, (E)EPROM, flash or SRAM programmability.

As a result of the continuous drive for increased density and performance, simple PLDs are losing their market share in favour of the high-density flexible PLD architectures. In this way, PLDs are moving closer and closer towards a gate array or cell-based design and are becoming a real option for implementing *systems on silicon*. Another piece of evidence for this trend is the fact that several vendors are offering libraries of embedded cores and megacells. In the following, several architectures are presented to show the trend in PLDs.

Field Programmable Gate Arrays (FPGAs)

FPGAs combine the initial PLD architecture with the flexibility of an *In-System Programmability (ISP)* feature. Many vendors currently offer very high-density *FPGA* architectures to facilitate *system-level integration (SLI)*. Current FPGAs are mostly SRAM-based and combine memory and *Look-Up Tables (LUTs)* to implement the logic blocks. Vendors offering LUT-based FPGAs include Xilinx (XC3000-XC5000 and Virtex families), Lucent Technologies (ORCA families) and ALTERA (FLEX families).

Initially, FPGAs were used to integrate the glue logic in a system. However, the rapid increase in their complexity and flexibility make them potential candidates for the integration of high-performance, high-density (sub)systems, previously implemented in gate arrays [5]. The potentials of an FPGA will be discussed on the basis of a generic FPGA architecture (figure 7.35).

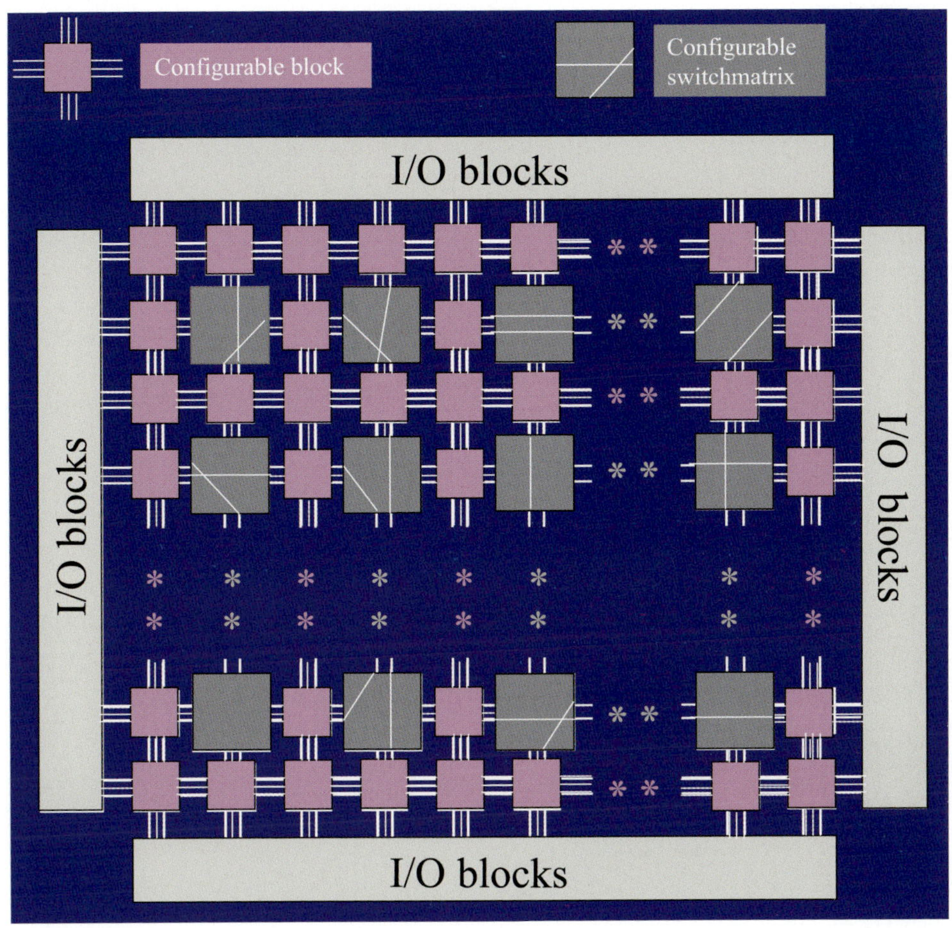

Figure 7.35: *General representation of an FPGA architecture*

Today, these architectures consist of a large array of hundreds of thousands of programmable (re)configurable logic blocks and configurable switch matrix blocks. A logic block generally offers both combinatorial and sequential logic. Figure 7.36 shows an example of a configurable block.

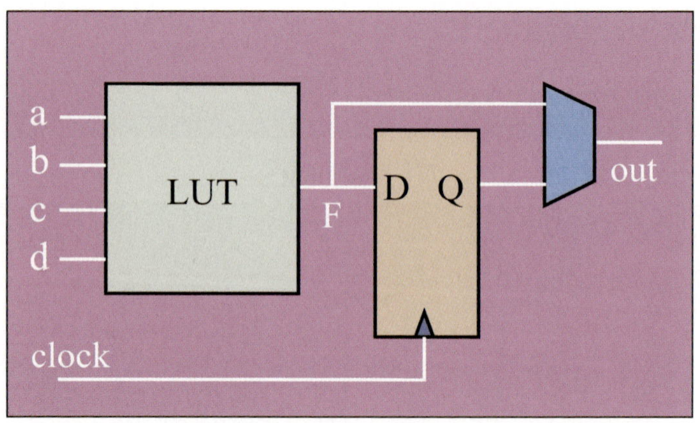

Figure 7.36: *Example of a 4-input configurable block*

In many FPGA architectures the configurable block includes one or more *look-up tables (LUTs)* , one or more flip-flops and multiplexers. Some also contain carry chains to support adder functions. The combinatorial logic is realised by the LUTs, which each may contain 3 to 8 inputs. Figure 7.37 shows an example of a 4-input LUT.

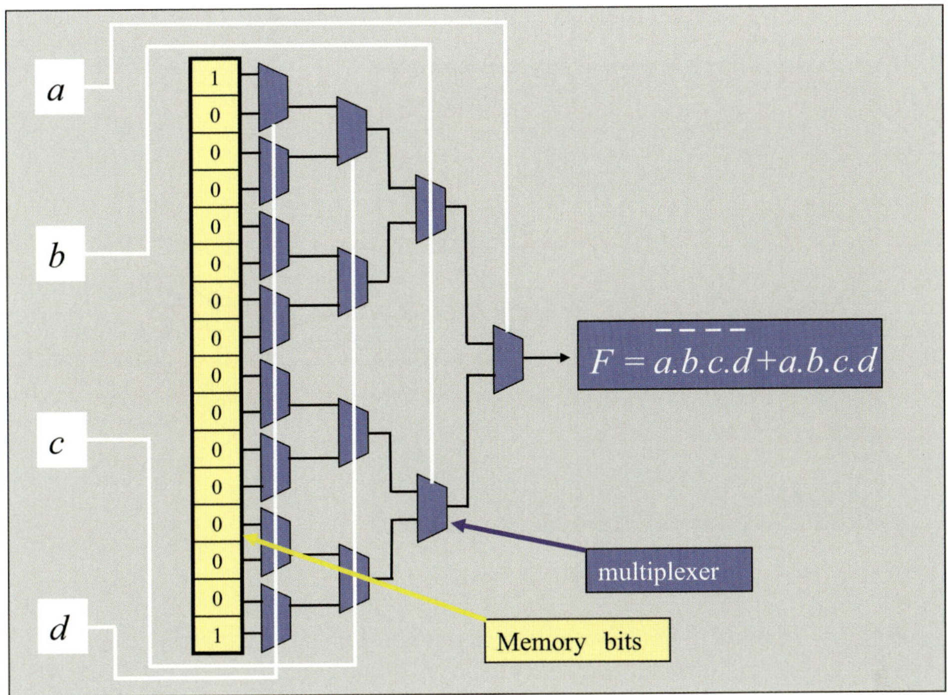

Figure 7.37: *Example of a 4-input LUT*

It is basically a small memory consisting of sixteen memory cells and a
couple of multiplexers. By changing the values in these memory cells
(when the application is loaded into the FPGA), any logic function (F)
of the four inputs (a, b, c, and d) can be created. The data stored in the
memory cells of the example represents the following logic function:

$$F = \overline{a} \cdot \overline{b} \cdot \overline{c} \cdot \overline{d} + a \cdot b \cdot c \cdot d$$

The LUT, however, can also serve as a distributed memory in the form
of synchronous or asynchronous, single or dual-port SRAM or ROM,
depending on needs of the application.

Many FPGAs contain short wire segments for local interconnections
as well as long wire segments for 'long distance' interconnections. The
logic blocks are connected to these wire segments by the configurable
switch matrix blocks. Figure 7.38 shows an example of such a block.

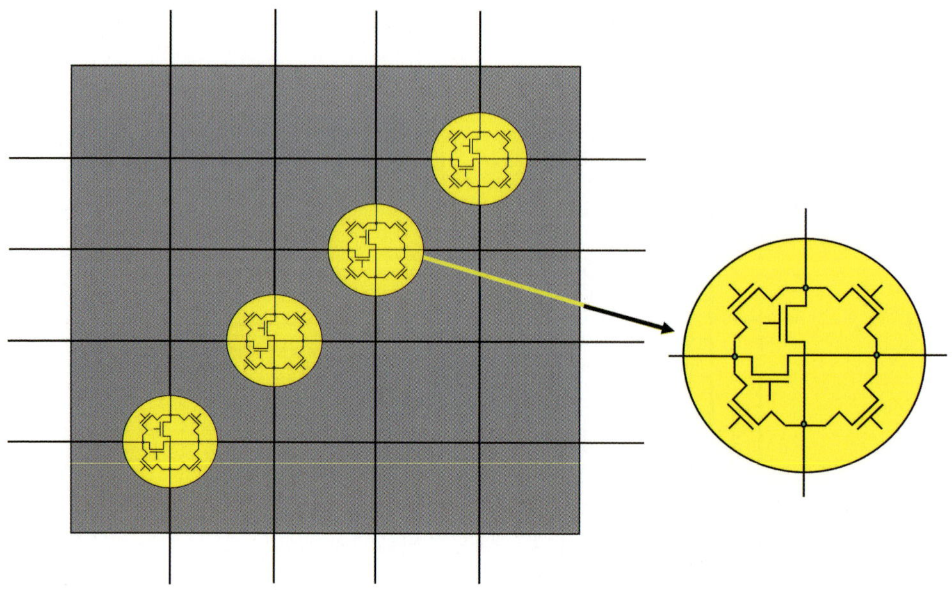

Figure 7.38: *Example of a configurable switch matrix*

The individual switches in such a block are controlled by the so-called configuration memory cells, whose data is also stored when the application is loaded into the FPGA. Most FPGAs use SRAMs to store the configuration bits, although there are also a few who store them in a non-volatile EEPROM or flash memory. All FPGAs that use SRAM for configuration storage need a shadow non-volatile backup memory on the board to be able to quickly download the application into the on-chip configuration memory. Downloading from a software program would lead to relatively large configuration times, whenever the application is started again after a power down.

Next to the configurable logic and switch matrix blocks, many FPGA architectures include dedicated IP cores, digital signal processors (DSPs), microprocessors such as ARM and PowerPC, single and/or dual port SRAMs and multipliers.

Finally most I/O blocks support a variety of standard and high-speed interfaces. Examples of single-ended interfaces are: LVTTL, LVCMOS PCI, PCI-X, GTL and GTLP, HSTL and SSTL. Examples of differential I/O standards are: LVDS, Extended LVDS (2.5 V only), BLVDS (Bus LVDS) and ULVDS, Hypertransport^{TM}, Differential HSTL, SSTL.

Of course also several dedicated memory interfaces, such as DDR, DDR-2, DDR-3 and SDRAM, are supported.

Among the state-of-the-art FPGAs are the Xilinx Virtex^{TM}-4 and the Altera's Stratix-III families. To get a flavour of the potentials of these FPGAs, some of the characteristical parameters of the Virtex^{TM}-4 architecture will be highlighted. It supports three different application platforms (figure 7.39): LX, which is optimised for logic applications, SX, for the high-end DSP applications and FX, which supports embedded processors and high-speed serial I/O.

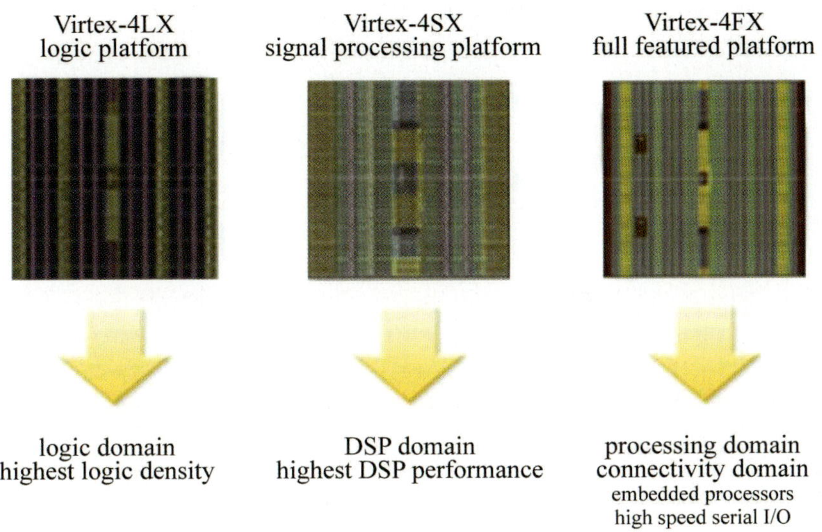

Figure 7.39: *Different platforms supported by the Virtex-4 FPGA (Source: Xilinx)*

The Virtex-4 FX architecture is very similar to that of the Virtex-II PRO, which is shown in figure 7.40.

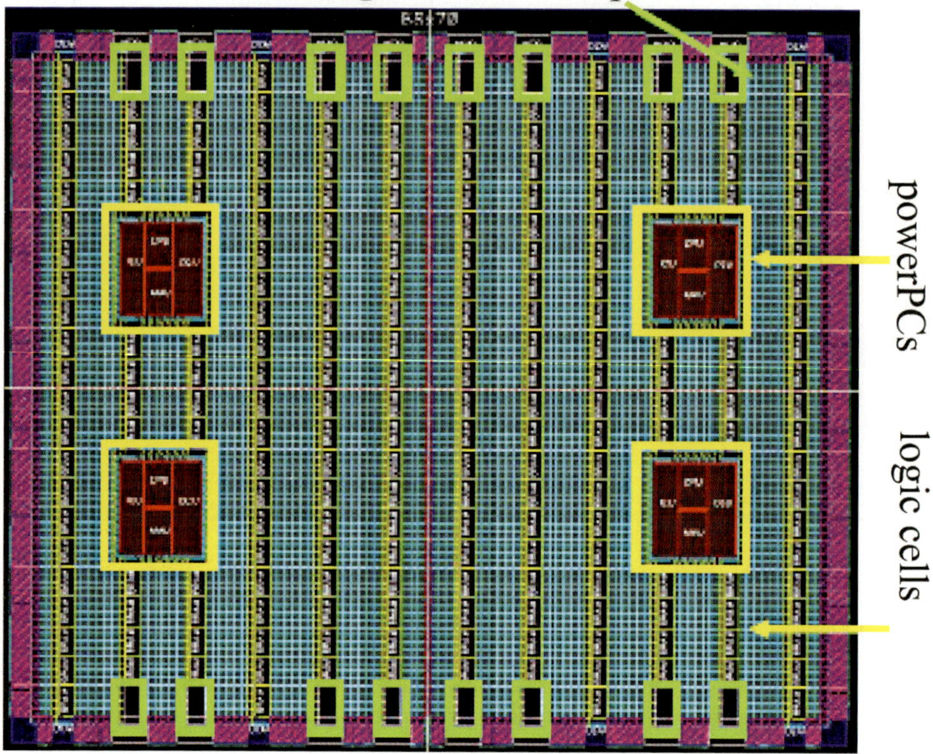

Figure 7.40: *Architecture of the Virtex-II PRO (Source: Xilinx)*

The Virtex-4 FX can have 63,000 configurable logic blocks (CLB) with 142,000 logic cells and around 10 Mb of block SRAM, 192 DSP slices, 2 PowerPC^{TM} embedded processors, 4 Ethernet MAC blocks and 24 high-speed serial interfaces [6].

The basic Virtex-4 building blocks are an enhancement of those found in previous Virtex-based products, allowing upward compatibility of existing designs. Virtex-4 devices are produced in a 90 nm copper process using 300 mm (12 inch) wafer technology.

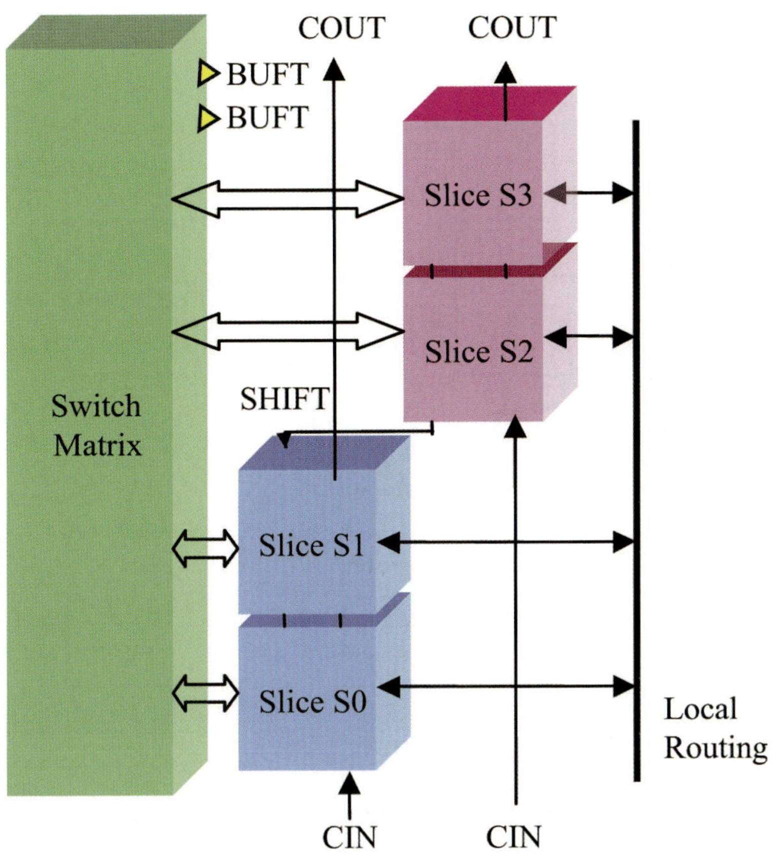

Figure 7.41: *A configurable logic block in the Virtex^{TM}-II FPGA architecture (Source: Xilinx)*

Figure 7.41 shows the architecture of a CLB. It consists of four so-called slices, two dedicated carry chains to support fast arithmetic, a 16:1 multiplexer (not shown in the figure). On one side, the slices are connected to the configurable switch matrix (as explained in figures 7.35 and 7.38), while on the other side it contains local routing to provide fast interconnect. Each slice (figure 7.42) contains two 4-input LUTs, two flipflops/latches, some carry logic to support arithmetic operations.

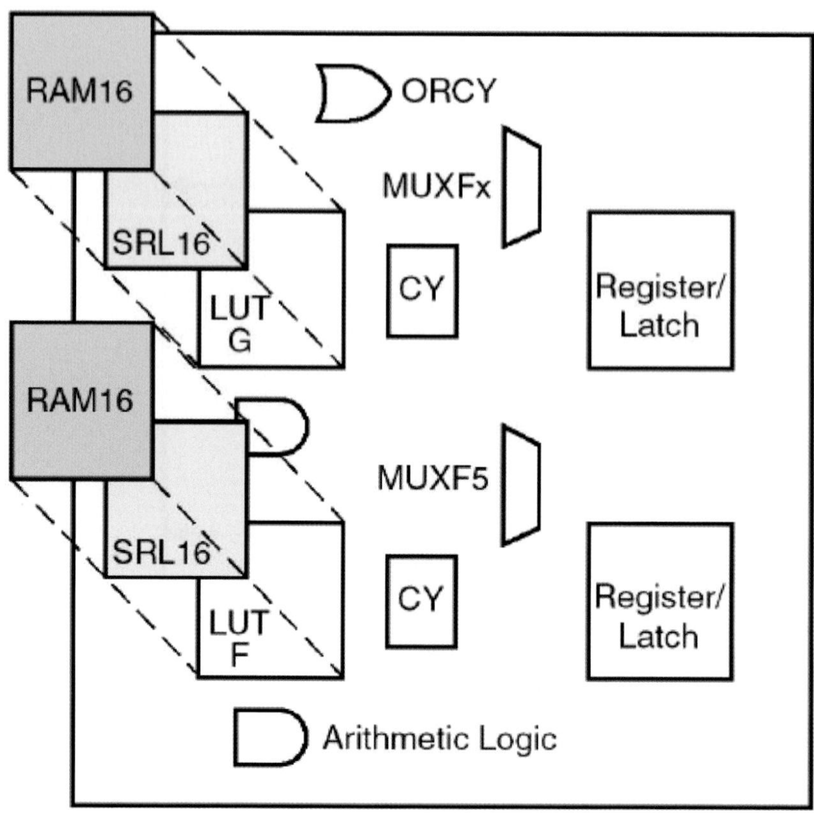

Figure 7.42: *The Virtex logic slice (Source: Xilinx)*

A 4-input LUT can also be used as a 16-bit synchronous RAM to create a distributed memory facility or as a 16-bit (or cascadable variable) shift register.

A presentation of more details on the Virtex-4 architecture and implementation is beyond the scope of this book. This section is meant to present a flavour of the potentials of current state-of-the-art FPGAs. As explained before, most FPGAs reconfigurability (logic as well as interconnect) is controlled by on-chip configuration SRAM memory bits and require additional non-volatile configuration back-up memory on the board.

Complex Programmable Logic Devices (CPLDs)

The structure of a PLD has evolved from the original PAL^{TM} devices, which implement sum-of-products (min terms), where the AND-array is

programmable and the OR-array is fixed (see section 7.6.4). Figure 7.43
shows an example of the basic PAL architecture, which implements three
logic functions.

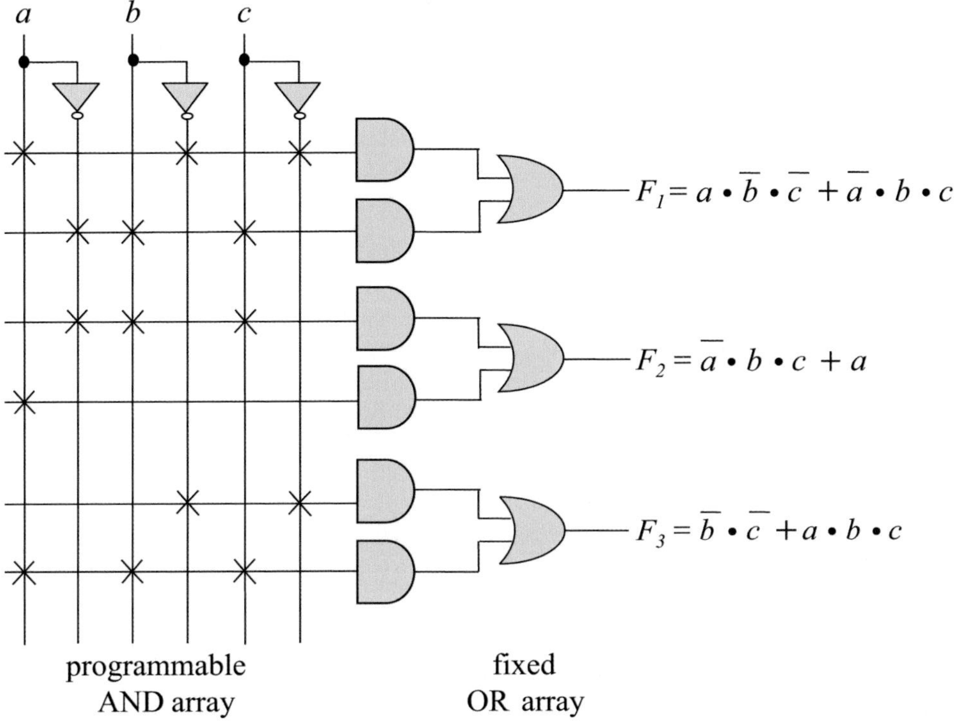

$$F_1 = a \cdot \overline{b} \cdot \overline{c} + \overline{a} \cdot b \cdot c$$

$$F_2 = \overline{a} \cdot b \cdot c + a$$

$$F_3 = \overline{b} \cdot \overline{c} + a \cdot b \cdot c$$

Figure 7.43: *Example of a basic PAL architecture implementing three different logic functions of three inputs*

The connections in the AND-array of the CPLD are commonly realised
by floating-gate transistors (figure 7.44), which means that it can be
(p)reprogrammed using in-system programmability and it will securely
retain its program, even when it is powered off.

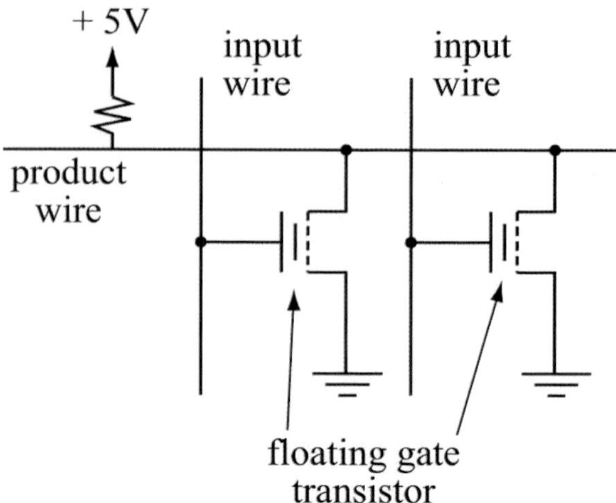

Figure 7.44: *Use of floating gate transistors to realise AND-array connections in CPLD (Source: IEEE Design & Test of Computers)*

There is no technical reason why the previously discussed FPGAs use SRAM or anti-fuse programming techniques in stead of non-volatile, except that the fabrication process will be cheaper.

The original simple PLDs only implemented some tens of logic functions. A large design had to be split to fit them into a couple of PLDs, which became a barrier for PLD usage. As a result, ASIC vendors started developing PLDs with much larger arrays and the *complex PLD* or *CPLD* was born. CPLDs are offered by a large number of vendors, including Altera (MAX®series), Xilinx (Coolrunner[TM] and XC9500[TM] series), Lattice Semiconductors (ispMACH4000Z), etc. Most CPLD architectures look very similar and are based on the previously discussed PAL AND-OR arrays. Since the logic depth of these arrays is relatively short, even wide-input PLD functions offer short pin-to-pin propagation delays. Many of them also include registers, but their total complexity in terms of equivalent logic gates and flip-flops is usually relatively low, compared to FPGAs.

One example, in which the CPLD architecture has departed from the original PAL array, allows more design flexibility and shows more similarity with an FPGA, is the Altera MAX-II CPLD family. Its architecture consists of an array of Logic Array Blocks (LAB's), a flash configuration memory, an additional flash memory for data storage, a JTAG and control block and supports a large number of I/Os through the staggered pads configuration.

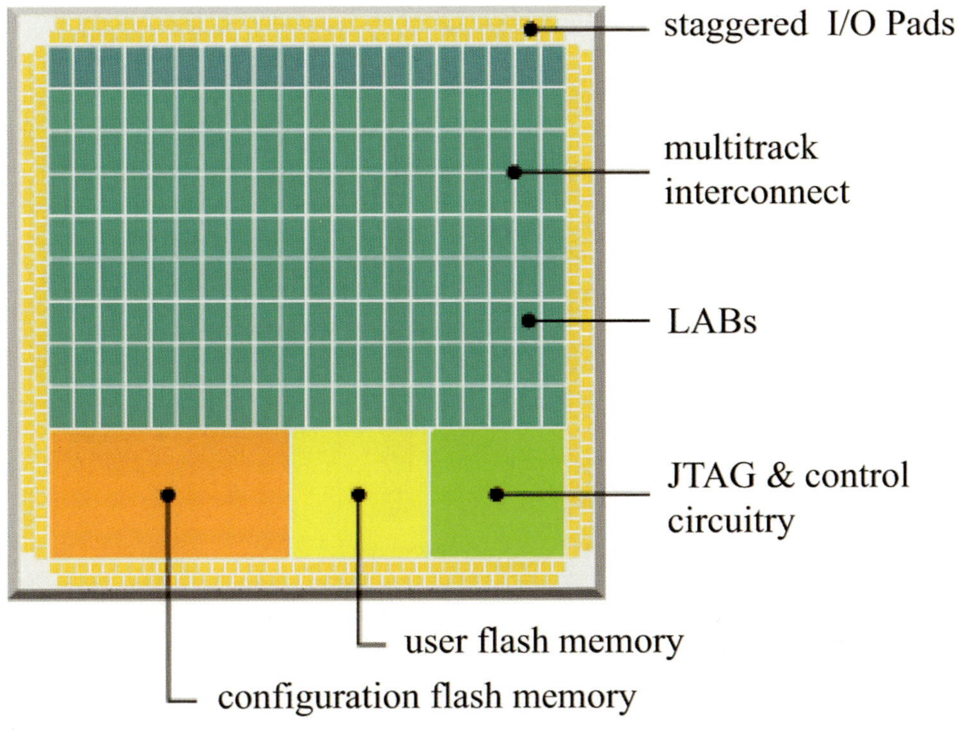

Figure 7.45: *Architecture of Altera's MAX-II CPLD (Source: Altera)*

Each LAB is built from 10 Logic Elements (LE's), LUT and carry chains, several LAB-control signals and local and register interconnections. One LAB can handle 26 independent inputs and 10 local (within the same LAB) feedback inputs. Figure 7.46 shows the global and local interconnections of a LAB. The DirectLink connections allow flexible and fast communication between adjacent LABs without using the row and column routing facilities.

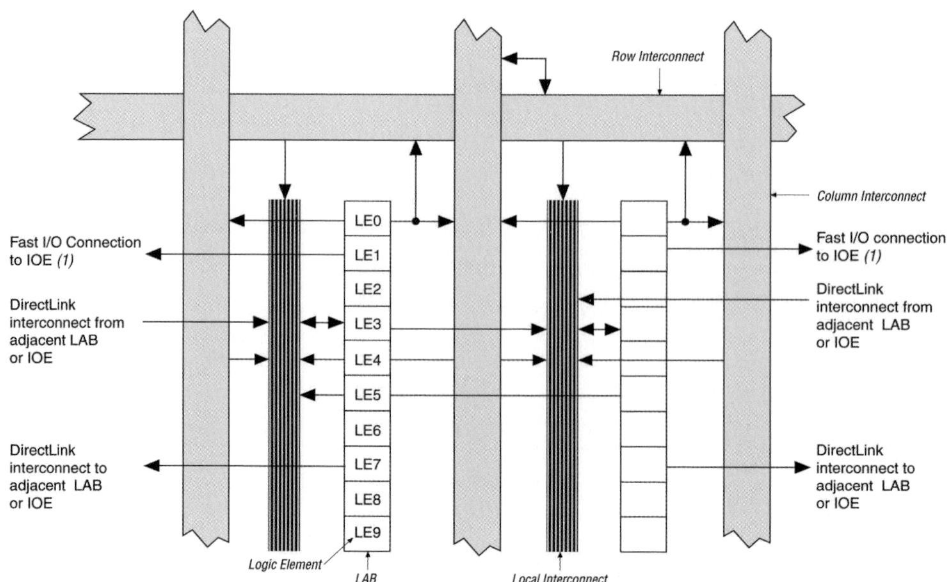

Figure 7.46: *Architecture of a LAB block together with its interconnect features (Source: Altera)*

Rather than a PAL array, which is most commonly used in a CPLD, the MAX-II CPLD uses a 4-input LUT as the basis for the generation of the logic function in an LE (figure 7.47)

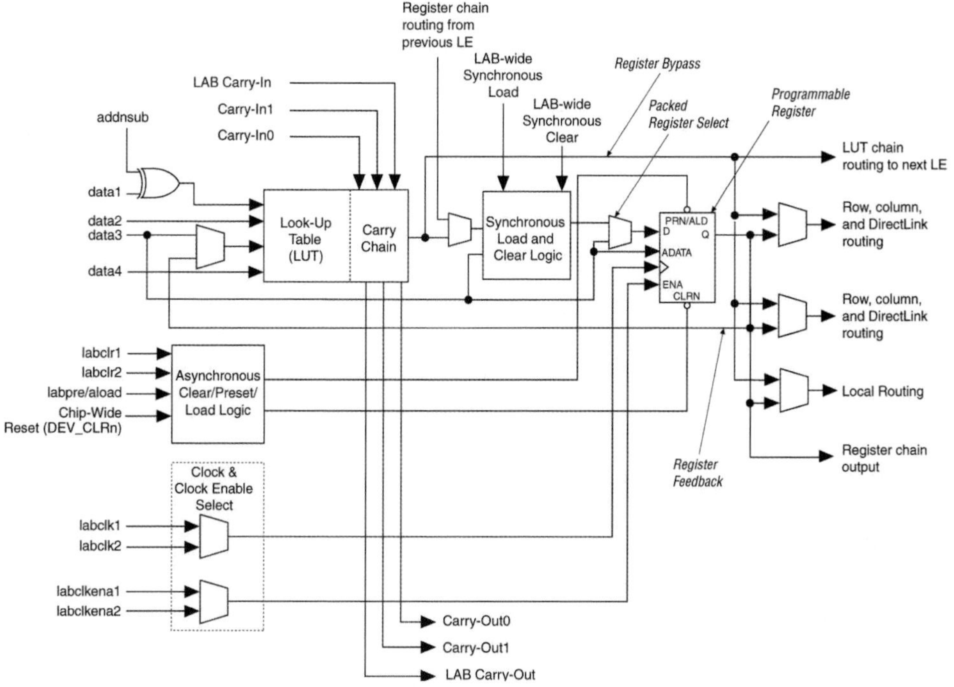

Figure 7.47: *Architecture of a Logic Element of the MAX-II CPLD (Source: Altera)*

Next to that, an LE contains a programmable register and carry chain with carry select capability [7]. It also supports dynamic single bit addition or subtraction mode selectable by an LAB-wide control signal. Each LE drives all types of interconnects: local, row, column, LUT chain, register chain, and DirectLink interconnects. An LE can drive up to 30 other LEs through these DirectLink interconnects.

It is beyond the scope of this textbook to discuss all features supported by an LE. Inclusion of the MAX-II CPLD architecture into this textbook is only meant to present a flavour of today's capabilities of CPLD ASICs.

As stated before, the total complexity of most CPLDs in terms of equivalent logic gates and flip-flops is relatively low, compared to FPGAs. They are therefore often used in small systems to implement complex finite-state machines, fast and wide decoders or high-performance control logic. Because the functionality is stored in a non-volatile way, most CPLDs are also suited for use in applications where they can be completely switched off during idle times, without losing their functionality as an SRAM-based FPGA would. The high-end (high-complexity) CPLDs applications show some overlap with the low-end FPGAs. Because of the large number of flip flops and the dynamic reconfigurability, FPGAs are much more flexible in use, compared to CPLD.

Programmability of FPGAs and CPLDs

The most important switch-programming techniques currently applied in FPGAs are SRAM, anti-fuse, and non-volatile memory cells. Figure 7.38 shows an example of a configurable switch matrix to configure the routing of signals through available interconnect patterns. SRAM cells or flip-flops are also used in a look-up table to configure logic functions (figure 7.37).

In the majority of current commercially-available CPLDs, the switches are implemented as floating-gate devices, like those in (E)EPROM and flash technologies (figure 7.44) [8]. However, CPLDs with SRAM programmability appear on the market. Here, the switches are used to program the AND and OR-array of the PAL, see figure 7.43. In 90% of the CPLDs, the connections are made through programmable multiplexers or full cross-point switches. If an input is not used in a product term (minterm) in an AND plane on a CPLD, the corresponding EPROM gate transistor is programmed to be in the off-state. Similar architectures can be built with EEPROM transistors.

Large complexity PLDs, based on anti-fuse programmability, are not discussed here, because the memory-based programmable PLDs/FPGAs are dominating the PLD market.

7.6.8   Embedded Arrays, Structured ASICs and platform ASICs

The previously-discussed cell-based designs (section 7.6.5) may include standard cells, macro cells, embedded memory blocks and IP cores, etc. A rather new development in cell-based designs is the inclusion of *embedded arrays*. In most cell-based designs that include an embedded array, all masks are customised, as in the cell-based designs. Embedded arrays combine a gate array-like structure and large cells such as microprocessor cores, memories and I/O functions. Cores can either be mapped onto the sea-of-gates array (see section 7.6.6) or can be implemented as a separate block. Figure 7.48 shows the architecture of an embedded array ASIC.

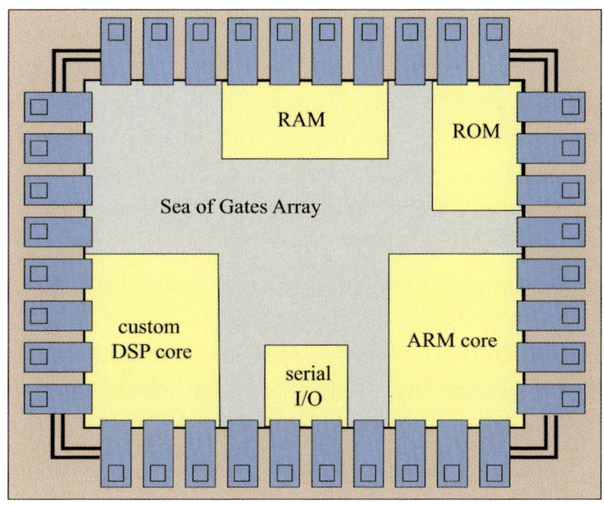

Figure 7.48: *Architecture of an embedded array ASIC (Source: ICE)*

The idea behind such an ASIC is to reduce the total turn-around time from spec definition to first silicon. During the first 20% of the spec development time, almost 80% of the system is defined. So, at that time, the engineers know already which memory type (SRAM, DRAM, flash, etc.), and how much memory is needed, what type of IP cores (CPU, DSP, ARM, analog IP, etc.) are needed and also what type of I/Os the application requires. Also a rough estimation of the required number of logic gates can be made at that time. These are then implemented as a kind of mask-programmable sea-of-gates array. The chip is then sent to the fab and is being processed up to the final back-end masks (metal layers and vias), in parallel to the design team defining the remaining 80% of the spec to come to the final spec definition. After completing the spec, only the final metal and via masks need to be defined and processed, thereby reducing the turn-around time and more specifically the time-to-market. Even last-minute design (spec) changes are allowed. Due to the very small lifetimes of today's products in many consumer and communication markets, it has become very important to have the ability to put prototype products quickly to the market, perform a fast customer product review and transfer it, if necessary, into a high-volume standard-cell design. Toshiba uses this embedded array concept in their 'universal arrays' ASIC architecture, where the customer can define his own ASIC, with a selection of various available IPs and I/Os, and with the logic implemented on a sea-of-gates array, available in 130 nm (TC280 series) and 90 nm CMOS (TC300 series) [9]. In normal standard-cell blocks, the empty areas are filled with *filler cells*, which do not contain any transistor, but are only used to extend the supply lines and nwells and pwells and allow routing in most metal layers. Due to the sea-of-gates approach in the universal array architecture, the 'empty areas', here, also contain unused transistors and offer additional flexibility for creating small design changes. The first product needs to undergo all mask and processing steps, but redesigns, or derivatives with small changes in the logic content, can be quickly realised by changing only the final metal and via masks and performing only the back-end processing. You need to do the design yourself, using the vendor's technology and design kit. The NRE costs for the first run may be in the order of 250-350 k\$ for a 120 nm CMOS design with a few million gates and a few Mb of embedded SRAM. It includes the mask costs and delivery of about 100 samples. A new run, with only minor metal mask changes, may cost 50 k\$. For a 90 nm design these NRE costs grow with about 50%.

Structured ASICs and platform ASICs

(Mask-programmable) gate arrays have suffered from a declined popularity over the last decade. This has increased the gap between the cell-based design ASICs and FPGAs. A *structured ASIC* or *platform ASIC* is a combination of the cell-based and FPGA design concepts, which targets prototyping applications and relatively low volume markets (10k-100k). It offers a large selection of IP cores, which can be customised through a limited number of masks. Basically personalisation can be done by customising all metal and via masks, by customising only a subset of the metal and via masks, or by customising only one via mask. NRE costs are relatively low (from 50k\$ to several 100k\$), but the price per individual chip can be four to six times the cell-based design version. In the following a structured array ASIC example is presented to show some capabilities of this category of ASIC products.

eASIC's Nextreme structured array ASIC Family This structured (array) ASIC is an example of customisation through only one top-level via mask. The Nextreme family [10] consists of six members, pre-processed up to metal6, offering from 350k to 5 million gates (also configurable to a maximum of 5 Mb distributed memory) and 416 kb to 5 Mb of dedicated block memory. Customisation is done, only through the VIA-6 mask, allowing very short production turn around times. Figure 7.49 shows its basic architecture.

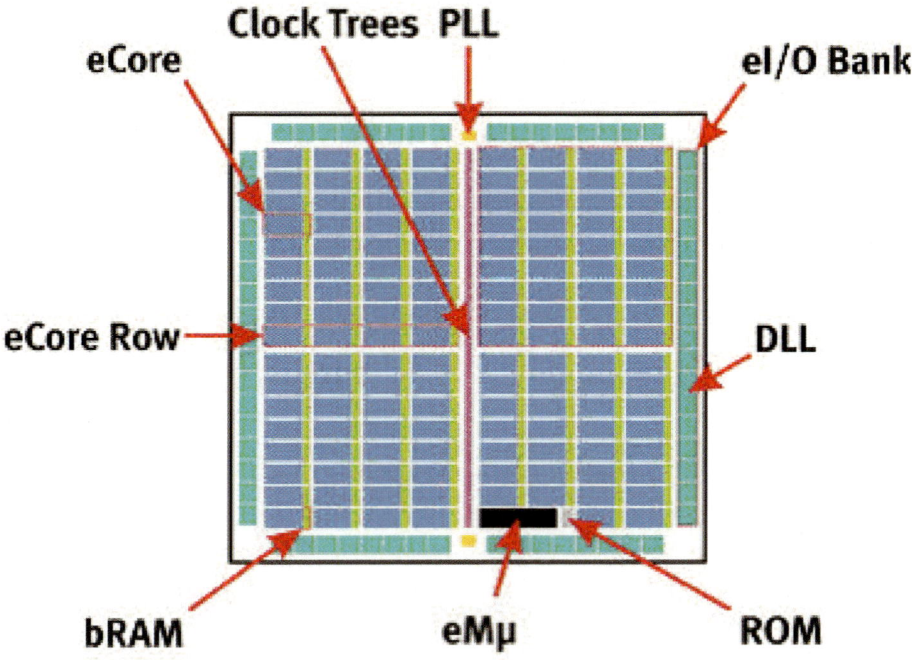

Figure 7.49: *Example of a 1-mask programmable structured array ASIC (Nextreme; eASIC)*

An eCore contains 2048 eCells that can be configured to logic or (distributed) memory. An eCell contains two 3-input LUTs and two 2-input NAND gates. A LUT and a NAND can form almost any combination of four inputs. It also contains a 2-input multiplexer, a scannable D-type flip-flop and buffers and drivers to drive high load nets. The dedicated memory (bRAM) may consist of multiples of single-port 32 kb memory blocks, which are VIA-6 configurable to 32kx1, 16kx2, 8kx4, 4kx8, 2kx16, 1kx32 architectures. The Nextreme architecture also contains a VIA-6 programmable 16 kb ROM. The 8051 microcontroller (eMµ) is used to initialize all distributed and block memories. Configurable PLLs and DLLs are embedded for clock generation and clock-phase shifting purposes. Next to a variety of I/O standards, also SERDES (serialiser-deserialiser), differential and DDR interfaces are supported through a library of input, output and bi-directional I/Os, which can be configured into a large variety of options and drive strengths.

For prototyping and other low-volume applications a direct-write eBeam machine is used to perform this VIA-6 customisation, to avoid the costly mask production. For high volumes the custom VIA-6 mask is generated from the same design data base.

Structured ASICs attack the low-end of the ASIC market. Although there has already been a 'structured arrays ASICs vendor' shake out, there are more vendors than the ones referred to in this section. The selection that has been made here presents a good flavour of the potentials of available products of this ASIC category.

7.6.9 Hierarchical design approach

The *hierarchical layout* design style is characterised by a modular structure, as shown in the example in figure 6 (in the preface section). The different modules are identified during the design path. With a complex system on chip, for example, the various required functional modules emerge from the specification. These modules may include microprocessor core, ROM, RAM and signal processors, etc.

A top-down design strategy generally leads to a satisfactory implementation of a hierarchical layout. The hierarchical division allows various designers or design teams to simultaneously produce layouts of the identified modules. Reasonable gate or bit densities are combined with a reasonable speed. The afforded performance renders the hierarchical layout design style suitable for most VLSI and ASIC designs. The design time for hierarchical layouts can be drastically reduced with good CAD tools. Available libraries may contain parameterised *module generators*. Also, IP cores (which are available from different vendors) can be "plugged in", see section 7.2 (definitions: IP) and section 7.4.2.

These (mostly) software descriptions are synthesised to produce netlists, which can be used to create layouts of required modules. Assembly of the resulting instances and bond pads leads to the creation of a complete chip layout. Even the assembly and interconnection is automated in *placement and routing* programs (*place & route tools*).

The hierarchical design style can, of course, include modules which are created by using different layout design styles, e.g., standard-cell or handcrafted module layouts. The hierarchical style was disadvantaged by the relatively large routing areas that could be necessary. However, with the present availability of six to nine metal layers, interconnections and buses can be routed across the logic blocks. In some cases, however, the chip area may not be optimum as a result of the *Manhattan skyline* effect, which results from different block shapes.

Figure 7.50 shows the *meet-in-the-middle strategy* used in the hierarchical design approach. This strategy was already introduced by Hugo de Man in the early eighties [11]. Here, the high-level system description is used to synthesise a design description comprising macro blocks at the implementation level. This implementation level lies roughly in the middle of the top-down design path. The choice of implementation form is still open at this level and possibilities may include a gate array or a cell-based layout. It must be possible to generate these macros from existing design descriptions. Sometimes, module generators are also used to generate a core. The (re)use of IP cores allows a fast "plug-in" of different functional blocks, which are standardised to a certain extent. Clearly, the results of design and layout syntheses meet at the implementation level.

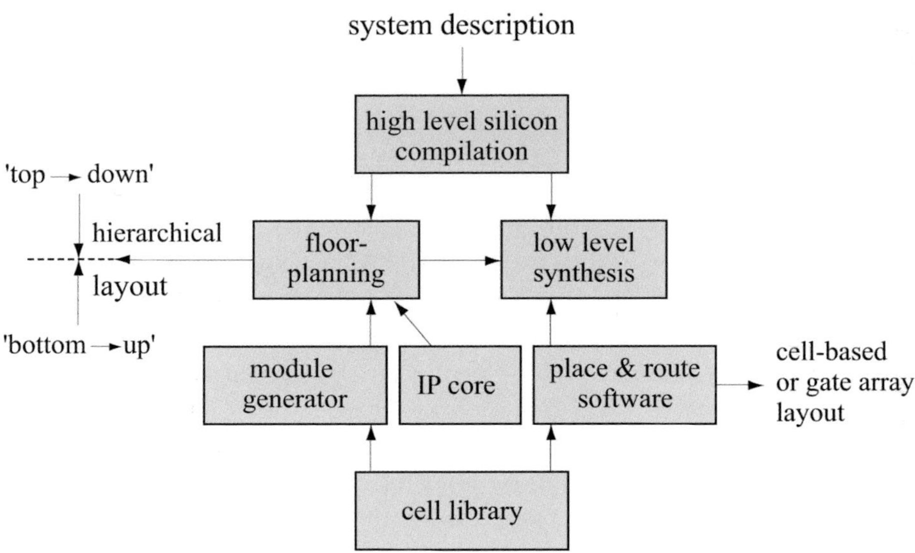

Figure 7.50: *Meet-in-the-middle strategy*

7.6.10 The choice of a layout implementation form

The unique characteristics of each form of layout implementation determine its applicability. The choice of implementation form is determined by chip performance requirements, initial design costs, required volumes and time-to-market requirements. Figure 7.51 shows a cost comparison of the different forms of layout implementation.

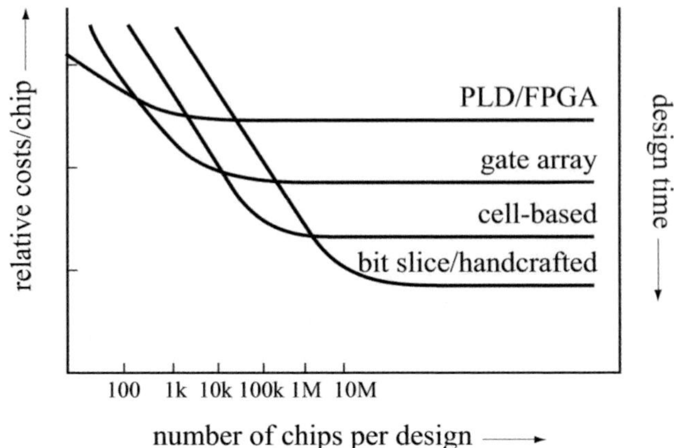

Figure 7.51: *Cost comparison of the different layout implementation forms*

A single chip may combine different implementation forms. The previously discussed embedded array ASICs and structured ASICs are examples of this. Figure 7.52 shows a photograph of a conventional microprocessor in which handcrafted, bit-slice and memory layout styles are combined. The nVidia GeForce 8800 is an example of a chip that combines data path layout with standard-cell, memory and full-custom design.

An implementation technique that was popular in the eighties and early nineties and is still used in some cases today, is the symbolic layout and compaction technique. A *symbolic layout* is a technology-independent design, which can be used for every layout implementation form. In a symbolic layout, transistors and contacts are represented by symbols whose exact dimensions are unspecified while wires are represented by lines whose widths are also unspecified. The abstract symbolic layout is transformed to an actual layout by a compaction program, which accounts for all of the design rules of the envisaged manufacturing process.

The symbolic-layout technique allows a short design time and relieves designers of the need to know specific layout and technology details. The technique is, however, disadvantaged by the associated relatively low gate density and low switching speed. These compare unfavourably with handcrafted layout results. Furthermore, the abstract nature of a symbolic layout only loosely reflects technological aspects. This may result in fatal design errors. Currently, symbolic layout and compaction are are only very rarely used.

Finally, the dimensions of all circuit components and wiring in an IC layout are scaled versions of the actual on-chip dimensions. This *geometric layout representation* is generally described in a *geometric layout description language (GLDL)*. Such languages are common to many CAD tools and usually serve as the data-interchange format between IC design and manufacturing environments. A GLDL has the following typical features:

- It facilitates the declaration of important layout description parameters, e.g., masks, resolution, dimensions

- It facilitates the definition of geometrical forms, e.g., rectangles and polygons

- It facilitates the definition of macros, e.g., patterns or symbols

- It enables transformations, e.g., mirroring and rotation

- It contains statements for the creation of matrixes.

Currently, *GDSII* is the de facto standard for physical chip design exchange in the semiconductor industry.

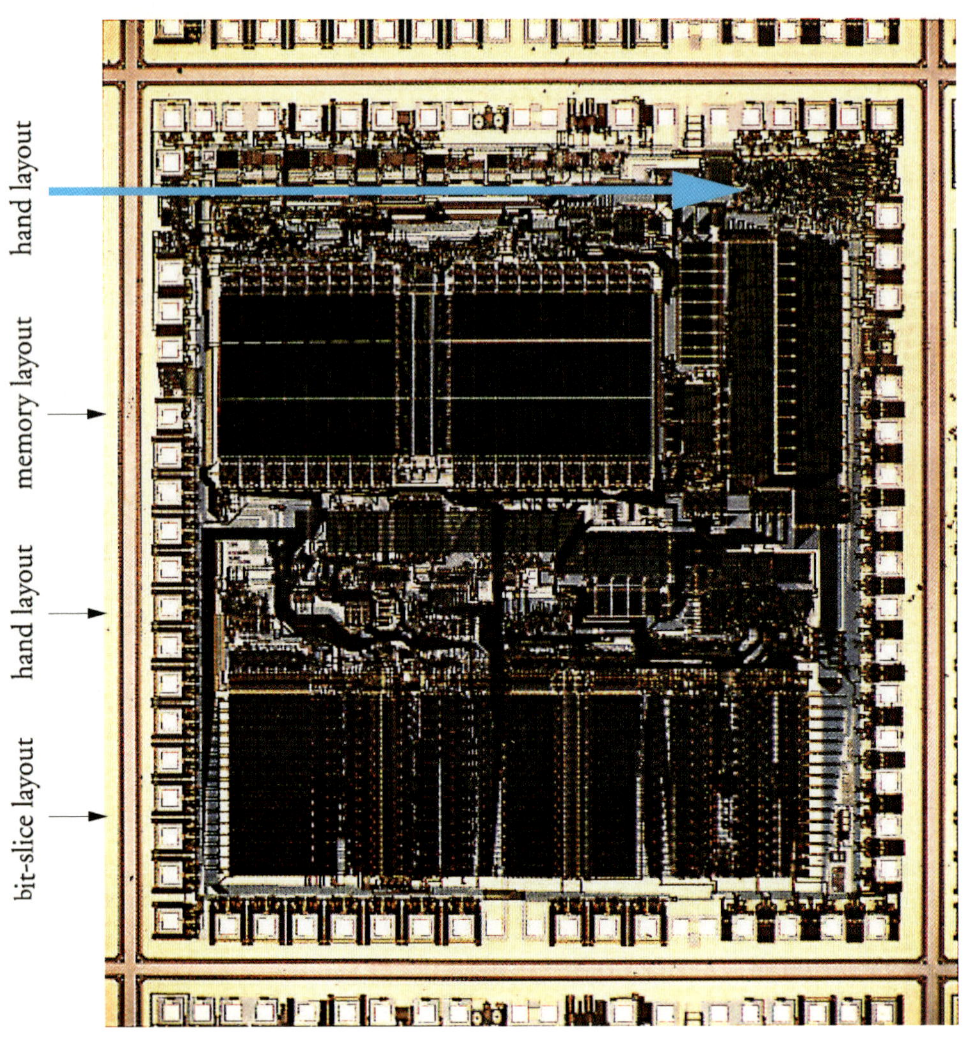

Figure 7.52: *A microprocessor chip which combines different layout implementation forms (Source: NXP Semiconductors)*

7.7 Conclusions

This chapter introduces various VLSI design and layout realisations and their characteristic properties. A top-down design approach, combined with a bottom-up implementation and verification through a hierarchical layout style appears suitable for most VLSI circuits. In practice, the design process consists of a number of iterations between the top-down and bottom-up paths, the aim being to minimise the number of iterations.

The use of IP cores that are available from different vendors is fuelling the reuse of existing functionality, such as microprocessor and signal processing cores and memories, etc. This reuse increases the problems with timing and communication between cores from different origins. Chapter 9 discusses these problems in detail.

During the last decade, the design complexity of an ASIC has dramatically increased and caused the design costs to increase about 25 times (see chapter 11). This has put a permanent pressure on the efficiency of the design process. Semiconductor companies have built application-domain specific platforms, which are key to a higher design productivity and improved product quality. Since IC production fabs are becoming extremely expensive, more companies will share the same production facility and production process and become *fab-lite* (outsourcing 40-50% of the manufacturing operations) or even *fabless*. So, semiconductor (design) houses can then only differentiate themselves to design better products faster and cheaper.

Various ASIC design and implementation styles have been presented. Standard-cell designs, mask-programmable gate arrays, field-programmable gate arrays and structured ASICs, all are different in the way they are designed, in the way they are fabricated and in the way they are used in an application. The choice of ASIC style largely depends on the required turn-around time and product volume.

A good IC design must be accompanied by a good test and debug strategy. Testability and debug are discussed in section 10.2 and require considerable attention during the design phase. The use of an extra 5 % of chip area to support testability and debug might, for instance, lead to a 50 % reduction in test costs.

7.8 References

[1] A.Abbo, et al.,
'XETAL-II: A 107 GOPS, 600 mW Massively-Parallel Processor for
Video Scene Analysis",
ISSCC Digest of Technical Papers, San Francisco, 2007

[2] R. Goering,
'Startup Liga promises to rev simulation', EE Times, 17-07-2006

[3] I. Okhura, et al.,
'A novel basic cell configuration for CMOS gate-array',
CICC 1982, pp 307-310, May 1982

[4] H.J.M. Veendrick, et al.,
'An efficient and flexible Architecture for High-Density Gate Arrays',
ISSCC Digest of Technical Papers, San Francisco, 1990

[5] Xilinx,
'The future of FPGAs', Xilinx Web-site, 1998

[6] Xilinx,
'Virtex-4 Family Overview', October 2006,
http://www.xilinx.com/bvdocs/publications/ds112.pdf

[7] Altera,
'Chapter 2. MAX II Architecture'
MAX II Device Handbook, Volume 1, August 2006,
http://www.altera.com/literature/hb/max2/max2_mii51002.pdf

[8] S. Brown, J. Rose,
'FPGA and CPLD Architectures: A tutorial',
IEEE Design&Test of Computers, Summer 1996

[9] 'ASIC 0.09 μm-TC300', 2006, www.toshiba.com, products
$\Rightarrow$ ASIC & Foundry $\Rightarrow$ ASIC 0.09 μm-TC300

[10] 'Nextreme features eASIC', 2006, http://www.easic.com

[11] H. de Man, et al.,
'An Intelligent Module Generator Environment',
Proceedings of the 23rd Design Automation Conference, 1986, pp.
730-735

7.9 Exercises

1. Why are abstraction levels used for complex IC designs?

2. What is meant by floor planning?

3. Explain what is meant by logic synthesis.

4. What does the term 'Manhattan skyline' describe in relation to a VLSI layout?

5. Assume that a standard-cell and a gate array library are designed in a CMOS technology. The libraries consist of logic cells with identical logic functions. Describe the main differences between the two libraries in terms of:
   a) Cell design
   b) Chip area
   c) Production time and cost
   d) Applications

6. Random logic functions can, for instance, be implemented using a ROM or a standard-cell realisation. Explain when each of these possibilities is preferred.

7. Draw a schematic diagram of a PLA which implements the following logic functions:

$$F_0 = \overline{\overline{x}\ \overline{y}} + xyz \quad F_1 = \overline{x\ \overline{y} + \overline{x}y + x\overline{z}} \quad F_2 = \overline{xyz + \overline{x}\ \overline{y}\ z}$$

8. Explain what is meant by mixed-level simulation.

9. Explain in your own words what is meant by IP. What is the cause of its existence? How can it affect design efficiency and what are the potential problems involved with it?

10. Explain the differences between an FPGA and a CPLD.

11. Explain the 'meet-in-the-middle' strategy.

12. Explain why a cell-based design implementation is much smaller than a design implemented with an FPGA.

Chapter 8

Low power, a hot topic in IC design

8.1 Introduction

Although already used in the seventies, it took until the mid-eighties before CMOS became the leading edge technology for VLSI circuits. Prior to that time, only a few designs were implemented in CMOS. At that time, only those applications that really required the low-power features of CMOS were designed in it. Most examples, then, were battery supplied applications, such as wristwatches (tens of millions per year), pocket calculators, portable medical devices (hearing aids and implantable heart controls) and remote controls.

From the 1970s until today, however, the number of transistors increased from only a few thousands to more than a billion, while chip frequencies, particularly in the high-performance processor category, increased from several Megahertz to several Gigahertz. In that period, the power consumption of these ICs increased from less than 1 W to above 100 W, while the power consumption of the ASIC category of ICs has reached the level of one to several Watts, which is in the range of the maximum allowed power consumption of a cheap plastic package. This is one of the main driving forces for low-power CMOS. It was also the reason for switching from nMOS to CMOS technology in the early eighties.

Currently, the requirement to also have access to powerful computation at any location is another driving force to reduce CMOS power dissipation.

The increasing number of portable applications is a third driving force for low-power CMOS. In the consumer market, we can find examples such as games, mp3 players, photo and video cameras, GPS systems, DVD players and flat screen TVs. In the PC market, an increasing percentage of computers is sold as notebook or laptop computers. Digital cellular telephone networks, which use complex speech and video compression algorithms, form a low-power CMOS application in the telecommunication field.

Finally, the emerging multimedia market will also show many new products in the near future. At the time of going to print, we see the portable full motion video and graphics as examples of such low-power applications. The personal digital assistant (PDA) was already available to the consumer during the 1990s. The development of these portable and hand-held devices has increased the drive for significant battery performance improvements. Therefore, the next section in this chapter will present a short summary on existing battery technologies.

Another important driving force for low power is the future system requirement. In a $22\,\text{nm}$ CMOS technology ($\approx$ year 2015), 50 to 500 billion transistors will be packed on a board of 20 by $20\,\text{cm}$ with very high-density packaging techniques (multi-chip modules (MCM), system in a package (SiP) and system on a package (SoP)).

Current power levels are not acceptable for these systems. In general, low power also leads to simpler power distribution, less supply and ground bounce and a reduction of electromigration and electromagnetic radiation levels. A low-power design attitude should therefore be common in every IC design trajectory, because it is beneficial for power consumption, robustness and reliability of current and future ICs and systems.

8.2   Battery technology summary

A *battery* is usually built from more than one cell, which can chemically store energy for a certain period of time. Based on the difference in the chemical process, we can distinguish two different types of batteries. *Non-rechargeable batteries* use so-called primary cells with a non-reversible chemical reaction and must be hand in as small chemical waste, when empty. These *primary battery cells* perform much better in terms of charge capacity, charge storage and charge leakage, but are less cost-efficient in high-performance systems or systems that are always on. *Rechargeable batteries* use *secondary battery cells*, which deliver energy by transforming one chemical substance into another. This transformation is reversible in that it can be put back into its original chemical state during recharging. In battery-operated systems that need frequent replacement of the batteries, rechargeable batteries would be a more economically viable solution. But, in applications which need a long battery lifetime, e.g., a year for an electronic clock, rechargeable batteries must be recharged at least every three months, while a non-rechargeable battery may "tick" for more than a year.

The growing diversity of battery operated systems, combined with the increasing performance and longer battery lifetimes, requires an improved *battery energy* efficiency, while smaller weight and shrinking dimensions require a reduced number of stacked battery cells. The performance of cells in series is substantially worse than that of individual cells. A single-cell battery with both high cell voltage and high energy efficiency is a real need in many applications.

Advances in rechargeable battery technologies are aimed at improving the battery capacity per unit of volume. Nickel-cadmium batteries have dominated the battery market, but they suffer from low cell voltage and low energy efficiency (see table 8.1).

Table 8.1: Characteristics of rechargeable batteries

Battery type	Nominal cell voltage [V]	Energy/ volume [Wh/l]	Energy/ weight [Wh/kg]	Self-discharge rate [%/month]
Nickel-cadmium (NiCd)	1.2	200	100	10-20
Nickel-metal-hydride (NiMH)	1.2	300	150	30
Lithium-ion/lithium polymer	3.7	400	200	5-10

The nickel-metal-hydride batteries are rapidly replacing the nickel-cadmium ones because of the higher energy capability. Both the NiCd and NiMH types of batteries suffer from the so-called memory effect. In many applications, these batteries are recharged before they are sufficiently or completely empty. After many of such recharge operations, the battery starts suffering from a so-called voltage depression, also known as the *battery memory effect*, which reversibly degrades its energy storage capacity. They therefore need a periodic deep discharge to prevent this memory effect.

During the last decade, single-cell lithium-ion and lithium-polymer (Li-pol) batteries have emerged as the more favoured choice. In a Li-pol battery the lithium electrolyte is a solid polymer medium as compared to the organic solution in a Li-ion battery. They both offer a higher cell voltage and a higher energy density (up to 400 Wh per litre). Because lithium is one of the lightest elements (third) of the periodic system, it helps to save weight, particularly in tiny handheld devices. Moreover, the *self-discharge rate* is only 5% per month and they hardly exhibit the memory effect. A major disadvantage of the lithium batteries is their sensitivity to over (dis)charge, or short circuit, because this can cause them to ignite or even explode. Li-ion and Li-pol battery packs therefore may contain internal protection circuits to monitor its voltage to prevent battery damage and its temperature to disconnect from the application, in case it gets too hot. These batteries should therefore not be used in applications in which they could be exposed to high temperatures.

As the world becomes more mobile, the demand for better battery technology will continue to increase. Most of these applications are in the range of 10 mW (jpeg encoding in a cell phone) to 10 W (peak power in a mobile device). However, the incremental improvements in battery technology do not keep pace with this increase in battery demand and, as such, it puts an additional burden on the shoulders of the IC design community by requiring a more intensive use of less-power design methods.

More information on battery technologies can be found in [1] and [2].

8.3 Sources of CMOS power consumption

During the operation of CMOS circuits, there are four different sources that contribute to the total power consumption:

$$P_{\text{total}} = P_{\text{dyn}} + P_{\text{stat}} + P_{\text{short}} + P_{\text{leak}} \tag{8.1}$$

where P_{dyn} represents the dynamic dissipation .

This is the power dissipated as a result of charging and discharging (switching) of the nodes, and can be represented by the following equation:

$$P_{\mathrm{dyn}} = C \cdot V^2 \cdot a \cdot f \qquad (8.2)$$

where C is the total capacitance,
V is the voltage swing,
f is the switching frequency and
a is the activity factor.

The activity factor represents the average fraction of gates that switch during one clock period. This number can be as low as 0.05 (low activity), for example, but it can also be as high as 2 to 4 (very high activity) because of hazards, see paragraph 8.5.3.

P_{stat} represents the static dissipation. This is the power dissipated as a result of static (temporary or continuous DC) current. In section 8.5, the basic causes of the different contributions are explained in detail.

The contribution of the short-circuit dissipation is represented by P_{short}. This is the power dissipated in logic gates as a result of short-circuit currents between supply and ground during transients.

Finally, the last contribution to the total power dissipation is made by the leakage dissipation P_{leak}. This is the power dissipated as a result of subthreshold leakage currents, gate leakage currents and substrate leakage currents. Both technology and design can affect several of these power dissipation contributors, see table 8.2.

Table 8.2: Power dissipation contributors

Contributor	Technology dependent	Design dependent
P_{dyn}	x	x
P_{stat}		x
P_{short}		x
P_{leak}	x	x

The following sections discuss the technological and design measures that can be taken to reduce the different power consumptions.

8.4 Technology options for low power

As can be seen from in table 8.2, technology can affect both the dynamic power dissipation and the leakage power dissipation.

8.4.1 Reduction of P_{leak} by technological measures

As a result of scaling the channel length over generations of technologies, we arrived at a point (when channel lengths became less than $0.5\,\mu\text{m}$), where we also had to reduce the supply voltage to limit the electrical fields inside a MOS transistor, see chapter 2. Between the $0.8\,\mu\text{m}$ CMOS technology node and the $120\,\text{nm}$ node, the supply voltage has been gradually reduced from 5 V to 1.2 V. Reducing the supply voltage means that the circuits become relatively slower. Therefore, the threshold voltage also has to be reduced.

This has severe consequences for the leakage currents as well as for the noise margin within digital circuits. Because of the subthreshold (weak-inversion) and gate leakage currents, as discussed in chapter 2, we will have a leakage current through an nMOS transistor when its gate voltage is at zero volt. The higher the threshold voltage, the less leakage current will flow at $V_{\text{gs}} = 0\,\text{V}$.

Let us define the subthreshold slope $s_{\text{subthr.}}$ to be the change in threshold voltage, causing a ten-fold increase of the subthreshold current at $V_{\text{gs}} = 0\,\text{V}$. In current technologies, $S_{\text{subthr.}}$ is between:

$$63\,\text{mV/decade}(I) < s_{\text{subthr.}} < 80\,\text{mV/decade}(I) \tag{8.3}$$

This means that a reduction of the threshold voltage of 100 mV leads to an increase of leakage current (at $V_{\text{gs}} = 0\,\text{V}$) of a factor close to 18. It should be clear that, for power and speed reasons, an optimum has to be found for the threshold voltages of both nMOS and pMOS transistors.

Example:
Assume a reference transistor with an aspect ratio $W/L = 240\,\text{nm}/60\,\text{nm}$. If $V_{\text{T}} = 0.42\,\text{V}$, then its leakage current might be 80 pA. Suppose the threshold voltage shifts to 0.3 V, now the current will increase to approximately 2 nA. Present standby currents in large RAMs can vary from nano amperes to milli amperes, depending on their storage capacity and application environment.

With decreasing channel lengths, the threshold voltage also decreases as a result of the small channel effects (threshold voltage roll-off; chapter 2). Consequently, the threshold voltage can be as low as 0.2 V for a minimum channel length transistor in a 65 nm CMOS technology. Also, for real low-voltage applications, the threshold voltage should be low to allow for a certain speed. However, at these low threshold voltages, the circuits suffer from a relatively large loss of power caused by leakage currents , especially in the standby mode.

There are several solutions to this problem. One is to vary the threshold voltage by applying a back-bias voltage during *standby mode* [3]. Depending on the K factor in the equation for the threshold voltage (equation (1.16)), the threshold voltage can be increased by about one hundred millivolt by applying a negative pwell bias, for a nMOS transistor, or a positive nwell bias, for a pMOS transistor, equal to the supply voltage. These additional *back-bias* voltages (both nMOS and pMOS need back-bias in the standby mode) can either be supplied by additional supply pads, or generated on the chip. The back-bias voltage can be offered to the complete chip, or only to distinguished cores (e.g., processors or memory cores) that need to be put in standby, while others remain active. To be able to apply a different bias voltage to a limited number of logic and/or memory cores, the pwell areas of these cores need to be isolated from pwell areas of the cores that remain active and don't need a pwell bias. This can be realised by a so-called *triple-well technology* that offers an additional third well [4].

Figure 8.1 shows a cross-section of a triple-well device. In this technology, the pwell and nwell areas can respectively be connected to V_{ss} and V_{dd}, or to separate pwell and nwell bias voltages. The nMOS transistors are isolated from the substrate. A triple-well technology has another important advantage. Because it physically separates all pwell areas from the p^--substrate, it is much more difficult for the noise induced into the nMOS substrates (pwells) to propagate through the triple well into the p^--substrate. This is of particular importance in designs that combine analog and digital circuits on one IC. In a triple-well technology the analog circuits are better isolated from the digital noise.

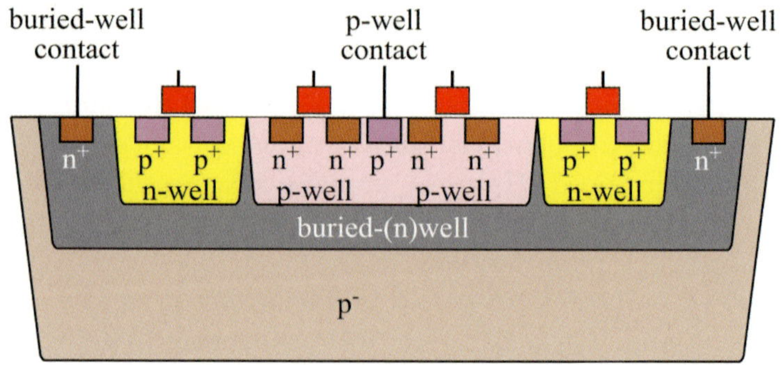

Figure 8.1: *Cross-section of a triple-well device*

Active well biasing for leakage power reduction

The previous discussions are not only limited to the assignment of a
fixed voltage to the substrate or nwell. Also dynamic regulation of the
well bias to vary the threshold voltage (V_T) for reduction of the *leakage
power* is applied. However, the continuous scaling of the device feature
sizes introduced short-channel effects (SCEs), in which the extended
depletion layers around the source and drain junctions lead to the so-
called V_T-roll off (chapter 2). This has required the implantation of
compensation dope, locally in the channel around the source and drain
areas (halos). The negative side effect of this local dope is an increased
junction leakage current, particularly when the junctions are reverse
biased, e.g., by using well biasing. Also in certain future technologies,
the small gate-oxide thickness causes the gate leakage to reach the same
order of magnitude as the subthreshold leakage current. The diagram
in figure 8.2 shows the leakage current in a 65 nm nMOS transistor as a
function of the gate and well-bias voltages.

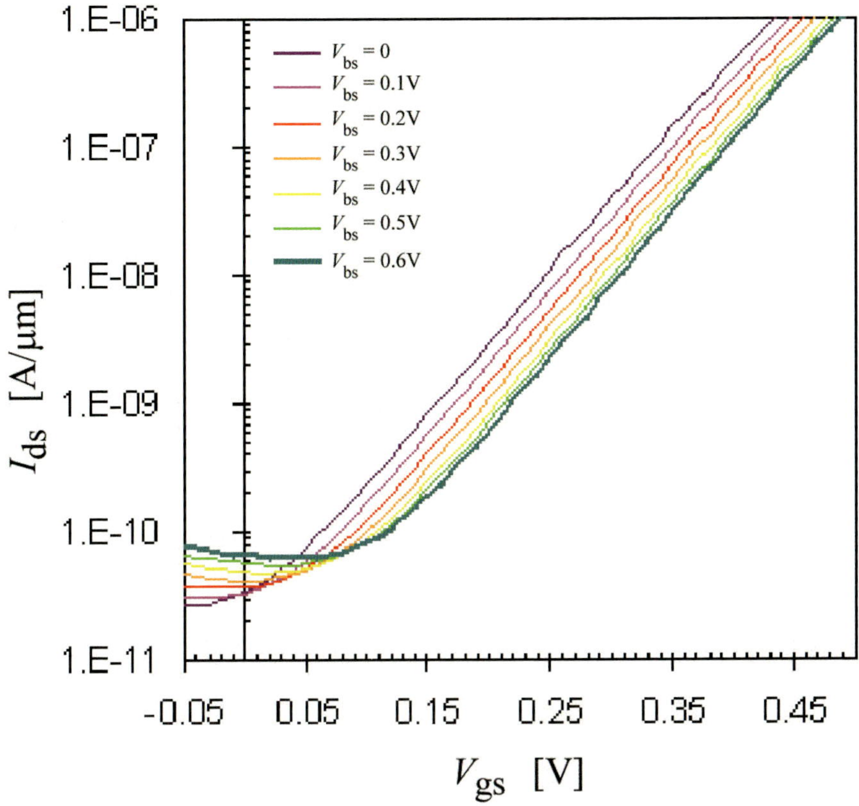

Figure 8.2: *Drain current versus gate voltage at different well bias voltages*

In this particular example, a negative well bias causes an increase rather than a reduction of the transistor off-current (V_{gs}=0). Similar diagrams can be derived for the pMOS transistor. This effect is dependent on the process and the parameters (V_T) and physical dimensions of the transistors (channel length and gate-oxide thickness) in that process. Devices in a general-purpose process show leakage properties different from those in a low-leakage process. A high-V_T device also behaves differently from a low-V_T device. So, the influence of a *well-bias* voltage on the *standby current* of a core is very much related to which device is used in which technology. The general trend is that this effect will only become stronger in smaller bulk-CMOS processes, since they require a continuous increase of the halo doping [5].

Consequently for the 65 nm CMOS node and beyond, *well biasing* is an increasingly less effective instrument for reducing the leakage current during the standby mode. Active well biasing is also seen as a means to compensate for process parameter spread. This can be done in an adaptive way. Both *adaptive body bias (ABB)* and *adaptive voltage scaling (AVS)* can be used to compensate for process variations [6,7]. The design implications of these techniques are discussed in subsection 8.4.2.

A second approach to reduce standby (leakage) currents is to use multiple thresholds [8]. Now, the power supply of the core (with low V_T circuits) is switched by a very large transistor with high V_T, see figure 8.3.

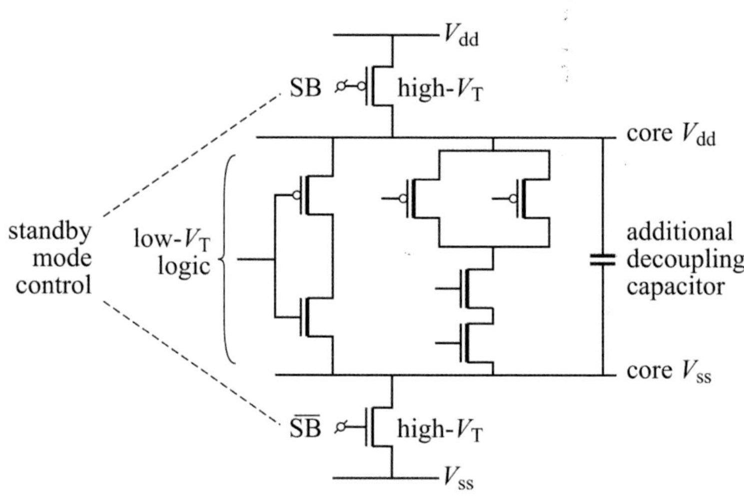

Figure 8.3: *Power supply switch in a multiple V_T environment*

The width of this transistor is such that there is only a marginal voltage drop across it. However, all storage cells and memories in the core must be connected to the permanent power supply and have a high V_T, in order not to lose the cell data. Although the figure suggests the use of both an nMOS and pMOS switch, usually only one *power switch* is used. An nMOS power switch has the advantage of occupying less area, due to their larger current capability. However, many CMOS circuits use the ground as a reference for signals and protection devices. An nMOS power switch would therefore need a change in related design concepts and introduces additional risks. Therefore pMOS power switches are more commonly applied. The use of high-V_T power switches also introduces additional noise peaks into the power supply network. When a large core is switched of, its intrinsic decoupling capacitance will fully be discharged after a certain period of time. When the core is switched on again, large current peaks flow through the supply network, to fully bring the core back to the supply voltage level. These peak currents can be reduced, by implementing the power switch as a combination of many smaller power switches in parallel and then successively switching them on one by one.

Another way of using a multi-V_T *(MTCMOS)* or dual-V_T technology is to design all library cells with both low-V_T and high-V_T transistors. A smart synthesis tool can then implement most logic paths with high-V_T cells, and only use low-V_T cells in the critical paths. For many designs this means that about 10% of the logic is built from low-V_T cells, rendering in a power reduction close to one order of magnitude.

An alternative way to reduce the subthreshold leakage currents is to use longer than minimum transistor channel lengths. Due to the V_T-roll off effect, as discussed in chapter 2, the threshold voltage increases with the channel length. This leads to a reduction of both the leakage current (I_off) and the on-current (I_on) of the transistor. So, depending on the application area, the designer may decide to use library cells with non-minimum channel lengths, when available.

The other major component in the total leakage current of a transistor is the *gate-oxide leakage current*, which is important for an oxide thickness below 2.5 nm (see chapter 2). Particularly for an oxide thickness below 2 nm [9] this leakage component may become larger than the subthreshold leakage. Current low-leakage processes may offer, next to the dual or multi-V_T option, also a dual t_ox option. These options can only be fully exploited when they are supported by the libraries and tools to efficiently reduce the leakage-power components in standby mode.

8.4.2   Reduction of P_dyn by technology measures

In the following formula for the dynamic dissipation, both capacitance C and voltage V are partly determined by the technology:

$$P_\mathrm{dyn} = C \cdot V^2 \cdot a \cdot f$$

Generally, the load (capacitance) of a logic gate is formed by the interconnection capacitance, the gate capacitance (fan-in of the connected logic) and the parasitic junction capacitances in the driving logic gate itself.

A reduction of the gate capacitance means a thicker gate oxide, which also affects the β and thus the speed of a MOS transistor dramatically. So, this is no alternative to reduce the capacitance.

The reduction of the interconnect capacitances depends on the thickness and the dielectric constant of the oxide and on the track thickness, see figure 8.4.

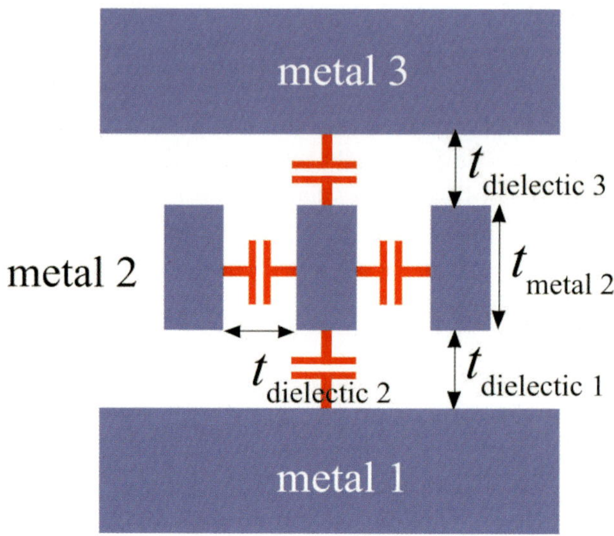

Figure 8.4: *Cross-section of multilevel metal to show capacitance*

As a result of resistive and electromigration effects, the thickness of the metal layers could only be reduced when other metals were used for routing. In this respect, copper was found to be a good candidate. The sheet resistance of copper is about 40% smaller than that of aluminium. However, copper diffuses through oxides and needs therefore to be fully encapsulated within a barrier (chapter 3). This reduces the effective resistance improvement to only 25%. This advantage is used to reduce the copper track thickness with 25%, so that they exhibit about the same resistance as aluminium tracks, but at a much lower mutual capacitance. This maintains the signal propagation across the interconnect, while at the same time the cross-talk and power consumption are reduced.

Thicker oxides require more advanced planarisation steps. The space between two metal tracks in the same layer cannot be increased much, as it will increase the chip area as well. It thus hardly affects the power dissipation, because the metal lines would become longer.

One way to decrease the dielectric capacitance is to find materials with a lower ε_r. The ε_r of SiO_2 is around 4, the ε_r of air is 1. Current values for ε_r are between 2.5 and 3. A value close to two may be achievable in the future.

Junction capacitances are formed by the depletion regions of the source and drain junctions of both nMOS and pMOS transistors. The thicknesses of the depletion regions and, therefore, the values of their capacitances, are determined by the dope of the n^+ and p^+ regions. A reduction of the junction capacitances is not expected, because of the increasing dope of halo implants needed for the suppression of the short-channel effects (chapter 2). An alternative to the main stream current CMOS processes for low power might be a Silicon on Insulator CMOS process, which is discussed in section 3.2.4.

8.4.3 Reduction of P_{dyn} by reduced-voltage processes

The decrease of the channel length over generations of technologies has increased the peak of the electrical field in the pinch-off region near the drain to unacceptable values. For a $0.7\,\mu$m technology, LDD structures (section 9.5.3) brought a satisfactory reduction of this electrical field, primarily to reduce hot-carrier effects. However, from about $0.6\,\mu$m technologies and beyond, these LDD structures are no longer sufficient. The only way to reduce the peak electrical field is to lower the supply voltage, see figure 8.5.

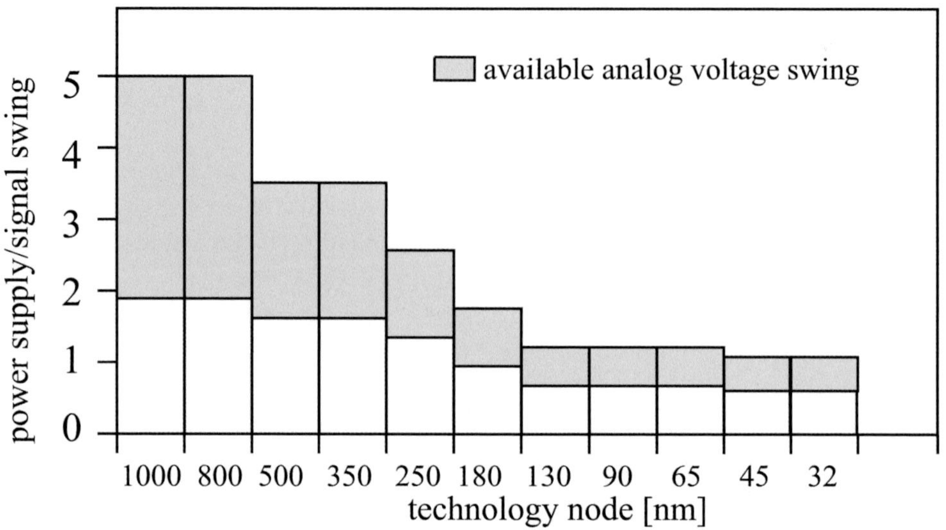

Figure 8.5: *Reduction of supply voltage as a function of the channel length*

Shorter channel lengths will require lower V_{dd} voltages. For performance reasons, the threshold voltage V_{T} had also to be reduced. However, this has led to an increase of the subthreshold (leakage) currents, see section 8.4.1. The *power-delay product* (τD-product; $\tau =$ gate delay and $D =$ dissipation) was the classic performance metric used in technologies above 120 nm, because the dynamic power consumption was the largest power component in those technology nodes:

$$\text{Energy} = \tau \cdot D = \text{delay} \cdot P_{\mathrm{dyn}} = \frac{1}{f} \cdot CV^2 \alpha f = CV^2 \alpha \qquad (8.4)$$

where α represents the average switching activity factor of the logic gates in a core. For traditional CMOS circuits the following two expressions apply:

$$\text{power}(D) \quad = \quad C \cdot V^2 \cdot a \cdot f \qquad\qquad\qquad (8.5)$$

$$\text{delay}(\tau) \quad = \quad \frac{2C \cdot V}{\beta(V - V_{\mathrm{T}})^2} \text{ from :} \quad \left\{ \begin{array}{rcl} Q = I \cdot t & = & C \cdot V \\[2ex] I & = & \frac{C \cdot V}{t} \\[2ex] \beta(V_{\mathrm{gs}} - V_{\mathrm{T}})^2 & = & \frac{2C \cdot V}{t} \end{array} \right\}$$

To reduce both the power and the delay, capacitance C must be reduced.

From the previous two equations, the $\tau \cdot D$ product will be equal to:

$$\tau \cdot D = b \cdot \frac{V^3}{(V - V_{\mathrm{T}})^2} \tag{8.6}$$

where b is a constant.

The minimum will exist for $\frac{\delta \tau D}{\delta V} = 0$, which results in: $V = 3V_{\mathrm{T}}$. Thus, when a ratio of three is used between the supply voltage and the threshold voltage, the process should allow for optimum performance.

The power-delay product assigns equal weight to the power and to the delay of a circuit. For circuits for which power has a higher priority than speed, we might give a higher weight to the power than to the delay and the metric becomes:

$$\mathrm{Power} \cdot \mathrm{Energy} = P_{\mathrm{dyn}}{}^2 \cdot \tau \tag{8.7}$$

For high-speed circuits we might give more weight to the delay instead of to the power. Then the metric becomes equal to the *energy-delay product*:

$$\mathrm{Energy} \cdot \mathrm{Delay} = P_{\mathrm{dyn}} \cdot \tau^2 \tag{8.8}$$

So, the chosen metric depends on the requirements of the application.

Most CMOS technology nodes (90 nm and below), today, support two different categories of ICs. The general-purpose variant is meant for those ICs that demand a reasonable speed for their circuits. The low-leakage variant is targetted at application areas with reasonably long standby times of the logic and/or memory cores, because it offers both high-V_{T} nMOS and pMOS transistors. This may be combined with a thicker gate oxide for reduced gate leakage. Particularly large memory cores would benefit greatly from this low-leakage technnology variant.

Another category of applications are those that are always on and therefore require *low-operating power (LOP)*. As an example, an MP3 player is either on, when selected, or completely switched off. For such applications or functions, the dynamic power consumption is the largest contribution to the overall power, which requires another power optimisation approach. For these applications a reduction of both the threshold and the supply voltage would render them to run at the same speed, consuming less active power but more leakage power. This optimisation allows the exchange of operational power (dynamic power; active power) and leakage power, without sacrificing performance. The choice of V_T determines the amount of on-current (I_{on}) and leakage current (I_{off}). It turns out that the optimum power is achieved when the operational power (P_{dyn}) is about equal to the leakage power (P_{leak}) [10]:

$$P_{leak} = I_{off} \cdot V_{dd} = C \cdot V_{dd}^2 \cdot \alpha \cdot f = P_{dyn} \tag{8.9}$$

with:

$$f = \frac{1}{T} = \frac{1}{L_d \cdot \tau} \quad \text{and} \quad C \cdot V_{dd} = I_{on} \cdot \tau \tag{8.10}$$

where L_d represents the average logic depth of the logic paths.

This results in:

$$I_{off} \cdot V_{dd} = I_{on} \cdot V_{dd} \cdot \tau \cdot \alpha \cdot \frac{1}{L_d \cdot \tau} = I_{on} \cdot V_{dd} \cdot \frac{\alpha}{L_d} \tag{8.11}$$

or:

$$\frac{I_{on}}{I_{off}} = \frac{L_d}{\alpha} \tag{8.12}$$

The optimum I_{on}/I_{off} ratio for a consumer IC, with a logic depth of about 40 gates and an activity factor of around 0.1 would be 400. For a high performance video processor with a logic depth of 15 gates and an activity factor of 0.3, the optimum I_{on}/I_{off} ratio would be equal to 50, requiring CMOS devices with a much lower V_T. To be able to use this optimisation concept in a given technology with fixed threshold voltages, synthesis tools have to be developed to match the L_d/α ratio of as many logic paths to the I_{on}/I_{off} ratio of the transistors.

8.5   Design options for power reduction

As shown in table 8.2, we can reduce the dynamic, the static and the short-circuit dissipation by taking measures in the design. Because the measures for the latter two are clear and compact, we start with these two first.

8.5.1   Reduction of P_{short} by design measures

During an input transition at a CMOS logic gate, there may be a temporary current path from supply to ground. The resulting short-circuit power dissipation can be relatively high if no attention has been paid to this [11]. Consider the example of figure 8.6, which is currently still representative for output drivers.

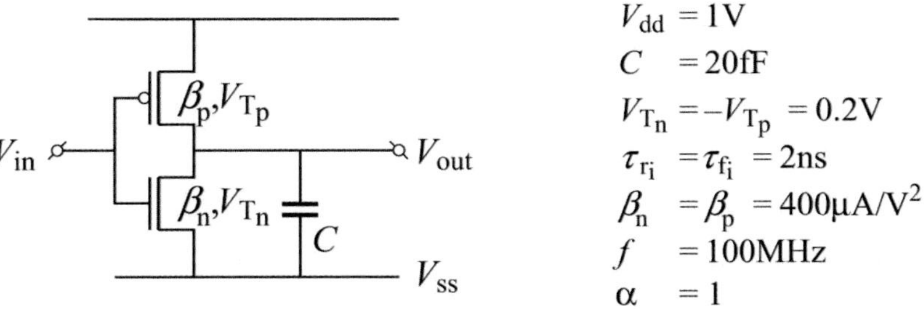

Figure 8.6: *Inverter example to illustrate the level of short-circuit power dissipation*

With these numbers, the dynamic power dissipation becomes:

$$P_{dyn} = C \cdot V^2 \cdot a \cdot f = 2\,\mu\text{W}$$

and the short-circuit power dissipation becomes [11]:

$$P_{short} = \frac{\beta}{12} \cdot (V_{dd} - 2V_T)^3 \cdot \frac{\tau}{T} = 1.44\,\mu\text{W}$$

Conclusion: either τ_f and τ_r on the inputs are much too large or the β of the pMOS and nMOS transistors must be reduced. For CMOS drivers (internal, clock and output drivers), this short-circuit power can be minimised when τ_f and τ_r are equal on all nodes. This requires tapering of the inverters in such a driver, see figure 8.7.

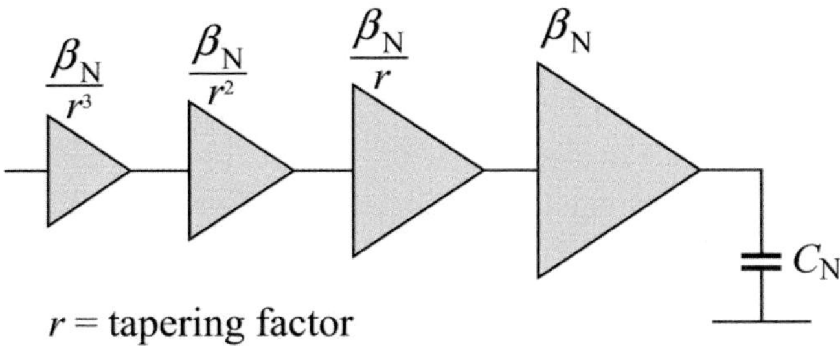

Figure 8.7: *Inverter chain with tapering factor*

In conventional CMOS technologies, a tapering factor between 8 to 16 usually resulted in a minimum short-circuit dissipation, which was less than 10% of the total dissipation [11]. In advanced CMOS processes (beyond 100 nm), the short-circuit power consumption is fully negligible (section 4.3.2).

An important remark to be made here is that the pMOS and the nMOS transistors can never conduct simultaneously during a transient when $V_{\mathrm{dd}} < V_{\mathrm{T_n}} + |V_{\mathrm{T_p}}|$, thereby eliminating the short-circuit dissipation completely.

8.5.2 Reduction/elimination of P_{stat} by design measures

In complex logic gates which require many pMOS transistors in series (four or more input NOR gates, address decoder in memories, etc.), pseudo-nMOS solutions are sometimes applied, see figure 8.8. When the output of such a gate is low, there is a continuous static current from V_{dd} to ground.

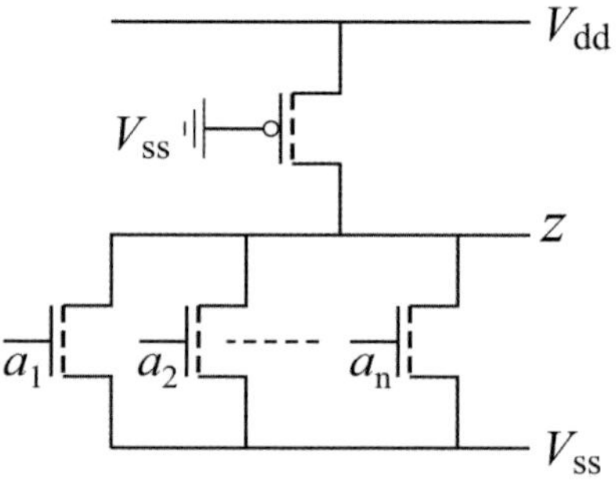

Figure 8.8: *n-Input pseudo-nMOS NOR gate*

For low-power applications, this is not an efficient way of implementation. In this case, the power can be reduced by replacing the grounded pMOSt by a clocked pMOSt. This may reduce the power by a factor equal to the clock duty cycle. For a low-power design, this is not a good solution, because a *pseudo-nMOS logic* gate consumes about 10 to 20 times that of a full static CMOS realisation. Therefore, to eliminate static power consumption, no pseudo-nMOS should be used at all.

8.5.3 Reduction of P_{dyn} by design measures

The dynamic dissipation was expressed by:

$$P_{\mathrm{dyn}} = C \cdot V^2 \cdot a \cdot f$$

By means of design techniques, we are able to influence all parameters in this expression. We will therefore present several alternative measures for each parameter to reduce its contribution to the power consumption. Examples of these measures are given at various hierarchy levels of design: algorithm/architecture, logic and transistor level. They show that the decisions taken at the higher levels have much more impact on the power consumption than those taken at the lower levels.

Power supply (V) reduction

A lower voltage generally means less performance and less chance for latch-up. Let's assume we have the following circuit on a chip, see figure 8.9.

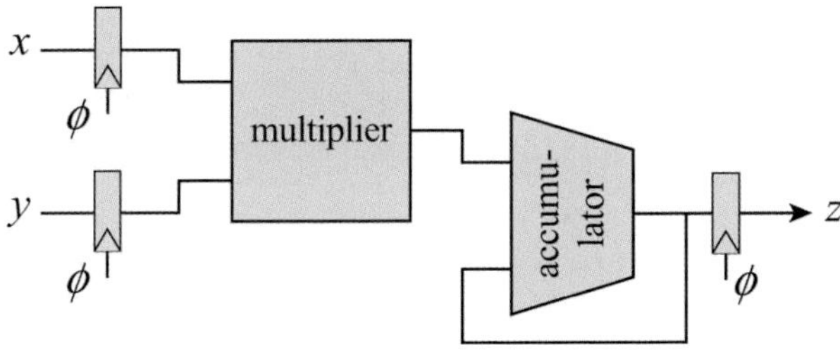

Figure 8.9: *A basic data path*

The total propagation delay time through the logic is equal to the sum of the propagation delays of the multiplier and accumulator. This total propagation delay determines the minimum duration T of the clock period. If we double this clock period, the propagation delay is allowed to be twice that of the original circuit. To achieve this doubling, we may reduce the supply voltage from 1.2 V to 0.95 V, for example, in a 1.2 V 65 nm CMOS technology. However, if the throughput is to be retained, two of these circuits can be connected in parallel and their inputs and outputs multiplexed (parallelism) or additional latches can be placed in between the logic functions to shorten the critical delay paths between two successive flip-flops (pipelining).

A) *Parallelism*

Figure 8.10 shows a parallel implementation of the circuit. As a result of demultiplexing and multiplexing the signals, the same performance can be achieved as in the original circuit of figure 8.9, which runs at twice the clock frequency.

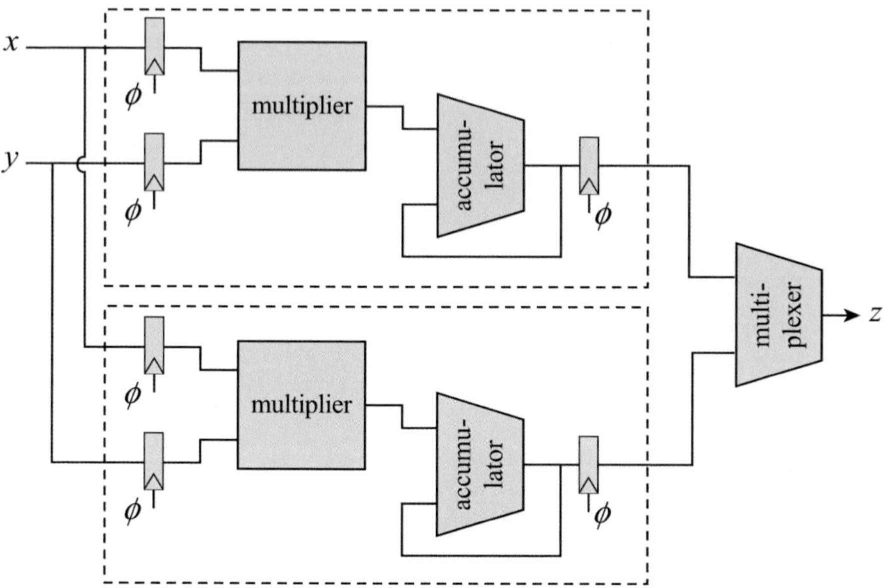

Figure 8.10: *Parallel implementation*

When we include multiplexers and additional wiring, this parallel architecture results in an increase of the total capacitance to be switched by a factor of about 2.25. The power dissipation comparison for the circuits of figure 8.9 and the parallel implementation in figure 8.10 then yields:

$$P_{\mathrm{dyn}}(\text{basic data path}) \;=\; C \cdot V^2 \cdot a \cdot f_{\mathrm{ref}} = P_{\mathrm{ref}}$$

$$P_{\mathrm{dyn}}(\text{parallel data path}) \;=\; (2.25C) \cdot \left(\frac{0.95}{1.2}V\right)^2 \cdot a \cdot \frac{f_{\mathrm{ref}}}{2} = 0.7 \cdot P_{\mathrm{ref}}$$

where f_{ref} and P_{ref} represent the frequency and power consumption of the reference circuit of figure 8.9, respectively.

Thus, the parallel implementation of the data path results in a power reduction of a factor of about 1.42, however at the cost of area overhead of more than a factor of two. This is sometimes not allowed, especially in the cheap high volume consumer markets.

Another way to maintain performance at a reduced power supply voltage is pipelining.

B) *Pipelining*
In figure 8.9, the critical path is equal to:

$$T_{\text{crit}} = T_{\text{mpy}} + T_{\text{acc}} \Rightarrow f_{\text{ref}}$$

where T_{mpy} and T_{acc} represent the worst-case delay paths (critical paths) of the multiplier and accumulator, respectively.

Let us assume that the propagation delays of the multiplier and the accumulator are about the same and that we put a pipeline in between the multiplier and accumulator. Figure 8.11 shows the circuit with the additional pipelines.

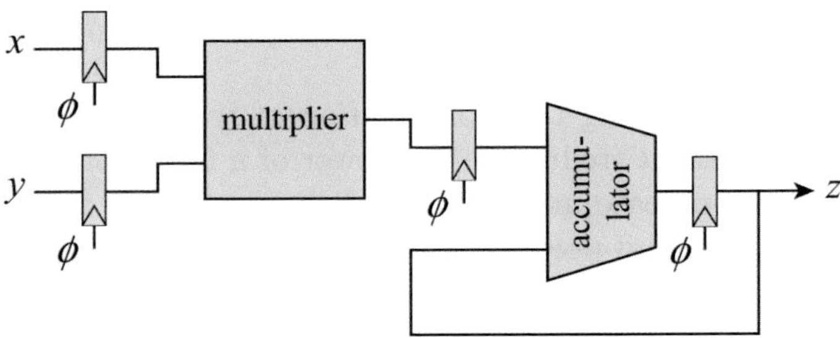

Figure 8.11: *Pipelined implementation*

Now, the critical path is:

$$T_{\text{crit}} = \max[T_{\text{mpy}}, T_{\text{acc}}] \quad \Rightarrow \quad f_{\text{max}} > f_{\text{ref}}$$
$$\text{if } T_{\text{mpy}} \approx T_{\text{acc}} \quad \Rightarrow \quad f_{\text{max}} \approx 2 \cdot f_{\text{ref}}$$

The additional pipeline allows a frequency which is about twice as high. Therefore, the voltage may reduce to about 0.95 V to maintain the same frequency again. As a result of the additional pipelines and multiplexer, the area increase will be about 20%. Comparing this pipelined architecture with the original one leads to the following

- When V_{in} becomes equal to $0.2\,V$, the pMOS transistor switches on and the output switches to $0.6\,V$.

- Finally, when $0\,V < V_{in} < 0.2\,V$, the pMOS transistor remains on and the output remains at $0.6\,V$.

Although these kinds of circuits ($V_{dd} < V_{T_n} + |V_{T_p}|$) are relatively slow, they have been used for a long time in battery-operated products, e.g., watches. One advantage of these circuits is that a short-circuit current can never flow, because one transistor always switches off before the other one switches on. Therefore, there is no short-circuit dissipation at all. Not every library is suited for low-voltage operation. This means that a new low-voltage library must be developed and characterised, including a RAM, a ROM and other generators. Moreover, because of the low-voltage supply, the threshold voltage (V_T) must be controlled very accurately, since the circuits are then much more sensitive to threshold voltage variations.

E) *Voltage regulators*

   Generally, ICs also contain low performance parts which could actually run at lower supply voltages. These can be supplied externally, or generated on chip by means of voltage regulators [12], see figure 8.13.

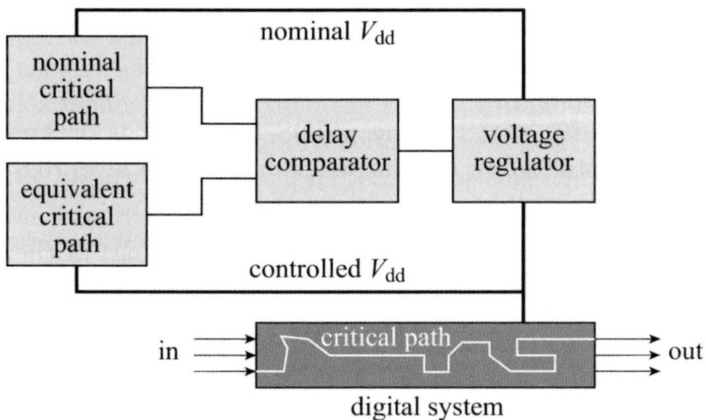

Figure 8.13: *Example of voltage regulator principle*

If such a voltage regulator is used, attention should always be paid to its power efficiency. A better alternative is to run the chip at the lowest required voltage and perform a voltage-up conversion only for the high-performance circuit parts. Such voltage-up converters are used in single cell hearing aids, for example. One can also use *DC-DC converters*. Here too, the power efficiency is an important factor in the use of such circuits. Currently, this efficiency is in the ninety percent range.

F) *Reduced voltage swing*

Bus widths, both on chip and off chip, are tending to grow to 32, 64 and even to 128 bits. This means that the number of simultaneously switching buses and/or outputs has already increased dramatically and this number will continue to increase. If the power dissipation becomes high with respect to other parts of the chip, then a lowering of the voltage swing on these buses (or outputs) has to be considered. As an example, we take the interface between a high-performance microprocessor and a 400 MHz DDR1 1 Gb SDRAM. Such an SDRAM may provide a 64-bit datapath, while another 24 bits are needed to access all 64-bit words in the memory. In addition, several ECC bits (when available), clock and control signals are needed in this interface, which may lead to a total of about 96 interface pins. DDR1 supports an I/O voltage level of 2.5 V. If we assume that all bits, representing 15 pF of load capacitance each, would switch at the fastest rate, the total power consumption of this interface would be:

$$P_{\mathrm{dyn}} = C \cdot V^2 \cdot f = 96 \cdot 15 \cdot 10^{-12} \cdot 6.25 \cdot 400 \cdot 10^6 = 3.5\,\mathrm{W}$$

By replacing this 2.5 V DDR1 interface with a 1.8 V DDR2 interface, this power dissipation would reduce to about 1.8 W. Reduced voltage swing techniques are frequently used to reduce the power dissipation of large 32-bit or 64-bit processors.

G) *Dynamic voltage and frequency scaling (DVFS)*

In case the application does require a further reduction of the power by dynamically adjusting the voltage to the performance needs, there are two possibilities to do so. The first one is to adaptively control the voltage (*adaptive voltage scaling; AVS*) by means of an analog feedback loop that continuously matches the required supply voltage to the performance needs of the running application. The second possibility is to have the voltage switched to one of the discrete (lower) voltage levels supplied to the chip.

To continuously match the supply voltage to the performance demands of the system, requires the integration of an on- or off-chip voltage regulator circuit into the system. On-chip voltage regulation can only be performed, by varying the resistance of a big transistor, which is positioned in between the core and the supply voltage (figure 8.14).

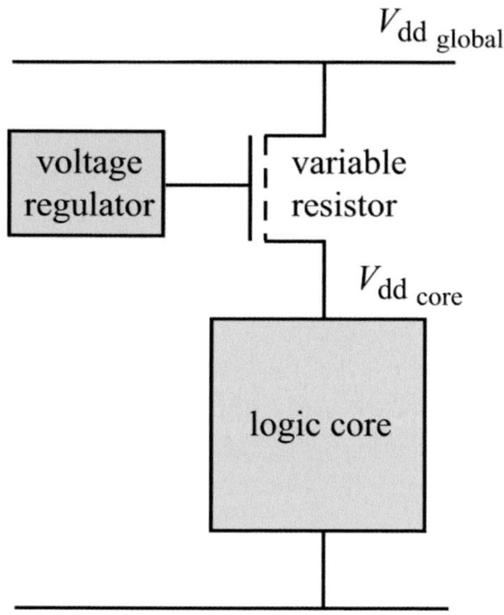

Figure 8.14: *On-chip voltage regulation*

However, with this type of linear voltage regulators, also called *low-dropout (LDO) regulators*, the power reduces only linearly with the V_{ddcore} instead of quadratically, since part of the power saving in the core is now consumed in the variable resistor. Until today, the main application of an LDO was to transform a relatively large battery voltage to the required operating voltage levels of an IC (e.g., 1.2 V in a 90 nm CMOS chip, derived from a 4.3-4.7 V lithium battery).

A more efficient power regulator uses a dynamically controllable off-chip DC-DC converter. Current DC-DC converters show power efficiencies close to 90%.

Fully adaptive voltage scaling is more applicable to certain CPU cores (e.g., on a Pentium) or for a single MPU core (e.g., ARM core) on an

ASIC, where relatively large design teams are assigned to the power management task, and where most critical delay paths in that core are exactly known due to a huge series of extensive simulations, based on accurate circuit models. For a dedicated core, like an ARM core, a replica of its worst-case delay path could be used.

Applying adaptive voltage scaling generally in the cores of an ASIC is not easy because it becomes very difficult to find a *"replicator"* that mimics the voltage behaviour of all these cores correctly. These cores may then run at (unknown) supply voltages, which requires full library characterisation across the full supply-voltage range. Communication between cores in different voltage domains needs the use of *level shifters*. However, if the voltages of the different domains are automatically adjusted, these level shifters need to be very flexible in that on one IC it needs to transfer signals from a low-V_{dd} to a high-V_{dd} domain, while on another IC from a different batch of wafers, it might need to do the opposite. Although the principle of using level shifters in itself looks easy, it can have far more design implications than seen at first sight. A level shifter introduces additional time delay in the communication path between two cores in different voltage domains. Particularly for high-performance communication there is no time budget left to do level shifting at all. In all other applications the level shifter have serious impact on the timing closure of the overall IC design.

Including the tolerances of the voltage regulators themselves, this makes synchronous communication between such cores very complex and sometimes even impossible. Moreover time delay, complexity and risks are added to the design, as well as debug and diagnose time. Next to this, the scheduling of, and elapsed time required for the new supply voltage to settle (often a hundred to a few hundreds of microseconds) is defined by the application.

The power management system needs direct knowledge about the current and future workload generated by (bursty) operations [13]. For non-periodic applications this is not an easy task and the information must be derived from monitoring the system load and by using appropriate models to predict future performance demands. Although the idea of dynamic voltage and/or frequency scaling exists for a long time, the technique is not widely used until now, mainly due to the problem of reliably predict the performance demands.

Next to that, this type of dynamic power management must also be fully supported by the libraries and design flow, such that it becomes completely transparent for the designers.

To a lesser extend, the above considerations also hold for the case in which the supply is not fully adaptively regulated, but where the power management system selects the appropriate voltage level from a few discrete voltages supplied to the chip.

It should be noted that switching the supply voltage to different levels, including a complete power on- or off switch, may introduce large current peaks in the circuit or even in the board, affecting the signal integrity and/or EMC behaviour of other on-chip cores or on-board devices.

Finally, it has been shown [14] that although dynamic voltage scaling renders the lowest energy dissipation for most microcontrollers, it is not always dramatically better than using a combination of dynamic frequency scaling and the built in power-down modes, which is much less complex and less expensive to implement.

For certain high-volume devices, this voltage assignment can be done after silicon realisation.. In this so-called *power-binning concept*, every chip is measured and the assignment of the voltage to the different supply domains on the chip is based on the real silicon performance measurement. The supply connections are then made by using polysilicon fuses, analogous to the selection of redundant columns in a memory. Power binning allows manufactures of power-critical devices to sell high-end and low-power versions at a premium price.

H) *Subthreshold logic*

Most low-power applications require power levels in the milliwatt range. However, there is an inceasing number of applications, e.g., RF-ID cards, some low-power processing in the idle mode in mobile applications, that require power consumption in the microwatt range. These *ultra-low power* applications can be realised with various CMOS circuit architectures. The most simple one, which also allows the use of the standard CMOS design flow, is to operate CMOS logic close to, or beyond the threshold voltage. This so-called *subthreshold logic* uses transistors that operate in the weak-inversion region, in which the current has an exponential relation with the voltage (section 2.7.1). Because subthreshold logic operates with very small supply voltages (most commonly between 0.2 V and 0.4 V), it is extremely sensitive to process parameter spread. It is therefore a major R&D challenge to create subthreshold logic circtuits with reduced sensitivity to this spread [15].

Capacitance reduction

The total capacitance to be switched on an IC can be reduced or limited at two levels: at system level and at chip level. The decisions taken at system level usually have more effect on the IC power than those taken at chip level. This is because a different architecture for an ALU/multiplier or for a filter structure can have more area consequences for the total hardware. This is shown in the following example:

A) *System level*
Suppose we have to perform the following algorithm:

$$y(n) = \sum_{m=0}^{k-1} x(n - m) \tag{8.13}$$

A possible hardware implementation is shown in figure 8.15.

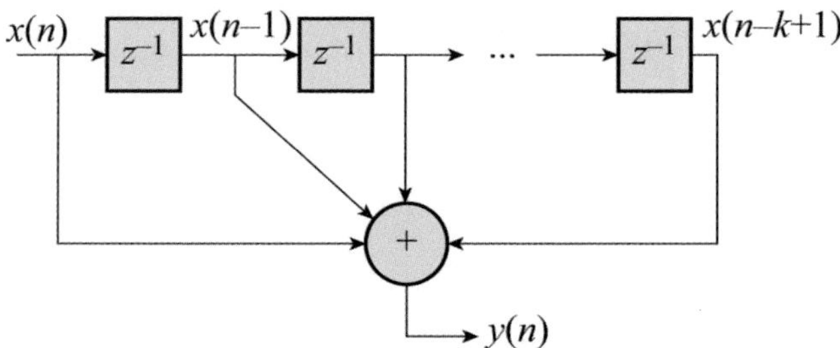

Figure 8.15: *Digital realisation of the running sum algorithm*

When k is large, many additions have to be performed. Here the hardware configuration will contain a lot of full adders to generate the sum and carry functions. The data has to ripple through a large number of full adders, leading to long propagation times and a limited clock frequency. A high-performance implementation would even require additional pipelines and/or carry-look-ahead techniques to improve speed. With regard to the power consumption, this implementation

is very disadvantageous. Figure 8.16 shows an alternative recursive
realisation:

$$y(n) = y(n-1) + x(n) - x(n-k) \qquad (8.14)$$

Although it consists of two adders, each adder here has only two
inputs, which means that much less hardware is involved.

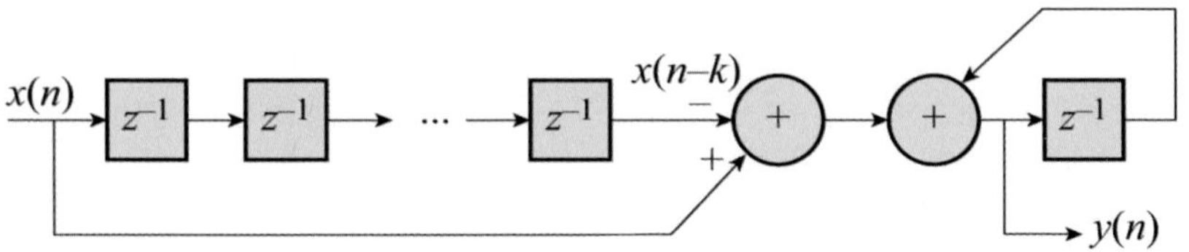

Figure 8.16: *Recursive realisation of the running sum algorithm*

From this example, we can conclude that the algorithm level is at
least as important as the chip level for a low-power realisation of a
certain function.

B) *Chip level*

   At chip level, there are many options for reducing the power consumption by capacitance reduction. This ranges from libraries, via
   tools and circuit techniques, to layout implementation.

   - Optimised libraries
     In many cases, low power implicates the use of a reduced voltage.
     This requires library cells with a low gate complexity (maximum
     three inputs). These cells suffer from less body effect and show
     a better performance than cells with a higher complexity. Most
     libraries were designed for high performance. They contained
     relatively large transistors which consume power accordingly.
     Using these libraries for a low-power design is an overkill in both
     power and area. In a real low-power library, the transistor and
     cell sizes must be limited, such that the fan-in capacitance, the
     parasitic junction capacitances of source and drain regions, and
     the total interconnect lengths after routing will also be reduced.
     Source and drain regions can be reduced by adapting a very
     regular layout style.

Flip-flops are probably the most frequently used cells of a library. In many synchronous chips, ten to fifty percent of the total layout area is often occupied by flip-flops. They therefore play a dominant role in the performance, the area, the robustness and the power consumption of a chip. It is clear that the flip-flops should be designed for low power, not only for their internal power consumption, but also for the clock driver power consumption. A low fan-in for the clock input combined with better clock skew tolerance (more output delay) allows smaller clock driver circuits, thereby reducing both power consumption and current peaks. Standard-cell libraries may be available with different cell heights. Usually this cell height is expressed in the number of metal grids (= minimum track width + minimum spacing). Cell heights of 12 to 14 grids are often used for common VLSI designs, while libraries with cell heights of 9 grids are often used for low-power designs. Current standard-cell libraries may contain between 1000 to 2000 cells, with different transistor drive strengths and threshold voltages to support both high-performance as well as low-power and low-leakage applications, as explained in section 8.4.1.

- Pass-transistor logic (transfer gate; pass gate; transmission gate) This logic already existed in the nMOS era. The most efficient circuits to be implemented in *pass-transistor logic* are multiplexers, half adder and full adder cells. The basic difference between this logic and conventional static CMOS logic is that a pass-transistor logic gate also has inputs on the source/drain terminals of the transistors. A major disadvantage of nMOS pass-transistor logic is the threshold voltage loss ($V_{\mathrm{out}} = V_{\mathrm{dd}} - V_{\mathrm{T_n}}$) at high output level. When such a signal is input to a CMOS inverter, a leakage current flows in this inverter when $V_{\mathrm{T_n}} \geq |V_{\mathrm{T_p}}|$. nMOS pass-transistor logic will thus not be an alternative for low-power design. For different reasons it is usually not feasible to control the threshold voltages (i.e., $V_{\mathrm{T_n}} \geq |V_{\mathrm{T_p}}|$) at the technology level. To compensate for the threshold voltage loss and for other disadvantages of nMOS pass-transistor logic, several pass-transistor logic styles have been presented in literature. The most important ones will now briefly be discussed.

Complementary Pass-Transistor Logic (CPL) [16]

A CPL gate (figure 8.17) basically consists of two nMOS logic circuits, two small pMOS transistors for level restoration and two inverters for generating complementary outputs. Without the cross-coupled pMOS pull-up transistors, CPL would also show the same problems as the above-discussed nMOS pass-transistor logic.

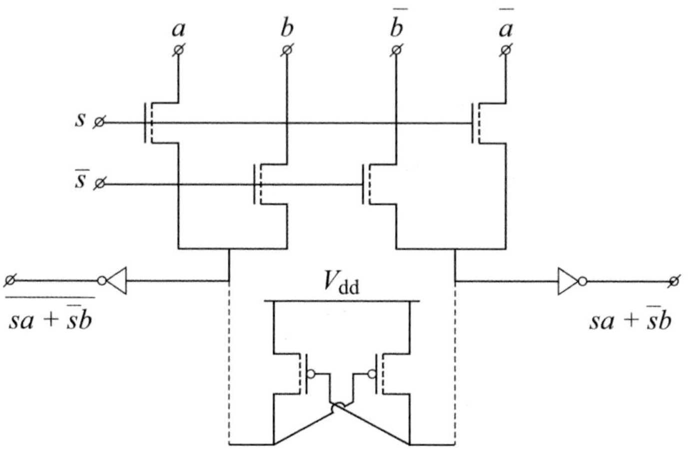

Figure 8.17: *2-input multiplexer in CPL*

Because of the complementary logic circuits, the fan-in and the complexity of a CPL gate approaches that of a conventional CMOS gate. Because of the availability and necessity of the complementary signals, much more routing area is required. Moreover, simple logic functions require a relatively high transistor count.

Double Pass-Transistor Logic (DPL) [17]
 A DPL logic gate uses both nMOS and pMOS logic circuits in parallel, providing full swing at the outputs, see figure 8.18.

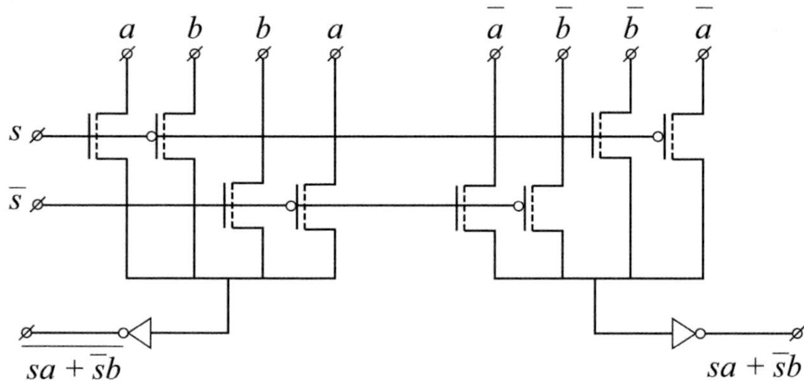

Figure 8.18: *2-input multiplexer in DPL*

Because of the necessity of complementary signals, this logic style has the same routing complexity as CPL. Although it needs no swing restore circuit, it contains more transistors, particularly in complex gates, and has a higher fan-in than CPL. This usually requires more switching power. It is therefore less attractive than other pass-transistor logic and standard CMOS logic.

Other styles of pass-transistor logic
There are several other styles of pass-transistor logic. One, *the Swing Restored Pass-Transistor Logic* (SRPL; [18]) is derived from CPL. Here, the output inverters are mutually cross-coupled (compare figure 8.17) and must be overwritten by the pass-transistor network. This makes this logic less robust for general implementation. In *Lean Integration with Pass-Transistor* (LEAP; [19]), single-rail pass-transistor logic is used. This basically contains an nMOS logic network and a level restore circuit, consisting of an inverter and a feedback pMOS pull-up transistor. This is slower than CPL. At supply voltages of $V_{\mathrm{dd}} < 2V_{\mathrm{T_n}}$, this pass-transistor style is no longer applicable because the output inverter can no longer be turned on.

Finally, new styles of pass-transistor logic are being invented continuously (DPTL[20]; EEPL[21]; PPL[22]). However, many are derived from previous styles with only minor changes. What-

479

ever style of pass-transistor logic will be invented yet, they will all have the same disadvantages: they will either suffer from threshold voltage loss and need a level restore circuit or they will need the double rail approach (complementary inputs and outputs).

Conclusions

Although different pass-transistor logic families are presented in literature, showing better performance in terms of power delay products than conventional CMOS logic, the opposite is also published [23].

Initially, pass-transistor logic showed equal functionality with low transistor count. However, with reduced voltages, complex ICs and low-power focus, this advantage has been undone by the necessity of a level restore circuit and/or dual rail implementation. Except for half and full adder functions, conventional CMOS circuits perform better than any pass-transistor style where both power and robustness are concerned. As a result of increasing process variations and extending application environments, the robustness will play an especially dominant role in the development of deep-submicron (standard) cell libraries.

- Synthesize logic functions into larger cells.

  Usually, logic functions are mapped onto library cells. This, however, is rather inefficient in terms of area and power. The full-adder function might serve as a good example, where S is the sum function and C represents the carry:

$$S = a\overline{bc} + \overline{a}b\overline{c} + \overline{ab}c + abc$$
$$C = ab + ac + bc$$

  In a standard cell library without a full-adder cell, the sum function would require four 3-input AND functions and one 4-input OR. With a dedicated full-adder library cell, the area will be roughly halved. Generally, a cell compiler, capable of optimising complex functions and creating logic gates, would be a good tool for optimising both area and speed. However, good characterisation tools must then also be available to generate accurate timing views of these compiled cells.

- Use optimised synthesis tools.

  Good tools are required for an optimum mapping of complex logic functions onto the library cells. These tools must include reasonably accurate timing models. Usually, the less hardware is used, the less power will be consumed.

- Use optimised place & route tools.

  Many current CAD tools for place & route are area or performance driven. Part(s) of the circuits can have different weights for high performance. These require priority in the place & route process. With a focus at low power, power driven (activity/capacitance) place & route tools are required, resulting in minimum wire lengths.

- Use custom design, if necessary.

  Reduction of the interconnection lengths can be achieved by different layout styles. Especially *cell abutment* is a way to optimise data paths in very regular structures, such as bit slice layouts and multipliers, etc. Custom design must only be applied if the additional design time can be retrieved. Practically speaking, this only holds for high volume chips, or for chips with very tight power specifications, which cannot be achieved with other design styles.

- Make an optimum floor plan.

  Although this sounds very commonplace, it is not self-evident. During floor planning, the focus should be on wasting less area and on reducing bus and other global interconnections. The cores that have intensive communication with each other should be placed at minimum distance, to reduce the wire length of the communication buses.

- Optimise the total clock network.

  Clock signals run globally over the chip and usually switch at the highest frequency (clock frequency f; data frequency $< f/2$). As discussed, the number of flip-flops and their properties are a dominant factor in the total clock network. The flip-flops should be optimised for low fan-in and a better clock skew tolerance so that smaller clock drivers could be used. Section 9.2.2 presents a robust flip-flop, which is also very well suited for low-power designs.

- Use well-balanced clock trees.

  Balanced clock trees are those in which drivers and loads are tuned to one another, such that equal clock delays are obtained, anywhere in the chip. This reduces the clock skew, which allows for smaller clock drivers.

- Dynamic versus static CMOS.
  Chapter 4 presents implementations of static and dynamic CMOS logic gates. With respect to capacitance, a dynamic CMOS gate generally has less fan-in capacitance. This is because the function is usually only realised in an nMOS network, while the pMOSt only acts as a (switched) load. Because every gate is clocked, we get very large clock loads. Moreover, as a result of the precharging mechanism, the average activity in a dynamic gate is higher than its static counterpart. A more detailed look into the activity of static and dynamic CMOS logic is presented in the following paragraph.

- Memory design.
  To reduce the total capacitance to be switched in a memory, the memory can be divided into blocks (block select), such that they can be selectively activated (precharge plus read/write). Divided word lines and divided bit lines means that less capacitance is switched during each word line and bit line selection. Wider words (64 bits instead of 32 bits) reduce the addressing and selection circuit overhead per bit.

  The precharge operation can be optimised by selectively precharging the columns (only those to be read or written) instead of all simultaneously.

Reduction of switching activity

Most of the switching activity of a circuit is determined at the architectural and register transfer level (RTL). At the chip level, there are less alternatives for lowering the power consumption by reducing switching activity.

This paragraph presents several of these alternatives, starting at the architectural level.

A) *Architectural level*
   Choices made at the architectural and RTL level heavily influence the performance, the area and the power consumption of a circuit. This subsection summarises the effect that these choices have on the activity of the circuit.

- Optimum binary word length.

  The word length must be not only optimum in terms of capacitance but also in terms of activity, which means that only that number of bits is used that is really required to perform a certain function.

- Bit serial versus bit parallel.

  Figure 8.19 gives two alternative implementations for a 16 by 16 bit multiplier: a bit serial iterative multiplier and an array multiplier.

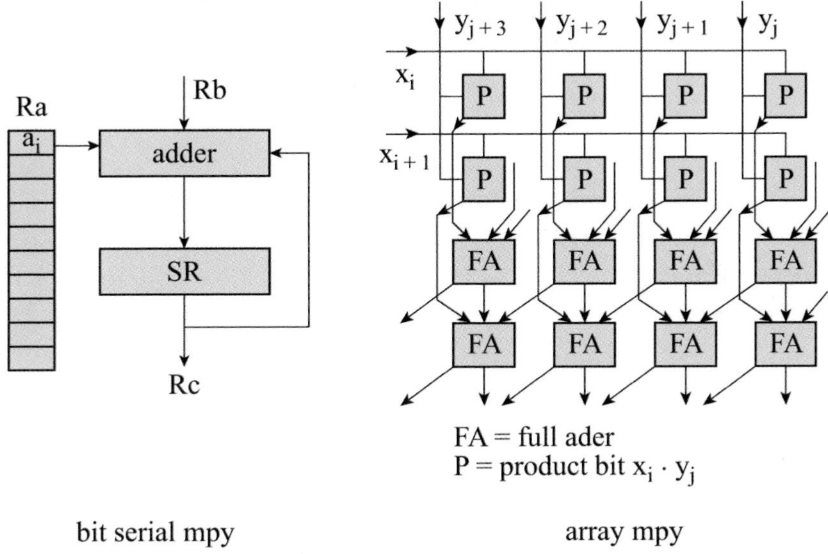

Figure 8.19: *Bit serial iterative and array multiplier*

The array multiplier only consists of logic that is really required for its function. In the bit serial approach, not only the required logic for multiplication is needed, but also the logic for additional control and registers. For a fair comparison, a complete multiplication must be taken. For the parallel multiplier, we have power∗1 (period); for the bit serial one, we have power∗16 (periods). This means that for a full 16∗16 bits multiplication, data has to go 16 times through the serial multiplier, while it only needs to go one time through the hardware of the parallel multiplier. From this example, we may conclude that a parallel

implementation generally has less overhead than a bit serial one
and will therefore consume less power.

- Optimise system power instead of chip power only.
  Complete systems use blocks such as DSP, A/D, D/A and memories, etc. As a result of the increasing communication bandwidth (data word length times frequency) of signals between
  these blocks, a lot of power would be wasted in the I/O circuit
  if each block was a separate chip. If possible, all functions should
  be on one chip. This will increase the chip power, but it will
  reduce the system power. A concentration of high-performance
  system parts and low performance system parts on different areas on one chip is attractive for power as well. The low performance parts could then run at lower frequencies and reduced
  voltages, to save power.

- Number representation.
  The choice of the number representation can also have an effect
  on the power consumption, see also figure 8.20.

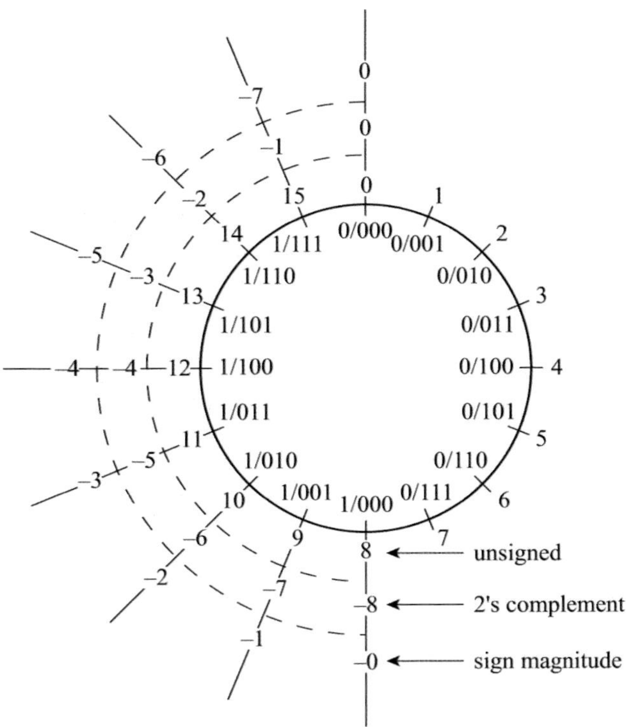

Figure 8.20: *Number representation diagram*

It is clear that unsigned code is only favourable for positive numbers. The most significant bit is then also used for magnitude representation. The two's complement notation shows problems (discontinuity) at the transition from $7 \rightarrow -8$. The diagram shows two discontinuities for the sign-magnitude notation: at the transition from $7 \rightarrow -0$ and also at the transition from $0 \rightarrow -7$. It is therefore more difficult when used in counters.

When small values are represented by many bits, the most significant bits in the two's complement notation adopt the value of the sign bit. If the signal is around zero, it will frequently switch from a positive to a negative value and vice versa. In the two's complement notation, a lot of bits will then toggle, while in the sign-magnitude notation only the sign bit will toggle, resulting in less power consumption. In the following example, the use of the two's complement notation and the sign-magnitude notation in adders and multipliers is compared.

Example:
8-bit adder/subtractor. The representation is shown in figure 8.21:

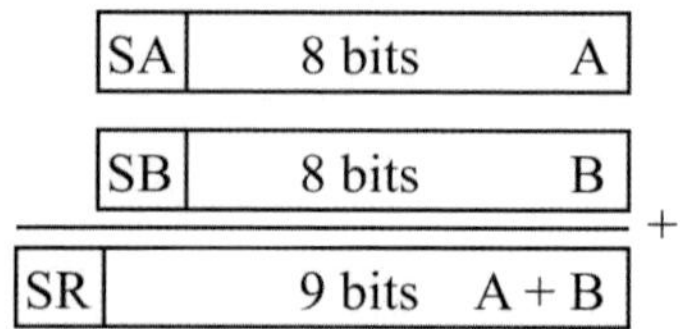

Figure 8.21: *Representation of an 8-bit adder/subtractor*

In the two's complement notation, the addition/subtraction operation does not give any problems. However, in the sign-magnitude notation, additional comparators must be used for a subtraction:

$$\text{if } A < B \quad \Rightarrow \quad sum = B - A$$

$$\text{else} \quad \Rightarrow \quad sum = A - B$$

Implementation with synthesis and standard-cell place & route tools reveals a difference in silicon area of a factor of about three in favour of the two's complement notation.

Example:
Two's complement multiplication:

$$X = -X_{n-1} \cdot 2^{n-1} + \sum_{i=0}^{n-2} X_i \cdot 2^i \tag{8.15}$$

$$Y = \underbrace{-Y_{m-1} \cdot 2^{m-1}}_{\text{sign}} + \underbrace{\sum_{j=0}^{m-2} Y_i \cdot 2^j}_{\text{value}} \tag{8.16}$$

The result of multiplying X and Y is:

$$X \cdot Y = X_{n-1} \cdot Y_{m-1} \cdot 2^{n+m-2} + \sum_{0}^{n-2} \sum_{0}^{m-2} X_i Y_j 2^{i+j} \tag{8.17}$$

$$-\left(\sum_{j=0}^{m-2} X_{n-1} \cdot Y_j \cdot 2^{n-1+j} + \sum_{i=0}^{n=2} Y_{m-1} \cdot X_i \cdot 2^{m-1+i} \right)$$

486

The realisation in an array multiplier requires the last two product terms to be skipped. A nice alternative is the Booth multiplier, in which half the number of full adders is replaced by multiplexers and where these two product terms are automatically skipped.

Example:
Sign-magnitude multiplication:

$$X \;=\; -1^{X_{n-1}} \cdot \sum_{i=0}^{n-2} X_i \cdot 2^i \tag{8.18}$$

$$Y \;=\; -1^{Y_{m-1}} \cdot \sum_{j=0}^{m-2} Y_j \cdot 2^j \tag{8.19}$$

and the product:

$$X \cdot Y = \underbrace{-1^{X_{n-1} \oplus Y_{m-1}}}_{\text{sign}} \sum_{i=0}^{n-2} \sum_{j=0}^{m-2} \underbrace{X_i \cdot Y_j \cdot 2^{1+j}}_{\text{magnitude}} \tag{8.20}$$

In this notation, the sign bit of the product is just a simple EXOR of the individual sign bits, while the magnitude is just the product of only positive numbers.

Conclusions on number representation

Although the sign-magnitude notation is convenient for multiplier implementation, the Booth algorithm array multiplier is more popular. Such a multiplier requires relatively little hardware and is thus suited for low power implementation.

The sign-magnitude notation is convenient for other applications. However, use is limited to representing absolute values in applications with peak detection, but even here it is still used more for number representation than for calculation. If only number representation is considered, the sign-magnitude notation shows less activity when the signal varies around zero. Note that, with compression techniques such as MPEG, a lot of zeros (000..00) are only represented by one bit. The use of compression techniques automatically reduces the power consumption.

- Optimum code.
  Even the code in which an operation is expressed can influence the power consumption. An example is shown in table 8.3

Table 8.3: Comparison of switching activity in a BCD
counter and a Gray code counter

Standard binary code (BCD)		**Gray code**	
number of changing bits			
0 0 0	3	1	0 0 0
0 0 1	1	1	0 0 1
0 1 0	2	1	0 1 1
0 1 1	1	1	0 1 0
1 0 0	3	1	1 1 0
1 0 1	1	1	1 1 1
1 1 0	2	1	1 0 1
1 1 1	1	1	1 0 0

2   1

↖↗

average/clock

Table 8.3 shows the switching activity of two 3-bit counters: a
BCD counter and a Gray code counter. The table also shows
that the BCD counter exhibits twice the switching activity of
the Gray code counter.

- Alternative implementations for arithmetic multiplier and adder
  circuits.

  Besides the previously-discussed options (bit serial versus bit
  parallel and number representations), there are many other alternatives that can influence the power consumption of arithmetic logic. Alternatives for multiplier implementation include:
  Booth multiplier, array multiplier and Wallace tree multiplier,
  etc. Alternatives for the addition process are carry select, carry
  ripple, carry save and carry look ahead techniques. With respect
  to power consumption, a general rule of thumb is: "every implementation that speeds up an arithmetic process will require
  additional power." The choice of an arithmetic implementation depends on the priorities in an application with respect to

488

speed, area and power consumption. Therefore, no fixed prescribed choice can be given for low power here.

- Microprocessor and microcontroller architecture.
  Many products use microprocessor cores: mobile phones, medical electronics, automotive and consumper products, watches and games. Maintaining or improving the performance while reducing the power consumption is a continuous challenge for the designers of new products in these fields. Generally, an instruction in a RISC architecture needs more execution cycles than in a CISC architecture. Pipelined RISC microprocessors use one or two cycles per instruction, while the CISC microprocessor often uses 10-20 cycles. However, complex algorithms mapped on a RISC machine generally require more instructions than a CISC machine. The CISC architecture may have too much hardware for only simple algorithms, which leads to a kind of overkill.
  In these cases, CISC power consumption may be more. From literature, it appears that each architecture (whether RISC or CISC) can in itself be optimised for low power. No real winner can be distinguished here because both architectures have many parameters to be adjusted for optimum low power.

- Limited I/O communication.
  In many applications, many I/O pins are used for communication between processor and memory and/or A/D or D/A converters. To reduce activity, these blocks have to be integrated on one single die. This may increase the chip power, but it certainly reduces the system power.

- Synchronous versus asynchronous.
  In synchronous circuits, the data transfer to, on and from the chip is usually controlled by a global clock signal. However, this clock signal does not contain any information. In contrast, asynchronous circuits proceed at their own speed. Here, the output of one circuit is immediately used as an input to the next. The relatively large difference in delay paths may lead to random operation and requires a special design style and test strategy. Actually, there are two kinds of asynchronous circuits: asynchronous subfunction(s) of synchronous designs and purely asynchronous designs (self-timed circuits).

- Asynchronous subfunction (of synchronous design).
  A synchronous chip is nothing more than a collection of asyn-

chronous circuits which are separated by flip-flops (registers). Thus, asynchronous blocks are embedded between registers. A 4-bit counter may serve as an example.

Figure 8.22 shows an asynchronous implementation and two synchronous alternatives of this counter. In the synchronous versions, each flip-flop is clocked at the highest frequency, which consumes a lot of power. The synchronous counter with parallel carry consumes the most power because it has more hardware than the ripple carry counter. In the asynchronous counter version, only the first flip-flop (LSB) runs at the highest frequency, whereas the others act as frequency dividers (divide by two). This version therefore requires much less power (about 1/3) than the best of the synchronous versions.

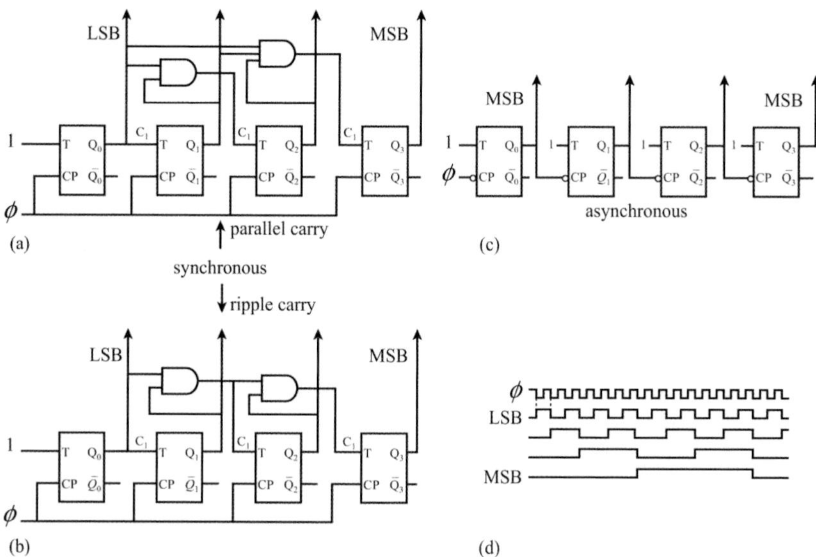

Figure 8.22: *Different versions of a 4-bit counter with timing diagram. a) synchronous with parallel carry b) synchronous with ripple carry c) asynchronous and d) timing diagram*

Asynchronous logic was already introduced in the early 80s [24], but has not been used intensively since then, mainly because many design concepts and flow are different from the synchronous standard cell design concepts and flow.

- Pure asynchronous designs *(self-timed circuits)*.
  A basic *asynchronous design* requires additional hardware to perform the necessary request (GO) and acknowledge (DONE) signals. Figure 8.23 shows a full-adder cell implemented as an asynchronous logic cell.

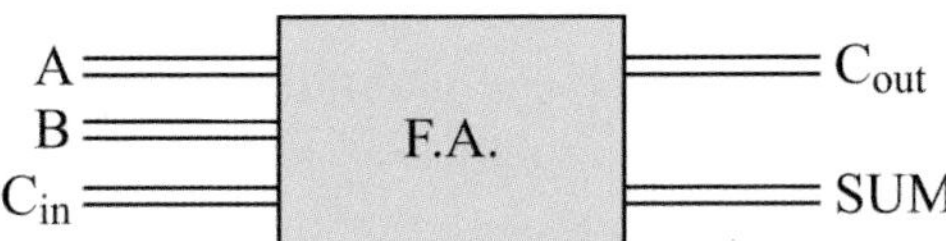

Figure 8.23: *Self-timed logic cell*

In this technique, an enormous area overhead must be spent to implement and route the additional logic that is associated with each request and acknowledge signal. This overhead is at least a factor two. An advantage is that no glitches can occur (see next subsection B). Another way of implementing self-timed circuits is to generate the request and acknowledge signals at a higher level of circuit hierarchy, see figure 8.24.

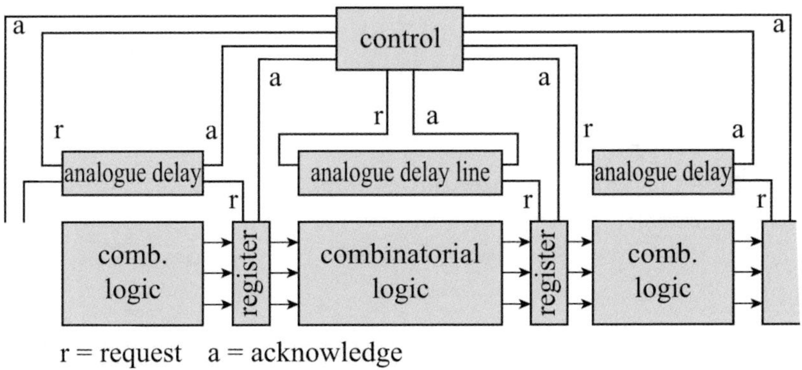

Figure 8.24: *Self-timed circuit by using analogue delay that mimics combinatorial logic delay*

After the data is stored in the register, it generates an acknowledge signal *a*, which is transformed by the control block into a request signal r, which then propagates through the analogue delay line at the same time that the logic signal propagates through the combinatorial logic block. Shortly after the data has become valid at the output of the logic block, the request signal arrives at the register, which then stores the data at the output of the logic block. If this data did not change, no new request data signal is generated. In this way, a higher component efficiency is achieved. A major disadvantage is that the dummy delay lines must be designed to be marginally slower than the slowest path through the combinatorial logic. This combinatorial logic also shows glitches as in synchronous logic, see next subsection B.

The key to this form of self-timed logic is the ability to model the combinatorial logic delay with a very small analogue delay line (inverter chain). Self-timed techniques are also used in synchronous systems, for instance, to generate the clocks needed in smaller parts of the chip. In RAMs, many self-timed clocks are generated on chip. A final discussion on power consumption of synchronous and asynchronous circuits leads to the following statement:

'Although asynchronous circuits are only active when necessary and thus operate at reduced power, these need not be *the* implementation for low-power circuits.'

Synchronous logic, optimised for low power, can achieve a power level that approaches that of asynchronous circuits. However, synchronous logic was mostly optimised for high speed (and, in some cases, for small area). Certain circuits are particularly suited for asynchronous implementation. But, for those that are not, the power consumed by the control circuit and the large test circuit can be greater than the advantage gained by having no clocks.

Several design houses are quietly replacing relatively small portions of their systems with asynchronous units. Already in the mid 1990s, Hewlett-Packard added an asynchronous floating-point multiplier to its 100 MHz RISC processor. These approaches are probably the wave of the future: asynchronous sub-units residing in a synchronous framework [25], or vice versa. More recently, asynchronous designs are used in an increasing number of application domains, e.g., smart cards [26], automotive, internet routers (switches) [27] and wireless products (ARM cores [28]) Particularly in the smart-card application, the reduction in power consumption directly leads to a performance increase, since these cards operate with a fixed power budget.

The increasing popularity of asynchronous design was caused after the successful combination of this design style with scantest, which resulted in stuck-at test coverage (see chapter 10) better than 99%. This asynchronous test methodology is based on applying a synchronous full-scan in asynchronous *handshake circuits* [29].

Another advantage of an asynchronous implementation compared to a synchronous one, is the general reduction of interference and noise. Figure 8.25 shows the result of a typical standard Dhrystone benchmark instruction set running on an ARM11 core. Dhrystone compares the performance of the benchmarked-processor core to that of a reference core, by measuring the average time the core takes to perform many iterations of a single loop containing a fixed sequence of the instructions of the benchmark.

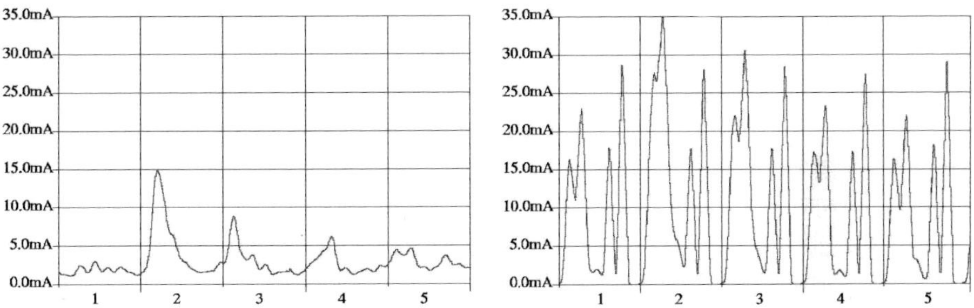

Figure 8.25: *Comparison of current peaks in an asynchronous (left) and synchronous ARM core (right) when executing the same benchmark operation*

The figure shows the current peaks in the supply lines of the asynchronous ARM core (left) during the execution of five Dhrystone loops within a period of $34\,\mu s$. Next, the clock period of the synchronous ARM core is adjusted such, that the same five

493

Dhrystone loops also last $34\,\mu s$, resulting in much larger current peaks (right), which also results in larger noise and emission levels. Although asynchronous designs exhibit a broader frequency spectrum of the generated supply noise, the amplitude is usually much less than that of their synchronous equivalents.

- Optimised memory design.
  The previously-discussed comparison can also be used in the realisation of memories. To reduce internal memory activities, self-timed techniques are used to generate a lot of different clocks or acknowledge signals which should be active according to some sequence. The alternative to performing one single operation (such as activate precharge, deactivate precharge, select word line, activate sense amplifier and select column, etc.) in one clock period means that a lot of clock periods are needed for only one read or write operation. This would be at the cost of increased power consumption.

B) *Implementation level.*

- Reduce glitching.
  Static CMOS circuits can exhibit *glitches* (also called dynamic hazards, critical races or *spurious transitions*) as a result of different propagation delays from one logic gate to the next. Consequently, a node can have multiple unnecessary transitions in a single clock cycle before it reaches its final state. Figure 8.26 gives an example.

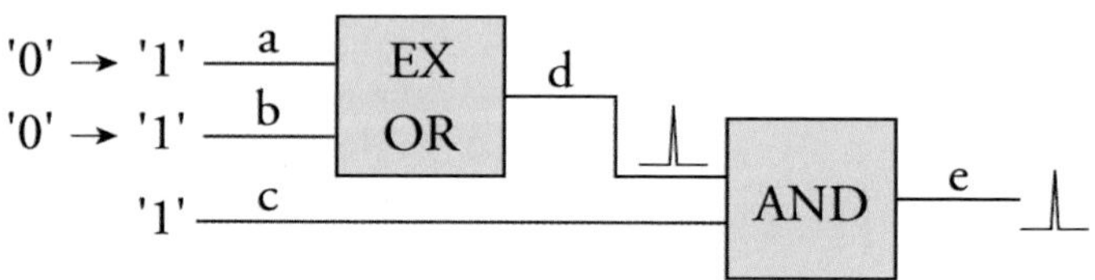

Figure 8.26: *Unnecessary transitions in a simple logic circuit*

Suppose the (a,b) inputs of an EXOR gate switch from (0,0) to (1,1). In both situations, output d of the EXOR should remain low ('0'). However, because of a different delay in the switching of the input signals, the state diagram of the (a,b) inputs might follow the following sequence $(0,0) \rightarrow (0,1) \rightarrow (1,1)$. Therefore, the (a, b) inputs are $(0,1)$ for a very short period of time, resulting in a temporary '1' at output d. This glitch also propagates through the next AND gate.

Such unnecessary transitions dissipate extra power. The magnitude of this problem is related to the kind of circuit to be realised. As a result of the occurrence of glitches, an 8-bit ripple carry adder with random input patterns consumes about 30% more power. For an 8*8-bit array multiplier, this number can be close to 100%, for a 16*16-bit array multiplier and for standard cell implementation of a progressive scan conversion circuit, it can be as high as 200%! Generally, the larger the logic depth, the larger the skew between the arrival times of input signals at a logic gate and the higher the probability of a glitch at the output of that gate. Therefore, a large power saving could be achieved in such circuits if all delay paths were balanced.

Different architectures can lead to a different percentage of unnecessary transients. A 16*16 bit Wallace tree multiplier has only 16% glitches, compared to the above 200% for a 16*16-bit array multiplier. The Wallace tree multiplier has far more balanced delay paths.

Finally, another way of reducing the number of glitches is to use retiming/pipelining to balance the delay paths.

- Optimise clock activity.
  There are two reasons why clock signals are very important with respect to power dissipation. The first is that clock signals run all over the chip to control the complete data flow on the chip in a synchronised way. This means that clock capacitance caused by both very long tracks and a large number of flip-flops can be very large. In complex VLSI chips, the clock load can be as high as one to several picofarads. The second reason is that the clock signal has the highest frequency (the maximum switching frequency of data signals is only half the clock frequency). The total power consumed by the clock network depends heavily on the number of connected flip-flops and latches.

  Figure 8.27 shows the relative clock power consumption as a function of the average activity on a chip. This is expressed as a fraction of the total power consumption.

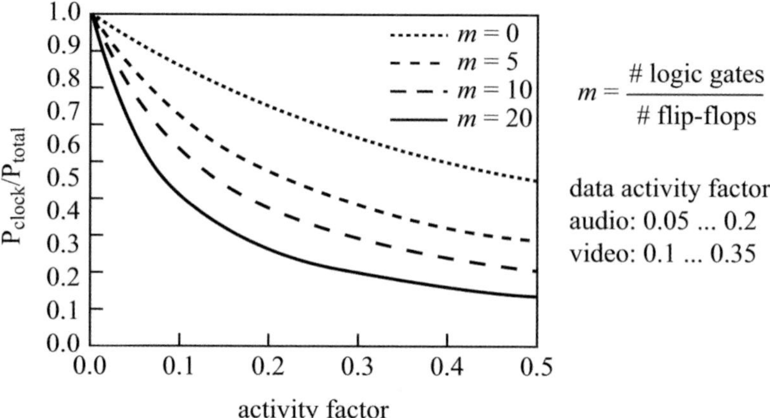

Figure 8.27: *Relative clock power consumption as a function of the activity*

Because the clock dissipation can be as high as 20-50% of the total chip dissipation, its activity should be reduced. This can be done because clock transitions carry no information. There are several ways to reduce clock activity. Including the use of *Dual-Edge Triggered (DET)* flip-flops. If a flip-flop could be triggered on both edges of the clock pulses instead of on only one edge, it would be possible to use a clock at half frequency for the same data rate, thereby reducing the power dissipation of the total clock network.

A flip-flop that acts at both edges of the clock pulse is called a dual-edge triggered flip-flop, whilst the conventional positive and negative-edge triggered flip-flops belong to the category of *Single-Edge Triggered (SET)* flip-flops. However, the use of DET flip-flops has been limited up to now by the high overhead in complexity that these flip-flops require and because they are not fully compatible with the current design flow. Both the SET and DET flip-flops have two latches. Basically, in a DET flip-flop (see figure 8.28) the two latches are arranged in parallel, while in a SET flip-flop, see figure 8.28(a), they are placed serially [30]. DET and SET flip-flops show comparable maximum data rates, however, DET flip-flops either require additional silicon area, or they are more difficult in use with respect to timing aspects [31,32].

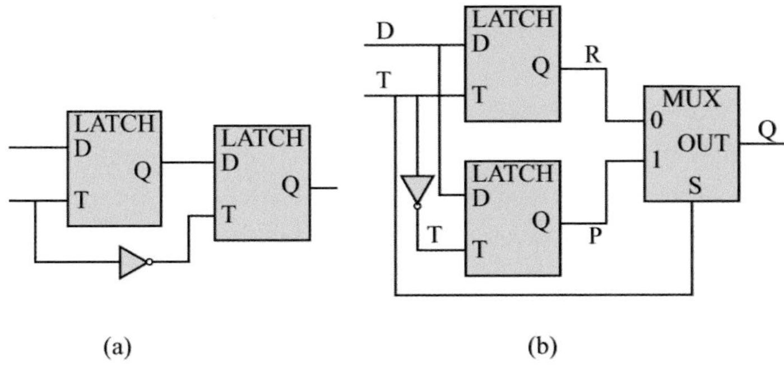

Figure 8.28: *Schematic diagram showing a SET and a DET flip-flop*

Since the clock contribution to the overall chip power consumption depends on the data activity, power savings of 10 to 20 percent are possible, at the cost of some additional flip-flop area (10 to 20%), when using DET flip-flops.

In conventional synchronous designs, the complete system is clocked at the highest frequency, even when some portions of the chip could operate on fractions of this frequency. In some cases, clock dividers are used to generate these lower frequencies. From a low-power point of view, we should start from the opposite direction.

This means that we supply the chip with the lowest required frequency and generate higher clock rates locally, if needed. This can be achieved by PLL-like circuits. In this way, the globally distributed clock would run at the minimum clock frequency and the higher clock frequencies would only be used where they are really needed. This might reduce the global clock activity drastically and also reduce the functional activity.

Another approach to reduce the total chip activity is to switch the clock off temporarily for certain functional blocks, or even for the complete chip during moments that no useful operations are executed. In this respect, different names are used for the same issue: gated clocks, stop-the-clock, sleep mode and power-down mode, etc.

A representative example is a coefficient ROM, whose power consumption can be relatively large. In many cases, such a ROM is often used for less than 1% of the time. Forcing this block to the power-down mode, e.g. by switching off its clock, saves 99% of its total power consumption.

When a signal processor enters the power-down mode, all its internal memory and register contents must be maintained to allow the operation to be continued unaltered when the power-down mode is terminated. Depending on the state of some control register(s), external devices can cause a wake-up of the DSP, e.g., when terminating an input operation. The processor enters the operating state again by reactivating the internal clock. The program or interrupted routine execution then continues.

A disadvantage of gated clocks (sleep modes, etc.) is that some logic operation has to be performed on the clock signal. This causes an additional delay for the internal gated clock, which may result in timing problems during data transfer between blocks that run at the main clock and those that run at a gated clock. Therefore, compensated delays must be used in those blocks that do not use a gated clock. Generally, gated clocks decrease the design robustness with respect to timing (see chapter 9).

- Dynamic versus static CMOS.
  The decision to implement a circuit in dynamic or static CMOS logic not only depends on power considerations. Aspects of testability, reliability, ease of design and design robustness are also very important here. In the comparison of dynamic and static CMOS realisations, several differences show up with respect to power. As precharge and sample periods in dynamic CMOS circuits are separated in time, no short-circuit dissipation will occur. Also, the absence of spurious transitions (hazards) reduces the activity of dynamic CMOS. However, precharging each node every clock cycle leads to an increase of activities.
  EXAMPLE:
  Let us assume that all input combinations in table 8.4 are uniformly distributed.

Table 8.4: Function table of a 2-input NOR and an EXOR gate

2-input NOR		EXOR	
a b	z	a b	z
0 0	1	0 0	0
0 1	0	0 1	1
1 0	0	1 0	1
1 1	0	1 1	0

Because each logic gate output in a dynamic CMOS chip is high during precharge, the output will be discharged in 75% of the input combinations of a 2-input NOR $\Rightarrow$ activity factor 0.75. For the EXOR: activity factor 0.5. In static CMOS, power is only dissipated when the output goes high:

$$\text{NOR}: \quad P_{0 \to 1} = P(0) \cdot P(1) = 3/4 \cdot 1/4 = 3/16$$
$$\text{EXOR}: \quad P_{0 \to 1} = P(0) \cdot P(1) = 1/2 \cdot 1/2 = 1/4$$

Usually, the logic function in dynamic CMOS is realised with an nMOS pull-down network, while a pMOS transistor is used for precharge. This leads to small input capacitances, which makes dynamic logic attractive for high-speed applications. Besides the higher activity factor, the additional clock loads to control the precharge transistors also leads to much higher dissipation. The use of dynamic logic is not as straightforward and common as static logic. In terms of design robustness and ease of design, static CMOS is favourable as well. Finally, when power reduction techniques (such as power-down modes, in which the clock is stopped) are being implemented, dynamic CMOS is much more difficult to apply because of its charge leakage. Generally, it can be stated that dynamic logic is not a real candidate for low-power (low-voltage) realisation.

- Connect high-activity input signals close to the output of a logic gate.

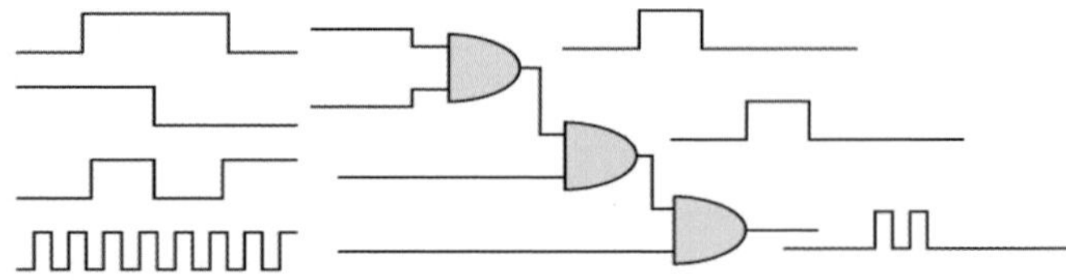

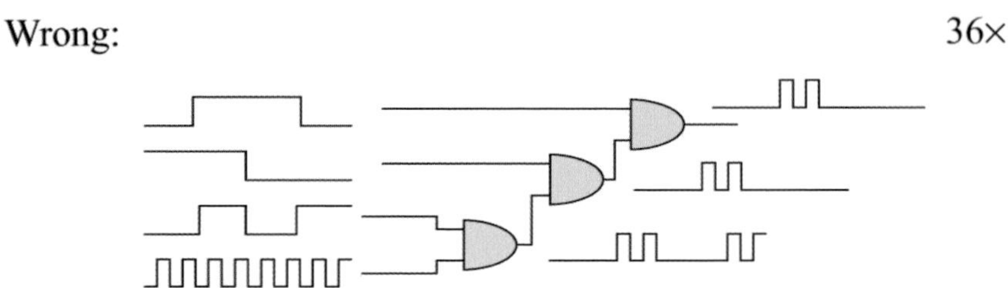

Figure 8.29: *Reduction of total activity by ordering signals*

Figure 8.29 shows that connecting signals with high activity close to the output of the propagation chain will reduce the total switching activity and so the total power consumption of that chain.

- Exploit the characteristics of library cells.
  Here again, when there are signals showing high activity, it is obvious that these will cause less power dissipation when they are connected to the low-capacitance inputs of logic gates. Figure 8.30 shows an example.

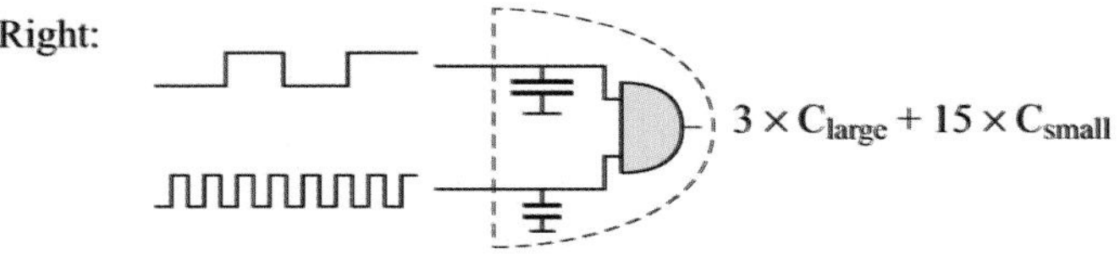

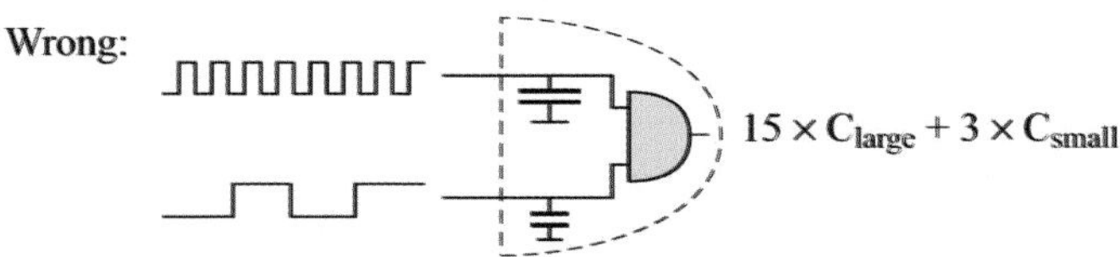

Figure 8.30: *Reduction of power dissipation by matching high-activity signals with low-capacitance inputs*

It should be clear that the power savings of these last two items can only be achieved by dedicated software programs, which perform some statistics on the signal activity inside a logic block.

8.6 Computing power versus chip power, a scaling perspective

The scaling process and its impact on the performance, reliability and signal integrity of MOS ICs is extensively discussed in chapter 11. However, the scaling process with respect to the system performance of digital signal processors (DSPs) requires a different approach.

An important parameter reflecting this system performance is the computing power of a DSP. Generally, this computing power (U) is proportional to:

$$U = n \cdot f \tag{8.21}$$

where n represents the number of transistors and f is the frequency.

The power dissipation of the DSP is proportional to:

$$P = n \cdot f \cdot C \cdot V^2 \tag{8.22}$$

From these two equations, it can be derived that the computing power per Watt dissipation is equal to:

$$U/[\text{W}] = \frac{1}{C \cdot V^2} \tag{8.23}$$

In the following discussion on scaling, V_T effects and velocity saturation are neglected. If the scaling factor between two successive process generations is s (usually $s \approx 0.7$), then the number of transistors will increase to:

$$n_s = n/s^2 \tag{8.24}$$

and the frequency to:

$$f_s = p/s^2 \cdot f \tag{8.25}$$

Where p equals the voltage scaling factor $(V_p = p \cdot V)$, as this factor may differ from s.
The capacitance C scales to:

$$C_s = s \cdot C \tag{8.26}$$

Combining the previous equations results in the following expressions concerning computing power and chip power impact: the computing power scales to:

$$U_s = n_s \cdot f_s = p/s^4 \cdot U \tag{8.27}$$

and the power dissipation per unit area with:

$$P_s = n_s \cdot f_s \cdot C_s \cdot V_p^2 = p^3/s^3 \cdot P \tag{8.28}$$

Therefore, the computing power per Watt after scaling increases to:

$$U_s/[\text{W}] = \frac{1}{s \cdot p^2} \cdot U/[\text{W}] \tag{8.29}$$

Remarkably, voltage scaling has more impact on the computing power per Watt than the process scaling. From the $0.8\,\mu\text{m}$ CMOS to the $120\,\text{nm}$ CMOS node, the voltage has continuously been scaled with a factor p that was about equal to the process scaling factor s. This means that, neglecting the second-order effects, the computing power per Watt for these generations of DSPs has increased according to:

$$U_s/[\text{W}] = \frac{1}{s^3} \cdot U/[\text{W}] \tag{8.30}$$

502

Because the subthreshold leakage current increases exponentially with a reduction in the threshold voltage, it has caused a slow-down in the reduction of the supply voltage, which has maintained almost constant from the 120 nm CMOS node to the 45 nm CMOS node. Although this has limited the increase in standby (leakage) power consumption, it has reduced the power efficiency improvement when moving to the next technology node.

Each DSP generation will therefore still become more power efficient, but to a lesser extent. Second-order effects have a more negative impact on the transistor performance and thus on the DSP efficiency. However, even after such a reduction in efficiency improvement, a lot of new DSPs are still expected to enter the market with improved power efficiency.

8.7　Conclusions

With respect to conventional CMOS processes and design styles, large power savings could be achieved because they were optimised for speed and area. Power can be reduced in different ways, but the largest power savings can be achieved by reducing the supply voltage. In this respect, the scaling process from $0.8\,\mu$m CMOS technologies to the 120 nm node, in which the supply voltage gradually reduced from 5 V to 1.2 V, has had a huge impact on the total power reduction. This was the basis for the integration of a large variety of functional features (camera, MP-3, phone, games, internet access, blue tooth communication, audio, video, GSM, etc.) into one single mobile gadget, which we still call a mobile phone.

In CMOS technology development, a few measures can be taken to reduce power: limit the leakage currents and limit the parasitic capacitances.

In the design, however, there are many options for reducing the total capacitance and activity on a chip. It has been shown that the decisions taken at the higher hierarchy levels have much more impact on the system power consumption than those taken at the lower levels of design (circuit and layout level).

A complete and clear set of design rules cannot be given, because the use of many of these options depends on the application. This chapter is meant to present a flavour of these options and to provide the designer with a low-power attitude.

Finally, although several alternative low-power CMOS design styles have been presented at conferences and magazines during the last decade, static CMOS logic is still favourable in many ways. It is very robust with respect to transistor scaling and supply voltage reduction. Besides this, design integrity is becoming a key issue in nanometer VLSI design, which also makes static (complementary) CMOS the best candidate for many process generations to come.

8.8 References

[1] J. Kopera,
'Considerations for the utilization of NiMH battery technology in stationary applications',
International Stationary Battery Conference 2005, pp. 4.1-4.10

[2] aan Frans Schoofs vragen

[3] K. Seta, et al.,
'50% Active-Power saving without speed degradation using standby power reduction (SPR) Circuit',
IEEE Digest of Technical papers, pp 318,319, Feb. 1995

[4] T. Kuroda, et al.,
'A 0.9 V, 150 MHz, 10 mW, 4 mm^2, 2D Discrete Cosine Transform Core Processor with variable Threshold Voltage (V_T) Scheme',
IEEE Journal of Solid-State Circuits, pp 1770-1779, Nov. 1996

[5] A. Montree, et al.,
'Limitations to adaptive back bias approach for standby power reduction in deep sub-micron CMOS ICs',
Proc.of the '99 European Solid-State Device Research Conf. (ESSDERC), Sept. 1999, pp.580-583

[6] T. Chen, et al.,
'Comparison of Adaptive Body Bias (ABB) and Adaptive Supply Voltage (ASV) for Improving Delay and Leakage Under the Presence of Process Variation',
IEEE Transactions on Very Large Scale Integration (VLSI) Systems, Vol. 11, No. 5, October 2003, pp. 888-899

[7] M. Meijer, et al.,
'Limits to performance spread tuning using adaptive voltage and body biasing',
International Symposium on Circuits and Systems (ISCAS), 2005, pp. 5-8

[8] M. Izumikawa, et al.,
'A 0.25 μm CMOS 0.9 V, 100 MHz, DSP Core',
IEEE Journal of Solid-State Circuits, pp 52-61, Jan. 1997

[9] D. Lee, et al.,
'Gate Oxide Leakage Current Analysis and Reduction for VLSI Circuits',
IEEE Transactions on VLSI Systems, Vol. 12, No. 2, February 2004

[10] C. Piguet,
'Design Methods and Circuit Techniques to Reduce Leakage in Deep Submicron',
Faible Tension Faible Consommation, FTFC 2003

[11] H.J.M. Veendrick,
'Short-Circuit Dissipation of Static CMOS Circuitry and its Impact on the Design of Buffer Circuits',
IEEE Journal of Solid State Circuits, Vol. SC-19, No. 4, August 1984, pp 468-473

[12] Von Kaenel, et al.,
'A Voltage Reduction Technique for Battery-Operated Systems',
IEEE JSSC. Vol. 25, Oct. 1990, pp 1136-1140

[13] Johan Pouwelse,
'Dynamic Voltage Scaling on a Low-Power Microprocessor',
Proceedings of the 7th annual international conference on Mobile computing and networking, 2001, pp. 251 - 259

[14] R. Ghattas, et al.,
'Energy management for commodity short-bit-width microcontrollers',
Proceedings of the 2005 international conference on Compilers, architectures and synthesis for embedded systems, pp. 32-42

[15] N. Verma, et al.,
'Nanometer MOSFET Variation in Minimum Energy Subthreshold Circuits',
IEEE Transactions on Electron Devices, Vol. 55, No. 1, January 2008, pp 163-174

[16] K. Yano, et al.,
'A 3.8 ns CMOS 16x16-b multiplier using complementary pass-transistor logic',
IEEE JSSC, Vol. 25, April 1990, pp 388-393

[17] M. Suzuki, et al.,
   'A 1.5 ns 32b CMOS ALU in double pass-transistor logic',
   Digest ISSCC, Feb 1993, pp 90-91

[18] A. Parameswar, et al.,
   'A swing restored pass-transistor logic-based multiply and accumulate circuit for multimedia applications',
   IEEE JSSC, Vol. 31, June 1996, pp 805-809

[19] K. Jano, et al.,
   'Top-down pass-transistor logic design',
   IEEE JSSC, Vol. 31, June 1996, pp 792-803

[20] J.H. Pasternak and C. Salama, 'Differential pass-transistor logic',
   IEEE circuits & Devices, July 1993, pp 23-28

[21] M. Song, et al.,
   'Design methodology for high speed and low power digital circuits with energy economized pass-transistor logic (EEPL)',
   Proc. 22nd ESSCIRC Digest, 1996, pp 120-123

[22] W.H. Paik, et al.,
   'Push-pull pass-transistor logic family for low-voltage and low-power',
   proc. 22nd ESSCIRC Digest, 1996, pp 116-119

[23] R. Zimmermann and W. Fichtner,
   'Low-Power Logic Styles: CMOS Versus Pass-Transistor Logic',
   IEEE JSSC, Vol. 32, July 1997, pp 1079-1090

[24] C. Mead and L. Conway,
   'Introduction to VLSI Systems',
   Chapter 7 by C.Seitz, Addison-Wesley, 1980

[25] C. Maxfield,
   'To be or not to be asynchronous that is the question',
   EDN, December 7, 1995, pp 157-173

[26] J. Kessels, et al.,
   'A design Experiment for a Smart Card Application consuming Low Energy',
   Chapter 13 in 'Principles of Asynchronous Circuit Design: A Systems Pespective', Kluwer Academic Publishers, 2001

[27] Ä. Lines,
'Asynchronous Interconnect for Synchronous (SOC) Design',
IEEE Micro Journal, Vol. 24, No. 1, pp.32-41, 2004

[28] A. Bink,
'ARM996HS, the first licensable, clockless 32-bit processor core',
IEEE Micro Journal, Vol. 27, 2007

[29] F. te Beest, et al.,
'Synchronous Full-Scan for Asynchronous Handshake Circuits',
Journal of Electronic Testing: Theory and Applications, Vol. 19,
pp.397-406, 2003

[30] R.Hossain, et al.,
'Low Power Design Using Double Edge Triggered Flip-Flops',
IEEE Trans. on VLSI, Vol.2, No.2, June 1994

[31] Jerry Yuang, et al.,
'New Single-Clock CMOS Latches and Flipflops with Improved Speed
and Power Savings',
IEEE Journal Solid State Circuits, January 1997, pp 62-69

[32] A.G.M. Strollo, et al.,
'Low power double edge-triggered flip-flop using one latch',
Electronics Letters, Vol. 35, 4 February 1999, pp 187-188

8.9 Exercises

1. Why must every designer always have a low-power attitude?

2. Which of the different power contributions is the larger and why?

3. How could the subthreshold leakage power dissipation be reduced?

4. In optimizing a complete library for low power, which of the library cells would you focus most of your attention to?

5. What is the greatest advantage of constant-field scaling with respect to power dissipation?

6. What would be the difference in activity factor between a static and dynamic CMOS realisation of the next boolean function: $z = \overline{abc}$

7. Repeat exercise 6 for $z = \overline{a + b + c}$

Chapter 9

Robustness of nanometer CMOS designs: signal integrity, variability and reliability

9.1   Introduction

With shrinking feature sizes and increased chip sizes, the average delay of a logic gate is now dominated by the interconnection (metal wires) rather than by the transistor itself. Most of the potential electrical problems, such as cross-talk, critical timing, substrate bounce and clock skew, etc. are related to the signal propagation and/or high (peak) currents through these metal wires.

Currently, complex VLSI chips may contain hundreds of millions to more than a billion transistors that realise complete (sub)systems on one single die. For the design of these ICs, a lot of different tools are used, see chapter 7. The sequence in which these tools are used, from the upper hierarchy levels down to the layout level, is called the "design flow".

IC design flows have been automated so much that "first time right silicon" is considered as natural. However, keeping control over all the tools used in the design flow (the high-level description language, the synthesis tools and the verification tools, to name a few) requires the complete attention of the designers. Thus, even when designers are familiar with the physical aspects of complex ICs, the potential physical and electrical problems do not get the attention that they require, particularly in nanometer technologies.

First silicon (especially of high-performance ICs) therefore shows first-time-right functionality but often at lower or higher supply voltages and/or at lower frequencies than required. Actually, at a time where designers are drifting away from the physical transistor level into abstract high-hierarchy levels of design, exactly the opposite would be required to get current and future VLSI chips operating electrically correctly. Many ICs are therefore no longer "correct by design" but are "designed by corrections".

This chapter deals with the robustness of digital circuits in relation with the continuous scaling process. It will cover most topics related to signal integrity (timing, cross-talk, signal propagation, voltage drop, supply and substrate noise, soft-errors, EMC, etc.), variability (systematic and random variability) as well as such reliability issues as electromigration, leakage, ESD, latch-up, hot-carrier effect and NBTI.

Because of the increasing clock frequencies and increasing chip complexity, e.g., multi-frequency and multi-voltage domains, timing closure has become one of the most time consuming activities in the total design trajectory. Therefore the next subsection will start with some of the related timing and clocking issues.

The reducing signal integrity is a result of two conflicting effects: the increase of noise and the reduction of the noise margins (V_{dd} and V_T). A relatively large section is therefore devoted to almost all aspects related to signal integrity and ways to maintain it at a sufficiently high level. A continuous reduction of the noise margins also has a severe impact on the quality of the IC test. The increasing discrepancy between chip operation during test and in the application will result in more customer returns and design spins. The section will therefore also include some remarks on the effect of scaling on test coverage and complexity.

As a result of the continuously reducing transistor sizes and voltages, the transistor behaviour is becoming much more sensitive to an increasing number of variability causes. Moreover, also the diversity and level of variations increases. The fourth paragraph presents a flavour of the impact of these variations on both analog and digital circuit design.

The continuous scaling of the devices and interconnects also has a severe impact on the reliability of the integrated circuit. Moreover, a necessary move to new materials may even have dramatic consequences for the overal IC reliability. The fifth paragraph in this chapter is therefore devoted to reliability and ways to maintain it. A robust design not only refers to a robust electrical operation with respect to the specified performance, it also takes great discipline to set up a robust database and create a complete design documentation. This is needed to enable quick redesign spins and support re-use. These aspects are discussed in the sixth paragraph. Paragraph seven presents some conclusive remarks. The negative impact of scaling on signal integrity and reliability depends on the way scaling is performed. The final chapter in this book discusses the effects of scaling in general, but will also come back to the influence of constant-voltage scaling, constant-field scaling and constant-size scaling on the robustness of operation of nanometer CMOS ICs.

9.2 Clock generation, clock distribution and critical timing

9.2.1 Introduction

The majority of today's integrated circuits are synchronous designs in which the data transfers to the chip, on chip and off chip are controlled by one or more clock signals. Clock frequencies may vary from a few tens of megahertz for extremely low-performance ASICs to a few gigahertz for high-performance microprocessors. The timing complexity of an IC is not only related to the frequency of its clock signals, but also on the diversity, complexity and number of cores integrated on a single chip, today. The overall timing complexity is so high, that full chip-level timing closure has become a real burden. Additionally, advanced power reduction techniques, such as clock disabling, different voltage and frequency domains, power switching and dynamic voltage and frequency scaling, in combination with increasing process parameter variations and circuit noise have made overall chip timing to become a nightmare. Next to the fact that the corresponding design solutions need to be supported by the models, the tools and the design flow, also the back-end designers must be well-educated so that they understand the issues and can manage the "global timing picture".

This subsection, however, will discuss only some of the basics of clock distribution and clock generation and is only meant to trigger and focus the attention of the design community to the huge challenges of timing closure.

9.2.2 Clock distribution and critical timing issues

Very complex designs may contain hundreds of millions of transistors on silicon die areas of less than half a square centimeter. Most VLSI designs contain synchronous logic, which means that data transfer on the chip is controlled by means of one or more clock signals. These clock signals are fed to latches, flip-flops and registers, which temporarily store data during part of the clock period.

Current VLSI chips may contain several hundred thousands of these latches or flip-flops and the total wire length of the clock signals may exceed several metres. To achieve high system performance, the clock frequency is often maximised. This combination (a large clock load and a relatively high clock frequency) is the cause of many on-chip timing problems.

The following sections discuss potential timing problems, most of which are related to the clock signals.

Single-phase clocking

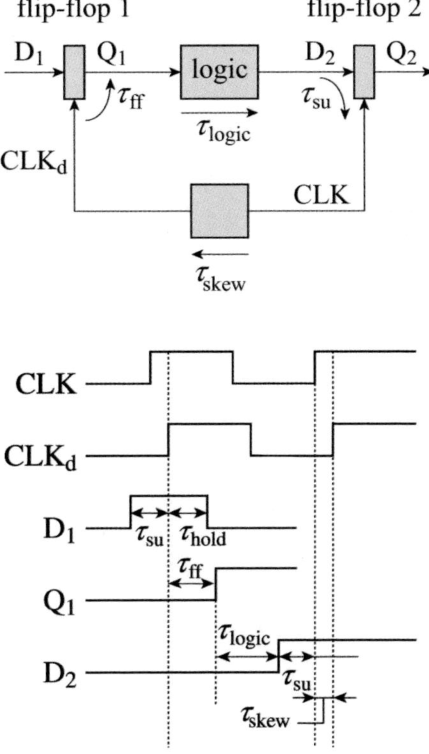

Figure 9.1: *Single-phase clock system and its timing diagram*

From figure 9.1, we can derive that the minimum cycle time is given by:

$$\tau_{\min} = \tau_{ff} + \tau_{logic} + \tau_{su} + \tau_{skew} \tag{9.1}$$

where
τ_{ff} is the flip-flop delay from clock to output,
τ_{logic} is the propagation delay through the logic and
τ_{su} is the setup time of the data of flip-flop 2.
τ_{skew} is the maximum amount of time that the clock of flip-flop 2 can be earlier than that of flip-flop 1.

Especially τ_{logic}, which is dominant in equation 9.1, must be carefully simulated to be sure that the required frequency (clock period) will be achieved. This "simulation" is usually performed by the static timing analysis tool, which adds the worst-case delay of each of the gates together to determine the total delay of the logic path. In combination with the synthesis tools it should guarantee satisfactory timing results. Sometimes the logic path between two flip-flops is absent. This is the case when pipe line and/or scan registers are implemented by using series connections of flip-flops (figure 9.2).

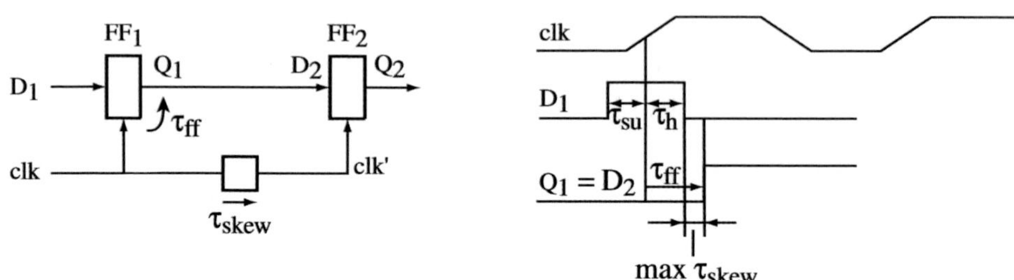

Figure 9.2: *Critical timing situation in case of direct communication between two flip-flops*

Especially in the scan mode during testing (see chapter 10), the logic paths are bypassed and flip-flops are directly connected to other flip-flops. In figure 9.3, a flip-flop of logic block 1 can be directly connected to a flip-flop of logic block 2.

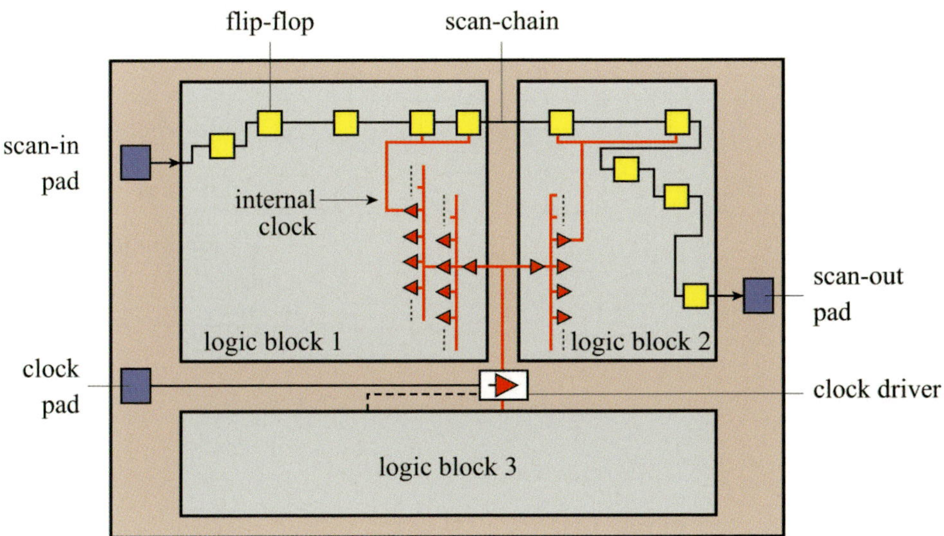

Figure 9.3: *Example of a scan-chain in a complex VLSI circuit*

With a direct connection, the propagation time of the data between these flip-flops can be very short. As the clock is routed through these blocks automatically, its time of arrival at the first flip-flop in the scan chain of logic block 2 can be later than the arrival time of the data. This will result in a race, which can also occur in registers. Therefore, each (scan) register should be carefully checked with respect to the above critical timing situation. If necessary, additional delay by using several inverters must be included at these critical positions in the scan chain.

Generally, there is a variety of single-phase clocked flip-flops in a library. As many of these flip-flops need two clock phases, one or both are generated inside the flip-flop by means of inverters.

Figure 9.4 shows a flip-flop [1] that can also be used in a low-power design: it has a small number of transistors, it is controlled by a single-phase clock and it has a relatively low clock load. Here also, the clock may be generated locally in the latch by means of an additional inverter.

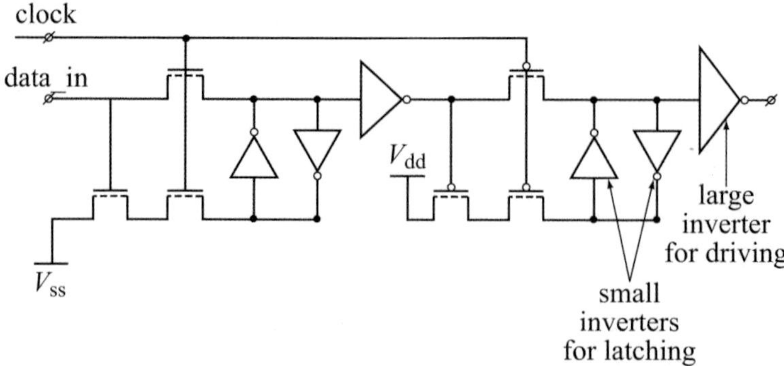

Figure 9.4: *Efficient single-phase clock flip-flop*

Clock skew and clock jitter

Clocks are routed locally in IP cores and globally across the chip, which may consist of different clock domains. For proper operation it is required that the clock arrives at all flip-flops within one clock domain almost exactly at the same time. However, due to many different causes this can not always be sufficiently guaranteed. The difference in clock arrival times at these flip-flops is called *clock skew*.
There are several causes of clock skew:

- different lengths of clock paths

- different loads of the clock drivers

- variations in clock driver delay due to

    - process variations
    - supply noise and IR-drop
    - temperature variations

The actual clock skew between two flip-flops is different from one chip to another and varies with the time.

The probability of occurrence of clock-skew is higher, when there is direct communication between flip-flops within digital cores, or between flip-flops in different cores. An example of clock skew is demonstrated in figure 9.3, where the clock signal in logic block 1 has to propagate through a different number of clock drivers in the clock tree than the clock signal in logic block 2, before it arrives at the respective flip flops.

Clock skew is also introduced in ICs that include cores whose clock may temporarily be switched off (*gated clocks*) to reduce power consumption when its functionality is not needed.

Particularly in this era of IP reuse, there is a large variety of digital cores to be integrated onto a single SoC. These logic and memory IP cores may origin from different design sites within one company, or even be created by different IP vendors. Also only one type of flip-flop is preferred: e.g., positive edge-triggered flip flop. Clock and further timing parameters must very well be specified in order to have these cores operated correctly with one another.

When the clock skew exceeds a certain limit, it may cause *timing violations. Set-up and hold-time violations* can play a dominant role in the operating frequency of the circuit. *Setup time* is the time a flip-flop requires its input data to be valid before its clock-signal capture edge. *Hold time* is the time a flip flop requires its input data to be stable after its clock-signal capture edge. A hold-time violation is caused by a so-called short-path delay fault.

Usually the set-up time is defined by the worst-case behaviour of the design, while the minimum hold time is defined by the best-case situation. Set-up time violations can be recovered if the application allows a reduction of the frequency. A hold-time violation means that you have to throw away the chip.

Within the cores, the clock skew is usually limited by applying a well-balanced clock tree approach. It is extremely important that the different branches in the tree are equally loaded (same number of flip-flops and same lengths of the clock wires). This must be verified by tools, particularly in high performance complex circuits. Current tools offer a well-balanced clock tree synthesis, which enhances the quality of clock timing. An important advantage of this clock tree approach is the distribution of the different small clock drivers over the logic blocks. The use of distributed clock drivers also puts the clock drivers right there where they are needed. Distributed clock drivers keep the current loops short and they also do not switch simultaneously, but distributed over a small time frame. Moreover, they can use the intrinsic decoupling capacitance which is available in a logic standard cell block. This reduces the dI/dt fluctuations, which are responsible for most of the supply/ground bounce in VLSI designs.

In many synchronous designs, the total dissipation of the clock-related circuit may vary from 10% to even 60% of the total IC dissipation. It is obvious, then, that the clock system will also generate a large part of the total supply bounce.

Today's semiconductor fabrication processes allow us to integrate complete systems onto one single die. Such an IC may contain a large variety of functions which may operate more or less independent from one another. The corresponding large currents introduce voltage drop across the on-chip power distribution network, which negatively affect the timing behaviour. Also other deep-submicron effects, such as cross-talk, supply and substrate noise, variability, etc., which are discussed in chapter 9, all affect the local and global timing behaviour.

Moreover, because power consumption has become one of the biggest concerns in the design of these systems, they need to support more and more state-of-the-art power reduction techniques: clock gating, power switching, voltage reduction, (dynamic) voltage and frequency scaling techniques, etc. Signals that cross different voltage domains require level shifters in their paths, which create additional delay. Reduction of the supply voltage causes a complete change in timing behaviour of the connected cores. In other words: it is not only the increasing IC functionality and performance that pose a serious threat to a secure and reliable timing closure; it is also these additional design measures to reduce power in both active and standby modes that create new timing constraints.

Next to clock skew, there is another major problem related to the propagation of the clock signal. Particularly as a result of supply voltage changes, e.g., due to supply noise, IR-drop or temperature variations, the clock period may vary from one clock cycle to another. This is called *clock jitter*. While clock skew represents the difference in clock-edge arrival times at different flip-flops in the same clock cycle, clock jitter is the difference in clock-edge arrival times and/or clock period at the same flip-flop in different clock cycles.

It will be clear by now, that both clock skew and clock jitter may have serious impact on the overall timing and functional behaviour of an IC. This section is only meant to review some of the most potential problems related to the clock, which is the most important signal on a synchronous chip and must be handled with care. More information on clock distribution, clock skew and clock jitter can be found in [2] and [3], respectively.

Other timing problems

Particularly in low-power CMOS ICs, some logic blocks (or sometimes even the complete chip) may often be inactive for certain periods of time. Such a chip may contain different clock domains, of which the mode of operation (active or standby) is controlled by a gated clock. In many cases then, the main clock is used as input to a logic gate which performs a logic operation on the clock signal (gated clock). Figure 9.5 shows an example:

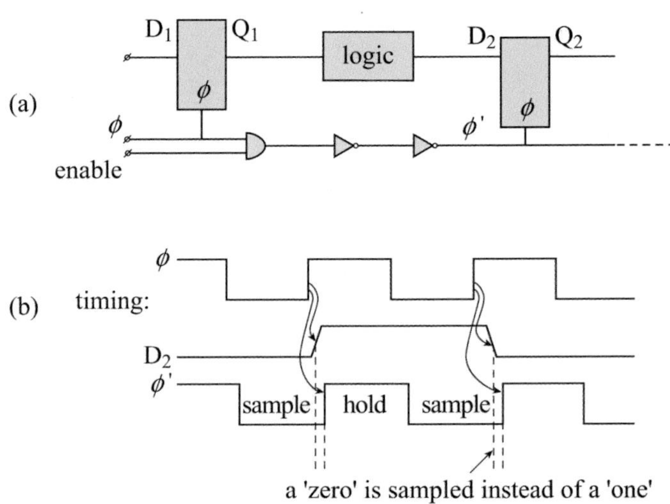

Figure 9.5: *(a) Example of a local clock-enabled circuit and (b) The corresponding timing diagram*

When the delay between the clock ϕ and then enabled clock ϕ' is longer than the data delay between the output Q_1 of one flip-flop in a certain core and the input D_2 of the next flip-flop in another core, this "new" data sample will be clocked into this flip-flop by the "old" clock and a race will occur.

Such clock-enabled signals are also often used in the design of memory address decoding circuits and are very critical with respect to timing margins.

Finally, timing problems could also occur when the data delay (caused by the logic and interconnection delay) between two successive latches or flip-flops becomes equal to or larger than one clock period. Figure 9.6

shows an example. When the total propagation time through the logic from Q_1 to D_2 exceeds the clock period, the data at D_2 can arrive after the sample period of flip-flop 2 has been terminated. It will then be sampled in the next clock period, resulting in incorrect output data. Timing simulation to find critical delay paths is therefore a must in CMOS VLSI design and is part of the design flow.

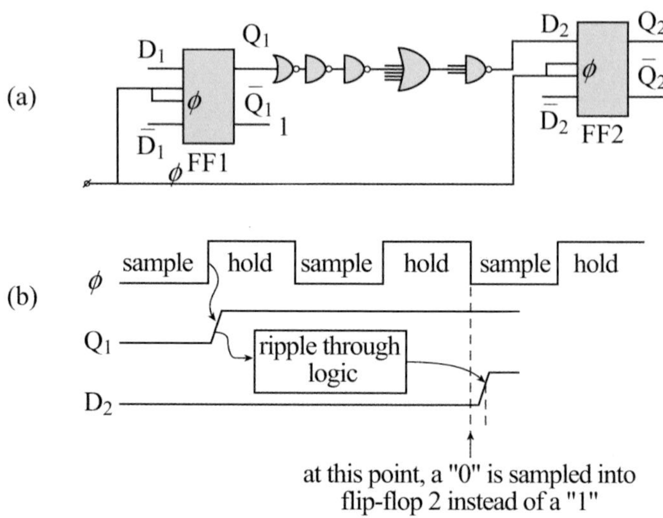

Figure 9.6: *(a) Example in which the data delay exceeds a clock period and (b) Its corresponding timing diagram*

Slack borrowing and time stealing

When a data path uses more than a full clock cycle in a single clock system, this is referred to as *cycle stealing*.

Slack borrowing refers to the case where a logical partition utilizes time left over (slack time) by the previous partition [4]. Important to note is that it can be used without the adjustment of circuitry and/or clock arrival times. This precludes the use of edge-triggered circuitry (dynamic logic and flip-flops). *Time stealing* refers to the case where a logical partition steals a portion of the time allotted to the next partition. This can only be obtained by adjusting the clock arrival time(s). Using one of these concepts to solve timing problems in (ultra) high-speed designs forces the designer to match certain design rule requirements. A well documented list of such design rules can be found in [4].

Source-synchronous timing (clock forwarding)

In a source-synchronous interface, data and clock signal propagation between transmitter and receiver are matched. This technique is currently used in high-performance microprocessors and SDRAM interfaces [5,6], but is also a potential candidate for on-chip chip time-of-flight compensation.

9.2.3 Clock generation and synchronisation in different (clock) domains on a chip

With IC complexities exceeding hundreds of millions of transistors, the total effort required to complete such complex VLSI designs is immense. This stimulates the reuse (IP) of certain logic blocks (cores) and memories. Current heterogeneous systems on chip may not only incorporate many clock domains, but can be built from cores, which are designed at different sites, with different specifications. Because each core has a different clock skew from the core's clock input terminal to the farthest away flip-flop, the clock phase of each core has to be synchronised with the main clock. This subsection discusses the generation of multiple clocks and the synchronisation of clocks in systems that use different cores running at different clock frequencies.

On-chip multiple clock generation

On-chip multiples of the clock can be generated by phase-locked loops (PLLs). Figure 9.7 shows a basic phase-locked loop concept.

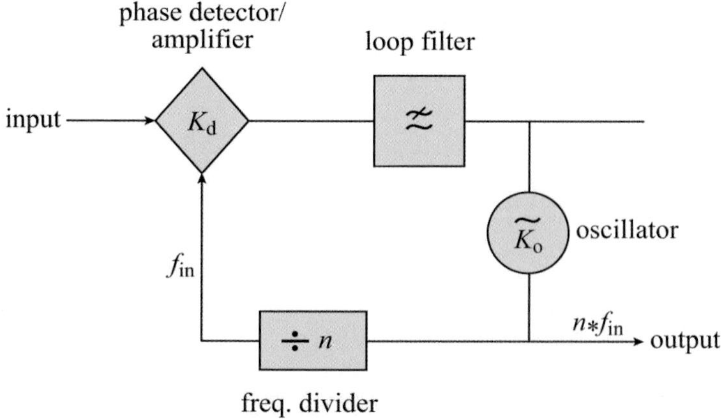

Figure 9.7: *Basic concept for a phase-locked loop*

The Voltage-Controlled Oscillator (VCO) - current-controlled oscillators
(CCOs) are also used - is basically an oscillator whose frequency is determined by an externally applied voltage. This frequency is a multiple
of that of the input. The phase detector is sensitive to differences in
phase of the input and VCO signals. A small shift in the frequency of
the input signal changes the control voltage of the VCO, which then
controls the VCO frequency back to the same value as that on the input signal. Thus, the VCO remains locked to the input. Based on this
principle, a PLL can be used to generate an output frequency which is
a multiple of the input frequency. The output frequency equals n times
the input frequency. A frequency divider $(\div\, n)$ is then used to create a
feedback signal with the same frequency (f_in) as the input signal.

As current complex ICs require many different clock domains, multiple frequencies must be generated on chip. Figure 9.8 shows an example
of a multi-clock generator based on PLL.

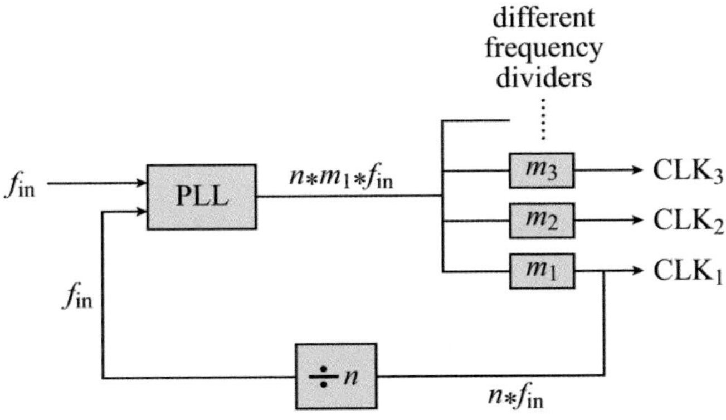

Figure 9.8: *Multi-clock generator, based on PLL*

In this example, the PLL output frequency equals $n \times m_1 \times f_{in}$. Using different divisions (m_i), many different clocks can be generated. The PLL, by nature, automatically locks these clocks in phase with the input.

Clock-phase synchronisation in multiple core environments

Because of differences in the clock arrival times at the flip-flops of different cores, these delays must be compensated for, to allow proper communication between different cores. There are several methods of synchronising the clock phase at the actual flip-flops in each core.

The first method is *adaptive skew control*. In this approach, the clock network of each core (domain) is extensively simulated. The clock skew in each core is then made equal to the worst case clock skew by using a chain of inverters. The length of this inverter chain is then adapted to the required additional delay in the specific core clock path.

The second method uses the PLL concept. The PLL property of locking one signal phase to the phase of another reference signal makes the PLL also suitable for the compensation of clock skew in different cores, see figure 9.9.

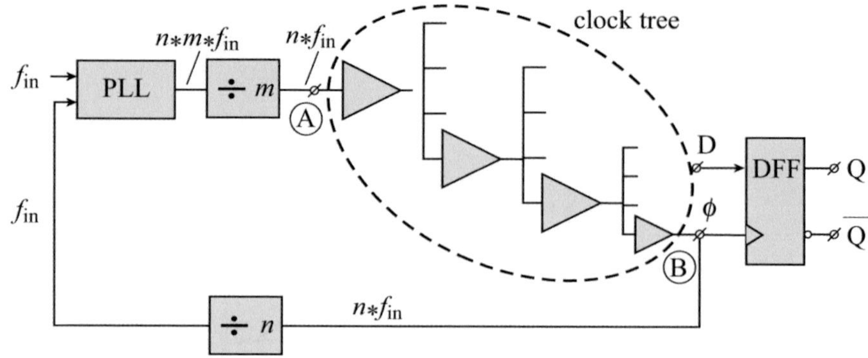

Figure 9.9: *Clock phase synchronisation to compensate for different clock skews inside different cores*

Node A represents the clock terminal of the core, and node B represents the clock terminal of the actual flip-flop in that core. The clock phase at the flip-flop will then be locked to the input reference signal, which is usually the chip's main clock. In this way, the clock tree delay (which might be different in all cores) can be compensated for. Moreover, when the frequency dividers in figure 9.9 are made programmable, then the same PLL can be used in all cores, even when they run at different frequencies.

Sometimes, reusable cores are only available with fixed instances and only in GDSII (layout description) format. In these cases, the clock tree must be thoroughly simulated and a delay chain, which mimics the core's internal clock delay path, replaces the clock tree between nodes A and B (figure 9.9) in the feedback path. The PLL must be placed outside the core. Disadvantages with the use of PLLs are:

- Because of high internal frequencies, PLLs can consume relatively high power.

- PLLs are difficult to start and stop.

- Especially the start-up takes a relatively long time.

- Multiple-clock concepts and the use of PLLs for clock generation and synchronisation makes testing very difficult. During testing, such PLLs must be set to the right mode first before the test procedure can be started.

Finally, to synchronise the clock phases to compensate for the different clock skews in different cores, delay-locked loops (DLLs) can also be used, see figure 9.10.

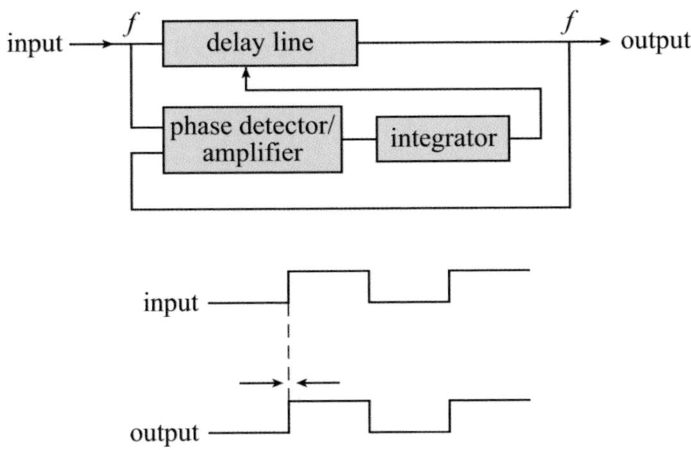

Figure 9.10: *Basic concept of a delay-locked loop and its timing*

The delay of the delay line can be controlled by the output voltage of the integrator. In this concept, the output signal is delayed over one complete clock period with respect to the input. If the delay is less, then the phase detector produces a signal which increases the delay of the delay line, via the integrator. The output signal in such a DLL has the same frequency as the input, and this concept of the DLL cannot be used to multiply the frequency.

Because the VCO or CCO in a PLL generates frequencies that depend on the supply voltage, clock jitter can occur when there is supply noise. Also, the delay in a DLL is susceptible to supply noise. Control of the clock jitter is therefore one of the most important constraints in the design of a PLL and DLL. For the synchronisation of the clock phases of all cores in a heterogeneous chip, each core needs its own PLL (DLL).

9.3   Signal integrity

Signal integrity indicates how well a signal maintains its original shape when propagating through a combination of circuits and interconnections. On-chip effects from different origin may influence this shape. Signals can be influenced by switching of nearby neighbours (cross-talk), by voltage changes on the supply lines (voltage drop and supply noise), by local voltage changes in the substrate (substrate noise), or when the signal node is hit by radioactive or cosmic particles (soft-error). In addition, the speed at which a signal propagates through bus lines is heavily affected by the switching behaviour of neighbouring bus lines.

The next subsections will focus on each of these signal-integrity topics individually and also present ways to limit the noise level or the influence of the potential noise sources that threaten the signal integrity.

9.3.1 Cross-talk and signal propagation

Due to the scaling of the transistors, their density has almost doubled every new technology node for more then four decades already. This forced the metal lines (width and spacing) to be scaled in the same order to be able to connect this increasing number of devices per unit of area. Per unit of area, however, the total length of the interconnections in one metal layer only increased with a factor of 1.4. This means that additional metal layers were needed to allow a high-density connection of all logic gates. The metal layers are also used to supply the current from the top metal layer all the way down to the individual devices. As will be discussed in the subsection on electro-migration, the current density also increased with a factor of 1.4 every new technology node, meaning that the thickness of the metal layers could not be scaled at the same pace as the width and spacing. Consequently the mutual capacitance between neighbouring signal lines has dramatically increased.

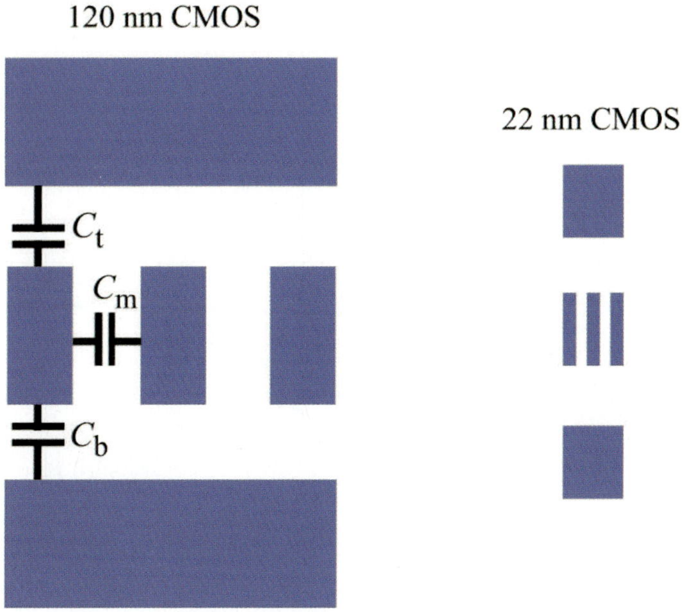

Figure 9.11: *Expected scaling of metal track width and spacing*

Figure 9.11 shows two cross sections of three parallel metal lines: one in a 120 nm CMOS technology and the other one in a 35 nm process. It clearly shows that the bottom (C_b) and top capacitances (C_t) reduce while the mutual capacitances (C_m) increase. This increase in mutual capacitance has dramatic effects on the performance and robustness of integrated circuits. The first one is the growing interference between two neighbouring interconnect lines, which is usually referred to as *crosstalk*. The second one is the growing signal *propagation delay* across the interconnect because of its increasing RC times. Third, the increased interconnect capacitances also affect the overall IC's power consumption. We'll discuss each one of these effects in more detail now. Figure 9.12 depicts the trend in the cross talk over several technology nodes.

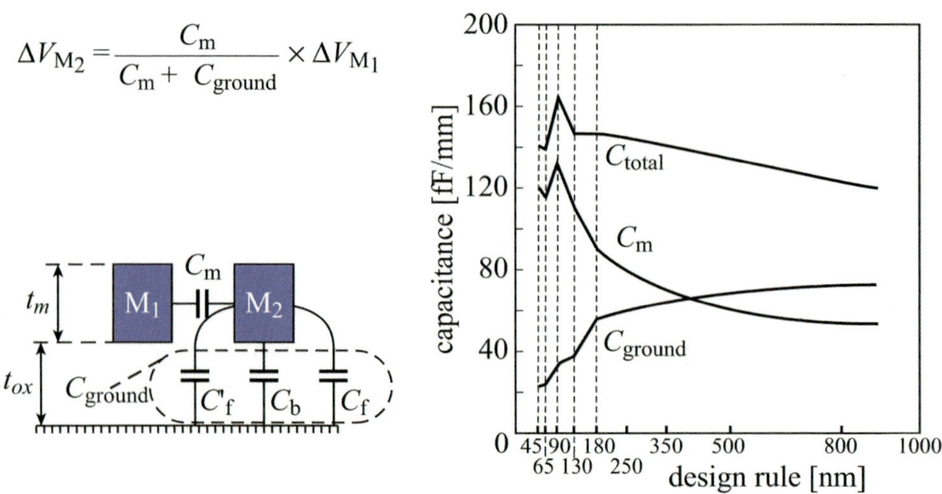

Figure 9.12: *Interconnect capacitances across various technology nodes*

The used model refers to two minimum-spaced interconnect wires in the same metal layer. A signal swing ΔV_{M1} on metal track M1, causes a noise pulse ΔV_{M2} on a floating metal track M2, as defined by:

$$\Delta V_{M2} = \Delta V_{M1} \cdot \frac{C_m}{C_m + C_{ground}} \tag{9.2}$$

Table 9.1 shows the capacitance values for various technology nodes. The bottom line in this table presents the amount that one signal propagates into the other one through cross-talk. For the 65 nm node this means that 84% of the switching signal propagates into its floating neighbours. Because of this, all floating lines (e.g., precharged bit lines in a memory and tri-state buses), are very susceptible to cross-talk noise.

Table 9.1: Capacitance values for second metal layer in different CMOS technologies

Node	180 nm	130 nm	90 nm	65 nm	45 nm
C_m	89 fF	110 fF	132 fF	115 fF	120 fF
C_{ground}	58 fF	36 fF	32 fF	21 fF	18 fF
C_{total}	147 fF	146 fF	164 fF	136 fF	138 fF
ΔV_{M2}	$0.6 \cdot \Delta V_{M1}$	$0.75 \cdot \Delta V_{M1}$	$0.8 \cdot \Delta V_{M1}$	$0.84 \cdot \Delta V_{M1}$	$0.86 \cdot \Delta V_{M1}$

Even non-floating (driven) lines in digital cores are becoming increasingly susceptible to cross-talk causing spurious voltage spikes in the interconnect wires. Traditional design flows only deal with top level cross-talk analysis in the back-end part, to repair the violations with manual effort, after the chip layout is completed. Because timing and cross-talk are closely related, they need to be executed concurrently with the place-and-route tools. The introduction of multi-V_{dd} and multi-V_T pose a challenge for the physical synthesis and verification tools because both design parameters affect timing and signal integrity.

In memory design, scaling poses other challenges to maintain design robustness. The layout of a static random-access memory (SRAM), for example, includes many parallel bit lines and word lines at minimum spacing in different metal layers. It is clear that these will represent many parasitic capacitances with an increasing contribution of mutual capacitances between the various contacts and vias (pillars) (figure 11.4 in chapter 11).

Memories in nanometer technologies therefore require very accurate 3-D extraction tools in order to prevent that the silicon will, unexpectedly, run much slower than derived from circuit simulations.

Next to the cross-talk between metal wires, the signal propagation across metal wires is also heavily affected by scaling. In a 32-bit bus, for example, most internal bus lines (victims) are embedded between two minimum-spaced neighbours (aggressors) (figure 9.13).

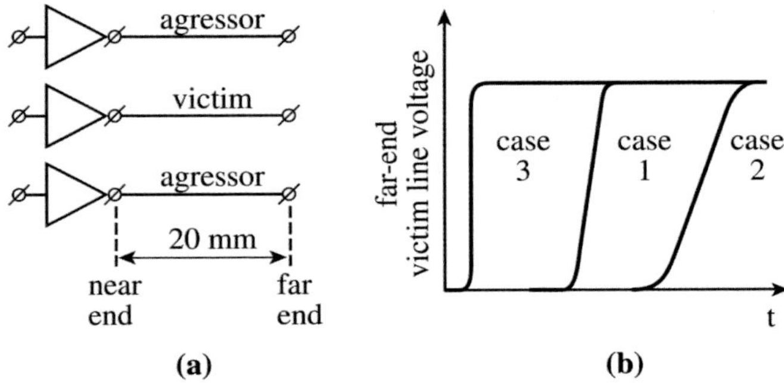

Figure 9.13: *Model for three neighbouring metal lines of an on-chip bus*

The switching behaviour of both aggressors with respect to the victim causes a large dynamic range in signal propagation across the victim line. In case both aggressors switch opposite to the victim (case 2), the signal propagation across the victim lasts about sixteen times longer than in case the aggressors and victim all switch in the same direction (case 3). Figure 9.14 shows this effect for 20 mm long bus lines in a 180 nm CMOS technology.

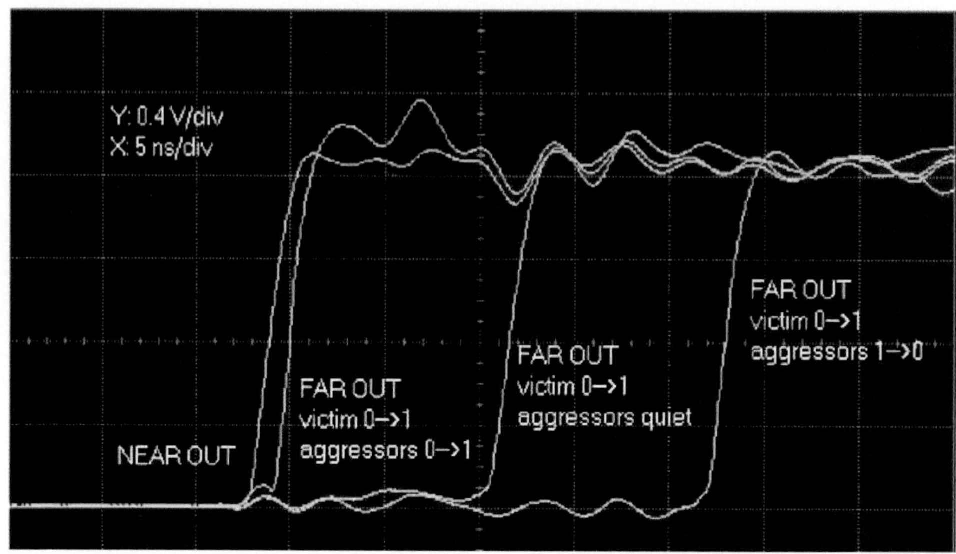

Figure 9.14: *Signal propagation across a 20 mm bus line in 180 nm CMOS technology in relation with the switching activity of both of its neighbours (at minimum distance)*

Figure 9.15 plots the increasing *propagation delay* (in nano-seconds) with the technology node for a 20 mm long bus line, embedded between two quiet (non-switching) aggressors.

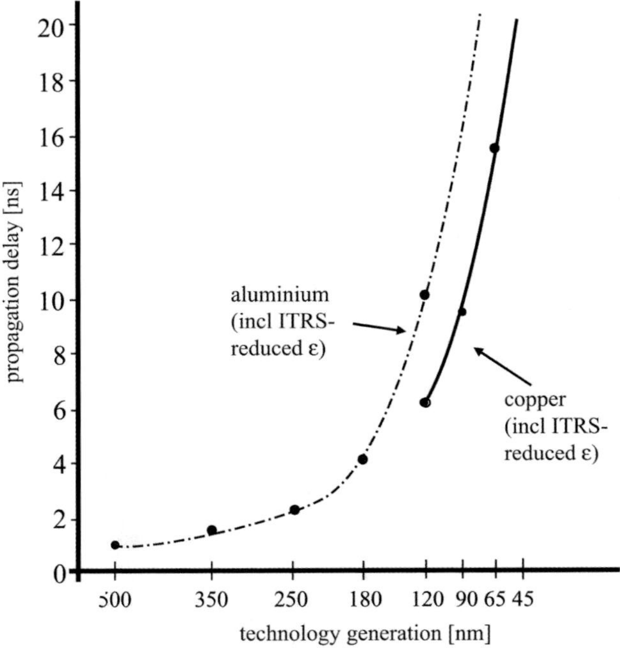

Figure 9.15: *Propagation delay versus technology node in case aggressors are quiet*

Although the introduction of copper with the 120 nm node shows some relief in the increase of the propagation delay, it only helped for about one technology node. This means that in the 120 nm node, with an aluminium backend, the interconnect propagation delay would have reached the same order of magnitude as the 90 nm node with a copper backend. The diagram also shows that the propagation delay will further increase. This requires different design architectures, in which the high-speed signals are kept local. Such architectures must allow latency in the global communication or communicate these global signals asynchronously (i.e., *islands of synchronicity; globally synchronous, locally asynchronous(GALS)*).

In the preceding discussions self- and mutual inductances where not taken into account. However, with the advances in speed and clock frequencies, the influence of these inductances becomes increasingly pronounced. The resistances of the metal lines in most of today's ICs still exceed the values of *inductance* by more than one order of magnitude. For one reason this is due to the fact that the *resistance* increases every technology node. The second reason is that the inductance is linearly proportional to the frequency (figure 9.16; [7]).

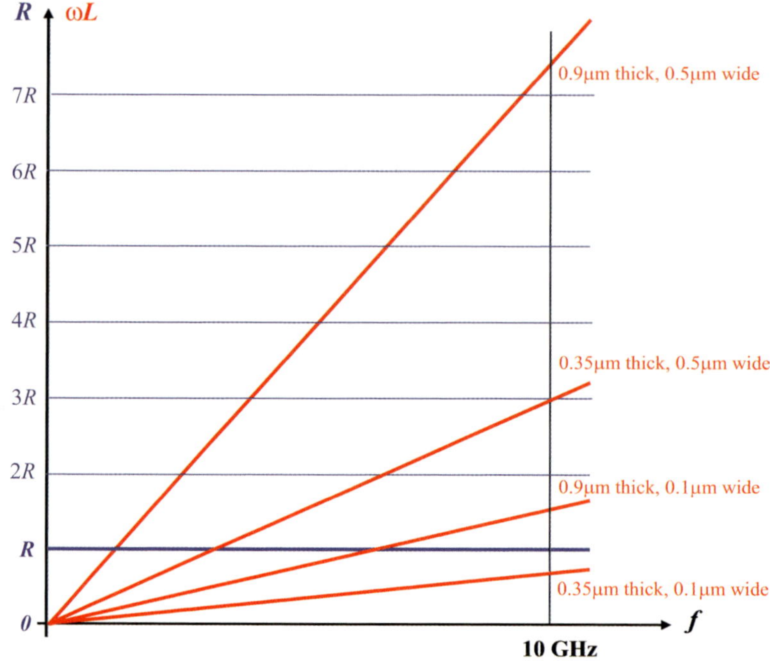

Figure 9.16: *Influence of the frequency on the inductance of 500 μm long on-chip metal lines*

At a frequency of 10 GHz, the inductance contribution (of a 350 nm thick signal line in a 65 nm CMOS process) to the total impedance of a metal wire reaches about two third of the resistance contribution. This means that we need to change from an *RC* interconnect model to an *RLC* model for designs that exceed 1 GHz (at this frequency the inductance value is about 10% of the resistance value and can thus no longer be neglected). Generally, there are two effects determining the difference in accuracy between an *RC* and an *RLC* model: the damping factor and the ratio between the input signal rise time and the signal propagation speed across the line. Therefore, even in designs that do not yet reach 1 GHz, the wider metal lines, with lower resistance (e.g., in clock distribution networks and upper metal layers) can exhibit significant inductive effects. The lines in figure 9.16 represent the relative ωL values with respect to their resistance values. So, the ωL value of a 0.9 μm thick and 0.5 μm wide metal track reaches the level of its resistance value already at a frequency of close to 1.5 GHz.

Because the rise times of signals on interconnect lines are also reducing with the advance of the technologies, RLC models need to be included in CAD tools soon, in order to avoid inaccurate performance predictions or underestimate signal integrity effects, which may also reduce the operating margins.

Finally a number of methods, depending on the requirements of the application, exist to reduce cross-talk and/or improve signal propagation. We will summarise them here, without discussing them in more detail:

- use fat wires to reduce track resistance

- increase spacing to reduce mutual capacitance

- use shielding between the individual bus lines

- use staggered repeaters to compensate noise

- use tools which can detect, replace and reroute critical nodes

- use current sensing or differential signalling for improved speed and noise compensation

Many of these solutions are described in various publications, which can easily be found on the internet.

9.3.2   Power integrity, supply an ground bounce

Every new technology node allows us to almost double the number of transistors. Next to this, the bus widths have also gradually grown over the last couple of decades: from 4-bit in the mid 1970s to 64-bit, or even 128-bit, today. The interface to a 1 Gb DDR-1 SDRAM, for instance, requires to communicate 64 data bits, about 30 address and control bits, totally adding up to some 96 parallel bits. In addition, due to the increased speed requirements, more flip-flops/pipelines are used within the logic blocks. All these individual trends contribute to a dramatic increase of simultaneously *switching activity* in an IC causing huge currents (i) and current peaks ($\mathrm{d}i$). These currents cause dynamic voltage drop across the resistance (R) of on-chip supply network, while the current peaks cause relatively large voltage dips and peaks across the self-inductances (L) in the supply path. As is discussed in the previous subsection, most of the *self-inductance* is still in the bond wires and the package leads, instead of in the on-chip metal supply lines.

Another trend that keeps pace with technology advances is the reduction in switching times (dt) of the logic gates and driver circuits. The combination of these two trends leads to a dramatic increase of di/dt, which term is mainly responsible for the *supply* and *ground bounce* generated on chip.

In total we can summarise the *dynamic voltage drop* by:

$$\Delta V = i \cdot R + L \cdot \frac{di}{dt} \tag{9.3}$$

The impact of this voltage drop on the behaviour of the chip is twofold. First, the average supply voltage throughout the complete clock period determines the speed of a circuit. Let V_{dd} be the nominal supply voltage of a chip. Most commonly this means that the chip is specified to operate within a 5% to 10% margin in this supply voltage. In case of a 1.2 V 65 nm CMOS design, this means that it should operate between 1.1 V and 1.3 V. So, in the application, the IC should operate correctly, even at 1.1 V. Because the logic synthesis is done using the gate delays specified at this lower voltage, an additional IR-drop within the chip could be disastrous for proper functionality. In other words, the designer should keep the total average voltage drop within stringent limits to assure the circuit operates according to the required frequency spec. It is commonly accepted that this *static IR-drop* is limited to just a small percentage of the supply voltage (around 1%). Second, ΔV introduces noise into the supply lines of the IC. The current is supplied through the V_{dd} supply lines and leaves the circuit through the V_{ss} ground lines. When the impedances of the supply and ground lines are identical, which is most commonly the case, the introduced bounce on the respective lines show complementary behaviour and are identical in level. The total inductance (L) consists of on-chip contributions of the supply and ground networks and off-chip contributions of the bond wires, package leads and board wires. Usually the damping effect of high resistive narrow signal wires reduces the effect of on-chip inductive coupling. To reduce the dynamic iR-drop in the above expression, however, the supply and ground networks require wide metal tracks in the upper metal layers with very low sheet resistance. Particularly for designs operating at GHz frequencies, inductance in IC interconnects is therefore becoming increasingly significant.

The supply noise can be reduced in several ways. When using n

supply pads for the supply connection, which are more or less homogeneously distributed across the IC periphery, the self-inductance will reduce to L/n. Both the use of a low-resistive supply network and multiple supply pads, however, contribute to a reduction of the overall impedance of the supply network. Because the bond wires, package leads and board wiring, all act as antennae, the resulting increase of the current peaks $(\mathrm{d}i/\mathrm{d}t)$ lead to a dramatic rise of interference with neighbouring ICs on the board and may cause EMC problems in the system. Therefore it is also required to keep the peak currents local within the different cores on the IC. In other words, it is necessary to lower the global $\mathrm{d}i/\mathrm{d}t$ contribution in the preceding equation as well. The use of staggered driver turn-on, to limit the amount of simultaneous switching activity, as well as encouraging the use of "slow" clock transients will directly contribute to a lower $\mathrm{d}i/\mathrm{d}t$. Another measure to limit the global $\mathrm{d}i/\mathrm{d}t$ is the use of decoupling capacitors within each of the different cores.

Figure 9.17 depicts two implementations of *decoupling capacitor* cells. Figure 9.17.a is a complementary set of transistors connected as an nMOS and pMOS capacitor, directly between V_{dd} and V_{ss}. Because the supply voltage in this cell is directly across the thin gate oxides, this cell needs some additional resistances either in the gate connection or in the source/drain connections, to limit the chance for ESD damage. Figure 9.17.b is a *tie-off* cell used as decoupling capacitor.

In several applications a tie-off cell supplies dummy V_{dd}' and V_{ss}' potentials to inputs of circuits, which, for reasons of electro-static discharge (ESD), are not allowed to be directly connected to the V_{dd} and V_{ss} rails. The channel resistances R_n and R_p (figure 9.17.c) of the nMOSt and pMOSt, respectively, serve as additional ESD protection for the transistor gates connected to the V_{ss}' and V_{dd}'. This advantage can also be exploited when we use this cell only as a capacitor cell between V_{dd} and V_{ss}, without using the dummy V_{dd}' and V_{ss}' terminals. When a supply dip occurs, the charge stored on the gate capacitance C_n (C_p) of the nMOSt (pMOSt) must be supplied to the V_{dd} (V_{ss}) in a relatively short time, which puts some constraints to the value of R_n (R_p). Therefore, *decoupling capacitor* cell b shows a better ESD behaviour compared to cell a.

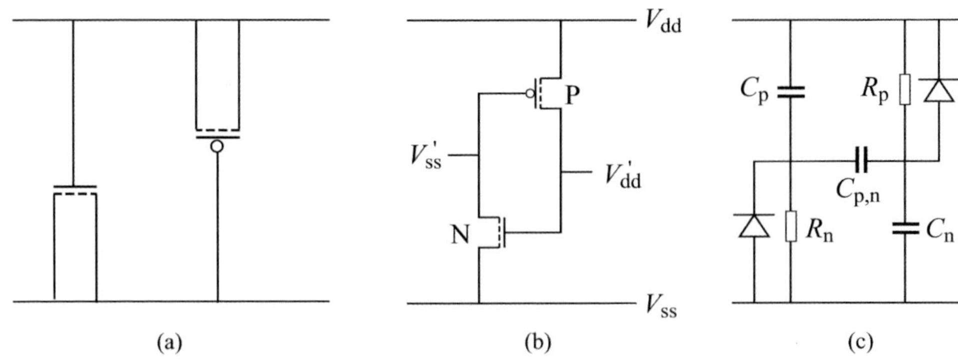

Figure 9.17: *Normal decoupling capacitor (a), tie-off cell decoupling capacitor (b), and equivalent circuit (c)*

These decoupling capacitors are charged during steady state, e.g., at the end of the clock period when the total switching activity has almost or completely come to an end. The additional charge, stored in these capacitors is then redistributed to the supply network during moments of intense switching, particularly at the clock transient that initiates the next signal propagation through the logic paths. These decoupling capacitor cells are designed as standard cells and are usually available in different sizes. The amount of decoupling capacitance that needs to be added in each core depends on the number of flip-flops in it and on the switching activity of its logic. The switching activity α is defined as the average number of gates that switch during a clock cycle. When a logic core has an activity factor of $\alpha = 1/3$, it means that the average gate switches one out of every three clock periods. Different algorithms require different logic implementations, which show different switching activities. It is known that average telecom and audio algorithms show less switching activity ($0.05 < \alpha < 0.15$) than an average video algorithm ($0.1 < \alpha < 0.4$), for example. These activity factor values are only meant as illustration, rather then an exact range for an application. As an example, the total additional decoupling capacitance in a logic block, performing a video algorithm, running at 1 GHz in a 65 nm CMOS core in a digital chip, may occupy about 10 to 20 percent of its total area. When the standard-cell block utilisation is less than 85%, most of this decoupling capacitance fits within the empty locations inside a standard-cell core. In certain mixed analog/digital ICs, however, this amount could grow dramatically, since the noise in these ICs is much more restricted by the sensitivity of the analog circuits.

Because of further scaling, i, R and $\mathrm{d}i$ (in equation 9.3) will increase, while $\mathrm{d}t$ will just do the opposite, potentially requiring an increasing number of design measures to limit the dynamic as well as the static voltage drop across the power network. Therefore, the power(-grid) integrity must be sufficiently guaranteed in order to enable correct chip behaviour. This power integrity must also be supported by the tools that can analyse (dynamic) voltage drop early in the design flow and will reduce the cost of chip debug and failure analysis and prevent the need for a respin. Static voltage-drop analysis targets at the average current (I) and on the resistive nature of the power supply network. Nanometer CMOS designs, however, also require a dynamic voltage-drop analysis, which focuses on the impact of instantaneous currents (i) and current peaks ($\mathrm{d}i$), early in the design cycle to avoid dangerous compromises between power grid design on the one hand, and power integrity, noise and timing requirements on the other.

9.3.3 Substrate bounce

Substrate bounce is closely related to the ground bounce. On a mixed analog/digital IC, usually the digital circuits are responsible for most of this bounce, while the analog and RF circuits are most sensitive to it (figure 9.18). The substrate bounce has several contributors. The transistor substrate current injection is responsible for only a few mV. Junction and interconnect capacitances account for several tens of mV. The highest noise levels (several hundred mV), however, are introduced through the current peaks in the supply network, also causing the previously discussed supply noise.

Figure 9.18: *Symbolic representation of a mixed analog/digital IC*

In most CMOS circuits it is common practice to connect the substrate to the V_{ss} rail, meaning that the ground bounce that is generated in the V_{ss} rail is directly coupled into the substrate. This is even a bigger problem, when the chip is realised on epitaxial wafers (see section 3.2.2 and 9.5.5) with a low-ohmic substrate, because it propagates the noise through the substrate to the analog part almost instantaneously and with hardly any loss of amplitude. Because the noise margins reduce with reducing supply voltages, the use of high-ohmic substrates is becoming increasingly important. Triple-well technology allows improved isolation of analog and RF circuits from digital cores. The level of isolation also depends on the frequency of the RF circuits. The use of a silicon-on-insulator (SOI) technology allows even a complete separation of the analog and digital circuits. Several other measures exist to reduce the level of substrate bounce. First, the measures that help reduce the supply and ground bounce, as discussed in the previous subsection, are also beneficial for substrate bounce reduction. Second, a physical separation of the core and I/O supply nets from the analog supply net, according to figure 9.19, prevents the relatively large noise introduced in these nets to propagate directly into the analog net [8].

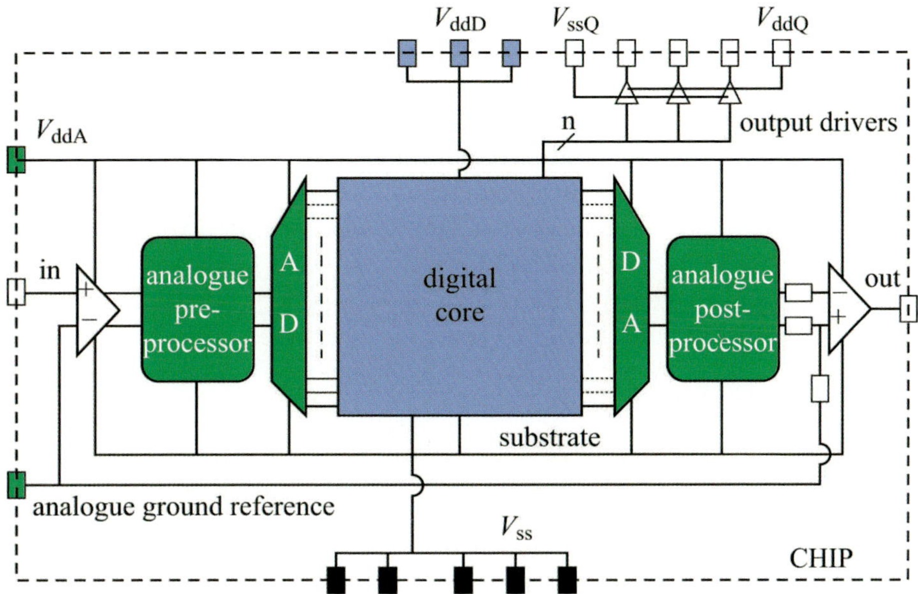

Figure 9.19: *Proposed supply connections in a mixed analog/digital IC*

The figure also illustrates that most digital and analog circuits share the same ground (V_{ss}), because it also serves as a reference for the communicated signals. Usually, the impedance of the internal and external V_{dd} and V_{ss} networks are almost symmetrical, meaning that they have equal widths and the same number of bonding pads. An increase in the impedance of the V_{dd} network with respect to the impedance of the V_{ss} network would increase the bounce in the V_{dd} supply network, while reducing it in the V_{ss} ground network. Because the analog and digital V_{dd}'s where separated anyway, this additional digital supply bounce is not coupled into the analog V_{dd}. Due to the fact that the analog and digital circuits share the same ground, the lower V_{ss} ground bounce also reduces the substrate bounce. Therefore, to increase the margins and robustness of mixed analog/digital ICs, it may be advantageous to dedicate more supply pads to V_{ss} and less to the V_{dd}. Advanced CMOS technologies offer an additional deep-nWELL (*triple-well technology*, which allows to isolate the pWELLs in digital cores from those in analogue cores. These cores may then also have separated V_{ss} pads.

Finally, particularly in the case of high-ohmic substrates, circuits with the highest switching activities and driving strengths, e.g., I/O pads, clock drivers, and drivers with a high fan-out, must be located as far away from the analog circuits as possible.

9.3.4   EMC

The problem of supply and ground bounce caused by large current changes is not restricted to on-chip circuits only. High current peaks may also introduce large electromagnetic disturbances on a printed-circuit board (PCB) because of the electromotive force and threatens the off-chip signal integrity. Because bonding pads, package, and board wiring act as antennae, they can "send" or "receive" an *electromagnetic pulse (EMP)*, which can dramatically affect the operation of neighbouring electronic circuits and systems [9].

When realising *electromagnetic compatible (EMC)* circuits and systems, the potential occurrence of EMP's must be prevented. The use of only one or a few pins for supply and ground connections of complex high-performance ICs is one source of EMC problems. Even the location of these pins is very important with respect to the total value of the self-inductance. The use of three neighbouring pins for V_{dd}, for instance, results in an electromagnetic noise pulse that is twice as large as when these supply pins were equally divided over the package. The best solution is to distribute the power and ground pins equally over the package in a sequence such as V_{dd}, V_{ss}, V_{dd} and V_{ss}. Bidirectional currents compensate each other's electro-magnetic fields in the same way as twisted pairs do in cables. Another source of EMC problems is formed by the outputs. They can be many (about 96 I/O pins for the address, data and control signals in a 1 Gb DDR SDRAM interface), contain relatively large drivers with high current capabilities and often operate at higher voltages than the cores. Actually, each output requires a low-inductance current return path, such that the best position for an output is right between a pair of V_{dd} and V_{ss} pads. This results in the smallest electro-magnetic disturbances at PCB level and reduces the supply noise at chip level. Because this is not very realistic in many designs, however, more outputs will be placed between one pair of supply pads. The limitation of this number is the designers' responsibility (simulation!) or defined by the characteristics of the library I/O or output cell. In this respect, the maximum number of *simultaneously switching outputs (SSOs)* per supply and ground pad is a combination of the characteristics of the output driver and the package. In addition, the di/dt, generated by these outputs, must also be limited to what is really needed to fulfill the timing requirements. Finally, all measures that reduce on-chip supply

542

and ground bounce, also improve the electromagnetic compatibility of
the chip and result in a more robust and reliable operation.

9.3.5 Soft errors

Because of the continuous shrinking of devices on an IC, the involved
charges on the circuit nodes have scaled dramatically. Ionising particles, independent of their origin, do have an increasing impact on the
behaviour of these shrinking devices. At sea-level, several categories of
particles can be distinguished, which all generate free electron-hole pairs
in the semiconductor bulk material [10]:

- *alpha particles*, originating from radio-active impurities (mainly
  uranium and thorium) in materials; these materials can be anything in the vicinity of the chip: solder, package or even some
  of the materials used in the production process of an IC (metals or dielectrics). These so-called α-particles can create a lot of
  electron-hole pairs along their track.

- *high-energy cosmic particles*, particularly neutrons, can even frac-
  ture a silicon nucleus. The resulting fragments cause the liberation
  of large numbers of electron-hole pairs.

- *low-energy cosmic neutrons*, interacting with boron-10 (^{10}B) nu-
  clei. When a ^{10}B nucleus breaks apart, an α-particle and a lithium
  nucleus are emitted, which are both capable of generating soft errors. This is only an issue if BPSG (chapter 3) is used in the
  processing, because it contains relatively large amounts of boron
  atoms.

In all cases, the generated electrons and holes can be collected by reversed-
biased pn-junctions in the circuit. This charge collection causes a current
pulse that can discharge capacitors (in dynamic logic and DRAMs) and
can flip states in both dynamic and static storage circuits (memories,
latches and flip-flops).

Figure 9.20 shows a cross section of an nMOS transistor which is part
of an SRAM cell. Assume that the drain of this transistor is connected
to the logic one (V_{dd}) side of the memory cell. An α-particle or neutron
creates electrons and holes in the silicon. Electrons are attracted to the
positive node (V_{dd}). If the number of collected electrons is large, it may
discharge the node so much that the SRAM cell can flip its state.

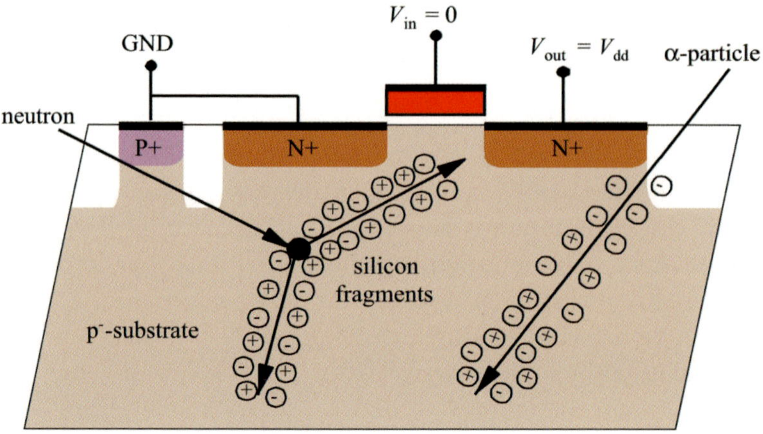

Figure 9.20: *Cross section of an nMOS transistor hit by an α-particle or neutron*

The resulting incorrect state is called a *soft error*, also known as a *single event upset (SEU)* because the flipped state has not caused permanent damage to any of the circuit devices. The rate at which a soft error occurs is called *soft-error rate (SER)* and is expressed in *failures in time (FIT)*. This rate reflects the number of expected failures in 1 billion operating hours. α-particles usually cause single-bit errors, while neutrons may cause both single- and multi-bit errors. The probability of a multi-bit upset is increasing with technology scaling. The total charge of a node is an important criterion for the probability of flipping its state after being hit by an ionising particle. The node charge Q_{node} representing one bit or one flip-flop equals:

$$Q_{\text{node}} = C_{\text{node}} \cdot V_{\text{dd}} \qquad (9.4)$$

where V_{dd} equals the supply voltage and C_{node} the total capacitance of the node.

In static storage cells (SRAM cell, latch or flip-flop), the critical charge is not only dependent on the capacitance of the nodes in these cells, but also on the drive strengths of the transistors that try to maintain the logic state. In this case, the critical charge varies with the width of the transient current pulse induced by a particle hit.

As a first approximation, the critical charge (Q_{crit}) needed to generate a soft error at a specific circuit node is given by:

$$Q_{\text{crit}} = Q_{\text{node}} + I_{\text{drive}} \cdot W_{\text{pulse}} \qquad (9.5)$$

where I_{drive} represents the transistor current needed for keeping the state and W_{pulse} is the width of the particle-induced current pulse.

Also as a first approximation, the SER of a single bit (or cell) can be represented by:

$$\mathrm{SER} \propto A_{\mathrm{diff}} \cdot e^{(-Q_{\mathrm{crit}}/<Q_{\mathrm{coll}}>)} \tag{9.6}$$

where A_{diff} represents the sensitive diffusion area and $<Q_{\mathrm{coll}}>$ the average collected charge. All three parameters in the above expression (A_{diff}, Q_{crit}, and $<Q_{\mathrm{coll}}>$) reduce with technology scaling. As a net result, the SER per Mb of SRAM is roughly constant. Because memory capacity (Mb per chip) is increasing with new technology generations, the SER per chip increases as well. Because of the shrinking memory cell sizes and distances, there is an increased probability that a high-energy particle hit may cause multiple bits to change state. The number of so-called multiple-event upsets (MEU) is therefore increasing [12].

As already mentioned before, latches and flip-flops are also sensitive to soft errors. It turns out that their SER per cell is also fairly constant. This also leads to an increasing SER per chip, because the number of cells (flip-flops and/or latches) per chip increases with new technology generations. The average soft-error rate for both SRAM cells and flip-flops in the $0.18\,\mu$m CMOS technology node is about equal and close to 1000 FIT/Mb. Table 9.2 shows the relative trend in soft-error rates for SRAMs and flip-flops (latches). Generally, an IC contains much more SRAM cells than flip-flops. Therefore, when the SRAMs are not protected against soft errors, they will dominate the SER per chip. However, when the SRAMs are protected with ECC, as discussed below, flip-flops (latches) may dominate it. Reference [12] more or less confirms the above described trend in SER sensitivity, although it is more optimistic regarding the trend in SER/bit for SRAMs, in that this rate even reduces every new technology node. It therefore claims that the SER at chip level is not expected to dramatically increase, but it also states that the SER remains an important point of focus for the quality of future semiconductor products.

Also logic gates become increasingly prone to soft errors, but their contribution to the SER per chip is usually (much) less than 10-15% of the total SER.

Table 9.2: *Relative trend in soft-error rate for different circuits (Source: Marc Derby (iRoC Technologies), IOLTS 2007) [11]*

	Technology node					
	180nm	130nm	90nm	65nm	45nm	Comment
Integration level (Mtransistor/device)	48	97	193	386	773	Source: ITRS 2005
Non-protected memory						
Memory integration (Mbit/device)	6.4	12.9	25.7	51.5	103.1	Hyp: 80% of T → 6-T mem cells
Memory SER per Mbit (norm. units)	1	0.78	0.91	1.03	1.10	Source: iRoC Technologies
Memory SER at chip level (norm. units)	1	1.58	3.66	8.28	17.71	**180nm to 65nm: ´ 8.3**
Non-protected sequential logic						
Flip-flop integration level (Mbit/device)	0.096	0.194	0.386	0.772	1.546	Hyp: 6% of T → 30-T mem cells
Flip-flop SER per Mbit (norm. units)	1	0.93	0.73	0.69	0.63	Source: iRoC Technologies
Flip-flop SER at chip level (norm. units)	1	1.88	2.94	5.55	10.15	**180nm to 65nm: ´ 5.6**

** SER numbers apply to high-energy-neutron and alpha-particle effects for an arbitrary device/technology (i.e., numbers do vary depending on source)*

** SER values given here include both single-bit upset (SBU) and multiple-cell upset (MCU) events; the SER values denote the bit-flip rate*

** 45nm SER values are estimates*

Source: Marc Derby (iRoC Technologies), IOLTS 2007

DRAM vendors keep the storage capacitance per memory cell at a level between 25 and 50 fF. As a result, the DRAM SER per bit has decreased dramatically because of the reduction of the sensitive diffusion are, which reduces the $<Q_{\mathrm{coll}}>$. However, at system level, the SER is almost constant because of the increasing memory capacity needs per system.

Finally, a high-energy particle hit may also introduce a *single event latch-up (SEL)*. Once such a hit creates sufficient charge it may trigger latch-up. This may be limited to a local region, but it may also propagate to larger parts of the chip, where the total induced current may become so high that it can even destroy the device. To recover the device operation after a non-destructive SEL condition, the power supply needs to be temporarily removed.

There are several measures to prevent or limit the occurrence of soft errors:

- Careful selection of purified materials (package, solder, chip manufacture, etc.) with low α-emission rates.

- Usage of a shielding layer, most commonly polyimide. This layer must be sufficiently thick (20 μm) in order to achieve about three orders of magnitude reduction of the soft-error rate (SER) caused by α-particles. This measure does not help to reduce the soft-error rate caused by the high-energy cosmic particles because they can pas through even half a meter of concrete.

- SER hardening of the circuits by changing memory cells, latches and flip-flops.

- Usage of process options or alternative technologies. Silicon-on insulator (SOI) circuits exhibit even an order of magnitude reduction of the SER because charges that are generated along a particle track in the main part of the wafer are physically isolated from the circuits in the top layer.

- Inclusion of error-detection/correction (ECC) circuits or making the designs fault tolerant. ECC is a very effective method to protect SRAMs and DRAMs against soft errors. In combination with design techniques such as physical interleaving of bit cells (scrambling), most multi-bit upsets may be considered as a multiple of individual single-bit upsets in multiple correction words and the ECC needs only to deal only with single-bit errors.

Currently, a lot of effort is being put into the evaluation and prevention of soft-errors, particularly in systems containing large amounts of densely packed memories [13].

9.3.6 Signal integrity summary and trends

From the previous subsections it can be seen that all noise components increase because of scaling and integrating more devices onto the same die area. At the same time that noise levels in digital CMOS ICs increase with scaling, the noise margins reduce due to reducing supply voltages (figure 9.21). Because they deal with large current peaks, high-performance ICs such as the PowerPC (IBM, Motorola), the Pentium (Intel) and the α-chip (DEC/Compaq/HP), have faced signal-integrity effects already in the early 1990s. The average application-specific integrated circuit (ASIC), however, consumes more than a factor of ten less power (and current) and therefore faces these problems a couple of technology generations later in time.

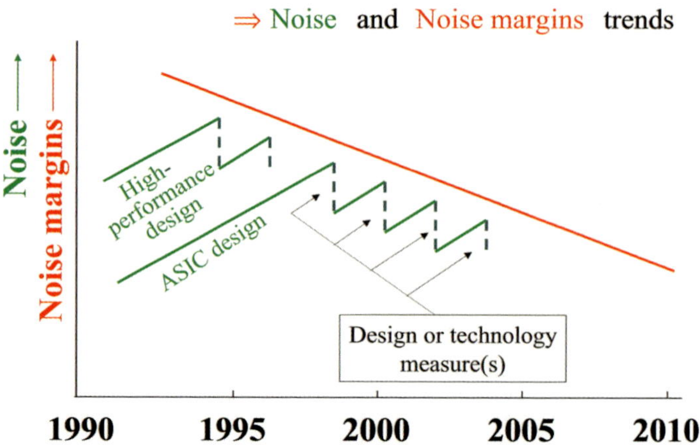

Figure 9.21: *Noise and noise margin trends over the past and current decade*

When a certain noise level has reached a maximum limit, a design or technology measure is required to reduce the noise level.
Examples of technology measures are:

- the use of copper instead of aluminium allowed a reduction of the metal height, thereby reducing the cross-talk (see section 9.3.1)

- the use of low-ϵ dielectrics in the back-end of the technology has the same effect

Examples of design measures are:

- the increase of space between long signal lines (buses) also reduces the cross-talk

- the use of on-chip decoupling capacitors reduces supply, ground, and substrate bounce

Whatever technology or design measure is taken, it only fulfills the requirements in that technology node. The next technology node offers twice the number of transistors, which can intrinsically switch faster. This results in a huge increase in the noise levels. In addition, the noise margin has reduced. Therefore, in every new technology node, it becomes more difficult to limit the noise within shrinking boundaries. In other words, the line (in figure 9.21) that represents the increasing noise must be bended in the direction of the line that represents the reducing noise margins. This can only be obtained by applying more and more design and/or technology measures. In example: in today's high-performance ASIC designs, the decoupling capacitors occupy between 5 to 10% of the total area within a standard-cell block. It is expected that this number will dramatically increase for extremely high-speed designs in the 32 nm node, which means that, by that time, a large part of all transistor equivalents on a chip is needed to support the other part in their functional and storage operations. This is yet another factor that adds up to the already rocketing semiconductor development costs.

Another increasingly important topic is the relation between signal integrity and test. Because noise has the tendency to increase, while noise margins reduce (again figure 9.21), there is not much room left for a reliable operation of an IC. Different operating vectors introduce different local and global switching activities. In many complex ICs, the operation and switching activity during testing are different from the operation and switching activity in the application. As a result, the noise, generated during a test, is different from the noise generated in the application. Because of the reducing noise margins, this increasing discrepancy between "test noise" and "application noise" cause products that were found correct during testing to operate incorrectly in the application. This is because, in many cases, scan tests are performed to verify the IC's functional operation. These tests are mostly performed locally and in many cases at different frequencies causing a lower overall switching activity and less noise than in the application. On the other hand, depending on the design, different scan chain tests may run in parallel, synchronous and at the same frequency, causing much more simultaneous switching and noise than in the application. These ICs may be found to operate incorrect during testing while showing correct functional behaviour in the application. Because of this and because of the lack of access to most internal signals, debugging a System-on-Chip (SoC) has become a very difficult and time-consuming task. In fact, first (and second) silicon debug has become a major component of time-to-market, as it may take up to 50% of designers' time.

A *Signal Integrity Self-Test (SIST)* architecture [14] allows real-time monitoring of different parameters (e.g., temperature, voltage drop, switching activity, supply noise, substrate noise, cross-talk, process parameter spread, clock jitter, and clock duty-cycle) that characterise the quality of operation of an IC, during test and debug or in the application. Moreover, even when first silicon is functionally correct, this SIST architecture allows the monitoring of signal integrity margins, in order to anticipate potential operating failures due to technology parameter variation in future silicon batches.

Since the margins continue to decrease, the additional design for debug measures will not be enough and provisions need therefore to be made in the designs to enhance operation robustness during the test as well as in the application. This poses additional challenges to the design, increases its complexity and also adds up to the total development costs.

9.4 Variability

This section presents a short overview on the increasing number and influence of different components of variability: process parameter spread, electrical variations, environmental variations, etc. The description starts by categorising these variability components by their nature in different ways. Next their influence on the behaviour of analogue and digital circuits is discussed.

9.4.1 Spatial vs. time-based variations

Spatial variations are variations due to the fact that identical devices can and will have a different physical environment, caused by a different orientation, a different metal coverage or other *proximity effects*, such as mechanical stress (e.g., *STI stress*), the position of a well in the vicinity of a transistor (*well-proximity effect*) [15], and/or pattern shape deviations as a result of imperfect lithographic imaging and pattern density variations.

Time-based variations include signal integrity effects, such as crosstalk, supply noise, ground bounce, and iR-drop, but also temperature variations over time, due to variations in workload.

9.4.2 Global vs. local variations

Device parameters can vary from lot to lot, from wafer to wafer, from die to die and from device to device. The first three are usually referred to as global or *inter-die variations*, which are more or less systematic and common to all devices on the same die. A mask misalignment, a smaller processed channel length or a V_T-shift are all examples that cause global variation. Advances in both the lithography and diffusion process have led to a much tighter overall process control, such that in mature processes, *global variations* are currently much better controlled than in the past. Most global variations are a result of systematic process imperfections. Nanometer CMOS technology nodes (e.g., 90 nm CMOS and beyond) show an increasing relation between design and systematic yield. Symmetrical and regular layout styles, such as used in *litho-friendly design* [16], can be applied to make the design less sensitive to these variations.

Variations between devices on the same wafer are usually referred to as local or *intra-die variations* or *mismatch*. Most *local variations* (*random variations*) are caused by stochastic processes during fabrication. Polysilicon gate *line-edge roughness (LER)* and the channel doping statistics are examples of local variations. The resulting device mismatch is particularly a problem in analog circuits, e.g., circuits with a differential transistor pair, clock generating circuits, current mirrors, operational amplifiers, etc. A common design approach is to simulate a circuit with respect to best-case and worst-case process corners (e.g., slow-nMOSt slow-pMOSt corner: snsp) [17]. The diamond in figure 9.23 connects these corners and represents the global variations in the voltages across the devices of figure 9.22. The clouds around every corner represent the random variations. The diagram clearly shows that, for individual devices, these random variations are in the same order of magnitude as the systematic variations.

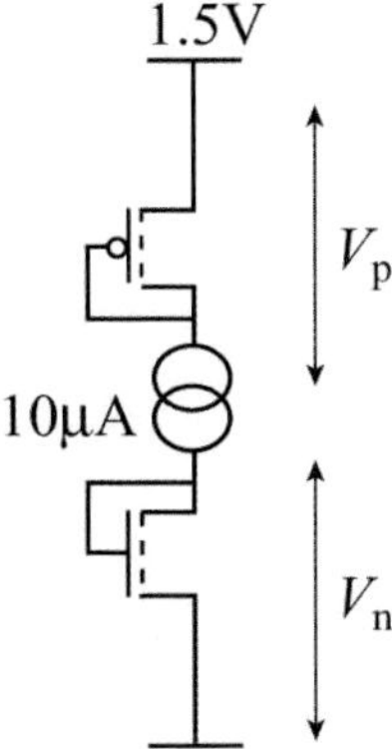

Figure 9.22: *Circuit used for the Monte Carlo simulation results of figure 9.23*

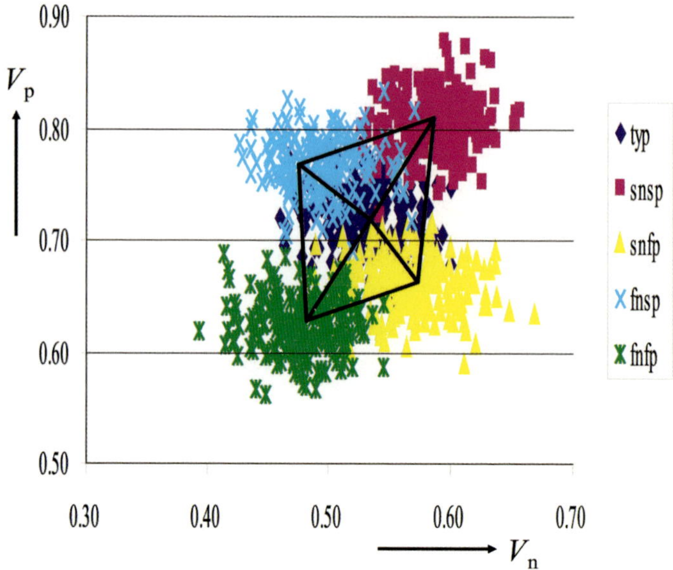

Figure 9.23: *Effect of global and local variations on the voltage across an individual device [18]*

Analog circuit performance is determined by the behaviour of the basic individual analog cells. Differential design eliminates the effect of the global variations in analog circuits, but they remain sensitive to local variations. The effects of global and local variations on the behaviour of a memory are comparable to that of analog circuits, because the behaviour of a memory is determined by the robustness of the individual cells.

In digital circuits these random variations become increasingly important due to tighter timing restrictions and reducing noise margins.

Since most local variations are of random nature, their influence on the total delay of a logic path relatively averages out with the depth N of the logic path (figure 9.24). The figure shows that the spread in the logic path delay, due to global variations, increases linearly with N, while the delay spread due to local variations "only" increases with $\sqrt{N}$. (To be able to depict all three characteristics into one diagram, the global and local variation have been scaled by the maximum length of the logic path (40) in the diagram.)

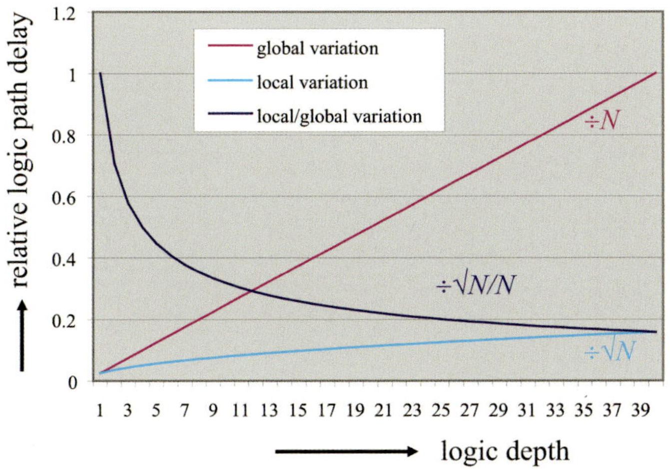

Figure 9.24: *Effect of global and local variations on the logic path delay vs. logic depth*

The effects of this relationship on the diagram of figure 9.23 is that the size of the diamond will increase linearly with N, while the size of the clouds will only increase with $\sqrt{N}$, showing a relative decrease of the local variability.

Figure 9.25 shows the influence of the local variability on the relative spread in logic path delay for different technology nodes.

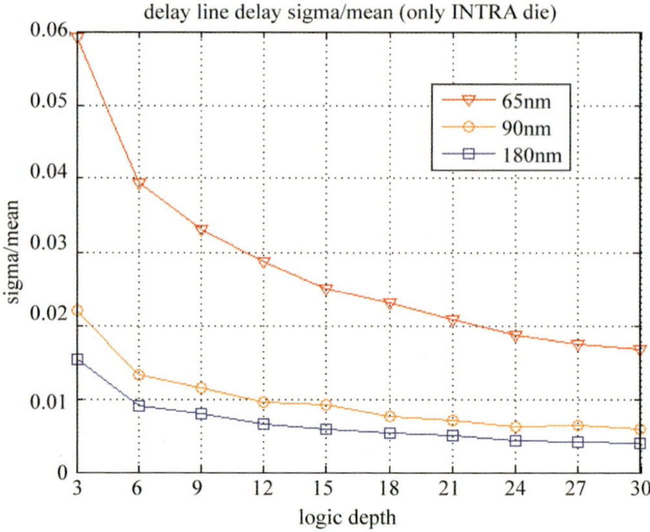

Figure 9.25: *Logic path delay (σ/μ) vs. logic depth*

In this analysis, based on simulation results, the relative performance spread defined as sigma/mean (σ/μ) was used as figure of merit. It shows the increase in random variability with the scaling of the technologies. It also shows that high-speed circuits, which only have a limited logic depth, face a larger influence of the random variability.

The conventional static timing analysis (STA) approach is no longer satisfactory for timing verification of this category of ICs.

9.4.3   Transistor matching

Matching of transistors means the extend to which two identical transistors, both in type, size and layout topology show equal device parameters, such as β and V_T. Particularly in analogue circuits (a memory is also an analogue circuit) where transistor pairs are required to have a very high level of matching [19], the spread ($\sigma_{\Delta V_\mathrm{T}}$) in the difference ($\Delta V_\mathrm{T}$) between the two threshold voltages of the pair results in inaccurate or even anomalous circuit behaviour. This spread is mainly caused by the doping statistics in the channel region of the MOS transistors. For bulk-CMOS devices this is defined as [18]:

$$\sigma_{\Delta V_\mathrm{T}} = \frac{A_{V_\mathrm{T}}}{\sqrt{WL}} \quad in \quad mV \qquad (9.7)$$

In which the *matching coefficient* A_{V_T} is defined as:

$$A_{V_T} = \frac{q \cdot t_{ox}}{\epsilon_0 \cdot \epsilon_{r,ox}} \sqrt{2Nt_{depl}} \qquad (9.8)$$

were N equals the number of active doping atoms in the depletion layer. The expression shows the proportionality of A_{V_T} with t_{ox}. Technologies with a good matching behaviour have their A_{V_T}/t_{ox} ratio close to $1\,\mathrm{V/mm^2}$. In [20] the following simple expression is derived:

$$\sigma_{\Delta V_T} = constant \cdot t_{ox} \frac{\sqrt[4]{N}}{\sqrt{WL}} \qquad (9.9)$$

Until the 45 nm node, N increased every next (bulk CMOS) technology node with a factor close to the reverse of the scaling factor s ($s \approx 0.7$), while beyond this node it is expected to be nearly constant with further scaling [20]. Therefore its impact on the spread is only marginal and we can simplify the relation to:

$$\sigma_{\Delta V_T} \propto \frac{t_{ox}}{\sqrt{WL}} \qquad (9.10)$$

It is clear from this relation, that the ΔV_T spread is inversely proportional to the square root of the transistor area.

For minimum transistor sizes (area), the ΔV_T spread increases every new IC process generation, such that both the scaling of the physical size and the operating voltage of analogue CMOS circuits lag one or two generations behind the digital CMOS circuits. Analogue designs are more prone to so-called proximity effects, such as *STI-stress* and *well-proximity* effects [19,22]. Different openings in the active areas lead to non-uniform compressive stress from the STI isolation into the active areas, influencing both the device saturation current and the threshold voltage. During the retrograde-well implant (see chapter 3) doping atoms may scatter laterally from the photoresist into the silicon, close to the edge of a well (figure 9.26).

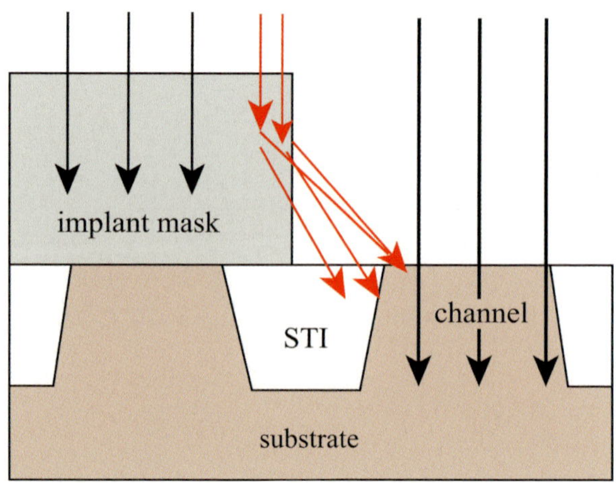

Figure 9.26: *Atoms scatter laterally from the photoresist into the silicon*

This causes a non-uniform well dope over a distance of one or more microns from the well-edge, and a change in threshold voltage and (saturation) currents as well. The influence of these effects is different for nMOS and pMOS transistors and require optimised and symmetrical layout design methods in order to minimise their influence on analogue circuit performance [22].

Also for logic circuits, matching of transistors is becoming an important issue, resulting in different propagation delays of identical logic circuits. Figure 9.27 presents two identical inverter chains (e.g., in a clock tree), but due to the V_T spread, they show different arrival times of the signals at their output nodes.

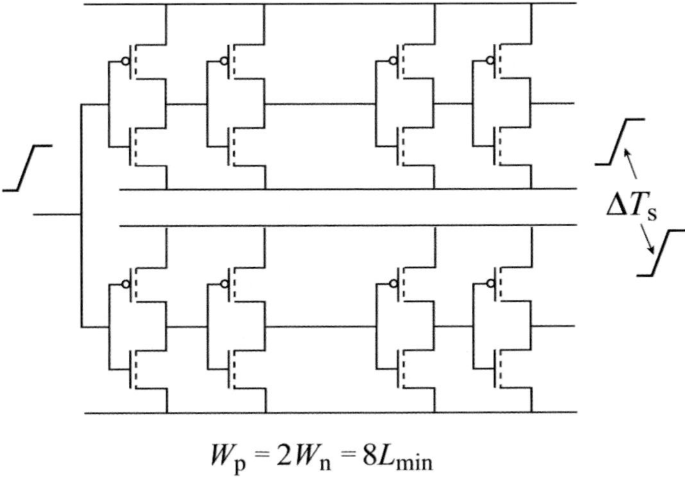

Figure 9.27: *Spread in signal arrival times due to transistor mismatch*

While the difference in arrival times at the second clock-tree stages may be 16 ps in a 0.25 μm CMOS technology, it can be as high as 88 ps in a 65 nm CMOS technology, which is in the order of several gate delays.

Table 9.3: Spread in signal arrival times for different technology nodes [23]

Technology node	250 nm	180 nm	130 nm	90 nm	65 nm
$\sigma_{\Delta T_s}$ ($C_{load} = 50$ fF)	16 ps	21 ps	38 ps	68 ps	88 ps
$\sigma_{\Delta T_s}$ (50,35,25,20,15 fF)	16 ps	16 ps	22 ps	33 ps	32 ps
Clock period T	10 ns	5 ns	2 ns	1 ns	500 ps

Table 9.3 shows the trend in this spread, simulated for a couple of technology nodes. It is right opposite from the ever increasing timing requirements.

Particularly for high-speed circuits, for which timing is a critical issue, transistor matching and its modelling is of extreme importance to maintain design robustness at a sufficiently high level.

9.4.4 From deterministic to probabilistic design

In the above subsection the influence of device parameter spread with respect to circuit performance is discussed. However, process-induced parameter spread in both the device and interconnect structures are also increasingly challenging chip-level timing behaviour and analysis. Transistors vary in relation to oxides, doping, V_T, width and length. Interconnects vary in relation to track width, spacing and thickness and dielectric thickness. So far, this spread was included in simulators in the so-called worst-case, nominal and best-case parameter sets in order to provide sufficient design margins. For example, in worst-case timing analysis it is assumed that the worst-case path delay equals the sum of the worst-case delays of all individual logic gates from which it is built. This produces pessimistic results, incorrect critical paths and over-design. *Static timing analysis (STA)* is a means to optimise and estimate timing across the chip. Current static timing analysis tools use the above-mentioned deterministic values for gate and wire delays, which is appropriate for inter-die parameter variations, but does not account for in-die variations. Particularly these in-die variations show significant impact on the overall timing behaviour. Delay faults caused by noise sources (cross-talk, supply noise, etc.) are also unpredictable with respect to the induced delay. *Statistical timing analysis* is therefore needed in order to cope with these local variations, which cause random gate and wire delays. These local variations are often dealt with by adding an artificial *on-chip variation (OCV)* factor. However, some companies use OCV for the compensation of unforeseen and unpredicted variability effects or parameter degradation over the lifetime of a chip. In either case, OCV introduces additional design overkill for most of the ICs.

An objective of statistical timing analysis is to find the probability density function of the signal arrival times at internal nodes and primary outputs.

However, characterising libraries for statistical timing analysis is not easy and the probability density functions are difficult to compute. Traditionally statistical timing analysis has suffered from extreme run times. Related research is therefore focused to reduce run times [24,25]. Statistical timing analysis is just taking off. For the 45 nm technology node and below, statistical timing analysis is considered to be a must, particularly for the complex and higher performance categories of ICs. The method needs to be simplified to make it a standard component of the verification process.

9.4.5 Can the variability problem be solved?

A lack of modelling accuracy in current IC designs turns variability into uncertainty and increases the risk of functional failures and reduced yield. Statistical timing analysis can only predict the probability of a circuit operating at a given frequency. A hold-time violation, however, only shows up in the process corner that causes it and still needs a complete set of simulations for all possible corners.

There are several measures in the design that can reduce the effects of variability, but these are only effective for a limited category of circuits. As already mentioned before, in many analog circuits the transistors do not use the smallest channel lengths and operate at the same (low) voltages as supplied to the digital cores. Also incorporating more regularity into the layouts of the library cells to support litho-friendly design, will reduce the variability effects of the lithographic process. A fully regular library, built from only one type of transistor, would allow the technologists to optimise the transistor architecture such, that it supports the lithography, reduces the variability and optimises the yield.

A lot of research is currently focussed on methods and tools to reduce the impact of variability on the performance, to reduce design overkill and to shorten design time, since it may lead to solutions that no longer need extensive full process-corner simulations. It will certainly take a while before these methods and tools will become mature instruments to effectively deal with the broad spectrum of causes of variability. But even then, new technology nodes may introduce new variability sources requiring a continuous R&D effort to create appropriate methods and tools.

9.5 Reliability

The continuous scaling of both the devices and interconnect has severe consequences for a reliable operation of an IC. Reliability topics, such as electro-migration, hot-carrier effects, Negative Temperature Bias Instability (NBTI), latch-up and ESD are all influenced by a combination of geometrical, physical, and electrical parameters: materials, sizes, dope, temperature, electrical field, current density, etc. Improving reliability therefore means choosing the right materials, the right sizes and doping levels and preventing excessive electrical fields, temperatures and currents. This section will discuss the effects of scaling on each of the aforementioned reliability issues.

9.5.1 Punch-through

The drain and source depletion regions of a MOS transistor may merge when a sufficiently large reverse-bias voltage is applied to the drain-to-substrate junction. This is particularly likely to occur in MOS transistors with very short channel lengths. The energy barrier, which keeps electrons in the source of an n-channel device, is lowered when the drain and source depletion regions merge. Consequently, many electrons start to flow from the source to the drain even when the gate voltage is below the threshold value and the transistor is supposedly not conducting. This effect is known as (sub-surface) *punch-through*. The drain-source voltage V_{PT} at which punch-through occurs is approximated as follows:

$$V_{PT} = \frac{q}{2\epsilon_0\epsilon_r} \cdot N_A \cdot L^2 \tag{9.11}$$

where N_A represents the substrate dope, L represents the transistor channel length and q represents the charge of an electron. The effect of this leakage mechanism can be reduced during processing by increasing the doping level of the substrate with an *anti-punch-through (APT)* implantation. The associated increase in the threshold voltage of the transistor can be compensated by reducing the oxide thickness. Punch-through is also regarded as a subsurface version of DIBL (see chapter 2). It is obvious that punch-through will not occur in the devices of a well-defined CMOS technology

9.5.2 Electromigration

The increase in current density associated with scaling may have detrimental impact, not only on circuit performance, but also on the IC's reliability. High currents, flowing through the metal lines, may cause metal ions to be transported through the interconnection layers due to the exchange of sufficient momentum between electrons and the metal atoms. For this effect, which causes a material to physically migrate, many electrons are required to collide with its atoms. This physical migration of material from a certain location to another location, creates open circuits or *voids* (figure 9.28.a) on locations where the material is removed, and *hillocks* (figure 9.28.b) on locations where material is added. This *electromigration* effect damages the layer and results in the eventual failure of the circuit. Electromigration may therefore dramatically shorten the lifetime of an IC. The impact of electromigration is eliminated by preventing excessive current densities. Electromigration design rules are therefore part of every design kit. These rules specify the minimum required metal track width for the respective metal (e.g., aluminium or copper) for a certain desired current flow at given temperatures. Electromigration effects increase with temperature because of the temperature dependence of the diffusion coefficient. This causes a reduction of the maximum allowed current density (J_{max}) at higher temperatures in on-chip interconnect. The required metal width for electromigration roughly doubles for every $10°C$ increase in temperature. Since many IC data sheets show a maximum ambient temperature of around $70°C$ or higher, the real worst-case junction temperature of the silicon itself may exceed $100°C$ in many applications. Therefore it is common design practice to use the value for J_{max} at $125°C$.

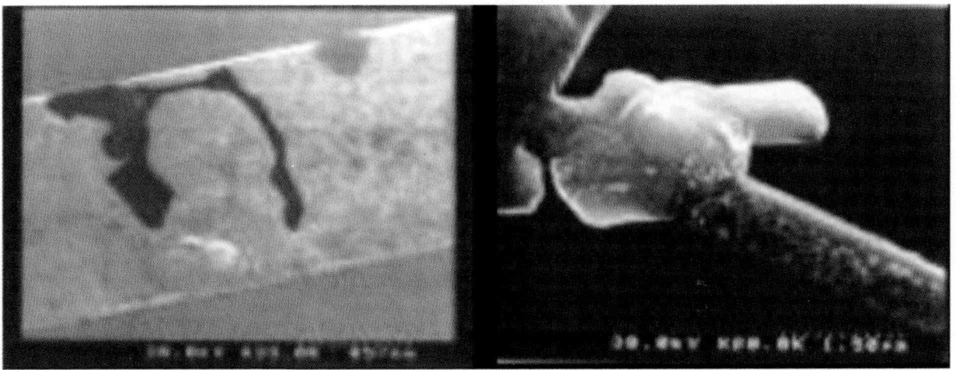

Figure 9.28: *Electromigration damage in metal interconnect lines: voids (a); hillocks (b) (Courtesy of University of Notre Dame, Notre Dame, Indiana)*

The minimum allowed width W_{em} of a metal wire with height H, to carry a current I, according to this electromigration requirement, is then equal to:

$$W_{em} = \frac{I}{J_{max} \cdot H} \tag{9.12}$$

Table 9.4 shows some parameter values, which are characteristic for metal layers in a $65\,\text{nm}$ CMOS technology.

Table 9.4: Metal characteristics for a 65 nm bulk-CMOS technology

Technology node and metal layer	R_{sheet}	H	J_{max}@125°C
lower metal layer (copper)	$85\,\mathrm{m\Omega}/\square$	350 nm	$3.2\,\mathrm{mA}/\mu\mathrm{m}^2$
top metal layer (copper)	$26\,\mathrm{m\Omega}/\square$	900 nm	$3.2\,\mathrm{mA}/\mu\mathrm{m}^2$

Since most of the currents in an IC flow through the supply lines, it is obvious that these are often implemented in the upper metal layer(s), which usually have a larger height. Because AC currents flow in both directions through a wire, the maximum value of these currents with respect to electromigration are about one order of magnitude larger than the maximum values for the (average) currents mentioned above. Similarly, currents through contact holes and vias must also be limited to eliminate electromigration-induced damage of the contact conductor. A typical maximum current density value for a 0.2x0.2 $\mu\mathrm{m}^2$ contact or via in a 65 nm CMOS technology is around $0.4\,\mathrm{mA}/\mu\mathrm{m}^2$ at 125°C. The increase in the aspect ratios of the contacts and vias, in combination with a reduction of maximum currents through them, makes them an incremental part of the overall IC reliability.

The continuous scaling of feature sizes and voltages (constant-field scaling) by about a factor of 0.7, for every new technology node, did not change the intrinsic power density of most standard-cell designs. However, due to the reduction in supply voltage, the supply current per unit area of logic increased with about a factor of 1.4 every generation. This has put severe constraints to maintaining electromigration reliability across complex designs.

Because of the expected increase in currents through the metal layers, more Joule heating is expected in these layers. This, in combination with low-ϵ dielectrics, which show a higher thermal resistance, made designers start worrying about the so-called *wire self-heating* mechanism. However, the width of a metal wire is not only specified by the appropriate electromigration requirements, but also by the maximum allowed voltage drop across the wire in order to limit speed loss of the connected circuit(s). Suppose an active logic block draws an average supply current of 100 mA. When this block is located nearby the supply pads of the chip, the width of the supply lines is determined only by the electromigration requirement for this $100\,\text{mA}$ current. When this block is near the centre of the chip, say at $5\,\text{mm}$ distance from the supply pads, the supply lines must be much wider in order to limit the voltage drop across it. So, above a certain distance from the supply pads, the width of the metal (and thus its cooling area) grows with its length, keeping the voltage drop across the line constant. As a result also the resistance of the line (and thus its total I^2R *Joule heating*) will then be constant. In other words: the maximum wire self-heating occurs in wires with length equal to a cross-over length L_{co}, which is defined to be the length at which the metal-width required by electromigration is identical to the width required by the maximum allowed voltage drop. In [29] it is shown that for $0.18\,\mu\text{m}$ and $0.12\,\mu\text{m}$ bulk-CMOS technologies, wire self-heating in supply lines causes only a limited temperature rise of the wires of just a few degrees. Also for the $65\,\text{nm}$ and $45\,\text{nm}$ technology nodes, this temperature rise is by far negligible compared to the temperature rise due to the power consumption of the silicon part of the chip. From this result it can be concluded that wire self-heating in supply lines should not be a real issue in current (and near future) properly designed bulk-CMOS VLSI chips.

9.5.3 Hot-carrier degradation

When carriers in the MOS transistor channel are given enough energy, they collide with the substrate atoms and generate electron-hole pairs. These, in turn, will also be accelerated and may also collide with substrate atoms. This so-called *impact ionisation* may cause large substrate currents, device breakdown and/or degradation of the silicon-to-gate-oxide interface. Electrons actually collide with the gate oxide. When electrons achieve sufficient energy, they may cross this silicon to silicon-dioxide (Si/SiO_2) interface barrier (with a barrier energy of about $3.1\,\text{eV}$ for electrons and $4.7\,\text{eV}$ for holes) and are injected into the gate oxide. Injected carriers lead to the degradation of the Si/SiO_2 interface (electrically active interface defects are generated), to the generation of defects in the gate oxide film and to charge trapping in the oxide interface (both pre-existing and newly generated). Oxide charge trapping and interface state generation induce a shift of the transistor threshold voltage and cause a degradation of the device drive current. This effect is called the *hot-carrier effect (HCE)* and leads to degraded device performance and reliability problems. Due to the lower mobility of holes with respect to

electrons in the transistor channel, impact ionisation in p-channel MOS-
FETs is less. Therefore, the hot-carrier effect is more severe in n-type
MOSFETs.

Graded drain and lightly doped drain structures are used to reduce
the maximum value of the electric field in small transistors and thus
prevent hot-carrier degradation. The *graded drain* transistor is a very
simple adaptation of the conventional transistor. The junction between
the drain and the substrate is made much more gradual by simply im-
planting phosphorous with a relatively low concentration in the highly
concentrated n^+ area. The phosphorous has a much higher diffusion coefficient than the arsenic in this area and therefore diffuses much further.
This results in a donor profile with a low gradient; an example is shown
in figure 9.29. The graded drain reduces the maximum electric field by
about 30 %. This implies that the operating voltage can be increased by
50 % for given transistor dimensions.

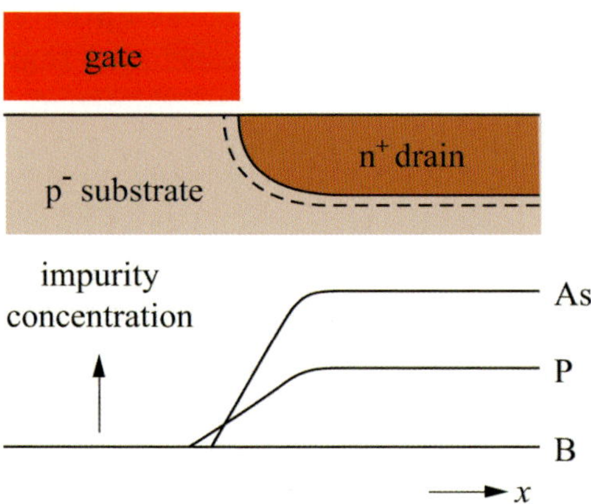

Figure 9.29: *Phosphorous (P) halo around arsenic (As) in the cross-
section of a graded drain transistor and the concentrations as a function
of the position* x

The *lightly doped drain (LDD)* method is a more difficult means of reducing the drain-substrate concentration gradient. It was introduced in
CMOS processes with channel lengths of around $1\,\mu$m, which exhibited
extremely large peaks in the electric field in the channel close to the
drain. The maximum electric field obtained by using LDD is lower than
that achieved with the graded drain. The various LDD process steps
are explained with the aid of figure 9.30. A conventional $0.35\,\mu$m nMOS

transistor with a gate oxide thickness of about 7 nm thickness is shown in figure 9.30a.

Conventional CMOS processing, which is described in chapter 3, is used to create the gate oxide. Phosphorous with a concentration that varies from 1×10^{18} to 4×10^{18} atoms per cm^3 is subsequently implanted. An oxide layer of about $0.35 \, \mu m$ thickness shown in figure 9.30b is then deposited. This is followed by an anisotropic etch, which leaves the *oxide spacers* shown on both sides of the gate in figure 9.30c. A subsequent highly concentrated implantation of arsenic and a drive-in diffusion produce the resulting n^- and n^+ areas shown in figure 9.30d. The magnitude of the transistor's horizontal electric field as a function of the channel position x is shown in figure 9.30e. Its maximum value is 50 % of that obtained in a comparable transistor with conventional arsenic drain and source areas. Two factors account for this significant reduction. The first is the relatively long region with a low donor (n^-) concentration. A depletion area will form much sooner in this area than in the n^+ area. A large proportion of the drain-source voltage drop is distributed over this area. The second factor is the extra separation between the gate and the n^+ drain area. This also reduces the influence of the second-order effects as discussed in chapter 2.

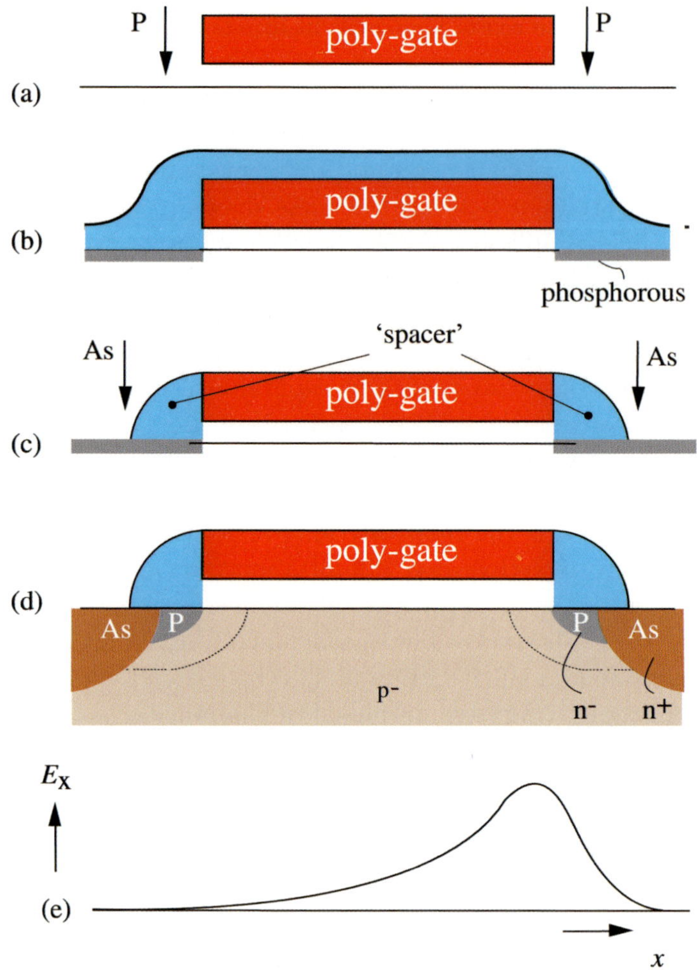

Figure 9.30: *Process steps for the creation of an LDD transistor and the resulting reduced horizontal electric field distribution*

The LDD transistor was difficult to create and has the added disadvantage of possible increased series resistance in the source and drain as a result of the n^- areas. Careful optimisation, however, yields small transistors with high operational voltages that can deliver high currents. LDD implants were included in CMOS technologies from about $1\,\mu$m down to $0.35\,\mu$m channel lengths. As discussed before, the required energy for carriers to cross the Si-SiO$_2$ interface barrier is at least $3.1\,$eV for electrons and $4.7\,$eV for holes. As supply voltages reduce with the advent of new process generations, these carriers can hardly ever reach such energies when the supply voltage is 2.5 V or less.

Theoretically, in a 0.18 μm CMOS technology with a supply voltage of 1.8 V, an electron can only get an energy level of 1.8 eV during its flow through the channel from source to drain. This is less than the above mentioned barrier energy to create hot electrons. However, due to multiple collisions, some electrons may acquire more energy than the required barrier energy and become "hot". From these considerations it was generally accepted that, when supply voltages are reduced, the chance to generate hot carriers in the transistor channel would reduce as well and the hot-carrier effect was expected to eventually disappear totally. As a result, LDD implants are no longer required in 0.25 μm CMOS processes and below. These are then replaced by a (lightly-doped) drain extension, as discussed in section 3.9.3.

With the continuous scaling process, critical-dimension (CD) control becomes more difficult leading to transistors with different channel lengths showing different hot-carrier behaviour. Shorter channel lengths easier introduce punch-through. Both the punch-through prevention and SCE suppression require different doping profiles around sources and drains, with increased doping levels. This has some negative effects on the hot-carrier behaviour.

When voltages across the transistor are scaled at the same pace as the transistor feature sizes, the electrical fields remain almost constant, and the chance for impact ionisation would hardly change. However, particularly with 90 nm and smaller CMOS technologies, the effective channel length is scaling faster than the supply voltage, so that the increase in electrical field may lead to increased impact ionization. Although these carriers do not acquire sufficient energy to cross the silicon-to-silicon dioxide barrier, they will still cause substrate currents. Hot-carrier effects may therefore manifest themselves again more in sub 100 nm technologies than in the technology nodes just above 100 nm, especially in the early process development phase due to bad transistor drain engineering. Assuming that the transistor is stressed under worst-case condition ($V_{ds} = V_{dd}$ and $V_{gs} = V_{dd}/2$) such that the substrate current is maximal), the hot-carrier lifetime is described by a well-accepted empirical expression (Takeda) as:

$$\tau_{drift} = A \cdot L_{eff}{}^{C} \cdot e^{B/V_{ds}} \tag{9.13}$$

where τ_{drift} represents the lifetime (usually at 10% degradation), L_{eff} the effective channel length and A, B and C are process-related coefficients. It is clear that the hot-carrier lifetime reduces with decreasing channel length and increasing voltage. So, when we scale the supply voltage with the same factor as the feature sizes, still this lifetime may increase, dependant on the constants A, B and C.

An additional effect is that for future technologies the silicon dioxide will be replaced by high-ε dielectrics. Most of them, however, have a significantly lower barrier [27] and the hot-carrier effects are not just slowly fading away because of reducing supply voltages below the barrier. Results from literature [28,29] stress the importance of a continuous attention for hot carrier degradation nanometer technologies, in order to maintain functional reliability at a sufficiently high level.

9.5.4 Negative bias temperature instability (NBTI)

Negative Bias Temperature Instability (NBTI) is a result of a negative bias applied to the gate of a p-channel MOS transistor with respect to the bulk. The mechanism is temperature activated. NBTI results in the degradation of many transistor parameters (threshold voltage, drive current and transconductance), but the threshold voltage appears to be the most degrading one. NBTI was first reported in 1967, but the attention devoted to this mechanism has been escalating since the millennium, due to the introduction of gate-oxide nitridation [30] that enhances NBTI and the fact that other oxide wear-out mechanisms, such as HCE and oxide breakdown, were expected to become less severe as the gate oxide scales down. NBTI is strongly process dependent. It has been reported that a higher nitrogen concentration in the oxide [30], boron penetration [31] and plasma processing can enhance NBTI, while fluorine incorporation in the gate dielectric is beneficial against NBTI [32]. The physical nature of the wear-out mechanism induced by NBTI is very difficult to identify. The most accepted models imply positive charge build-up in the oxide bulk and at the Si/SiO_2 interface (donor-like interface states) [32,33].

Whilst hot-carrier injection mostly affects n-channel MOSFETs and depends on the transistor channel length, NBTI mostly affects the pMOS transistor and is only slightly dependent on the transistor geometry, although it has also been reported that in shorter channel devices NBTI can be more severe [34]. Furthermore, the NBTI does not imply a current flow in the transistor channel and can occur at zero drain to source bias. This would mean that NBTI stress could even occur in the standby mode. Design configurations in which matched p-channel MOSFET pairs are subjected to unbalanced stress are reported as most sensitive

to NBTI degradation, since the threshold voltages of the transistor pair
change differently with the stress [35]. Also matched p-channel MOS-
FET pairs operated symmetrically can lead to reliability fails due to
NBTI when the transistors are subjected to different biases in power-
down mode. Burn-in can also be a source of NBTI-induced circuit fails,
due to the involved high temperature. The NBTI effect is more severe
for pMOS transistors than for nMOS transistors (PBTI) because of the
difference between holes and electrons in interacting with oxide states.
The NBTI (PBTI) degradation almost always recovers after the stress
is removed. This requires a quick engineering test to demonstrate the
impact of this reliability mechanism.

Even when an IC is produced in different fabs that run the same process, it may perform differently with respect to NBTI, because not all
individual processing steps are completely identical. NBTI is therefore
a technology issue, but critical design configurations, such as matched
p-channel MOSFET pairs subjected to unbalanced stress, either in operation or power-down mode, should be avoided.

The physical understanding of NBTI is continuously improving, leading to the development of various NBTI models. A description is beyond
the scope of this book, but many of them can be found on the internet.
Assuming a power-law dependence on the stress voltage (field), then the
change in V_T is proportional to:

$$\Delta V_\mathrm{T} = D \cdot {E_{ox}}^{m} \tag{9.14}$$

where D is a process dependant parameter, E_{ox} represents the electrical
field across the oxide, and m a coefficient dependant on e.g., the dielectric
material and the dielectric thickness (an approximate value is $m \approx 4$).

V_T shifts of $50\,\mathrm{mV}$ and more have been reported, so designers need
to be convinced to build enough tolerance in their designs. The occurrence of NBTI can be lowered when a device (chip) is not subjected to
voltage overshoot and/or high temperatures, either from its own heat
dissipation or from its application environment. Therefore, a reduced
power consumption would also be beneficial to reduce the chance for
NBTI stress.

9.5.5 Latch-up

The presence of nMOS and pMOS transistors in a CMOS process leads
to the creation of parasitic thyristors, as shown in figure 9.31. In this

figure R_1 and R_2 represent the substrate and n-well resistances, respectively.

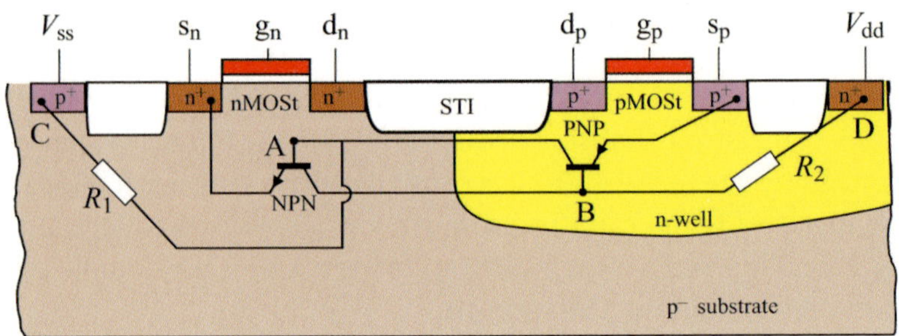

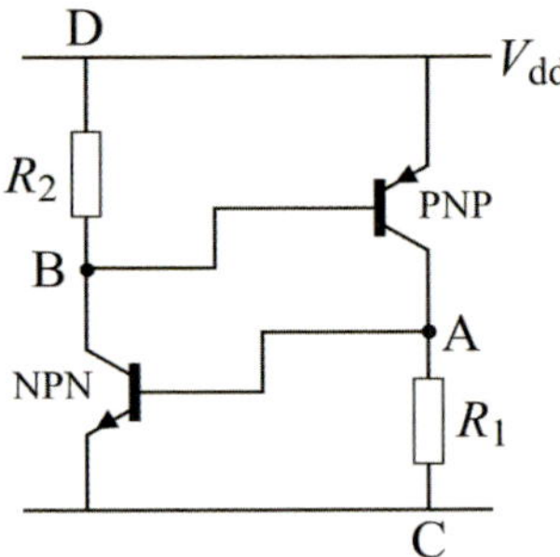

Figure 9.31: *Parasitic thyristor in CMOS and its equivalent circuit diagram*

Relatively high currents through the bipolar transistors will create relatively high voltages in the substrate and/or n-well. When a sufficiently high positive voltage is present somewhere in the substrate (e.g., at position A), it will turn on the parasitic NPN transistor, or when a local voltage (e.g., at position B) within the n-well that is sufficiently lower than the V_{dd}, it will turn on the parasitic PNP transistor. When both bipolar transistors conduct, they are connected into a feed-forward loop, which means that they enhance each other's conduction state, which will finally be latched (maintained) in the thyristor. This state can only be recovered when the supply is completely switched off. This undesirable effect is called *latch-up* and leads to incorrect circuit behaviour or even damage. Also inductive effects or coupling capacitances may cause the node connected to the drain to have overshoots and/or undershoots, thus forward biasing the drain substrate junction, which may initiate latch-up. This requires a controlled start up of ICs.

Latch-up in CMOS circuits can be avoided by applying the following technological and/or design remedies:

- Minimise the substrate and/or n-well resistances. This can be done in two ways. One is the use of many substrate and n-well contacts in the design, which will reduce the values for R_1 and R_2, respectively. The parasitic thyristor is then unlikely to turn on. Reducing both resistances by increasing the substrate and n-well doping is not an option, since it also changes the threshold voltages and overall transistor behaviour. A good alternative is the use of so-called epitaxial wafers (figure 9.32).

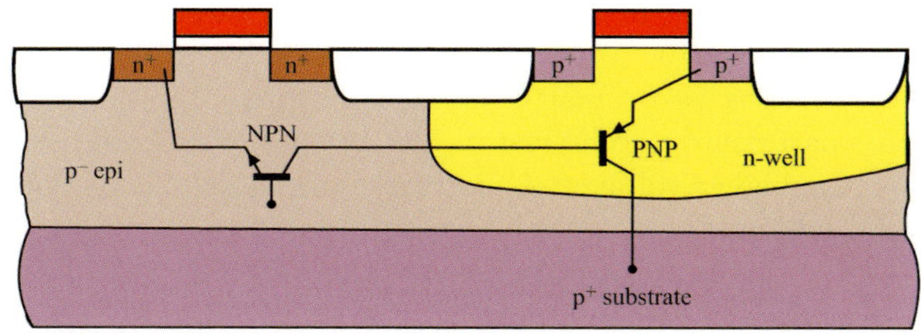

Figure 9.32: *Cross-section of a wafer with a thin p^- epi-layer on a thick p^+ substrate*

Epitaxy is a layer of single-crystalline silicon deposited/grown onto a single-crystalline silicon wafer (see section 3.2.2). The crystalline structure of the substrate is reproduced in the growing material. This epitaxial layer, in which the devices are formed and whose thickness is usually between 1 to 5 μm, can be doped, as it is deposited, to the required doping type and concentration (usually with a resistivity of $\approx$ 10-20 Ωcm) while continuing the substrate's crystalline structure. So, we can create a thin p^--epitaxial layer on top of a p^+-substrate. Because the current (unground) wafer thickness is typically 750 μm, the p^+-substrate is relatively thick and has a low resistivity ($\approx$ 5-10 mΩcm). Such low-ohmic substrates show very low values for R_1. A large part of the PNP collector current will therefore flow through this substrate and only a small part will flow into the base of the NPN transistor. This transistor can no longer be turned on easily and is then largely excluded from the latch circuit. *Epitaxial wafers* with low-ohmic substrates have been massively used for CMOS products in $0.18\,\mu$m technologies and above. Due to decreasing supply voltages and increasing noise levels, the combination of analog and digital circuits onto one single chip has made its design a difficult and cumbersome task. Particularly the substrate noise sensitivity of analog circuits requires a good isolation from the digital noise 'generators', which is why a high-ohmic substrate is preferred for mixed analog digital circuits. Because the latch-up effect decreases with reducing voltages, CMOS in 120 nm and beyond are most commonly processed on high-ohmic wafers (section 3.2.2).

- The use of guard rings is another way to make strong (low-ohmic) connections of local substrate and/or nwell areas to V_{ss} and V_{dd} respectively. Moreover, the distance between n-type and p-type areas is also a matter of concern during the design phase and is particularly of interest in I/O circuits, which are usually supplied by higher voltages. Guard rings are more effective on high-ohmic substrates.

- Apply a back-bias voltage to the substrate. When the p^--substrate in figure 9.31 is connected to a negative voltage instead of to V_{ss}, the base voltage V_A of the NPN transistor will be lowered. Therefore, this transistor can no longer be turned on easily. This technique is more a theoretical option and is not frequently used for latch-up prevention.

- Use Silicon-On-Insulator technology to completely isolate the nMOS transistors from the pMOSTs. In this technology the NPN and PNP transistors are completely isolated from one another and so the connections to create latching thyristor circuits are missing. The amount of "electrical isolation" depends heavily on the applied frequencies.

The application of one or more of the above remedies has increased latch-up immunity to a very high level. The highest chance of occurrence for latch-up is during testing. Standard testing requirements include immunity to 100 mA or more, depending on what the IC can and should withstand from an application point of view. This means that with epi-wafer material, 100 mA can be supplied to the output of an output buffer (driver) even though no output transistor is conducting. This current, then, directly flows into the substrate, thereby raising the substrate voltage and possibly turning the thyristor on (figure 9.31). In practice, some latch-up tests are done with 150-200 mA at a maximum ambient rated temperature for the device, depending on the target application area.

In future technologies, the latch-up phenomenon is likely to disappear inside electronic circuits, as the supply voltages will be reduced every new technology node. However, at the chip I/Os, the requirements on latch-up remain relatively high, since many applications still require a higher interface voltage (1.8 V, 2.5 V, 3.3 V). More on latch-up basics can be found in [36].

9.5.6   Electro-Static Discharge (ESD)

Integrated circuits are exposed to many possible sources of damage, both during and after the manufacturing process. The principle cause of damage is *electrostatic discharge (ESD)*, due to the transfer of charge between bodies at different electrical potentials. ESD pulse durations are very short and normally range from 1 to 200 ns, but they may introduce very large power spikes. The high impedance of MOS input circuits makes them particularly vulnerable to physical damage when they are exposed to these spikes. This may result from operations during the fabrication process or from handling (un)-packaged dies and bonding. It may also occur during testing and maintenance or in the application. While only a few devices or connections may be severely damaged, many more may suffer damage that is not immediately apparent. These latent failures will result in customer returns, which is one of the biggest worries of semiconductor vendors. Thus ESD is one of the most important factors that determine the reliability of an IC. It may also trigger the parasitic thyristor in figure 9.31 resulting in the occurrence of latch-up.

The damage caused by ESD is irreversible. The human body is one of the main sources responsible for ESD. Just by walking on a carpet on a low-humidity day, for instance, a person, wearing shoes with highly insulating soles can build up a voltage in excess of 30.000 V. The resulting charge can then be transferred via an ESD to an electronic circuit during touching. It is also very important that precautions need to be taken to prevent ESD damage during IC fabrication. In addition, protective measures must be included in an IC's design to ensure that it can withstand acceptably large ESD pulses. On-chip MOS protection circuits are used to increase the immunity of an IC to ESD pulses. These circuits are designed to provide input and output circuits with low-impedance shunt paths, which prevent excessive voltages to arrive at the IC's input, output, and core transistors.

ESD test models and procedures

ESD sources are emulated in several different ways. The *human-body model* is currently the most popular industry model and simulates the direct transfer of electrostatic charge from the human body to a test device. It is internationally accepted as a standard (JEDEC Standard No.22-A114-B). Figure 9.33 shows a human-body test set up. The basic requirement for this model, in combination with the parasitics (L) of the tester interface cables, is to generate ESD pulses with rise times between 10 to 15 ns.

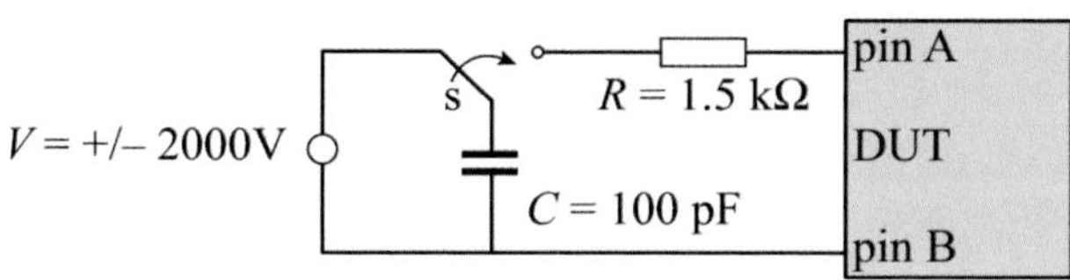

Figure 9.33: *A typical equivalent circuit based on the human-body model*

The test is normally done on an ESD tester. This human-body model has not changed much over the last few decades. Basically, a 100 pF capacitor is charged to the test voltage, and then discharged through a 1.5 kΩ resistor across any combination of pins A and B (table 9.5) of the DUT (device under test). The chip may consist of several supply (V_{dd}) and ground (V_{ss}) domains. Each domain may be supplied by more than one pin. The V_{ss} and V_{dd} in the table below refer to just one of the respective pins of a supply domain. In other words: each pin is then tested with respect to all grounded V_{ss} and V_{dd} domains and not to all grounded V_{ss} and V_{dd} pins, to save test time. Each signal pin is also tested with respect to all other grounded signal pins. The maximum test voltage ranges from 2 kV to 8 kV and depends on the application area of the chip. Since production environments are well controlled, a maximum voltage of 2 kV is usually required. However, because more and more IC pins can be touched in daily life (plug-ins like USB ports, chip cards, SIM cards, memory sticks and flash cards, etc.), the ESD-test requirements tend to increase. The 8 kV requirement is

therefore no exception anymore. The devices are classified when meeting
a particular sensitivity criterion. A class-2 device, for instance, has
passed the 2 kV, but fails after exposure to an ESD pulse of 4 kV (see the
above mentioned standard: www.jedec.org). There are also discussions
led by the Industry Council on ESD Target Levels to target for a safe
Human Body Model of 1 kV for many applications (e.g. automotive
and consumer), because the 2 kV level is really and over-design for these
applications.

Table 9.5: Different ESD test states

State	DUT	
	pin A	pin B
1	input	V_{ss}
2	V_{ss}	input
3	input	V_{dd}
4	V_{dd}	input
5	output	V_{ss}
6	V_{ss}	output
7	output	V_{dd}
8	V_{dd}	output
9	input	output
10	output	input
11	input	input
12	output	output
13	V_{dd}	V_{ss}
14	V_{ss}	V_{dd}

The first ESD tests start at 100 V. Generally three to five positive and
negative pulses are applied at 300 ms intervals in all test states. Stressed
pins are tested after application of each ESD pulse series. If no failure
is observed for a sequence through the pins, then the ESD voltage level
is increased by 100 V and the sequence is repeated. The ESD test is
complete when a failure is observed or when all pins on the DUT have
been stressed until the required maximum voltage is reached. Generally,
the following (example) criteria may be used to determine failure:

- Incorrect functional operation or a violation of the device specifications.

- A change of more than 5% in the forward voltage drop and breakdown voltage in the diode characteristic.

- An increase of more than 10% in the I_{ddq} leakage current (see
  chapter 10).

Another standardised and popular ESD test model is the *machine model*,
which emulates the rapid direct transfer of electrostatic charge, from a
charged conductive object (tool or equipment), to a test device. Compared to the human-body model of figure 9.33, the machine model spec-
ifies a discharge of a 200 pF capacitor through a 0.75 mH inductor. Due
to the absence of the current limiting resistor this model was seen as
more severe and the tests are done at lower voltages. The *charged-
device model* is an alternative ESD test set up, which is most commonly
used to emulate rapid electrostatic charge transfer during e.g., packaging
and assembly. More details on the latter two models can be found in
[37] or directly from the JEDEC website: www.jedec.org.

On-chip ESD protection circuits

Although much ESD and ESD-protection knowledge has been built over
the last couple of decades, the design of on-chip ESD protection circuits
is both scientific and experimental. This is due to the fact that in every
new semiconductor node, device architectures and feature sizes (e.g.,
width, spacing, oxide thickness, etc.) have changed with respect to the
previous node, which requires new protection solutions. Usually several
alternative protection circuits are explored in each new technology node
and often semiconductor process development goes hand in hand with
ESD protection development.

The purpose of a protection circuit is that it provides a low-ohmic
shunt path in parallel with the MOS input and output transistors during
the occurrence of an ESD pulse. In its simplest form, a protection circuit
consists of a spike filter and a set of diodes (figure 9.34).

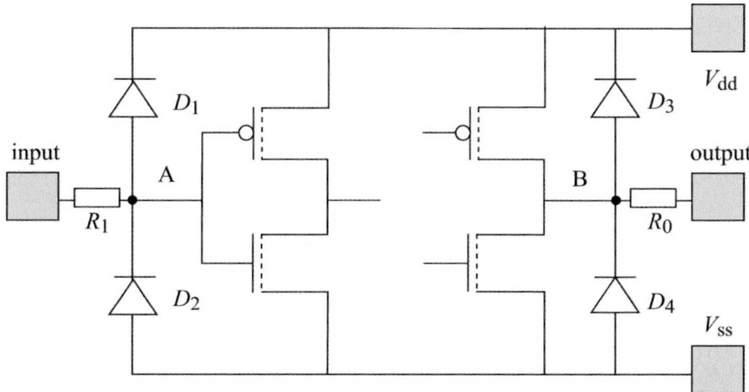

Figure 9.34: *Simplest form of a MOS input and output protection circuit*

Resistor R_i, in combination with the parasitic capacitance at node A, forms the input voltage spike filter, while diodes D_1 and D_2 clamp the input voltage to a junction voltage above the V_{dd} and a junction voltage below the V_{ss}, respectively. R_o, D_3 and D_4 do the same, but then for the output node. In many cases also MOS diodes are used for clamping the input and output voltages. Values for the input and output resistors are in the order of $100\,\Omega$ and $5\,k\Omega$, respectively.

Since MOS inputs are connected to high-ohmic transistor gates, the protection of input circuits is more critical than that of output circuits. Output pads are connected to drain areas. Usually these drain areas are relatively large, because outputs usually have to drive large capacitances (10-50 pF) and the complementary drain junctions act as intrinsically available diode clamps. Of course also the outputs must fulfill ESD design rules.

The behaviour of MOS protection circuits depends very much on their size and layout and on various process parameters. Each manufacturing process has its own specific design rules for ESD protection circuits. Therefore, the design of such circuits is done in co-operation with specialists in the field of protection devices.

Future technologies, particularly those for high-performance designs, may require different substrates, such as SOI and/or silicon germanium (SiGe). SOI technologies need a different approach for the development of ESD protection devices, since their devices are built on an isolating substrate. The implementation of ESD protection diodes on SOI needs to change from the high-perimeter bulk CMOS diodes to an SOI lateral-gated diode structure. SiGe technology has become another important alternative for high-speed communications and wireless applications. Since the change in material and mobility will also influence ESD, developing an ESD strategy for SiGe circuits will be very challenging. More about ESD and ESD models can be found in [38] and [39].

9.5.7 Charge injection during the fabrication process

Many IC processing steps use plasma or sputter-etching techniques, which introduce a flow of charged particles towards the wafer surface. These charge particles are then collected by conducting surface materials (polysilicon, metals). This, so-called *antenna effect* can create significant electrical fields across the thin gate oxides which can be stressed to such an extend that the transistor's reliability can no longer be guaranteed. It can also cause a threshold-voltage shift, which affects the matching behaviour of transistors pairs in analogue functions. It is industry practice to introduce additional "antenna design rules" to limit the ratio of antenna area to gate-oxide area (see chapter 3). The back-end design tools can handle these design rules by limiting the maximum wire (antenna) length in the different metal layers. Also, protection diodes can be used in the library cells to shunt the transistor gates. Due to the trend in gate-oxide thickness scaling the appearance of the antenna effect is expected to have greater consequences for the design. The use of high-ε gate dielectrics in building the transistor stack would therefore also be beneficial to reduce this antenna effect.

9.5.8 Reliability summary and trends

Most of the previously discussed reliability topics depend on size, doping profiles and levels, voltages, temperatures and device materials. Scaling requires a change in many of these parameters and will therefore have severe effects on the reliability of CMOS devices and circuits. Moreover, in technologies with channel lengths below $45\,\text{nm}$, the transistors are expected to be built from a completely different stack of materials as compared to today's high volume products. The bulk-silicon substrate will probably be replaced by SOI and/or SiGe; due to the high leakage current, the SiO_2 gate oxide is expected to be replaced by a high-ε dielectric and because of gate depletion the polysilicon gate may be replaced by a metal or fully silicided gate. This has an additional impact on the reliability of the devices and vice versa. Maintaining reliability at a sufficiently high level will put severe demands on this new transistor stack and makes the choice for the right materials a very difficult and cumbersome one.

9.6 Design organisation

Robustness of nanometer CMOS designs not only impacts the level to which the chip operates with respect to the functional and electrical requirements. A robust design intrinsically includes a decent organisation of its database.

A very important requirement for increasing the integrity of the database is to enable quick design changes (be it in the final design stage or during a redesign). There are two requirements for the database with respect to design changes:

- it should take minimum effort

- it should not introduce new errors caused by:

  - unintended modifications
  - forgetting something.

These requirements even hold for design changes after one or more years. Current VLSI chips often reuse existing building blocks, such as multipliers, memories or microprocessor cores. This requires a structured and well documented database set-up and design documentation.

Ten to thirty percent of test engineers time is lost as a result of incomplete documentation of the design. What really is required is:

- good and complete specification

- complete (sufficient) test vectors

- mixed signal ICs: the test engineer must understand the complete IC.

Generally, the best solution for a database set-up is a hierarchical approach, in which one has:

- directory hierarchy = design hierarchy (easy to find your way through)

- good version management; what object (netlist and layout, etc.) is generated from which source (e.g., VHDL or Verilog description), and which one is the latest; use of make files!

A first-time-right design requires a full hundred percent discipline, not only to fulfill the requirements of the specification, but also to create a structured and logical database in order to reduce the change of failures during (re-)design or during the creation of design derivatives.

9.7 Conclusions

After almost five decades of scaling, the robustness of integrated circuits has dramatically been affected by the consequences of the continuous increase of device and interconnect density. There are three major effects that contribute to the negative trend in the robustness of operation. The first one is the increase in IR-drop, supply and substrate noise and ground bounce due the increase of average current and more on-chip simultaneous switching and cross-talk, due to the increasing interference of the signal wires. The second one is the fact that the reduction of physical sizes of the devices and of the supply voltages have reduced the operating margins of the ICs and made them more sensitive to both internal and external influences. The third one is the increasing parameter variation, which is another source of non-ideal circuit behaviour and also has a negative impact on the robustness of operation. The reduction of the device sizes in combination with the increased density has also made ICs to become more vulnerable to soft errors. All these effects tend to move in the wrong direction and will have severe consequences for the design methods and flow to keep robustness at a sufficiently high level.

Until the change of the millennium, the reliability of ICs has not suffered dramatically from the continuous scaling of the technologies. This, however, is expected to change, when the semiconductor industry is required to move to new device concepts and new interconnect strategies. The impact of hot-carrier effects, NBTI and ESD is very much dependant on the physical sizes and material properties of the transistor stack. In this respect the alternative MOS transistors, such as multi-gate devices and FinFETs, are very challenging regarding reliability topics. An accurate prediction of device behaviour with reliability models turns out to be very difficult. The combination of new materials and device stacks may introduce new reliability problems, for which the models still need to be built.

Higher device densities and increased frequencies have led to a continuously increasing demand in IC current. At the same time, the metal widths and heights reduce, particularly of the lower metal layers. The corresponding expected increase in current density requires the permanent attention of the designers, to prevent electromigration to occur anywhere in the chip. Power estimation tools, capable of calculating the required currents, are therefore needed early in the design flow, to create a proper global supply network, while also paying attention to the local supply wire widths.

From these considerations it is clear that it will be a continuously challenging task for the designers to maintain the robustness and reliability level in a "shrinking environment".

A robust design not only requires the integrity of the electrical and physical operation of the chip, it also includes the set-up of a very well organised database. This allows easy, correct and rapid design modifications when redesigns or different versions of the design are required.

9.8   References

[1] B. Barton, et al.,
ESSCIRC, low-power workshop 1997, Southampton

[2] P.J. Resle, et al.,
'A Clock Distribution Network for Microprocessors',
IEEE Journal of Solid-State Circuits, Vol. 36, No. 5, May 2001,
pp.792-799

[3] S. Rusu,
'Clock Generation and Distribution for High-Performance Processors',
SoC 2004, http://www.tkt.cs.tut.fi/kurssit/8404941/S04/chapter5.pdf

[4] Kerry Bernstein, et al.,
'High-Speed CMOS Design Styles',
Kluwer Academic Publishers, 1999

[5] S. Rusu,
'Circuit Design Challenges for Integrated Systems',
Workshop on Integrated Systems, European Solid-State Circuits Conference, September, 1999

[6] H.A. Collins and R.E. Nikel,
'DDR-SDRAM high-speed, source-synchronous interfaces',
EDN, September 2, 1999

[7] H. Basit, et al.,
'Practical Multi-Gigahertz Clocks for ASIC and COT Designs',
DesignCon 2004

[8] B. Nauta and G. Hoogzaad,
'How to deal with substrate noise in analog CMOS circuits', European
Conference on Circuit Theory and Design, Budapest, September 1997

[9] H.B.Bakoglu,
'Circuits, Interconnections and Packaging for VLSI',
Addison-Wesley, 1990

[10] E. Dupont, et al.,
'Embedded Robustness IPs for transient-error-free ICs',
IEEE Design & Test of Computers,
Vol. 19, No. 3, pp. 56-70, May/June 2002

[11] Marc Derby,
'Soft-error impacts on design for reliability technologies',
Keynote talk at IOLTS, July 2007

[12] T. Heijmen, et al.,
'Soft-Error Rate Testing of Deep-Submicron Integrated Circuits',
Test Symposium, 2006. ETS '06

[13] N.Seifert, et al.,
'Radiation-induced Soft Error Rates of Advanced CMOS Bulk Devices',
IEEE 44th Annual International Reliability Physics Symposium, San Jose, 2006, pp. 217-225

[14] V. Petrescu, et al.,
'A Signal Integrity Self Test (SIST) concept for the debug of nanometer CMOS ICs',
ISSCC 2006, Digest of Technical Papers, session 29

[15] P. Drennan, et al.,
'Implications of Proximity Effects for Analog Design',
IEEE 2006 CICC conference

[16] J.M. Brunet,
'modelling Process Variability in the Design Flow',
Chip Design Magazine, Issue Dec 2005/Jan 2006

[17] M.Vertregt,
'The analog challenge of nanometer CMOS',
IEDM 2006, Digest of Technical Papers, pp. 11-18, December 2006

[18] M.Pelgrom, et al.,
'Transistor matching in analog CMOS applications',
International Electron Device Meeting (IEDM) 1998, pp. 915-918

[19] M.Vertregt,
'Embedded Analog Technology',
IEDM short course on System-On-a-Chip Technology,
December 5, 1999

[20] P.Stolk, et al.,
'Modeling Statistical Dopant Fluctuations in MOS Transistors',
IEEE Transactions on Electron devices, Vol. 45, No. 9, September 1998

[21] International technology Roadmap for Semiconductors,
2005 edition and 2006 update,
www.itrs.net/reports.html

[22] T. Kanamoto, et al,
'Impact of Well Edge Proximity Effect on Timing',
ESSCIRC 2007, Digest of technical papers, pp. 115-118

[23] M. Pelgrom, et al,
'Digital circuit insights from analog experiences',
ISSCC 2007, Special Topic Evening Sessions

[24] A.Agarwal, et al,
'Statistical Timing Analysis using Bounds',
DATE 2002

[25] Jing-Jia Liou, et al,
'Fast Statistical Timing analysis By Probabilistic Event Propagation',
DAC 2001, June 2001, Las Vegas

[26] H.J.M.Veendrick,
'Wire Self-heating in Supply Lines on Bulk-CMOS ICs',
ESSCIRC 2002, Digest of Technical Papers,
pp. 199-202, September 2002.

[27] G.D.Wilk, et al,
'High-k dielectrics: Current status and materials properties considerations',
Journal of Applied Physics,
Vol. 89, No. 10, pp. 5243-5275, May 2001

[28] Anit Kottantharayil,
'Low-Voltage Hot-Carrier Issues in Deep-sub-micron MOSFETs',
http://137.193.200.177/ediss/kottantharayil-anil/inhalt.pdf

[29] S.Mahaptra, et al,
'Device Scaling Effects on Hot-Carrier Induced Interface and Oxide-Trapped Charge Distributions in MOSFETs',
IEEE Transactions on Electron Devices, Vol. 47, No. 4, April 2000

[30] K. Kushida-Abdelghafar, et al.,
'Effect of nitrogen at SiO_2-Si interface on reliability issues negative

bias temperature instability and Fowler-Nordheim stress degradation',
Appl. Phyiscs Letters, 81 (23) (2002)

[31] Y. Hiruta, et al.,
'Interface state generation under long-term positive-bias temperature stress for a p+ poly gate MOS structure',
IEEE TED 36, p. 1732 (1989)

[32] T. B. Hook, et al.,
'The effect of fluorine on parametric and reliability in a $0.18\,\mu$m $3.5/6.8$ nm dual gate oxide CMOS technology',
IEEE TED, 48 (7), p. 1346 (2001)

[33] Ogawa, et al.,
'Interface-trap generartion at ultrathin (4-6 nm) interfaces during negative-bias temperature aging',
JAP 77 (3) (1995)

[34] A. Scarpa, et al,
'Effect of the Process Flow on Negative-Bias-Temperature-Instability',
Proc. 8th International Symp. on Process- and
Plasma-Induced Damage, p. 142, 2003

[35] P. Chaparala, et al.,
'NBTI in dual gate oxide PMOSFETs',
Proc. 8th International Symp. on Process- and
Plasma-Induced Damage, p. 138, 2003

[36] R.R. Troutman,
'Latchup in CMOS Technology',
Kluwer Academic Publishers, 1986, ISBN 0-89838-215-7

[37] http://www.esdlab.com/others.htm

[38] A. Ameraskera and C. Duvvury,
'ESD in silicon integrated circuits',
John Wiley, 2002, ISBN 0-471-95481-0

[39] M.D. Ker, et al,
'ESD Test Methods on Integrated Circuits; An Overview',
IEEE website

9.9 Exercises

1. Explain why the internal chip latch-up sensitivity will decrease every new process generation.

2. What are the main causes of supply noise inside a VLSI chip?

3. Explain why the power supply lines to a large driver circuit (e.g., clock driver or output driver) should be wider than the output signal track.

4. When we decide to reduce the thickness of the copper wires in a CMOS process, explain what would be the advantages and disadvantages in terms of signal integrity and reliability, if we would not adapt our design to this reduction?

5. What is generally the best place to position the clock drivers and why?

6. What are the main causes of clock skew and what are the measures to reduce it?

7. Explain how the back-end of the manufacturing process is dominating the IC behaviour.

8. Mention several reasons for increasing di/dt. What are the consequences?

9. What is the impact of an increased di/dt on the signal integrity?

10. Explain why the use of a good database management system is required during the design of a VLSI chip.

11. Why would the implementation of an ESD protection be more problematic on SOI than on bulk-CMOS?

12. Explain the use of decoupling capacitors and why they are needed.

13. Assume a certain IP-core is consuming $100\,\mathrm{mA}$ at $1\,\mathrm{V}$. Assume also that the maximum allowed combined average voltage drop (IR-drop) across the V_{dd} and V_{ss} supply lines in the top-level metal layer to this core is only 2% of the supply voltage. What would be the distance of the block to the supply bonding pads at which both the electromigration as well as the voltage drop requirements

would exactly be fulfilled? Use the electromigration number of copper in the section on electromigration at 125°C. The square resistance of copper is $R_\square = 22m\Omega/\square$ for the top level metal ($R = R_\square \cdot (L/W)$).

14. What would be the difference between a synchronous and an asynchronous implementation of a logic core in terms of signal integrity?

15. What would be the effect of a small forward body bias on the performance of a transistor? What would be the risks of using forward body bias with respect to reliability?

Chapter 10

Testing, yield, packaging, debug and failure analysis

10.1 Introduction

Although this is almost the final chapter in this book, it does not mean that the topics discussed here are less important than those of the previous chapters.

Testing, debugging, yield and packaging have a substantial influence on the ultimate costs and quality of a chip. Relatively short discussions of these topics are therefore included in this chapter.

An integrated circuit can fall victim to a large variety of failure mechanisms. Ideally, the related problems are detected early in the manufacturing process. However, some only show up during the final tests, or even worse, they might not be identified before the chip is soldered on a customer's board.

The next paragraph starts with an overview of different test methods currently in use, and continues with the measures that a designer can implement to improve the testability and support the debug of his design.

The engineering and evaluation of first silicon until it is considered to be "error free" happens to be a tough job. Programmable processors, for example, may be used in an almost unlimited number of different applications. It is almost impossible to guarantee even "fifth-time-right" silicon for these kinds of ICs.

Even when a failure is detected during the testing of first silicon, it might take a considerable time before the cause of failure is located and proven. This is because complex ICs contain up to several hundreds of millions of transistors and up to several hundreds to a thousand of I/O pins. It is therefore very complex to locate an internal failure via a limited number of external pins (I/O). Moreover, because of the increased number of interconnection (metal) layers, physical probing of signals has almost become impossible. Design for test and debug should therefore be adopted as a general design approach, to enhance controllability and observability, to ease the detection of design bugs and other failure mechanisms during the engineering phase of first silicon.

Eventually, the results of all tests determine which of the chips pass and which fail. The related yield is function of the test coverage. A flavour of the most important yield topics is therefore part of the discussions.

Many of the chip characteristics in terms of performance, robustness and reliability are co-defined by the way the chip is designed and packaged. This chapter, therefore, also contains a summary of the most commonly used packages. It includes a few presentations that describe the influence of the package on some of the electrical and thermal characteristics of the chip. Also trends in SoC and SiP integration solutions are identified.

Finally, this chapter is concluded by a presentation of failure analysis methods that support the detection of failures and their diagnosis to enable a fast identification of the failure mechanism during first silicon debug. This should prevent customer returns and will shorten time-to-market.

10.2 Testing

Testing is done to bridge the gap between customer requirements and
the quality of the design in combination with the manufacturing process. Testing thus helps to increase the quality of an IC. The yield is
determined by testing and can be influenced by the complexity of the
test: a simple test may lead to a higher yield but can lead to more *customer returns*. The yield for large and complex ICs can be relatively
low and can dominate the ultimate costs. How extensive a design needs
to be tested depends on many different factors. Three major different
test categories can be distinguished.

Characterisation tests, which are mostly executed manually, are developed for characterising the chip operation and/or operation area with
respect to different operation conditions. These tests are more focussed
on the accuracy of the test than on the speed of testing. These tests are
very much related to which parameters and what type of circuits are to
be characterised:

- Design errors, design margins, manufacturing defects.

- The on-chip circuits: pure logic, pure analog, pure memory, static
  circuit, dynamic circuit or a mix of analogue, RF and digital circuits.

- An increasing number of chip failures is related to dynamic effects
  such as cross-talk, charge sharing, critical timing and noise. Most
  automatically generated tests detect the 'stuck-at-one' and 'stuck-
  at-zero' faults, which cause circuit nodes to remain at '0' and '1',
  respectively. Nanometer CMOS technologies, however, result in
  lower supply voltages and, consequently, reduced noise margins.
  This will produce faults that are much more difficult to classify
  than the traditional stuck-at faults.

Production tests, which are performed in an automated mode, provide a way to reject those ICs that do not meet the required specification
criteria or performance limits. Production tests include a large number
of different tests to achieve the best possible test coverage and depend on
the quality requirements of the target product and/or target application
area:

- Consumer

- Computer

- Aviation: aircraft and space craft

- Automotive

Due to safety requirements it will be clear that the latter two require
higher test coverage.

Reliability tests, which are mostly performed manually, challenge the
chip operation during and after the exposure to extreme electrical and
environmental conditions:

- Electrical stress, burn-in.

- Temperature cycles, thermal shock and high-temperature storage.

- Increased humidity levels.

- Mechanical vibrations and shocks.

- Other reliability tests include electrostatic discharge (ESD) and
  latch-up, but since these are very much design related, they are
  discussed in chapter 9.

Because of safety requirements, aviation, navigation and medical applications usually require very exhaustive testing particular with respect
to reliability standards. For the same reason, also automotive products require exhaustive testing. Moreover, they operate in more "hos-
tile" environments which may include large supply transients or interference caused by switching of heavy or inductive loads such as lamps
and starter motors and require specific protection and more stringent
reliability tests.

When all tests are executed properly, only a few of the chips that
pass all tests still may be returned by the customer (*customer returns;
escapes*) because of a failure, which displayed itself in the application
either directly or after a while (day, week, month, year). The number of
customer returns is expressed in *ppm* (parts per million), which represents the ratio of customer returns per million supplied chips. This ppm
level has become representative for the quality of a delivered product.
Ppm acceptance levels are related to the quality requirements of the application domain. While typical automotive applications allow 1 ppm,
consumer applications and microprocessors may show ppm levels of 100
and 300, respectively.

Tests can be performed both on the wafer and on the final packaged product. Pre-tests, also known as *e-sort* (early sort), are usually performed directly on a wafer to prevent unnecessary assembly costs. The final tests are performed on the packaged die. There is often a lot of overlap between the pre-tests and the final tests. As a result of the associated additional costs, the number of redundant tests must be limited. During pre-test, the individual ICs are tested on the wafer by probing the bond pads of the chip. Figure 10.1 shows a photograph of an example of a probecard with more than 120 probes.

Figure 10.1: *Example of a probecard (Source: MICRAM Microelectronic GmbH)*

A probe station brings these small needles into contact with the IC's bond pads. A test system provides pre-determined stimuli for the IC and compares actual output signals to expected responses (figure 10.2). The stimuli should ensure that a large percentage of possible faults will result in discrepancies. This percentage is called fault coverage with respect to the applied fault model and is most commonly targeted at above 99%.

The fault coverage, however, is always related to the fault model used (stuck-at, bridging, stuck-open, transition (only once), gate delay and path delay). Redundancy can also be reason for reduced fault coverage. The test stimuli and response signals are transferred through a connector that provides a bi-directional link between the probes and the test system. The test system, *Automatic Test Equipment (ATE)*, can also be used to control the wafer prober or handler to automatically step from one circuit to another so that a number of ICs can be tested in rapid sequence.

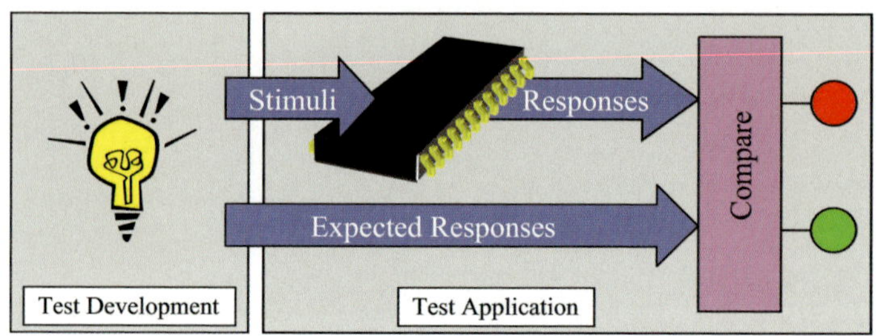

Figure 10.2: *Basic principle of IC testing*

It was relatively easy to manually determine test stimuli (vectors) for complete SSI (small-scale integration) circuits. For VLSI circuits, however, this is impracticable and has led to the development of computer programs that generate test vectors. A complete test program may consist of several subtests, of which the most important ones are discussed in the next paragraph.

The quest for high bit or gate densities consumes much design effort aimed at the realisation of a maximum amount of electronics on a minimum area. However, designers must ensure that their circuits are testable. For VLSI circuits, an increase in testability may, for instance, result in a chip area 'sacrifice' of 5% and a 50% reduction in test costs.

10.2.1   Basic IC tests

This section discusses the most important different tests that are applied to a chip to achieve sufficient test coverage. Some of these tests are done at chip level, some at core level, using so-called test protocols. A test protocol is a detailed description of how the test stimuli must be supplied (to which terminals of the core and in which time slots) and how the responses of the core to these stimuli must be captured.

Contact test

As discussed before, pre-tests are performed directly on the wafer to prevent unnecessary assembly costs. To perform the test, the needles of the probe card are brought into contact with the bond pads of the chip. The larger the number of bond pads, the bigger the chance that one or more needles make bad contact with the pads. During the contact test a voltage is applied to all pads. When a needle makes good contact to its corresponding pad, a current will flow through the (ESD) protection diodes (see chapter 9) connected to that pad. If no current is measured, the connection fails and the probe station may try to reconnect to the pads, or simply step to the next chip, without performing any of the other tests.

Functional test

Due to the large amount of different IP and memory cores on a chip, only a few of them will have direct access to the pins and can be functionally tested.

A functional test re-uses stimuli from the design simulation phase which exercise the function of the circuit.

Preferably, the whole function of the IC should be tested similar to the way it is used in the application. This full functional test should be performed at application speed, since only then the IC will show its real application behaviour. Some IC problems only manifest themselves if all circuits operate simultaneous and identical to the application:

- voltage drop along supply lines

- supply and ground bounce due to peak currents through the supply lines, bond wires and package leads

- cross-talk between neighbouring signals

- EMC behaviour

The on-chip noise is maximal when all circuits are active, just like in the application. When a circuit fails, it may not directly be an indication for a bad chip or bad design. It may also lead to the conclusion that the simulation is not correct. In many cases, however, the malfunction may be caused by a timing-critical failure, a noise related operation failure, a process related defect causing (resistive) shorts or opens or by a design error. Debug and failure analysis techniques are used to locate the failure and support the diagnosis. These are discussed in the final subsection in this chapter.

A test is not only used as a method to verify correct behaviour of an IC. It is also used as a quality measure for its operating area. A Shmoo plot can also be used for this purpose. A Shmoo plot is a graph that represents when a certain test passes or fails with respect to a large number of parameter settings at which the chip is repeatedly tested. Figure 10.3 shows an example of a Shmoo plot in which both the supply voltage and frequency are varied over certain ranges around the spec area. The shape of the Shmoo plot may contain information about the cause of the failure and will be discussed in more detail in section 10.6.3.

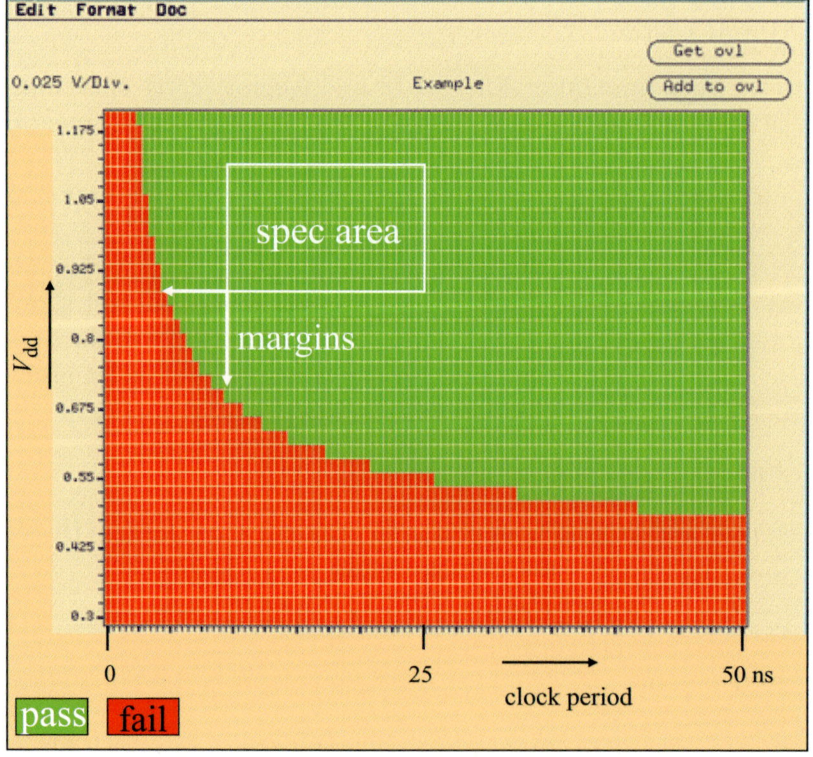

Figure 10.3: *Shmoo plot example, showing the margins between the measured operating and the spec area*

The Shmoo plot must also be measured for the worst and best-case temperatures and the way it changes its shape is representative for the operation of the chip (see section on failure analysis).

Let us assume that the Shmoo plot in this figure represents that of first silicon. When the margins of the operating area, with respect to the spec area, are small, there is a big chance that one or more of the following batches will show failing circuits. In other words, this kind of test allows you to anticipate already on what problems may show up in the (near) future and debug and failure analysis can already be started to determine the cause and location of the of the circuit that causes the small margin.

Delay-fault test

For a digital circuit to operate correctly, it is required that the propagation delay along all of its logic paths is more than, or less then a specified limit. Random process parameter variations and/or defects may cause propagation delays to exceed these specified limits, leading to operation faults. These faults can only be detected by a form of so-called at-speed testing. At-speed tests were only very scarcely applied and usually carried out with functional test patterns. The increasing IC complexity, both in terms of components and speed, has made at-speed functional testing unattractive mainly due to the test development costs and the cost of a complex test system with a very high timing accuracy. *Structural test* techniques, such as scan tests, which were originally developed for relatively slow stuck-at faults testing, are forming today a viable cost-efficient alternative to at-speed testing. Delay fault testing offers a structural approach to at-speed timing tests, while keeping test hardware costs limited.

A *delay fault* usually refers to a single logic gate or a logic path exceeding its maximum specified propagation delay. The output of a logic gate responds to a transition of one or more of its inputs. The time such a response takes is called *gate delay*. *Path delay* represents the accumulated gate delay plus the interconnect delays within that path.

There are two fault models related to delay fault testing: the transition fault model and the path delay model. Transition faults, also called gate delay faults, model defects which occur at the inputs or outputs of a logic gate and which lead to a gate delay outside its specified range. Path delay faults model defects that cause signal propagation along the path that takes longer than the specified maximum time (usually defined by the clock period).

A *transition delay fault* is caused by random variations occurring both in the devices and interconnections. Examples are: threshold voltage shifts, CMOS opens, highly resistive vias, narrow metal lines, signal integrity causes such as cross-talk, supply noise and IR drop.

Test pattern generation for delay fault testing (for transition delay and path delay faults) is based on stuck-at procedures and require two tests. The first test puts the targeted faulty circuit path in a certain state, while the second test introduces an input transition, such that it propagates to one or more primary outputs or scan flip-flops. Scan chains are normally used to guide the stimuli patterns to the faulty path, which may have been identified through static timing analysis. The responses of the faulty path to these input stimuli are captured by the output scan chain. It is virtually impossible to detect all possible delay faults. Therefore, several techniques have been proposed to enhance delay fault coverage for standard scan designs [1]. However, a discussion of these techniques is beyond the scope of this book.

A complete functional test of an IC or of parts of the IC is often complex and time consuming. Therefore a structural test approach is applied to most of today's ICs. This type of test is also called scan test.

Scan test (structural test)

As already stated before, the complexity of today's ICs is so high that most cores have no direct access to the pins. To be able to test whether these cores show correct behaviour, we have to create access to them by artificial means. A synchronous design is built from logic gates and flip-flops, which are used to control the data flow through a chip. The flip-flops are controlled by a clock signal and temporarily store a logic value and let it go on clock demand. So, in normal operation, the flip-flops are an essential part of the total function of a core. But, during test mode, these flip-flops are put in series, thereby bypassing the logic path in between them. As such, the flip-flops form a so-called *scan chain* to guide the data to target nodes in a logic core. Each *scannable flip-flop* is therefore equipped with an additional multiplexer to enable connection to the logic path during normal operation and connection to the previous flip-flop in the chain during test (figure 10.4).

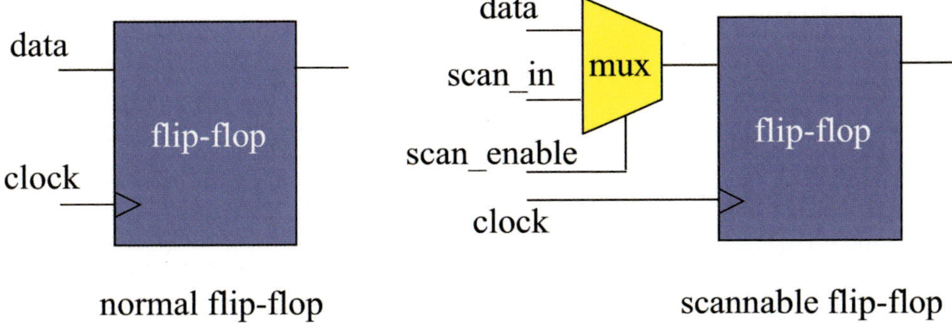

Figure 10.4: *Example of the use of a scannable flip-flop*

This multiplexer is controlled by the *scan_enable* signal, which needs to be routed through the chip as well. Different cores can be connected to the same scan chain (see figure 9.3 for a scan chain example). To test one of the cores in a chain, the following sequence of operations is executed:

- put the chip in the scan-test mode: *scan_enable* goes high to form the scan chain

- guide data through the scan chain to the flip-flops in the target core

- put the core in functional mode for performing a single clock cycle

- put the chip in test mode again

- scan out the results of the functional operation

- compare these results with the expected data (most commonly from simulations)

A test may activate many scan chains (10-200) in parallel to reduce test time. Their number is limited by the number of available (re-used) I/O pads of the application. Overall accessibility is ensured because each flip-flop in each logic block is part of a scan chain. The total chip area overhead to support scan test (the multiplexer in each flip-flop, the routing of the scan data, *scan* and *scanb* signals and the test-control block (TCB)), is typically less than 5%.

The I_{ddq} test, which is discussed next, uses the scan-test infrastructure to put the chip in different states, in order to detect defects or faults.

I_{ddq} and ΔI_{ddq} test

During the eighties, the IC testing was based on *stuck-at fault* models, which could detect failures at logic gates and flip-flops when their outputs were short circuited to V_{dd} or ground (stuck-at-one or stuck-at-zero, respectively). However, with these simple models, it was not possible to cover all process-oriented defects. I_{ddq} tests are particularly good at detecting bridging faults, power supply short circuits and punch-through failures.

In a normal static CMOS logic gate, either the pMOS pull-up network is conducting, keeping the output at high level (logic "1"), or the nMOS pull-down network is conducting, keeping the output at low level (logic "0"). In the steady state, no current usually flows through such a logic gate, except for a negligibly small subthreshold leakage current of the logic gates. Above $0.25\,\mu$m CMOS technologies, the magnitude of this leakage current was usually below $1\,\mu$A. At such a level of background current, larger steady-state currents, caused by different process-defect mechanisms, can easily be detected by measuring, as these currents are several orders of magnitude higher than the leakage current.

For example, common gate oxide defects may result in current values in the order of micro-amperes to several milli-amperes, depending on the size of the defect and on the size of the transistor involved. A drain-source bridging defect can easily cause steady-state currents up to several milli-amperes as well.

However, a *defect* is not always leading to a fault. It may cause a *structural fault* when it is large enough to connect two neighbouring conductors or disconnect a continuous pattern. Only faults lead to yield loss. Some defects do not lead to structural faults, but only to *parametric faults* which may affect the circuit performance.

During the measurement of the *steady-state current*, the chip has to be put in the steady-state mode. In many CMOS ICs, this state can be achieved by just switching off the clock. However, the chip often has to be put in a special I_{ddq} test mode before switching off the clock. In this way, defects are detected by the level of the supply current during the steady state. This is called I_{ddq} testing. I_{ddq} test pattern generation is only needed to put the chip or different parts of the chip in a certain mode (controllability). Observability need not be supported, as results of the test are simply measured via I_{ddq} currents. Because the current needs to settle during the measurement, I_{ddq} testing is a relatively slow process.

Especially in circuits that contain non-static CMOS circuits, such as PLLs, A/D, D/A and other analogue circuits, floating nodes (e.g., tri-state buses), dynamic and pseudo-nMOS circuits need additional attention during the design to make the total chip I_{ddq} testable.

At which I_{ddq} level the chip should be considered defect depends on many things. The number of gates is one important parameter, while the level of the threshold voltages of the nMOS and pMOS transistors is also dominant in determining the critical I_{ddq} level. Because of scaling, the threshold voltage is reduced every process generation, to maintain or increase the speed of each new generation of ICs. As shown in chapter 8, the subthreshold current in a transistor is defined by the *subthreshold slope* (chapter 2) of the device. A typical value of $80\,\text{mV/dec}$, for bulk-CMOS devices, leads to an increase of about a factor eighteen for each threshold voltage reduction of $100\,\text{mV}$.

CMOS technologies beyond $120\,\text{nm}$ exhibit even more leakage mechanisms. Next to subthreshold leakage also gate-oxide and junction leakage start playing a role. Because of the increased leakage levels, I_{ddq} testability is therefore no longer possible for most ICs made in $120\,\text{nm}$ CMOS technologies and beyond. An alternative, in this respect, is the ΔI_{ddq} test. During this test, the chip is put into several different states, by scanning a variety of test vectors through the scan chain. After every new test vector, the chip is put into the corresponding quiescent state and the I_{ddq} is measured. Next all I_{ddq} values are compared with each other and so the ΔI_{ddq} is determined. Figure 10.5 shows some measurement results for three ICs.

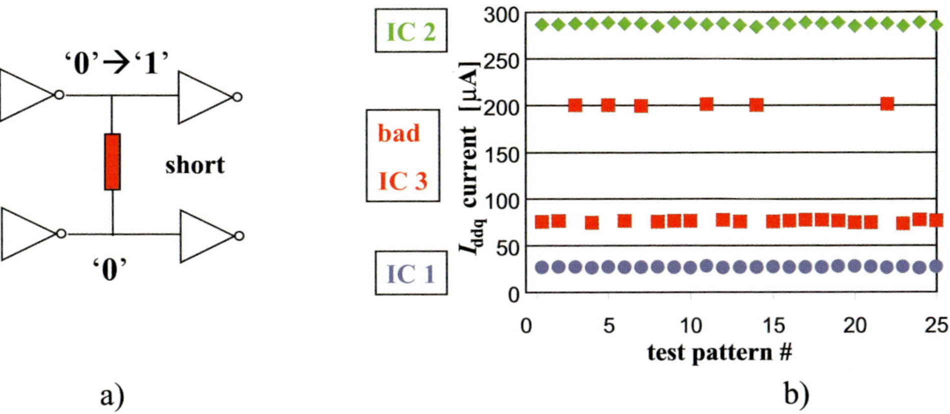

Figure 10.5: *Leakage measurements for different test patterns in different*

These ICs may originate from the different batches. A small threshold-voltage shift can cause a large change in leakage current. Different test vectors are needed to put the chip in such a state that a defect is detected. In figure 10.5.a the short will only lead to a defect-oriented current when the logic levels across the short are different. Although IC 1 and IC 2 show a relatively large difference between their I_{ddq} values, these values are independent of the state (test vector) of the chip. The bad IC 3, however, shows different values of the I_{ddq} current, which means that certain test vectors bring the chip in a state in which it manifests a short. The rejection criterion is not the absolute value of I_{ddq}, but the difference between I_{ddq} values of several measurements. Therefore, this test is called ΔI_{ddq}.

Very low voltage (VLV) testing

Bridging, *gate-delay*, and *path-delay* faults may not always be detected. These *delay faults* are mostly caused by opens (e.g., a bad via, which causes too much contact resistance). In non-critical delay paths, these faults may not be detected, but, then, they may not lead to functional errors and can thus be tolerated. When we define the *golden device* to be a product that operates perfectly according to the specification and even at voltages well below the worst-case specified levels, then *VLV (very low voltage)* tests can be performed on the other devices to detect weak (high-ohmic) shorts. At such low-voltage levels, these shorts can easily cause a much longer delay, leading to a detectable fault. Figure 10.6 shows an example of voltage-based testing of shorts.

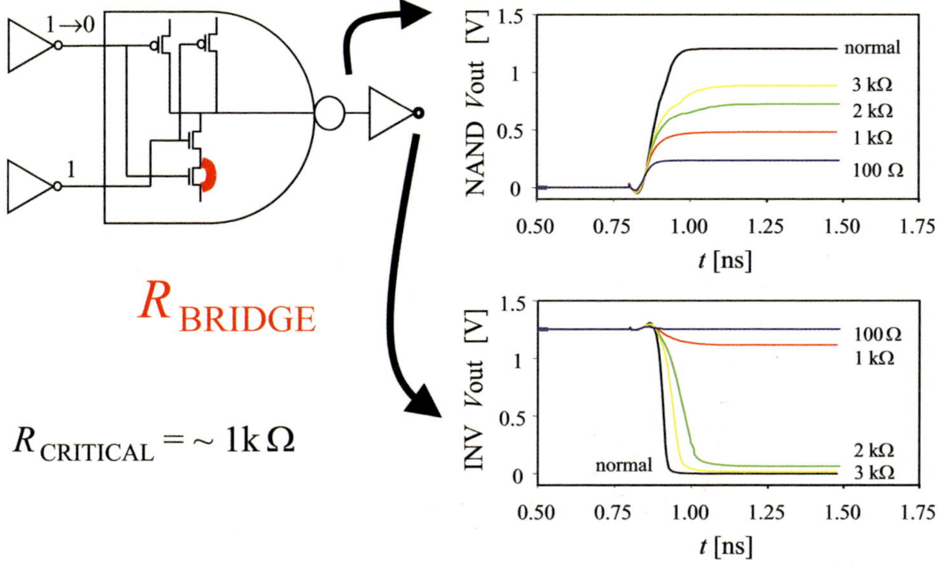

Figure 10.6: *Example of voltage-based testing of shorts*

Assume that, due to a manufacturing defect, there is a high-ohmic short (or bridge) between the drain and source of the transistor as indicated in the figure. When switching the gate of this transistor from a logic 'one' to a logic 'zero', it will not switch completely off. With no bridge, the output of the two-input NAND gate would switch to the supply voltage, indicated as 'Normal' in the upper diagram. However, dependent on the resistance of the bridge, this output will not reach this level. Shorts with more than $2\,\text{k}\Omega$ resistance will only manifest them selves as additional gate delay at the output of the inverter connected to the NAND (lower diagram). The fault will manifest itself as an additional path delay.

Figure 10.7 shows an experimental Shmoo plot measurement using VLV testing of a resistive short in an inverter.

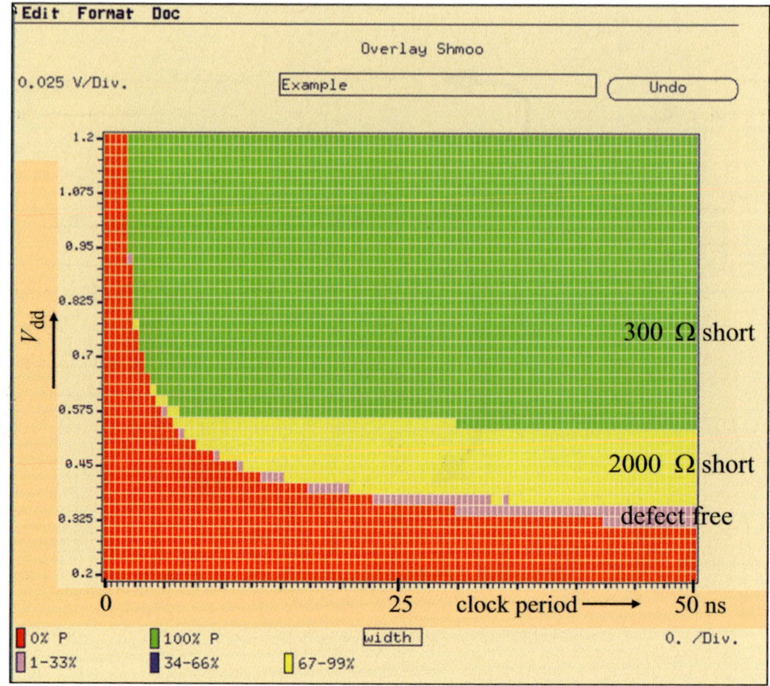

Figure 10.7: *Shmoo plot measurement used during a VLV test of an inverter*

The Shmoo plot clearly shows that the operating area reduces when the short becomes stronger (less resistive). At a resistive value of $300\,\Omega$, the circuit does not operate at voltages below $0.525\,\text{V}$, even at frequencies of only $20\,\text{MHz}$.

BIST

The costs of testing will dramatically increase as a result of the increase in the speed of the circuits, the reduction of the voltages (smaller noise margins) and the increase in the number of bond pads. The cost of a tester will increase from from a few million to more than 10 million US$ in the next decade. *Built-in Self Test (BIST* techniques are currently used in several (embedded) memories. Figure 10.8 shows an example of BIST in an embedded memory: *memory BIST*. To reduce the cost of overall chip testing, BIST techniques must also be included in the design of digital and analogue blocks.

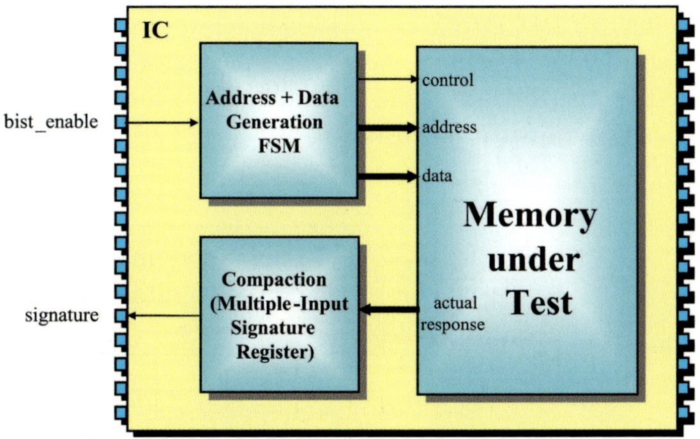

Figure 10.8: *Example of built-in self-test in an embedded memory*

Although the embedded memories in a VLSI chip contain most of its transistors, it is relatively easy to achieve a large memory test coverage, because of the regular architecture of its memory array. To be able to detect defects between neighbouring bit lines or word lines, they have to be set in different logic states. First the complete memory is loaded with '1's and then read. Next it is loaded with '0's and then read. Then a checkerboard (1 0 1 0 1 0 1) pattern is loaded into the memory, such that every '1' is surrounded by '0's and then read. Next the inverse checkerboard pattern is loaded and read. These tests only contain very regular patterns of '1's and '0's, which can easily be generated by a finite state machine (FSM) and an address sequencer. All output data (read back data) is sequentially stored in a *multiple-input signature register (MISR)*. A MISR basically contains a shift register with a built-in linear feedback loop. It generates a signature which is dependent on all bits that are fed into it. If one or more bits are wrong, the signature does not match its expected value, meaning that it has detected a fault. A MISR actually compresses the output data to a single signature to save test time.

Because of its simplicity in both the on-chip generation and comparison of the test vectors, memory BIST has already been incorporated in many IC designs and has become more or less standard practice, today. Currently, BIST is often combined with a repair action. This technique is called *built-in self-test and repair (BISTAR)* in which faulty columns are replaced by correctly operating redundant columns, by blowing polysilicon fuses.

Introducing self test into logic cores is much more complicated. This so-called *logic BIST* (LBIST) measures the response to random test patterns, fed to the different scan chains in the logic core. Figure 10.9 shows the basic architecture of LBIST.

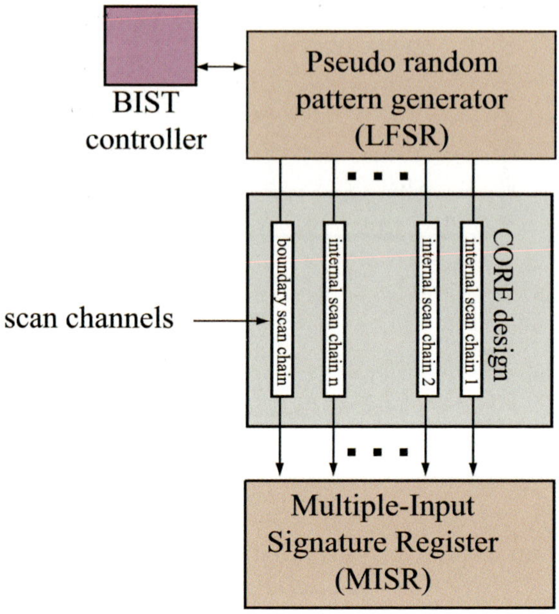

Figure 10.9: *Basic architecture of logic BIST*

The BIST controller generates a sequence of activities to perform the execution of BIST:

- The MISR is first put into a known state

- A pseudo/random pattern generator (PRPG), implemented as a linear feedback shift register (LSFR), generates the input stimuli to the different internal scan chains on the chip.

- Then a functional cycle is started to capture the responses to the input stimuli

- The MISR compresses these responses and at the end of a BIST execution, the final state that is captured in the MISR is called the signature.

- This signature is compared with the expected signature, known from simulation. A mismatch between these signatures is an indication for the occurrence of defects in the logic circuit.

There are a few remarks to be made here. A disadvantage of LBIST is the associated low fault coverage. *Deterministic LBIST (DLBIST)*, which uses an LBIST decompression/compression architecture capable of applying deterministic test patterns, shows a reasonable test coverage, however at the cost of a relatively large area overhead.

LBIST has therefore not yet become a mainstream test solution. It was and still is not an integral part of the synthesis tools and design flow. However, with the rapidly growing test cost, LBIST may become more generally accepted as a standard design for testability methodology. LBIST is already in use for some time in special applications, particular in security applications where a scan test would enable unwanted read out of the security key, and in applications that require field tests and where there is no tester nearby.

Boundary scan test

Advances in semiconductor and packaging technologies lead to such densely integrated modules that overall system accessibility is reduced. Also, the need for shorter time-to-market requires flexible and fast in-system testability. In 1990, a breakthrough in system test methods was made with the standardisation of the so-called *Boundary Scan Test (BST; IEEE 1149.1, JTAG)* method. BST reduces the overall test costs and simplifies board and system level testing.

Although BST increases chip and board costs (additional area dedicated to *design-for-testability* circuits), this is recovered by the advantages mentioned in this section. BST also supports system production efficiency and in-field serviceability. With BST, interconnection failures during the assembly of ICs and in between ICs on a board, such as the open circuits, short circuits and stuck-at faults, can be detected. In the BST approach, a boundary cell, which contains a flip-flop, is positioned between every pin to core connection. Each cell is also connected to its two neighbours, see figure 10.10. In the BST test mode, these cells form a scan register, which is able to serially scan in and scan out test data.

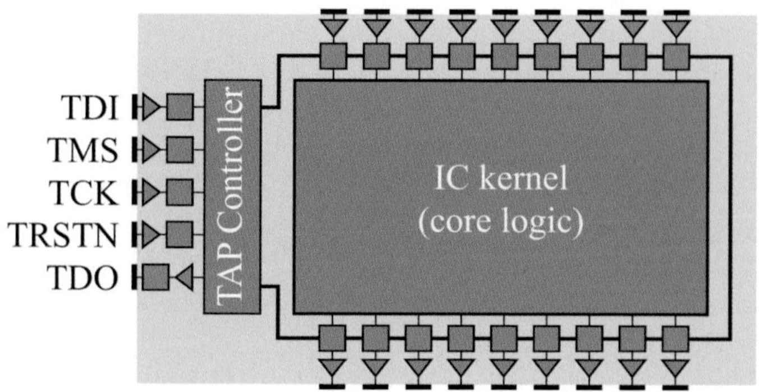

Figure 10.10: *The Boundary Scan Test approach*

Independently of the cores, such a scan chain can drive and monitor the pin connection of each chip in the system. A test clock and an additional test pin control the BST test mode of the system. BST supports three basic tests: interconnection tests between BST chips, IC core tests and function monitoring during normal circuit operation.

Because of the BST standard, ICs from different vendors supporting BST can be placed on the same board in a system to allow overall system testability. BST replaces the conventional 'bed of nails' test technique, in which a tester is connected through numerous wires to an array of pins on an acrylic substrate, whose footprint matches that of the test points on the PCB to be tested. Ideally, all components on a board are equipped with BST. However, even if some components do not have BST, there are still substantial benefits. IEEE 1149.1 mandates a few instructions to support board level interconnection testing, but is open for private instructions. Many companies make dozens of such private instructions, e.g., for IC production testing, silicon debugging, emulation and application debugging, etc.

10.2.2 Design for testability

The previous subsection discussed the different tests that are currently applied to achieve the best possible test coverage. This section discusses what a design team needs to do to support test: *design for testability* (DfT):

- Make the design I_{ddq} or ΔI_{ddq} testable.
  There are several generally accepted guidelines for this test. Usually test patterns are created at core level or at chip level by an

automatic test pattern generator (ATPG) tool (e.g., Tetramax, Fastscan, EncounterTest, and AMSAL). These test patterns are capable of putting all cores into various different states for making defects visible through varying I_{ddq} currents. The amount of test vectors needed to create a sufficient number of different states depends on the application domain of the chip. For certain IC categories, no I_{ddq} test is performed at all, because of test-costs savings. Other categories may require around ten to twenty well-chosen test vectors, which may create 98% I_{ddq} test coverage. Some ICs might even require several hundreds to a thousand test vectors. These then require the usage of an I_{ddq} monitor on the load board, to speed up the test.

- If possible, subdivide the chip into separately testable functional blocks.
  The possibility of executing full functional tests, allows to mimic the real application. Preferably the whole chip should be fully functional tested, because only then the real application conditions are created on the chip, including all current peaks, cross-talk and supply noise sources. Although the complexity of today's ICs does not allow full functional test, it is a very important that the test conditions match closely with the real application conditions, because there are an increasing number of cases in which the chip passes all tests, but still fail in the application, due to the different noise conditions. There are even examples in which the chip fails in the test mode, but still shows correct behaviour in the application due to the possibility of a higher switching activity in the test mode than in the real application!

- Add self-test logic to suitable cores of the chip.
  This is sufficiently discussed in the previous subsection and need no additional guidelines here.

- Make the design scan-testable.
  The scan test improves accessibility and observability and enables to guide signals to the relevant cores on the chip that are not directly accessible through the pins of the chip. This holds for almost all logic blocks on the chip, today. Compressed scan data, combined with on-chip decompression techniques may lead to a reduction of five to ten times in test time at the costs of only 5% in area overhead.

- Include boundary scan test (BST) for enhanced system testability. Most of the current PCBs are very densely packed with a lot of components and interconnect, which makes direct test access to the relevant areas on the PCB impossible. BST verifies the operation at pin level of every device in the system and checks the connection from the device pads through the leads of package to interconnections on the board.

Methods for testability improvement are meant for production testing of ICs. Prior to the computer test phase, however, design problems may appear during *IC characterisation*, debug and *engineering*. On-chip *waveform measurements* are essential when timing errors, noise margin problems or other non-stuck-at errors are suspected. These measurements facilitate functional checking of different IC parts and local verification of timing specifications.

10.3 Yield

The current diameter of wafers used in modern IC production is mostly 8 to 12 inches. The size of an IC determines the number of dies per wafer. Most IC sizes range between $25\,\mathrm{mm}^2$ and $200\,\mathrm{mm}^2$ and their number per wafer therefore ranges from a few hundred to a few thousand. The ultimate price of an IC is determined by the number of *Functionally Good Dies per Wafer* (FGDW). This number is not only dependent on the number of dies per wafer but also on the *yield*. Quite a lot of dies on a wafer do not meet their specified requirements during testing. An additional number of dies is lost during packaging. The yield observed during wafer probing depends on the quality of the manufacturing and on the sensitivity of the design to process-induced defects. The production of nanometer CMOS ICs places very high demands on the factory building, the production environment and the chemicals. Disturbances in the production environment may be attributed to the following parameters:

- *Temperature*: Fluctuations in temperature may cause the projected image of the mask on the wafer to exceed the required tolerances. Also several processing steps are done at elevated temperatures.

- *Humidity*: High humidity results in a poor bond between the photoresist layer and wafer. This may result in under-etching during the subsequent processing step (delamination).

- *Vibrations*: Vibrations that occur during a photolithographic step may lead to inaccurate pattern images on the wafer and result in open or short circuits.

- *Light*: The photolithographic process is sensitive to UV light. Light filters are therefore used to protect wafers during photolithographic steps. The photolithographic environment is often called the '*yellow room*' because of the specially coated lamps used in it.

- *Process induced or dust particles*: Particles that contaminate the wafer during a processing step may damage the actual layer or disturb a photolithographic step. This can eventually lead to incorrect circuit performance. For this reason, manufacturing areas are currently qualified by the class of their *clean room(s)*. Modern advanced clean rooms are of *class-one*. This means that, on average, each cubic foot (≈ 28 litres) of air contains no more than one dust particle with a diameter greater than 0.1μm. In contrast, a cubic foot of open air contains 10^9 to 10^{10} dust particles that are at least 0.1μm in diameter. The standard applied in *conventional clean rooms* required a class-one room to have no more than one dust particle with a diameter greater than 0.5μm per cubic foot. This was because smaller particles could not be detected. A conventional class-one clean room is comparable to class 100 in the currently-used classification.

A lot of effort is done to keep the contamination level as low as possible.

Clean room operators need to wear special suits to maintain high quality standards of the clean room with respect to contamination.

Silicon wafers are subjected to many process steps to build a complete circuit. Each step requires physical treatment performed with a dedicated tool. Feature size reduction has constantly increased the requirements with respect to the purity of the chemicals, gases and environments that contact the wafers during processing. The exposure of the wafer surface to the less pure clean room environment introduces defects and results in yield loss. Modern clean rooms have class 10 - 100 for the overall environment. A mini environment, with controlled airflow, pressure and much less particles (e.g., better than class 1) is used to transport

the wafer to the various process tools. Such a mini environment
is called a standard mechanical interface environment, a *SMIF
environment* or *SMIF pod*. It protects the wafers from particle
contamination and provides an automated and standardised interface to the process tools. The wafers remain either in the SMIF
pod or in the tool and are no longer exposed to the surrounding
airflow.

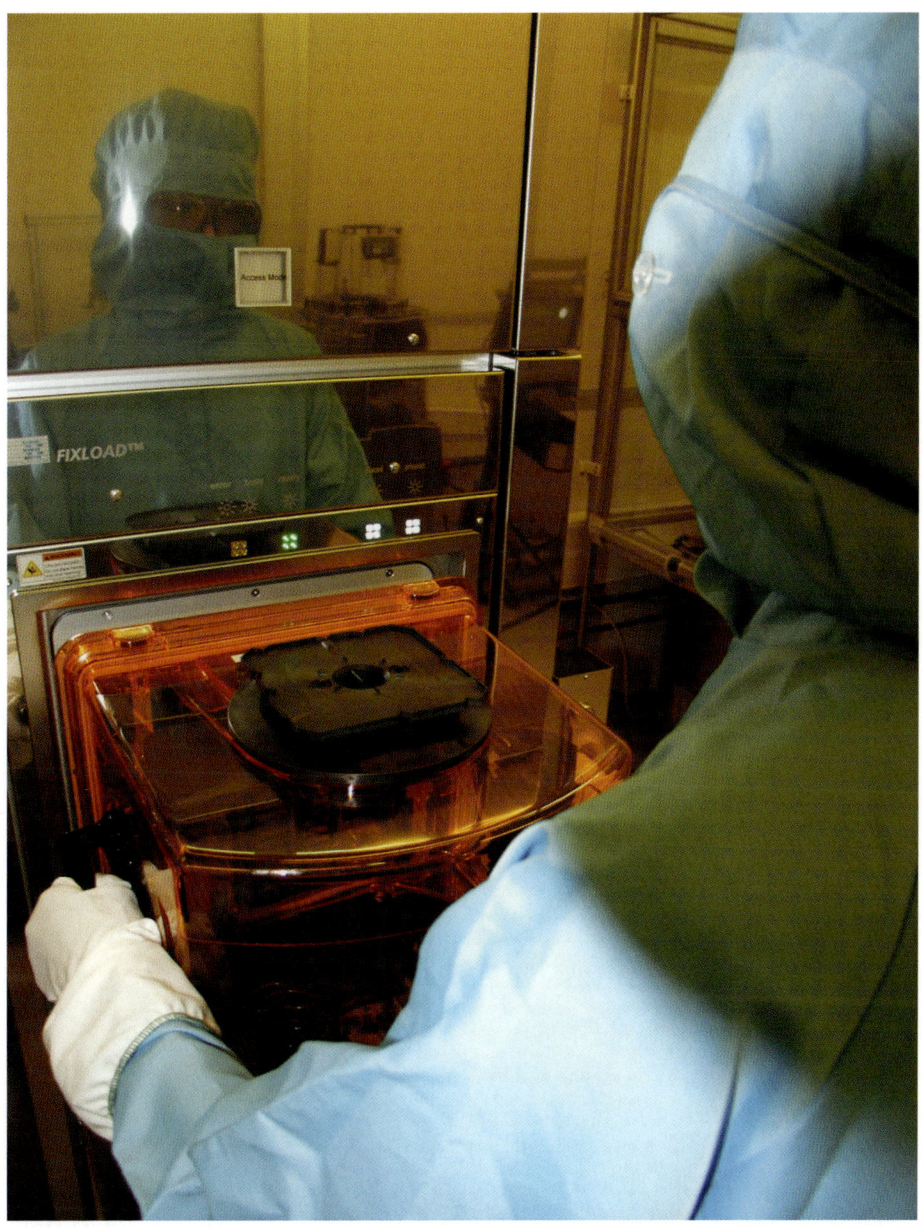

Figure 10.11: *Example of the use of a FOUP mini environment in a modern clean room (Source: Entegris)*

SMIF pods are usually used for wafer sizes up to 200 mm. The front opening unified pod or *FOUP mini environment* figure 10.11

was particularly developed for the constraints of the 300-mm generation. FOUPs may have RF-identification tags for automatic handling in the cleanroom.

- *Electrostatic charge*: Electrostatic charge attracts small dust particles. Very high charge accumulation may occur at a low humidity. This can lead to a discharge which damages the electronic circuits on ICs.

- *The purity of the chemicals*: The chemicals used must be extremely pure to guarantee the high grade of reproducibility and reliability required for ICs.

10.3.1   A simple yield model and yield control

The above parameters, the complexity of the process and the size of an IC determine the yield. Disturbances anywhere during wafer processing may cause defects. In order to control the production costs and predict the product's performance, yield loss mechanisms must be very well understood and accurately modeled. The basic cause of yield loss can be threefold. *Systematic yield loss* is usually caused by the sensitivity of process variations, process or lithography steps to certain pattern topographies in the layout. These are usually spatially or temporally correlated. *Parametric yield loss* is often caused by marginal operation of the design e.g., critical timing, too much switching noise or small noise margins. Finally, *random yield loss*, which is typically associated with physical mechanisms, such as metal shorts and opens due to defects (particles) or contaminants, or open contacts and vias due to misalignment or formation defects. These are usually characterised by the absence of any kind of correlation.

There exists several yield models today. Each model assumes a particular defect density distribution: exponential in the Seeds model, triangular in the Murphy model, gamma in the Negative Binomial model and random in the Poisson model. IC producers compare for a specific process, yield data versus die size with results from the selected model to achieve the best fit.

The overall die yield can generally be described as a product of parametric/systematic limited yield Y_s and random-defect limited yield Y_r. To keep the explanations simple, we will use the Poisson model. According to this model, the yield Y is expressed as:

614

$$Y = Y_s \cdot Y_r = Y_s \cdot e^{-D_0 A} \qquad\qquad (10.1)$$

where Y represents the pre-test yield, D_0 the defect density (#defects/cm^2) in diffusion and the product defect susceptibility, and A the chip area. The yield Y_s incorporates the wafer Area Usage Factor, stepper wafer layout definition, stepper alignment marker areas or other drop-in structures (if applicable), sytematic and parametric yield loss. Today's production lines use electrostatic clamping devices for wafer handling, which offers significant advantages over the conventional mechanical clamp ring by increasing wafer edge utilisation and yield.

The parametric yield is determined by the match of the product design and process window. Especially in the early phase of process development, yield loss is dominated by parametric/systematic issues. Such defects are the result of structural failure mechanisms, which may be caused either by physical process defects or by an incorrect or process sensitive design, and are relatively easy to find.

Most non-uniformly distributed defects originate from 'critical' processing steps. Particularly the steps that involve masks with very dense patterns are considered to be potentially critical. These masks include those used to define patterns in thin oxide regions, polysilicon layers and in metal layers.

The factor Y_{s}, which is area independent, does not include the unusable wafer area close to the wafer edge. The usable wafer area (see figure 10.12) is defined by the total area occupied by complete dies, with the exclusion of a circular edge area (with a width of several millimetres) and a bottom flat side. Current wafers (8" wafers and larger) no longer contain a flat side, but only a notch (section 3.2.3). The total number of dies within this usable area is called *Potential Good Dies per Wafer* (*PGDW*).

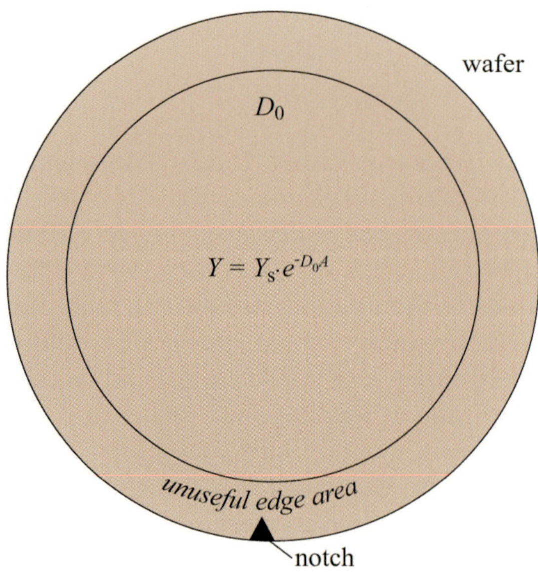

Figure 10.12: *Useful wafer area for PGDW*

The defect density D_0 in equation (10.1) represents the density of defects causing uniformly distributed failures. These are uncorrelated and randomly distributed over the wafer. Examples include dust particles which may affect each process step.

The number of *Functionally Good Dies per Wafer* (*FGDW*) is:

$$FGDW = PGDW \cdot Y \qquad (10.2)$$

The eventual production cost of a chip is determined by the cost of a fully processed wafer and FGDW:

$$Cost/chip = wafercost/FGDW \qquad (10.3)$$

Clearly, the best way to reduce the fabrication cost per chip is to increase the yield.

Particularly in the early phase of process development, Y_s will be relatively low and D_0 will be relatively high. Figure 10.13 shows an example of the yield Y according to equation (10.1) as a function of the die area A for two cases for a 65 nm CMOS process. Case 1 shows the situation during an early development stage of a new process, when $Y_s = 0.6$ and $D_0 = 2$ [defects/cm^2]. Case 2 may represent the situation after a year ($Y_s = 0.85$ and $D_0 = 0.5$ [defects/cm^2]). For more mature processes, typical values for $Y_s = 0.97$ and $D_0 = 0.25$ [defects/cm^2] (case 3).

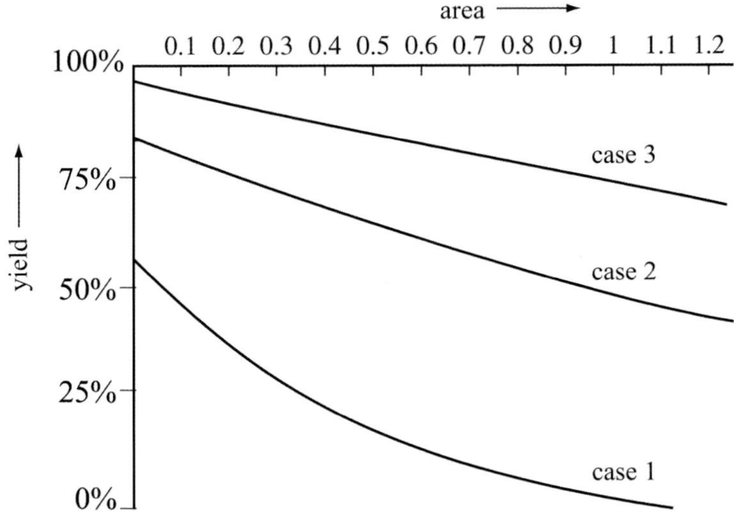

Figure 10.13: *Yield curves at different stages of process maturity*

Traditionally, during a wafer test, an ink dot was deposited on every die that failed the test. Today, the distribution of correct and failing dies across the wafer, a so-called *wafer map* or *wafer bin map*, is stored in the tester's memory. Some prober control tools allow real-time monitoring of the wafer map during testing, with the X-Y coordinates displayed relative to the reference die. The test results are put in bins, presented by colour-coded dies on the wafer. Figure 10.14 shows an example of such a wafer bin map.

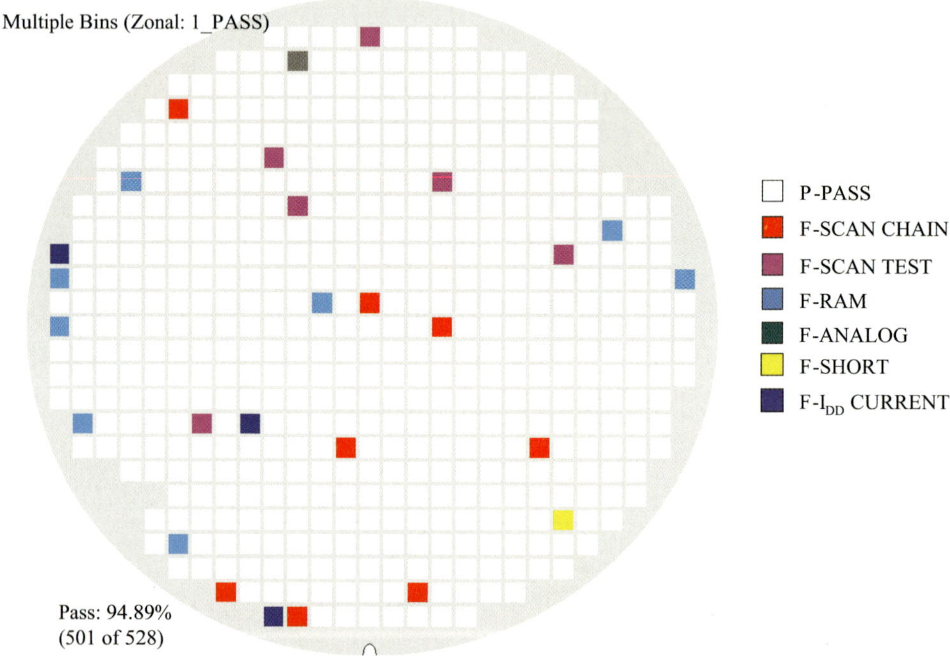

Figure 10.14: *Example of a wafer bin map showing correct and failing dies (Source: NXP Semiconductors)*

All white dies pass all tests. The red dies fail the scan-continuity test, which test correct behaviour of all scan chains. The pink dies fail the full scan test of the logic cores on the die. The light-blue dies have at least one failure in the SRAM. Dies with a failure in the analog circuits are coloured yellow. Finally the dark-blue dies represent dies, which fail the supply current I_{ddq} test. Wafer bin maps can also be used to aggregate data from multiple wafers and stack them for cross wafer or lot analysis. Specific patterns in a bin map are usually an indication for equipment problems or process variations. Several tools exist that can automatically recognize wafer bin map patterns and can provide valuable information for the diagnosis of failure causes. This supports the designers and the foundries to ramp up yields in shorter time.

For the purpose of yield control, *Process Control Modules (PCMs)* are included on wafers. Traditionally, a wafer contained about five PCMs reasonably distributed over its surface area. Today, these PCMs are positioned within the scribe lanes between the dies (figure 10.15). There will be many of them on a 12 inch wafer, but usually still only a limited number per wafer is measured.

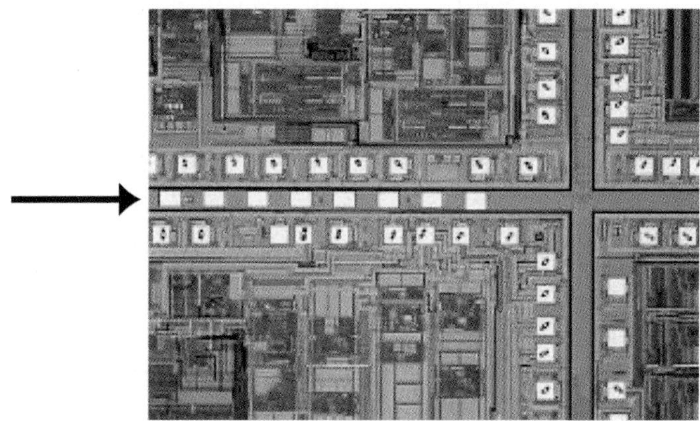

Figure 10.15: *Example of process control modules within the scribe lane between dies*

A PCM often contains transistors of various sizes (W, L) for the electrical characterisation of parameters such as β and V_T. PCMs also usually contain relatively large structures that facilitate the measurement of possible shorts and opens through meander structures, for example. These structures are sampled during and at the completion of the wafer fabrication. Often, more than a hundred parameters can be measured on a PCM.

During the introduction of a new process, the PCMs on all wafers are often measured. When a process becomes mature, usually a few wafers per lot and a few devices per wafer are measured. The measurement results are used as an early feedback to control the process.

Finally, when the correct dies are packaged, the final tests are done, which, besides functional, structural and reliability tests, also check the connections between package and die. These final tests, in combination with the pre-test (wafer test), must limit the number of customer returns to a minimum.

10.3.2 Design for manufacturability

Over the last decade, design costs for an average complex ASIC have started to explode, from approximately $1million in 1998 to approximately $25 million in 2006. This, combined with reducing product life cycles and manufacturing yields has increased the drive to reduce the number of respins and to ramp up the yield in shorter time, to meet time-to market, quality and cost targets. Design rules form the real link between process technology and design. In conventional CMOS technologies, "absolute" design rules *(DRC-rules)* were sufficient to create circuits with relatively high yields. From 90 nm to 65 nm and 45 nm, these absolute design rules are no longer sufficient. Additional rules *(DfM-rules)* are required to make the designs tolerant to photolithography and process deficiencies, in order to maintain a sufficiently high yield level. In current nanometer CMOS technologies, extensive yield evaluation must be performed before a design is sent to the fab. Particularly layouts are adapted to increase this yield. This so-called *design for manufacturability (DfM)* can reduce the design sensitivity to defects (opens or shorts), but it may also support the lithographic process (litho-friendly design; chapter 3). Figure 10.16 depicts some examples of random failure.

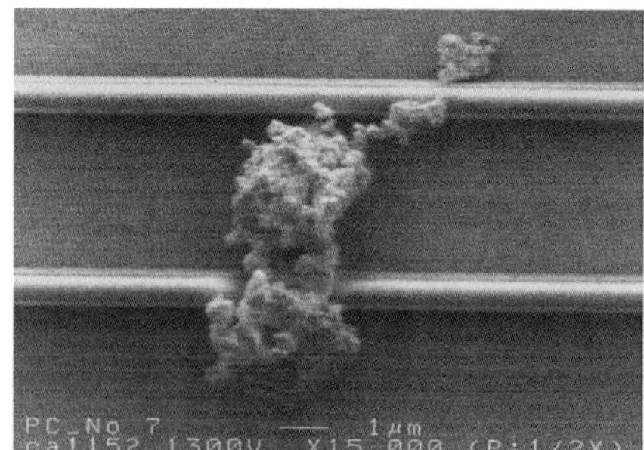
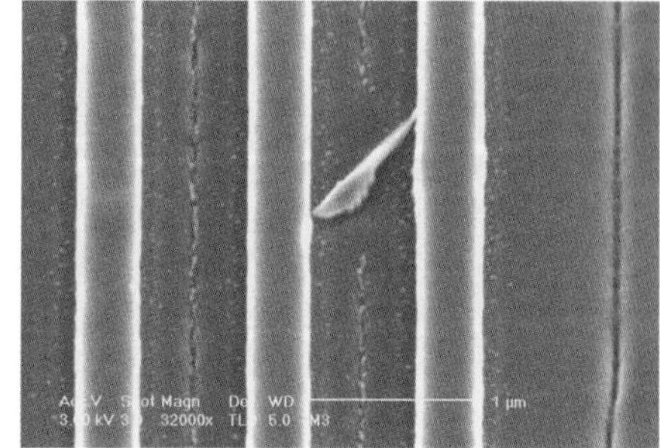

Figure 10.16: *Example of random failures: particles causing a potential short (Source: NXP Semiconductors)*

There exists no uniform definition for DfM. Some include all effects that are potential candidates to reduce the yield: defects, shorts and opens, lithographic variations, process variations, power integrity, substrate noise, electromigration, leakage currents, reducing noise margins, etc. Many of these effects are discussed in the previous chapter, since they also influence the design robustness and product reliability [2].

DfM includes a set of guidelines to make designs more robust against systematic, parametric and random yield loss and create more easy producible products. DfM is a way of anticipating on critical features or critical areas in the layout early in the design phase. Figure 10.17 shows an example of a systematic failure: a short between the two polysilicon areas. The photo has been taken after de-processing of the metal and via layers.

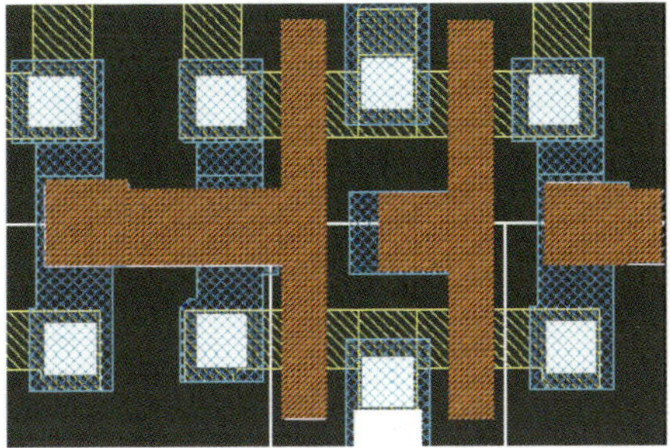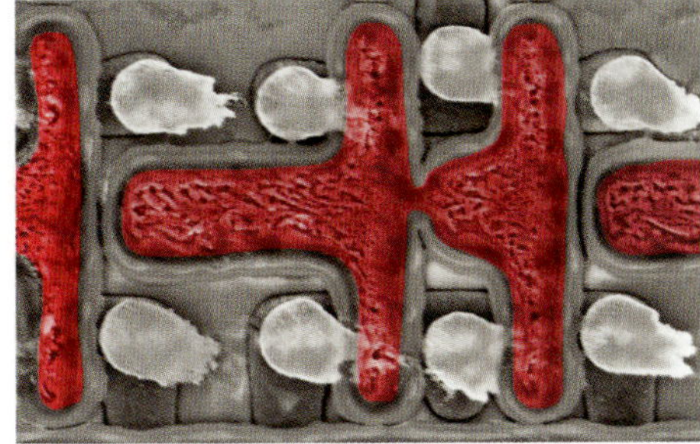

Figure 10.17: *Example of a systematic failure, showing the layout and a photograph of a polysilicon short*

Particularly at product introduction, when the design rules and process are not yet mature, the operating margins can be low and may cause parametric yield loss. Consequently, DfM rules may change as the process technology becomes more mature [3].

A few DfM rules have already become commonplace, such as antenna rules and rules for tiles (area fills to improve the CMP planarisation process (chapter 3). Rules for wire widening and improved wire distribution (*wire spreading*; figure 10.18) were introduced around the turn of the millennium.

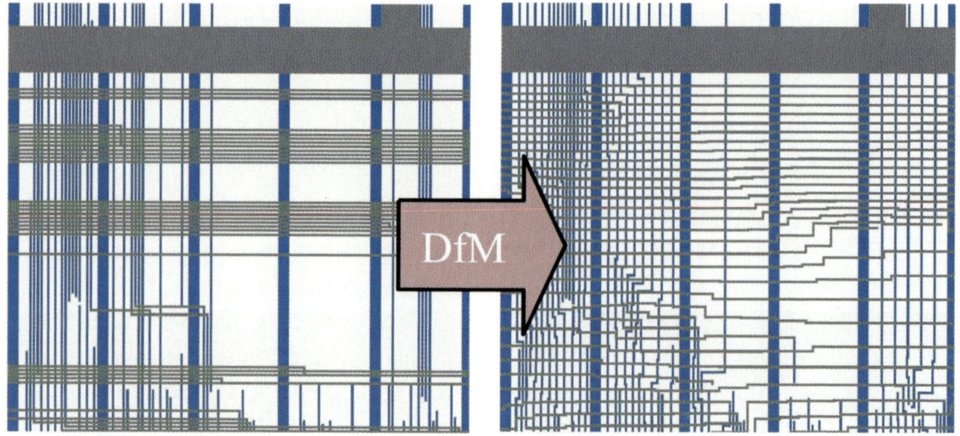

Figure 10.18: *Example of wire spreading to improve yield (Source: NXP Semiconductors)*

Wire spreading was particularly an issue in those areas of the chip, where many wires were routed at minimum width and spacing, while there is ample room for wider wires at (much) larger than minimum spacing. Over the last couple of years via doubling has entered the scene to reduce the number of opens on a chip. This is due to the fact that the number of contacts and vias with minimum metal overlap has dramatically increased. Doubling every via in a design is not possible, since it would have a serious area impact. The current approach is to double only the vias that have sufficient white space around them and which causes no area increase.

These additional DfM rules require the development of intelligent tools. An important requirement for applying these tools to achieve maximum yield improvements is that they are supported with adequate yield models that have the potential to recognize and fix critical layout areas without area increase. Approaching the end of Moore's law has forced the designers to also deal with many nanometer CMOS effects such as shrinking process windows, increasing process variability, changing defect mechanisms, increasing lithographic effects, increasing noise levels, reducing noise margins, etc, which are not all covered by DfM. It is better to use the term *design for anything* or *DfX*, which includes: DfT, DfM, design for robustness, design for reliability, litho-friendly design, design for debug, design for failure analysis, etc. Each of these "design for" topics requires additional design resources and increases the design complexity and costs.

10.4 Packaging

10.4.1 Introduction

The development of the IC package is a dynamic technology. Applications that were unattainable only a few years ago are now common place thanks to advances in package design. Moreover, the increasing demand for smaller, faster and cheaper products is forcing the packaging technology to keep pace with the progress in semiconductor technology. The huge diversity of application areas, e.g., automotive, identification, mobile communications, medical, consumer and military, to name a few, combined with an exponentially growing device complexity and the continuous demand for increased performance has generated a real explosion of advanced packaging techniques.

Packaging is no longer a final step in the total development chain of a semiconductor product and as such, it has become an integral and differentiating part of the IC design and fabrication process.

The package supports various important functions:

- Allow an IC to be handled for PCB assembly and protect it during further PCB production

- Mechanical and chemical protection against the environment

- Mechanical interface to the PCB

- Good electrical connection (signals and power supply) between PCB and chip

- Enhance thermal properties to improve heat transport for IC to environment

- Allow standardization

Currently, ICs may contain hundreds of millions to more than a billion transistors. With such high integration densities, the IC package has become increasingly important in determining not only the size of the component, but also its overall performance and price. Higher lead count, smaller pitch, minimum footprint area and reduced component volume all contribute to a more compact system implementation. As the package directly affects factors such as heat dissipation and frequency dependency, choosing the right package is essential in optimising IC performance.

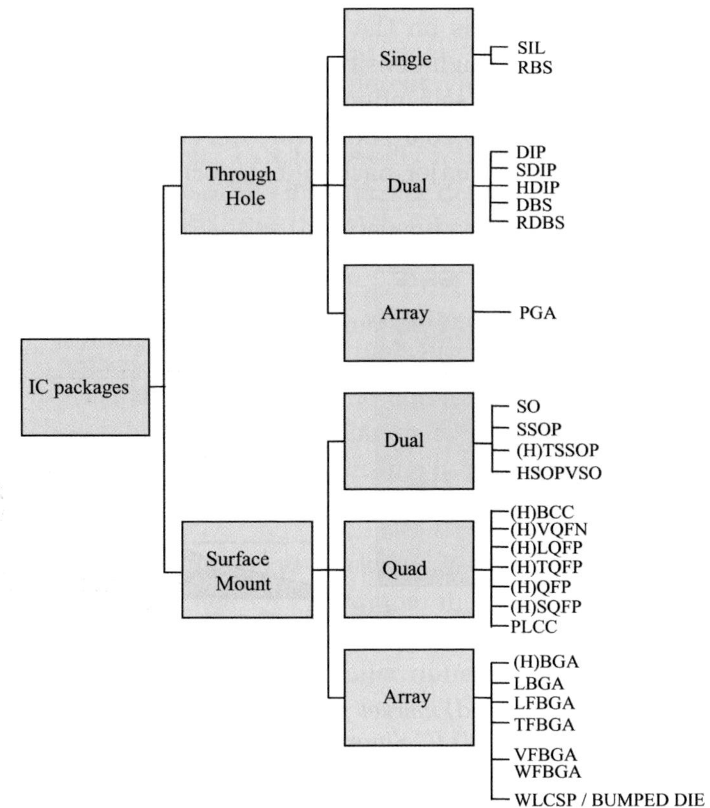

Figure 10.21: *Overview of the most commonly used packages*

Different versions have been developed for each of the package categories. The ceramic versions show better thermal behaviour and are meant for ICs that consume more power. The increase in power densisty, combined with a limited temperature budget requires creative approaches to thermal management. Dependant on the application demand, several alternative technologies can be applied to uniformly cool an IC, e.g., normal airflow, heat sink, heat spreader, thermally enhanced interface material, fan.

These versions can be categorized with three characteristics: overall package height (L, T, V, W), lead/ball pitch (S, F) and thermally enhanced (H). This naming convention is standardized by JEDEC, an industrial standardization committee.

Finally, figure 10.22 shows the trend in use of the various package categories.

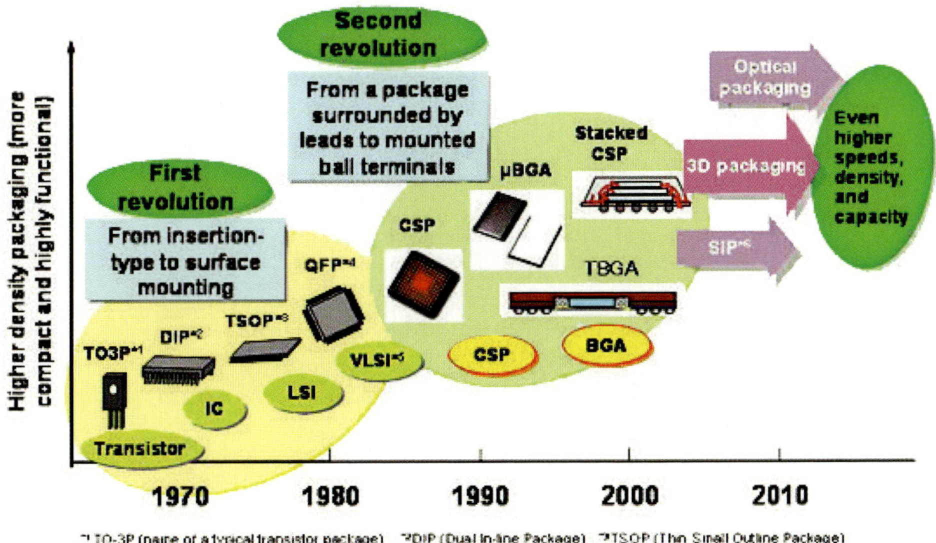

Figure 10.22: *Worldwide IC package trends (Source: HITATCHI)*

10.4.3 Packaging process flow

When a wafer leaves the waferfab, it first needs to be functionally and electrically tested before it can be assembled. This electrical testing, which is often referred to as probing, is done by means of metal needles that physically contact the bond pads on each die. Three main technologies can be identified: cantilever probing, as depicted in figure 10.23, membrame probing, often used for RF solutions and vertical probing, the preferred technology for bumped dies. IC's that do not pass the functional and electrical tests are marked by either an ink dot, in conventional processes, or identified as fail on an electronic wafermap (figure 10.14), which ensures that they will not be used during the die-attach/die-placement process for packaging.

627

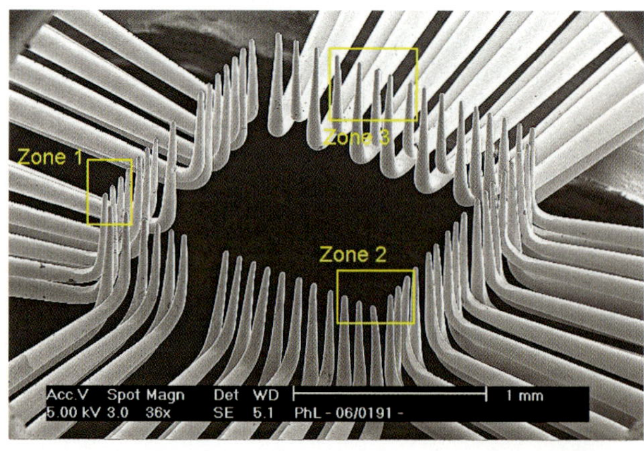

Figure 10.23: *Picture of cantilever probecard (Source unknown)*

Backgrinding and sawing

Before the actual packaging starts, each wafer will have to be back grinded to the optimal thickness. Typical thicknesses are $280\,\mu$m and $380\,\mu$m, while $100\,\mu$m is more common for very thin packages.

The physical backgrinding is typically a two-step process, in which the wafer is first back grinded to about $20\,\mu$m above the required thickness, with a coarse grinding wheel. The remaining last $20\,\mu$m are then grinded with a much finer grinding wheel. For very thin wafers ($< 150\,\mu$m) an etch step can be added for stress relief.

Once back grinded to the right thickness, the wafer has to be separated into individual dies. This is typically done by means of a diamond saw, although laser separation is an upcoming trend. *Laser dicing* has a couple of advantages compared to diamond sawing. Its is faster, it causes less material stress, it requires a smaller scribe lane and is able to dice devices with different form factors on the same wafer [4].

To allow this dicing scribe lanes of $50\,\mu$m to $200\,\mu$m are designed around each die.

Dicing with the conventional diamond saw can be performed in two different techniques. In the first technique, a single cut is made directly through the complete wafer. This process is typically used for conventional production processes. The more advanced processes, e.g., the ones that use low-ϵ dielectrics, use a so-called step-cut process. This process first uses a wide blade to only cut through the active area (1 & 2 in figure 10.24) into the bulk silicon. Next, a finer blade in the second step will separate the dies (3 & 4).

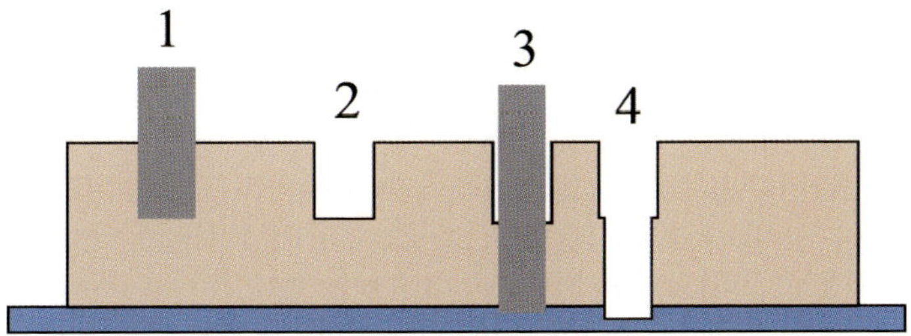

Figure 10.24: *The step-cut process for singulation of dies (dicing)*

Packaging

Once the wafers are back grinded and diced, the actual packaging can start. The package choice is very much related to the electrical, thermal and size requirements dictated by the application domain. Two main interconnect technologies can be identified to realize the electrical connections. The most common one is *wire bonding (WB)*, which is still responsible for about 90% of all chip interconnects. Before the actual wire bonding can take place, the die is first mounted on a carrier (lead-frame or substrate) in a process called *die attachment*. During this die attachment, an adhesive is deposited on the carrier and the functional good dies are picked from the wafer and placed in the adhesive. Which dies are being picked is determined by the wafermap that has been generated during probing (see figure 10.14), or, in conventional processes, by the ink dots placed during the wafer test. Adhesive materials are typically a mixture of epoxy and a metal (aluminium or silver) to ensure a low-electrical and low-thermal resistance between the die and package. For thermal enhanced applications also solder can be used to attach the die to the carrier.

IC reliability is strongly influenced by the quality of bonding wires. Diameters of the wires range from $15\,\mu$m for fine pitch applications to $150\,\mu$m for high power devices.

There are two common wirebond processes, depending on the applied wire material: Au ball bonding and Al wedge bonding. However, copper is gaining more popularity because of its reduced electrical and thermal resistances. But copper is more readily oxidized than aluminium and copper oxidation may cause reliability problems, in the form of poor adhesion or in the introduction of cracks at the bond interfaces.

During wirebonding the bonding tool is guided to the bondpad on the die. This so-called first bond (figure 10.25) is achieved by using thermal and ultrasonic energy. Next, the wire is stretched to the corresponding finger of the leadframe on the carrier and, again by using pressure and ultrasonic energy, the opposite end is welded to the leadfinger to form a stitch bond (also known as wedge bond). At the formation of this second bond the wire is also automatically cut in preparation for the next bond. After all the pads have been bonded, the die is encapsulated. Figure 10.25 shows an overview of the wire bonding process.

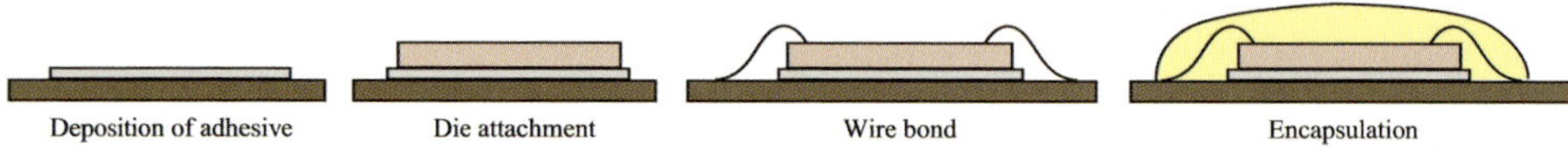

Figure 10.25: *Overview of the wire bonding process*

Because the number of transistors grows quadratically with the scaling factor, while the number of pad positions only grows linearly with it, there is an increasing demand for multi-row bond pad (staggered bond pads) connections to support a variety of applications. Figure 10.26 shows examples of a double and a triple-row wire bond interconnections.

Figure 10.26: *Examples of a double- and a triple-row wire bond inter-connects (Source: NXP Semiconductors)*

Very high current variations in high-speed VLSI circuits can cause an increased voltage drop ΔV across the bonding wires. This is because of the inductance (L) of the wires and is expressed as follows:

$$\Delta V = L\frac{di}{dt} \tag{10.4}$$

The above voltage drop may become critical in analog/mixed signal and high-speed digital circuits, unless suitable design measures are taken. This topic is addressed in chapter 9.

In *flip-chip bonding (FCB)*, which is the second interconnect technology, the die is assembled face down directly onto the circuit board (figure 10.19) with solder-, Au- or Au/Ni-bumps. Compared to wire bond, this technology comes with less area overhead, because there is no additional area needed for contacts on the sides of a chip. It enables the final packaged chip to be only marginally larger than the original die (*chip-scale package*).

In the example of the controlled-collapse chip connection (C4) soldering process, first, solder bumps are deposited on the die bond pads (figure 10.28), usually when they are still on the wafer, and at the corresponding locations on the substrate. Figure 10.27 shows a wafer-level CSP with two rows of dies sawn from the wafer. The zoom-in shows that also a redistribution layer is used, which is covered with a polyimide layer for passivation and stress relieve. Then holes are etched in this passivation layer and a direct ball drop method produces the balls needed for connection to the board.

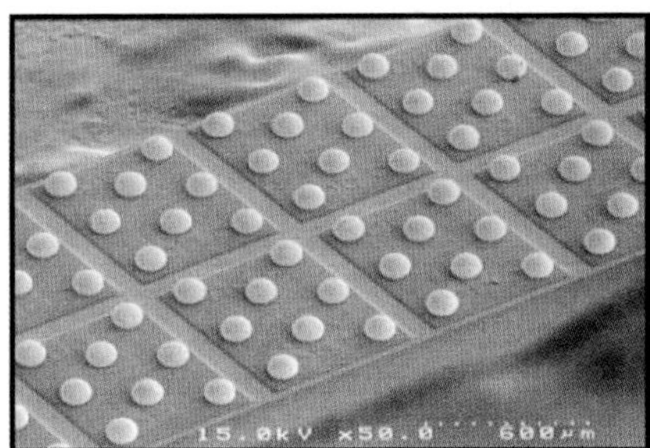
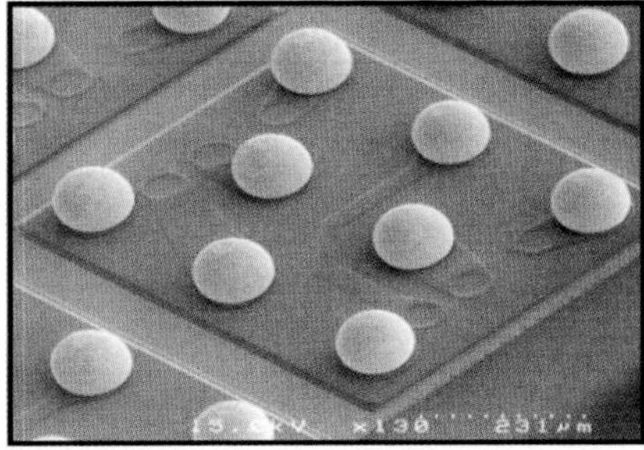

Figure 10.27: *A wafer-level chip-scale package with direct ball drop*

During placement, the array of balls on the die is aligned with the array on the substrate. Depending on the FCB technology, it is then either pressed and reflowed (melted), or the complete embodiment is reflowed (melted) in a furnace, to create all electrical connections. During this reflow step the chip is self-aligned to its exact position on the substrate (figure 10.28). If no material was deposited before, there will always be a gap between the die and substrate. In the next step, the die is encapsulated with good isolating material (epoxy) to fill this gap (underfill). This underfill serves to protect the chip from humidity and impurities, but also improves reliability in terms of mechanical and thermal stress.

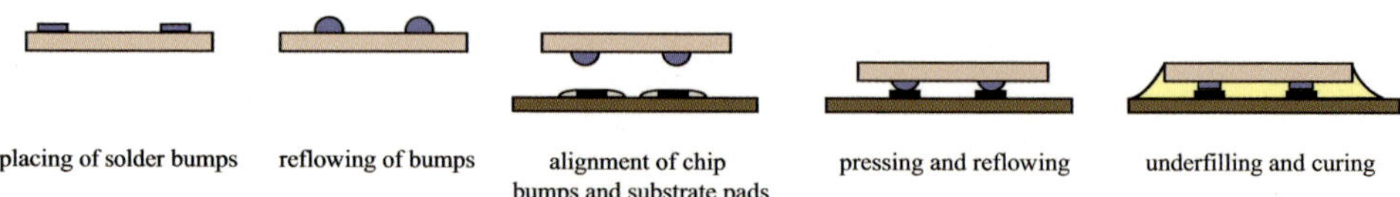

Figure 10.28: *Overview of flip-chip bonding process*

Finally the FCB process is completed by curing (hardening) the underfill material by using heat or light at a certain wavelength, depending on the underfill material.

Flip-chip bonding accomodates dies that may have several hundred bond pads placed anywhere on their top surface. In many cases an additional redistribution metal layer is required to transfer periphery wire bonding pad connections into an area array of connections for flip-chip bonding. Compared to other interconnection techniques, FCB results in very short connections and exhibits improved performance in high-speed applications.

Flip-chip connection can be found both in a silicon to substrate and silicon to silicon bumping. Figure 10.29 shows a combination of wire bonding and flip-chip bonding in a stacked-die application.

Figure 10.29: *Package with both flip-chip (FC) and wirebond (WB) interconnections (Source: NXP Semiconductors)*

The substrate with the array of balls, in the lower middle of figure 10.29, shows such a redistribution layer. Currently, the redistribution layer is part of the top aluminium layer, which is needed for coverage of the bondpads to support reliable bonding.

10.4.4 Electrical aspects of packaging

The drive for higher bandwidths and the resulting increase in signal speed, combined with the ever present demand for area and cost reduction, has caused the package development to become a significant and integral part of the total development process of integrated circuits. Package costs are mainly driven by the size of the package and is closely related to the number of package connections. A package with poorly controlled electrical characteristics (such as resistance R, inductance L, capacitance C and impedance Z) needs more pins than a package that meets the electrical signal interface design criteria.

The electrical characteristics of a package are determined by its construction. The type of signal interface indicates the desired $RLCZ$ of the package interconnects. A high speed differential interface may want $100\,\Omega$ differential impedance between metal tracks, while minimizing the inductance of the power and ground pins is more important for a single-ended interface. The signal type will determine the desired electrical characteristics of the package interconnect.

Generally, a small value of L is advantageous for both signal integrity and speed. Fast current changes may introduce large voltage changes (ΔV), see also chapter 9, according to expression (10.4). Certain package types, however, can only support certain ranges of electrical parameters. Conventional package types using lead frames have relatively large lead inductances (1-50 nH), because of longer lead lengths. They also tend to have a high mutual coupling. Ceramic multi-layer packages, such as pin grid arrays (PGAs) have better performance due to the presence of power or ground planes, but are relatively expensive. As previously discussed, the use of flip-chip bonding can also improve electrical performance by minimizing the lengths of the connections between the die and the substrate, resulting in inductances of 0.5-1 nH. Recent developments in package technology however, allow more flexibility in the design of package interconnects to meet cost targets. Some package types allow the design of specific impedances while others try to minimize them altogether. *Ball grid array (BGA)* packages use inexpensive laminate substrates allowing the inclusion of power and ground planes and therefore the design of transmission line structures. On the other hand, so called no-lead packages, such as QFNs, have virtually no pins and sometimes not even bond wires thereby minimizing the influence of the package on the overall electrical behaviour of the chip. It is therefore extremely important to understand that the requirements of the interconnect are dictated by the application. Particularly in high-speed applications, a reasonably accurate electrical model of the package is an essential part of the final performance simulations of the integrated circuit.

In the packaging of ICs, we distinguish several hierarchy levels of interconnections:

- first level of interconnection: chip to package connection

- second level of interconnection: package to PCB connection

- third level of interconnection: PCB wires

- fourth level of interconnection: PCB to system (back-planes) connnection

It should be clear that not only the package-to-board connections must be optimized for high-performance ICs, but to achieve maximum system performance, all levels of interconnections must be optimized and adapted to each other. This becomes even more important with the state-of-the-art packaging technologies, such as multi-chip modules (MCMs) and stacked dies (figure 10.34), system-in-a-package (SiP) and system-on-a-package (SoP) (figure 10.36). These packaging technologies are discussed in section 10.4.8.

10.4.5 Thermal aspects of packaging

Another dominating parameter in the performance and reliability of an integrated circuit is the physical temperature of the die inside the package, which is determined by the power consumption of the IC in combination with the *thermal behaviour* of the package. This requires a strong interaction between the IC, the package, the system design and its application. The most commonly used, but simple model for IC packages includes two thermal resistance parameters. For a given power dissipation P, the junction-to-air thermal resistance R_{JA} represents the ability of a package to conduct heat from junction (die) to ambient and is expressed as follows:

$$R_{JA} = \frac{(Temp_J - Temp_A)}{P} \quad [^\circ\text{C/W}] \tag{10.5}$$

Where $Temp_J - Temp_A$ represents the temperature difference between the chip (junction) and its environment (ambient).

R_{JA} is often determined corresponding to the JEDEC [5] requirements for standard test boards and in different air conditions, including still air. Table 10.1 shows some values for R_{JA} under still-air conditions.

Table 10.1: *Thermal resistance values for different packages under still-air conditions (Source: NXP Semiconductors [6])*

Package-Pins	Package Designator	Package Outline Code	RTH(J-A) ⁰C/W	RTH(J-C) ⁰C/W
DQFN-14	BQ	SOT762	61	32
DQFN-20	BQ	SOT764	50	23
HVQFN-16	BS	SOT629	40	18
HVQFN-24	BS	SOT616	40	18
LFBGA-96	EC	SOT536	60	16
LFBGA-114	EC	SOT537	55	15
PLCC-20	A	SOT380	80	32
PLCC-28	A	SOT261	73	29
PLCC-52	A	SOT238	48	13
SSOP-20	DB	SOT339	136	40
SSOP-24	DB	SOT340	125	37
SSOP-28	DB	SOT341	98	35
SSOP-48	DL	SOT370	88	25
SSOP-56	DL	SOT371	84	24
TSSOP-8	DP	SOT505	120	30
TSSOP-16	PW	SOT403	160	39
TSSOP-24	PW	SOT355	128	32
TSSOP-48	DGG	SOT362	104	23
TSSOP-64	DGG	SOT646	110	26
QFP-52	BB	SOT379	62	15
VFBGA-56	EV	SOT702-1	80	21

In many applications the maximum junction temperature is defined as $125\,°C$. If we assume a consumer application with an ambient temperature of $70\,°C$, the maximum allowed power consumption of an IC, packaged with a 48 pins SSOP (see table 10.1) under still-air conditions (on a reference board of the supplier), is then equal to:

$$P = (Temp_J - Temp_A)/R_{JA} = (125 - 70)/88 = 625\,\text{mW}$$

If the power consumption is more than this calculated maximum, either a heat spreader is required or an air-flow must be introduced, using a fan.

The other parameter is defined as the junction-to-case thermal resistance R_{JC}, and represents the ability of a package to conduct heat from the junction (die) to the surface (top or bottom) of the case (package) and is expressed as follows:

$$R_{JC} = \frac{(Temp_J - Temp_C)}{P} \quad [^\circ C/W] \tag{10.6}$$

This parameter is only applicable if an external heat sink is used and the heat is only conducted through that surface that connects to the heatsink.

If we assume an IC, consuming 1 W, which exceeds the above calculated maximum allowed power under the same conditions, then the required thermal resistance of the device must be equal to:

$$R_{JA} = (Temp_J - Temp_A)/P = (125 - 70)/1 = 55\,^\circ C/W$$

This can either be achieved by introducing an airflow or by using an external heatsink.

This model, however, only describes the steady-state heat conduction capability and does not account for the dynamics in power behaviour of the product in a real application. Heatflows are rarely one dimensional. Different application boards, or stacking packages change the environment of the product and can have a huge impact on its thermal behaviour. However, the value for R_{JA} can very well be used to compare thermal capabilities of different packages.

To obtain an accurate model for a particular thermal situation including two and three dimensional heat conduction paths therefore soon leads to a complex network. This has led to the development of compact thermal models, describing the thermal behaviour with an accuracy of 5% by using a thermal network with seven or more nodes connected by thermal resistances. A discussion of such compact models is beyond the scope of this book. An example of such a compact model is described in [6].

10.4.6 Reliability aspects of packaging

There are a few packaging aspects that are related to the reliability of the chip. First of all, the trend to reduce the dielectric constant of the *inter-level dielectric (ILD)* layers (low-ε dielectrics) in semiconductor fabrication processes makes these dielectrics more porous, less robust and more sensitive to physical pressure during test (probing) and bonding. Secondly, in a copper-backend CMOS process, copper is used for all metal layers including the one(s) used to create the bond pads. However, copper oxidises quickly and the oxidation prevents the creation of a good and reliable contact between the bond wire and the pad. Therefore, during an additional re-metallisation step, a so-called *Al cap* (*aluminium cap*) is formed above the pad area to create a good electrical contact with the bond wire. But, this does not solve all reliability aspects. Particularly the drive for finer pad pitches and smaller pads requires probe cards with smaller and sharper needles, which increases the probability to punch through the Al cap and expose the underlying copper. Also these exposed copper areas oxidise quickly, showing the same problems as described above. A solution to this problem is to increase the Al cap area such that the probe needles do not land in the wire bond region (figure 10.30) and can no longer damage the underlying copper layer because it is separated by the passivation layer.

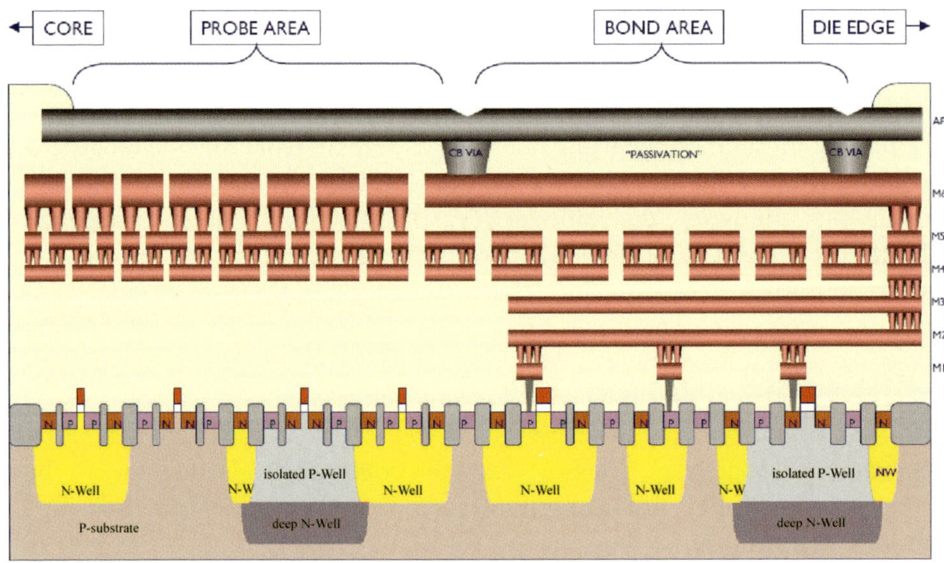

Figure 10.30: *Example of extended Al cap area to prevent pad damaging during probing (Source: NXP Semiconductors)*

The increasing number of pads, combined with the drive for smaller chip areas has forced the semiconductor industry to create *bond-over-active (BOA)* layout techniques, in which bond pads are not only located at the chip's periphery but also on top of active silicon areas at the periphery of the die core area on top of diodes, power and ground lines, I/O transistors and ESD protection circuits. BOA may lead to a significant reduction in die size [7].

In summary, any change in pad-related design and technology concepts may have severe consequences for the reliability of the bonding process and for an overall reliable chip operation.

10.4.7   Future trends in packaging technology

While during the 1980s and 1990s *surface-mount device (SMD)* technology became very popular at the cost of through-hole packages, a new trend towards miniaturisation is observed. Products like mobile phones get smaller and thinner every year, which automatically requires the same shrink for the components they are built from. This means that conventional leaded parts, such as quad flat packs (QFPs), will increasingly be substituted by leadless parts like QFN's or even bare dies (wafer level *chip-scale package (CSP)* (WLCSP)).

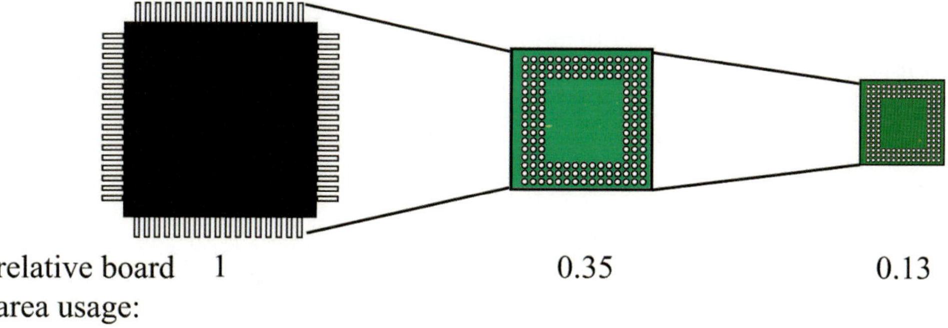

Figure 10.31:   *Miniaturisation in packaging (reduced board area's)* *(Source: OKI)*

Because the complexity of nanometer ICs has reached an incredibly high level and will continue to increase, it puts severe demands to the density of die pads and package connections (pins or balls). This drives the trend towards area array packages, e.g., ball grid arrays (BGAs), in which the connection between the package and the application board is formed by solder balls.

Figure 10.32 shows a variety of BGA packages for different applications. They mainly differ in the ball-pitch, package height and/or ther-

mal capability. Also a differentiation can be made between the bonding
technologies used within the BGA packages, e.g., wire bonding and flip-
chip bonding.

Figure 10.32: *Different flavours of BGA packages*

In literature BGAs are frequently combined with CSPs, which are usually referred to as packages whose sizes are less than 20% larger than
the die itself. Most CSPs are wafer level packages, where the peripheral bond pads are redistributed and rerouted to an area array of pads,
using a thin film like technology. This can be executed as an extension
of the wafer fab process, or the wafer can be sent to a bump supplier
who creates the redistribution directly on the wafer. On those rerouted
pads balls are placed, by means of direct ball attach, to create the CSP
(figure 10.33). Next, flip-chip bonding technology is applied, when the
CSP needs to be attached to a kind of laminate carrier.

Figure 10.33: *Rerouted Wafer Level CSP*

10.4.8 System-on-a-chip (SoC) versus system-in-a-package (SiP)

Over the last couple of years, bulk CMOS processes have enabled the integration of digital, RF and mixed-signal functions onto one single die. Time has come to make a trade-off between traditional board design with discrete passive devices and a fully integrated system solution like a *system-on-a-chip (SoC)*. A SoC is an extremely integrated single chip solution built from in-house and/or external IP. It contains the computing engine (e.g., μproc and/or DSP core), logic and various memories on the same chip. The key benefits of a SoC realisation are:

- better performance due to the smaller on-chip delays, compared to the delays caused by going on- and off-chip

- small physical size

- reduced overall system costs, due to a reduction in the number of components

- less power consumption

- increased reliability due to a reduced number of system components

However, there are also some critical remarks to be made here. For many applications, the time between inception and high-volume production of a SoC, may take several years. Most SoCs are therefore expensive, custom-designed products for high-volume market segments with a relatively long lifetime expectancy. The increasing diversity of the system's applications requires the development of more sophisticated IP.

Today, system complexity is growing at a faster rate than that of a SoC and a printed-circuit board (PCB). In many applications, Moore's law only deals with the integrated fraction of the system, leaving the largest part to relatively large discrete passive components such as antennas, filters, capacitors, inductors, resistors and switches. In the example of a cell-phone, only one tenth of the system consists of ICs, with the remaining part being passives, boards and interconnections and switches. An alternative to SoC integration is the use of *system-in-a-package (SiP)* technology, which usually refers to a single package that includes a multiple of interconnected integrated circuits and/or passive devices. SiP technology enables hybrid systems built from sub-functions that may have been created by different designs methods in different technologies. Some people see a SiP and *multi-chip module (MCM)* as the same system solution, however, an MCM is usually referred to as the integration of different dies on the same plane on the same substrate in one single package, while SiP also refers to stacked dies and/or passives in one single package. SiPs may use a combination of different packing technologies including wire bond, flip-chip, wafer-level packages, CSPs, stacked dies and/or stacked packages. Figure 10.34 shows examples of both an MCM and of wire-bonded stacked dies.

Figure 10.34: *Example of an MCM and of wire-bonded stacked dies*

Compared to a SoC realisation, a SiP approach offers much more flexibility in adding new functions and features to the system. As is the case with all packaging technologies that combine one or more naked dies, also SiP technology faces the challenge of *known-good-die (KGD)*, which is a chip that has been extensively tested before being placed into its package. When an expensive processor is to be combined with a cheap peripheral chip onto one substrate or into one package, an almost 100% guarantee is required that this peripheral chip will operate fully according to its spec. This is to prevent to throw away the total substrate, including the expensive processor, if only the cheap peripheral chip does not work properly. To avoid this problem, a new upcoming trend can be identified: *package-on-a-package (PoP)*. In this concept (figure 10.35) the expensive processor/ASIC is separately packaged in the POP bottom package, while the memory is packaged in the top package. Each of these two can be tested separately, while for the memory even burn-in can be applied. Once both are proven to be fully functional the parts can be mounted on the application board.

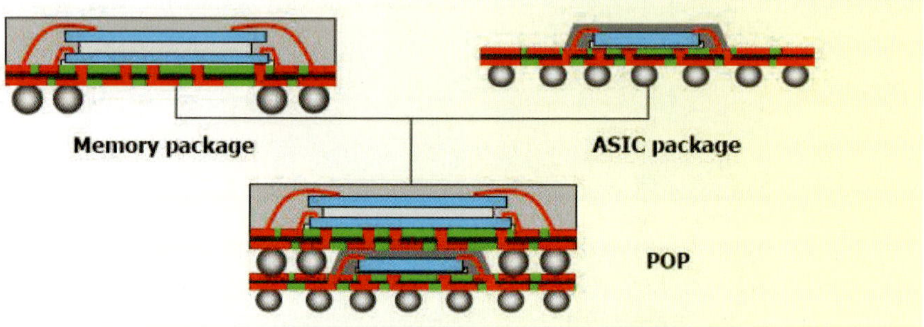

Figure 10.35: *PoP concept (Source: Amkor technology)*

Future systems, however, will incorporate features and functional complexity that will be even beyond today's imagination. They will combine the potentials of physics, optics, biology and chemistry with analog and digital signal processing and storage capabilities packed onto one composite substrate. This is usually referred to as 'more than Moore'. A target application may be a device that could be encapsulated below the human skin to permanently monitor a person's health. This could be done by checking vital organs through the monitoring of breath, heartbeat, blood pressure, blood glucose level, etc. The results could then wireless be communicated through the Internet to a medical advisor or physician, which can then propose the appropriate medical treatment, when necessary. In this case the sensors (monitors) may be attached on top of a substrate. Figure 10.36 shows an example of such a *system-on-a-package (SoP)*, which may combine optical circuits with passives, MEMs, SoCs and SiPs inside or attached to a composite substrate [8].

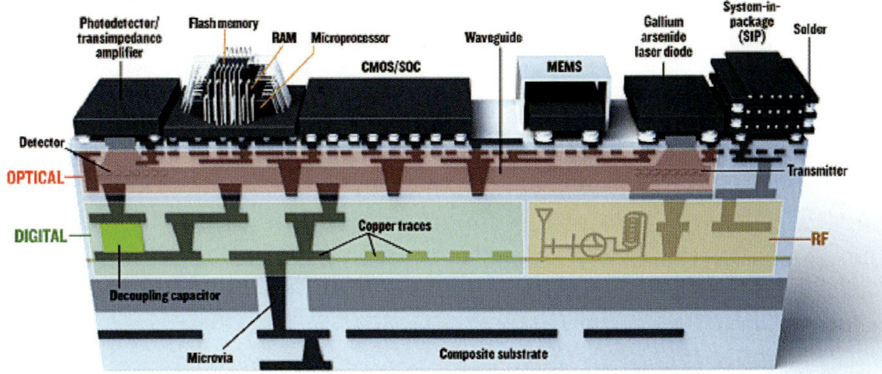

Figure 10.36: *Example of a system-on-a-package (SOP), consisting of optical circuits and devices, resistors, inductors, antennas, decoupling capacitors within a layered substrate and a combination of bare and packaged such as flash memory, CMOS/SoC, optical circuits, laser diode and SiP devices attached to the top layer [8] (Graphics design by www.bryanchristiedesign.com)*

In conclusion: SoC and SiP are competing technologies, but this does not mean that there will be a winner in the end. SoC and SiP systems will live together, but each in its specific application domains, where their properties are exploited to their full advantage.

10.4.9 Quality and reliability of packaged dies

Various quality and reliability tests are applied to packaged ICs before they are approved for sale or for application in high volume production. Many of these tests are standardised. An insight into the background to these tests and their implementations is provided below.

Quality

Vulnerability to *electrostatic discharge (ESD)* and sensitivity to *latch-up* are two important quality criteria on which chips are tested.

Parasitic bipolar devices in all CMOS chips form a *thyristor* between its supply and ground terminals. Activation of this thyristor results in latch-up. The result is a dramatic increase in current consumption and a chip malfunction. A chip's latch-up sensitivity can be tested by sequentially applying a voltage of one-and-a-half times the maximum specified voltage to each pin, while limiting the available current to, for example, 500 mA. The actual current consumption is observed for signs of latch-up.

Since ESD and latch-up sensitivity can be influenced by the design, these topics are discussed in detail in chapter 9. In addition, chapter 9 describes technological and design measures which can be taken to reduce the chances of failure in the associated tests. ESD tests and the related quality requirements are also discussed in that chapter.

Reliability

The increasing complexity of ICs means that their reliability has a considerable effect on the reliability of electronic products in which they are applied. Reliability is therefore an important property of an IC and receives considerable attention from IC manufacturers. Related tests subject an IC in active and non-active states to various *stress conditions*. This facilitates rapid evaluation of the IC's sensitivity to external factors such as temperature changes and humidity. The most important *reliability tests* are as follows:

- *Electrical endurance test*: This test exposes an IC to a high temperature (125 °C to 150 °C), while its supply voltage exceeds the specified maximum. Constant and varying signals are applied during the test, which may last for 1000 hours. These stress conditions will make the weak devices fail, which is detected by applying normal tests that monitor whether the ICs still show correct functional behaviour.

  The electrical endurance test reveals the following:

  - *Infant Mortality*, i.e., faults which are likely to arise in the early months of an IC's normal application;
  - *Early Failure Rate*, i.e., faults which are likely to arise after half a year;
  - *Intrinsic Failure Rate*, i.e., the probability of a fault occurring during a specified number of years;
  - *Wearout*, i.e., effects of prolonged use on the product.

  Faults that are observed during the electrical endurance test can usually be traced to errors in the manufacturing process which preceded IC packaging.

- *Temperature-cycle test*: This test emulates practical temperature changes by exposing the product to rapid and extreme temperature variation cycles. The minimum temperature in each cycle is between $-55\,^{\circ}$C and $-65\,^{\circ}$C. The maximum temperature is $150\,^{\circ}$C. The number of cycles used is typically five hundred. The test is carried out in an inert gas and/or an inert liquid. The main purpose of the temperature-cycle test is to check the robustness of the package and the robustness of the connections between the package and its die. The test should reveal possible incompatibilities between the temperature expansion coefficients of the various parts of an IC, e.g., the die, the lead frame and the package material.

- *Humidity test*: This test exposes an IC to a relative humidity of $85\,\%$ in a hot environment ($85\,^{\circ}$C to $135\,^{\circ}$C). The test reveals the effects of corrosion on the package and provides an indication of the quality of the *scratch-protection layer*. Usually, the corrosion process is accelerated by applying different voltages to consecutive pairs of pins, with $0\,$V on one pin and V_{dd} on the other. Most humidity tests last 1000 hours.

The required specifications of an IC depend on its application field, envisaged customer, status and supplier. It can therefore take a relatively long time before the quality and reliability of a new IC in a new manufacturing process reaches an acceptable level.

10.4.10   Conclusions

While packaging, in conventional IC designs, was seen as a necessity to be tackled at the end of the design process, today it is more and more a critical factor towards the success of an IC development. The combination of increased IC complexity, the drive towards miniaturization and the continuous pressure on cost reduction will not make the design process easier in the near future.

In the coming years it is expected that packaging technology will get closer to its limitations. For wire bonding, reductions in bond pad pitches will slow down or maybe even stagnate. Alternative bondpad layouts need to be explored and new technologies for die to package connections will have to be developed. To follow the fab technology miniaturisation, flipchip will gain in popularity, while on the other hand the clear distinction between fab and assembly will fade. Within semi-conductors vertical integration will start and new packaging trends will appear, where fab technologies will be used in combination with assembly techniques, as well as substrate manufacturing technologies will be combined with assembly techniques.

SoC or SIP will stay competing technologies, without a winner in the end. SoC and SiP systems will live together, but each in its specific application domains, where their properties are exploited to their full advantage. One thing will be sure; SIP will be extended towards non-conventional technologies integrating MEMS based applications, biosensors and/or optics.

Finally, the package choice has a huge impact on the overall quality and reliability of the chip.

10.5 Potential first silicon problems

When first silicon, either on a wafer or mounted in a package, is subjected to the first tests, one or even all tests might fail. Passing a test means that everything must be correct: the technology must be within specification, the tester operation must be correct, the test software (vectors and timing) must be right, connections between tester and chip (interface and probe card) must be proper and, finally, the design must be right. Therefore, passing a test means the logical AND of correct processing, correct tester and interface operation, correct software and, finally, correct design.

Especially in the beginning of the engineering phase of first silicon, problems may occur with the tester, its interface or the test software. Also, problems may arise from marginal processing or marginal design.

The following subsections discuss each of the different categories of failure causes.

10.5.1 Problems with testing

Very complex ICs contain hundreds of millions to several billions of transistors and can have several hundreds to more than a thousand bond pads. It is therefore a tough job to locate the failure somewhere in the chip, when, for instance, one output signal fails. The relation between an incorrect signal on one of the output pins and the location of an internal failure is very vague. Dedicated advanced testing techniques are already included in the design to support testing. Because not all functional blocks have (direct) access to output pins, they will be part of a scan chain (see section 10.2.1). In many cases, these scan chains run (and are tested) at lower frequencies. A potential problem is that such blocks are found to operate correctly on the tester (at a lower frequency) but may show failures when the chip is put in the application (board; speed check). Therefore, the chip should run at the same speed during scan test as in the application.

Test data, such as test vectors and expected output data from simulations are also subjected to failures. Testing of complex high-performance VLSI chips requires a lot of different test vectors to be applied to the chip at the right time. Normally, the test response is compared with the "expected data", most of which is generated during the simulation of the silicon at the verification phase of the design. To reduce the number of test pins and test time, large parts of the chip are simultaneously tested via scan chains.

A reduction of the number of test vectors is often achieved by the implementation of *Multiple Input Signature Registers (MISRs)* which allow compression of data over a number of clock cycles. The final data is then scanned out. Because such tests are not functional tests, they may not yet have been simulated thoroughly during the design phase, leading to incorrect test pattern generation or incorrect comparison data. Moreover, when a bit failure occurs in compressed test data (signature), it is very difficult to locate the cause of the failure. This requires a lot of simulation. Data compression techniques during testing must only be used if other techniques are not satisfactory.

Other causes of test errors are timing errors. Sometimes, the switch from a functional test to a scan test or vice versa may take more time on the chip for the multiplexers to adopt the new state. Waits must then be included in the test programs to properly test the chip. Even set-up and hold times for input pins or the amount of load that the tester offers to a chip output pin must be thoroughly verified. In some cases, even the tester hardware might show problems.

An important, but not yet discussed, source of initial test failures is the probe card, which is used in the initial test phase during failure analysis on the wafer instead of on packaged dies. In such a test environment, limited ground bounce can only be achieved by taking several measures. These measures are all related to preventing or limiting current slew rates $(\mathrm{d}i/\mathrm{d}t)$. Placing decoupling capacitances close to the supply pads is one measure. Another measure is to prevent large (ground) current loops. This can be achieved by using star grounds instead of serial grounds, see figure 10.37.

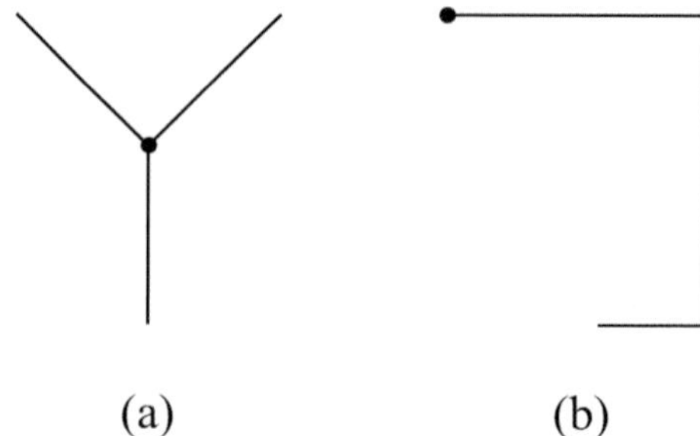

(a)   (b)

Figure 10.37: *Limiting large (ground) current loops by using (a) star grounds instead of (b) serial grounds*

Especially outputs can generate large current slew rates. The measurement of V_{OL} and V_{OH}, for instance, will often be done sequentially instead of testing simultaneously for all outputs. In conclusion, failures may arise during the development of the tests, during the development of the test boards and during the testing itself. Passing these test phases carefully can save a lot of time and frustration during the evaluation of first silicon.

10.5.2 Problems caused by marginal or out-of-specification processing

Each batch of wafers is processed under different environmental conditions: dust, temperature, humidity, implanter energy, etching time and doping levels, etc. This means that dies from different batches may show different electrical behaviour. The number of dust particles, for example, is one of the dominating factors that determines the yield, see section 10.3. In the following, we describe the influence of the most important technology parameters on the electrical behaviour of the chip.

Gate oxide thickness
The gate oxide thickness is the smallest dimension in the manufacture of MOS devices. It controls the gain factor β and the threshold voltage V_{T}, and it can also affect the IC's reliability.

When the gate oxide is thin, β will be high and an increased current capability of the transistors will be the result. In some circuit blocks, especially in memories, signals have to arrive in a certain sequence and they therefore propagate through different delay paths. However, when transistors become faster, the difference in delay paths may change, or may even become negative. This may cause a race, resulting in malfunctioning of the circuit.

Thin gate oxide may also lead to *pinholes*. These are oxide imperfections at locations where the oxide thickness is locally reduced (figure 10.38).

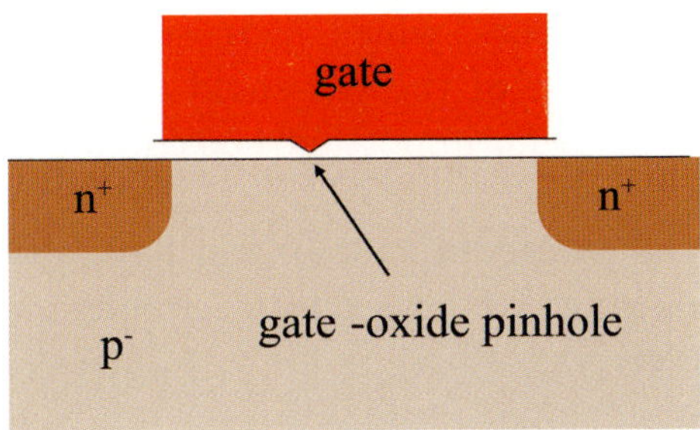

Figure 10.38: *Example of a pin hole in the gate oxide*

Sometimes, the oxide thickness at such a pinhole may be so thin that the voltage across it might cause carriers to tunnel through this oxide. The resulting leakage current increases slowly over time and eventually, as a result of this oxide breakdown mechanism, the chip no longer functions correctly. This process might take an hour, a week, a month or even a year. The sooner it is detected, the better. However, detection after shipping the device to customers will be disastrous and lead to customer-returns. Therefore, a bad gate oxide reduces reliability and can often be detected by means of I_{ddq} testing.

Polysilicon width

The distance between the source and drain of a transistor (called the channel length) is determined by the polysilicon width, forming the gate of the transistor. The wider the polysilicon, the larger the transistor channel lengths will be and the slower the transistor becomes. Signals that propagate through a combination of metal tracks and transistors will show different timing diagrams when polysilicon widths are wider than expected. This may lead to timing problems as a result of slowly operating circuits. On the other hand, narrow polysilicon leads to fast transistors. This may again result in timing problems such as races.

Threshold voltage

A change in threshold voltage can have different effects on the electrical behaviour of the chip.

A high threshold voltage, caused by a different channel dope, a thicker gate oxide or a larger body factor (K-factor) results in slower operation of the transistors. Especially a high body factor may lead to problems in pass-transistor logic and latches that use pass transistors. This may cause these circuits to operate significantly slower.

In contrast, a low threshold voltage results in somewhat faster circuits. Subthreshold currents, which increase by a factor of about 18 for every $100\,\text{mV}$ decrease of the threshold voltage, may cause larger standby currents. This is an important consideration in battery-driven applications.

A variation in the threshold voltage can have severe impact on the performance of analogue, RF and digital circuits, while it also determines their noise margins.

Substrate (p-well) and/or n-well dope

All together the n^+ diffusions of an nMOS transistor, the p^- substrate, the p^+ diffusions of the pMOS transistors and the n-well form parasitic thyristors.

When the p^- substrate is pulled to more than a junction voltage ($\approx 0.6 - 0.7\,\text{V}$) above the n^+ diffusion, such a thyristor might switch on, see also section 9.5.5. Because of the positive feedback in such a thyristor, it operates like a latch and the current may increase to unacceptably large values. This effect is called latch-up and can only be eliminated when the power supply is switched off.

Low substrate dope allows the thyristor to switch on much earlier and makes the circuit more susceptible to latch-up. The doping levels of substrate and n-well also determine the threshold voltages of the nMOS and pMOS transistors, respectively, as well as the thickness of the depletion layers across their source and drain junctions. The latter, in turn, determines the parasitic junction capacitances.

Next to these examples of how process technology can impact circuit behaviour and reliability, there are few relatively new physical mechanisms (such as STI stress, well-proximity effects and NBTI) that may cause variations in β and V_T, which are not yet completely understood and predicted by the models and the tools. These also form potential causes of performance reduction, which may lead to incorrect chip behaviour.

10.5.3 Problems caused by marginal design

Currently, verification software for integrated circuits has evolved to mature tools that are part of every design flow. Especially the verification on Register Transfer Level (RTL) and logic level (gate level) offers the potential of designing chips in which no logic error can occur. These tools almost guarantee that everything on the chip is connected correctly according to the specification. It is therefore important to first verify the specification, either by simulation or by emulation. Sometimes, in an application, the chip does not perform the function it was meant to execute. In many cases, it later appeared that the specification was insufficiently verified.

A hardware failure in very complex programmable chips can sometimes only be detected during very dedicated application tests. The number of different applications (and thus programs) of such chips is almost unlimited and extremely hard to simulate within an acceptable time.

Currently, most ASICs are designed in a mature process via a mature design flow and run at medium clock frequencies. First-time-right ASICs therefore should be the rule rather than the exception. However, modern technologies (90 nm CMOS and below) offer small feature sizes and thus the ability to integrate hundreds of millions to more than a billion transistors on one single chip. This, combined with the trend of increasing chip area, challenges the designer with many potential electronic problems that are not yet (or only partly) dealt with by the tools. Chapter 9 focuses on the underlying physical effects and on the measures that a designer can take to maintain the IC's reliability and signal integrity at a sufficiently high level.

10.6 First-silicon debug and failure analysis

10.6.1 Introduction

Current VLSI chips may contain hundreds of millions to more than a billion transistors, with only several hundred I/O pins. This means that hardly any logic block has direct access to output pins.

Without a direct access to the output pins, the other blocks must be accessed through a scan chain and tested as such (see section 10.2. In many cases, these scan tests run at a lower speed. This might lead to problems that show up only when the blocks are used in the real application because only then are all circuits running at full speed.

Logical (design) errors are easy to locate, both in scan test or in full functional test. On the other hand, identification of timing errors is much more complex!

When failures show up during the debug and engineering phase of an IC, it is important to know their source: whether it is logical, short circuit, latch-up or timing, etc. I_{ddq} testing is a means to quickly detect leakage currents and floating nodes, etc.

For circuits that can be tested at full functional speed, Shmoo plots can be drawn to gather information about the behaviour of the IC. Afterwards, different *failure analysis (FA)* techniques can be applied to locate the failure: laser scan, photo-emission, and in-circuit probing. The conventional failure analysis techniques such as picoprobing, liquid crystal, and electron beam, usually need access to the circuits from the frontside of the wafer or chip. Current FA techniques also use the backside to get access to the circuits, e.g., Time-Resolved Photo Emission and many *scanning optical-beam (SOM)* techniques. There are several other techniques that support these analysis tools and allow a quick repair of only a few samples.

10.6.2 I_{ddq} and ΔI_{ddq} testing

I_{ddq} and ΔI_{ddq} tests are described in section 10.2. In the following text with I_{ddq} we intrinsically mean I_{ddq} and ΔI_{ddq}. When a synchronous chip has been completely designed in static CMOS, hardly any current should flow when the clock is switched off. The only currents that flow are leakage currents caused by subthreshold and gate leakage mechanisms. However, in some cases, local higher-amplitude currents can flow. I_{ddq} testing, which is extensively discussed in section 10.2 is therefore a good means of locating certain defects or unusual behaviour which cause increased current levels during steady state.

10.6.3 Traditional debug, diagnosis and failure analysis (FA) techniques

This section discusses debug and FA techniques that were already in place during the last century. They may still be in use, either for designs in conventional CMOS technologies with only a few metal layers, or for designs in which special arrangements have been made, e.g., taking critical nodes up to the top metal so they can still be probed from the frontside of the wafer or chip or after de-processing of the IC.

Diagnosis via Shmoo plots

When a complete chip or part of a chip can be functionally tested, and an insight about operating margins with respect to the specification is required, then a Shmoo plot can be made.

A Shmoo plot shows the operating area of the software, the tester, the interface between tester and chip, and the chip itself, with respect to different parameters. When a Shmoo plot is not according to expectation (specification), the failure does not necessarily need to be in the IC design. It can also be in the technology, tester software or interface or the tester itself.

A Shmoo plot, which shows the operating area of a chip is, in fact, a quality measure. It shows whether the chip is marginal with respect to its specification (see figure 10.3). Measurements of Shmoo plots can be repeated at different temperatures to see how the margins shift.

Once the environment (tester, tester interface, connections, etc.) has proven to be correct, then, if the small operating areas of the chip are found to be too small, several different Shmoo plots must be measured to find dependencies: supply voltage, frequency, set-up time, temperature and I/O levels, etc.:

- If delay paths between flip-flops are too long:
  → frequency versus supply voltage Shmoo plot: lower frequency
  → better operation and higher voltage → faster circuits
  Conclusion: ⇒ use frequency versus supply voltage Shmoo plot at a fixed temperature.

- If races, which are independent of the frequency, occur:
  $\rightarrow$ supply voltage versus temperature Shmoo plot: higher voltage
  $\rightarrow$ faster circuits and higher temperature $\rightarrow$ slower circuits
  Conclusion: use supply voltage versus temperature Shmoo plot
  at a fixed frequency. For this test, often Shmoo plots with error
  count are made. Such Shmoo plots show bands of errors, which
  are independent of the frequency (figure 10.39).

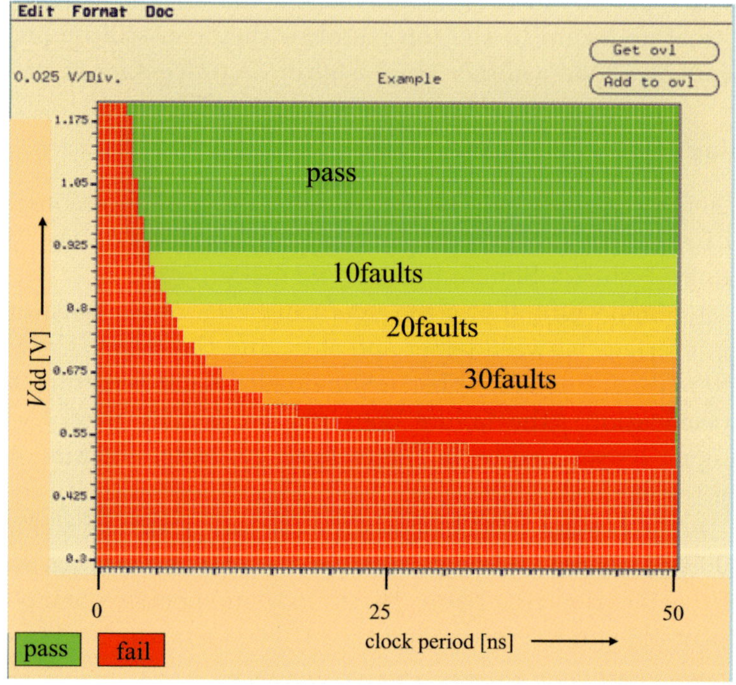

Figure 10.39: *Shmoo plot showing bands of faults, which are independent of the frequency*

- A Shmoo plot diagnosis may take a lot of time. Once a diagnosis
  has been made, it must always be verified by other techniques (such
  as probing). This is shown by the following example. Although
  this example refers to a chip in a conventional $0.25\,\mu$m CMOS
  technology, it is still very well suited for educational purposes.

EXAMPLE:

A certain signal processor contained two separated V_{dd} supply connections: V_{dd_1} and V_{dd_2}, which should have been connected together on the chip, but they were not. Figure 10.40 shows the Shmoo plot of the operating area of the memory on that chip:

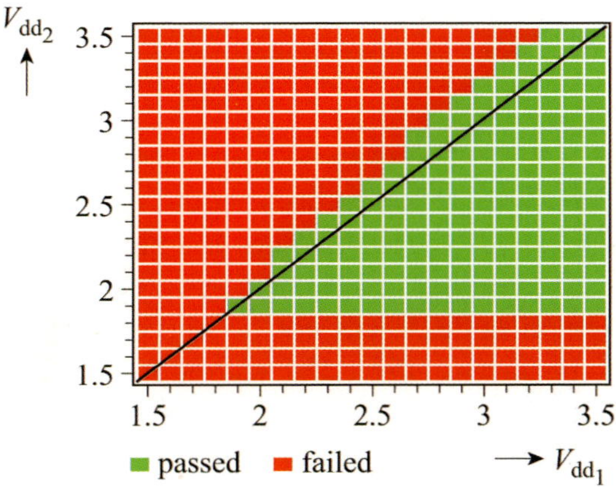

Figure 10.40: *Example of a very critical Shmoo plot*

Both supplies are connected to the same V_{dd} on the board. The $45°$ line in the plot shows the operating points for which V_{dd_1} equals V_{dd_2}. The Shmoo plot shows that when there is only a small on-chip supply noise in one of the supply domains, the chip would no longer operate. After a first inspection, the input registers of the memory were suspected. Because the ϕ and $\overline{\phi}$ clocks are generated by a clock buffer which is supplied by V_{dd_1} and the latches (figure 10.41) of the input register were supplied by V_{dd_2}, these latches could be the cause of the problem.

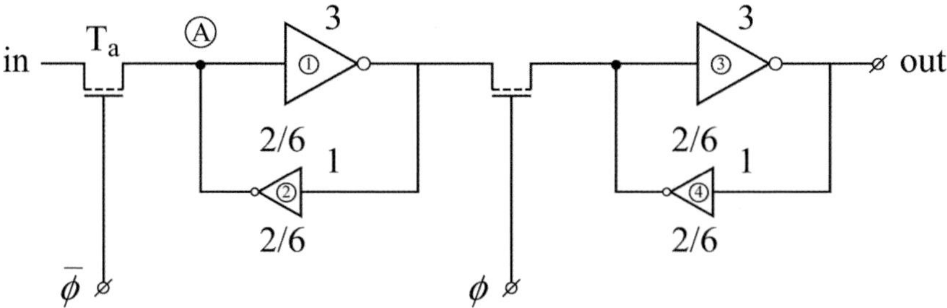

Figure 10.41: *Circuit for potential cause of the problem in the video signal processor*

Inverters 1 and 3 have a switching point equal to about $V_{dd_2}/2$, because the pMOS transistor width is three times the nMOS width (so $\beta_n \approx \beta_p$). Clocks ϕ and $\overline{\phi}$ are supplied via power supply V_{dd_1}. When $\overline{\phi}$ is high (V_{dd_1}), the voltage on node A will not be higher than: $V_{dd_1} - V_{T_a}$.

Because of the back-bias effect, V_{T_a} will be relatively high. Therefore, if $V_{dd_1} - V_{T_a} < V_{dd_2}/2$, the flip-flop will fail to switch to a logic "1". The results of a circuit simulation using worst-case process parameters are shown in figure 10.42:

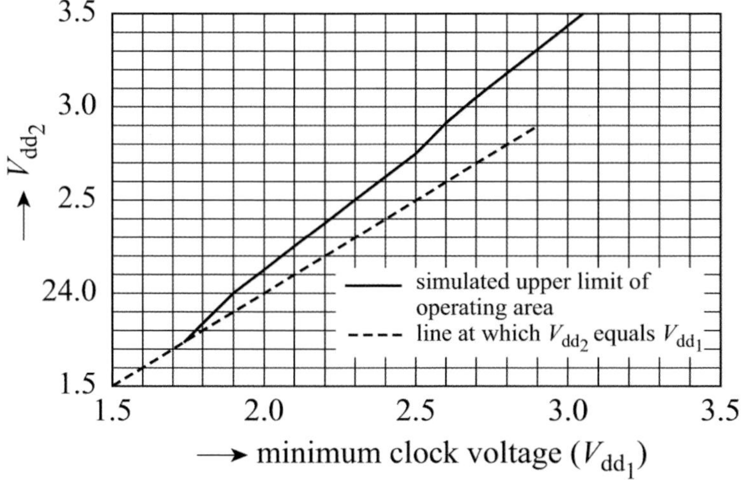

Figure 10.42: *Circuit simulation of the operating area of the latch of figure 10.41*

Below the solid line, the circuit operates correctly; above the line, it does not. If we compare this with the Shmoo plot of figure 10.40, we see that they are almost completely identical, and one would believe that the flip-flop is the real cause of the problem.

However, before changing the flip-flop design, the outputs of the flip-flops were probed to check the diagnosis. These flip-flops happened to operate much better than simulated and thus the real cause of the problem had to be found elsewhere. The process was certainly not in the worst-case corner.

Although a legacy technique with picoprobes was used for further analysis, this example is still representative for the necessity to check a potential cause of a failure.

Diagnosis via probing

Probing is a method that allows us to measure any node that is available at the top level metal when there is still no passivation layer (scratch protection) on the wafer, or when this layer has been removed locally (by etching, laser cutting, or Focused Ion Beam: FIB).

Conventionally, picoprobes were used, figure 10.44. They consist of needles as thick as a hair and with a very thin tip, less than several tens of a micron. This needle is connected to the input of a FET to reduce the capacitance. Values of 10 to 1000 fF for such a FET probe are available and are so low that they can be used within a digital IC almost without affecting the probed signal itself.

This technique was a reliable method for analysing incorrectly operating VLSI chips in semiconductor technologies with up to three metal layers. However, with the advent of multi-level metal technologies, it is becoming increasingly difficult to probe a signal that is only available in the lowest metal layer(s). During the design phase, additional metal stacks could be placed at the critical nodes to create probe pads. Another way to cope with this problem is to adapt the design style to design for debug, see section 10.6.7.

In the previous example, picoprobes were also used to try to locate the failure further. After a while, the real cause of the failure was found. Figure 10.43 shows the corresponding schematics:

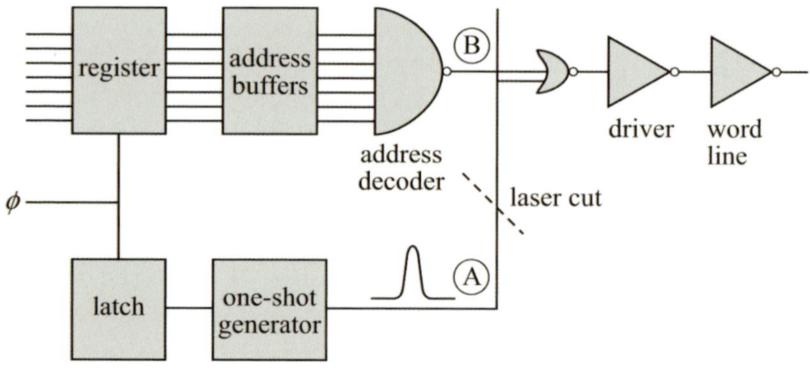

Figure 10.43: *Schematic of the real cause of failure*

Figure 10.44: *Picoprobes were used to measure a chip's internal signals (Source: NXP Semiconductors)*

The one-shot pulse on node A came 100 ps too early. At that moment the row address of the previous clock cycle was still available at the address decoder pins. This resulted in reading the wrong word from the memory. By cutting track A with a laser, probing the one-shot signal right before the cut and forcing it back via a pulse generator with variable delay right after the cut, the correct pulse could be found and a new Shmoo plot was measured. Figure 10.45 shows the result.

Thus, probing identified the exact location of the failure and also a way to solve the problem. A redesign (a one-mask change only) was made and the devices in the next batch operated correctly.

Besides picoprobing, there were several other conventional techniques to accommodate failure analysis, such as liquid crystal and electron beam. Because of the large number of metal layers which shield the lower signal lines and devices, these techniques are only applied occasionally and have been replaced by such techniques as laser-scanning and time-resolved photo emission. We will therefore no longer focus on these legacy techniques.

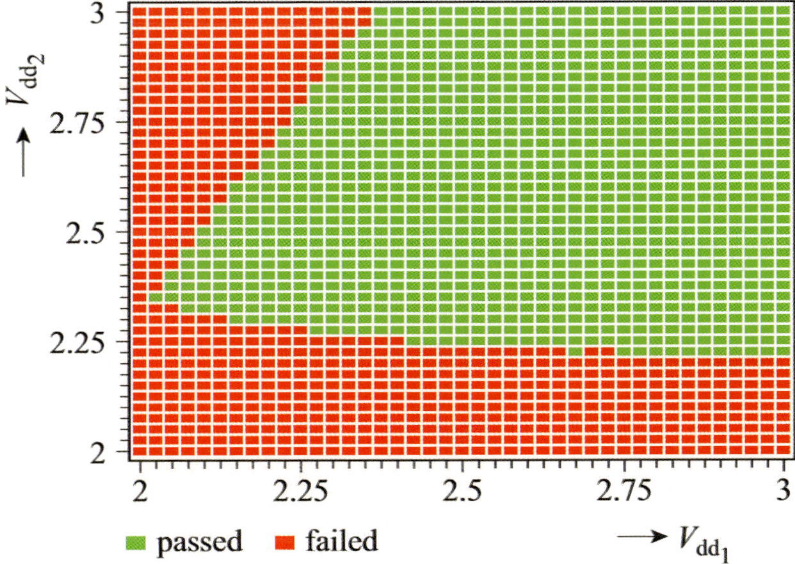

Figure 10.45: *Shmoo plot after correcting the one-shot pulse by forcing it via picoprobes*

661

Diagnosis by photon emission microscopy (PEM)

When charge carriers decay to a lower state of energy, the energy surplus is converted to photon emission (PE). This occurs, for instance, when electrons are accelerated in the transistor channel to excessive velocities, when carriers cross a potential barrier, or during breakdown, resulting in an avalanche of carriers [9]. It is therefore a good tool in identifying hot electrons. The light that is given off by operating ICs is captured by a microscope and used for imaging. Figure 10.46 shows several hot spots on a CMOS chip layout:

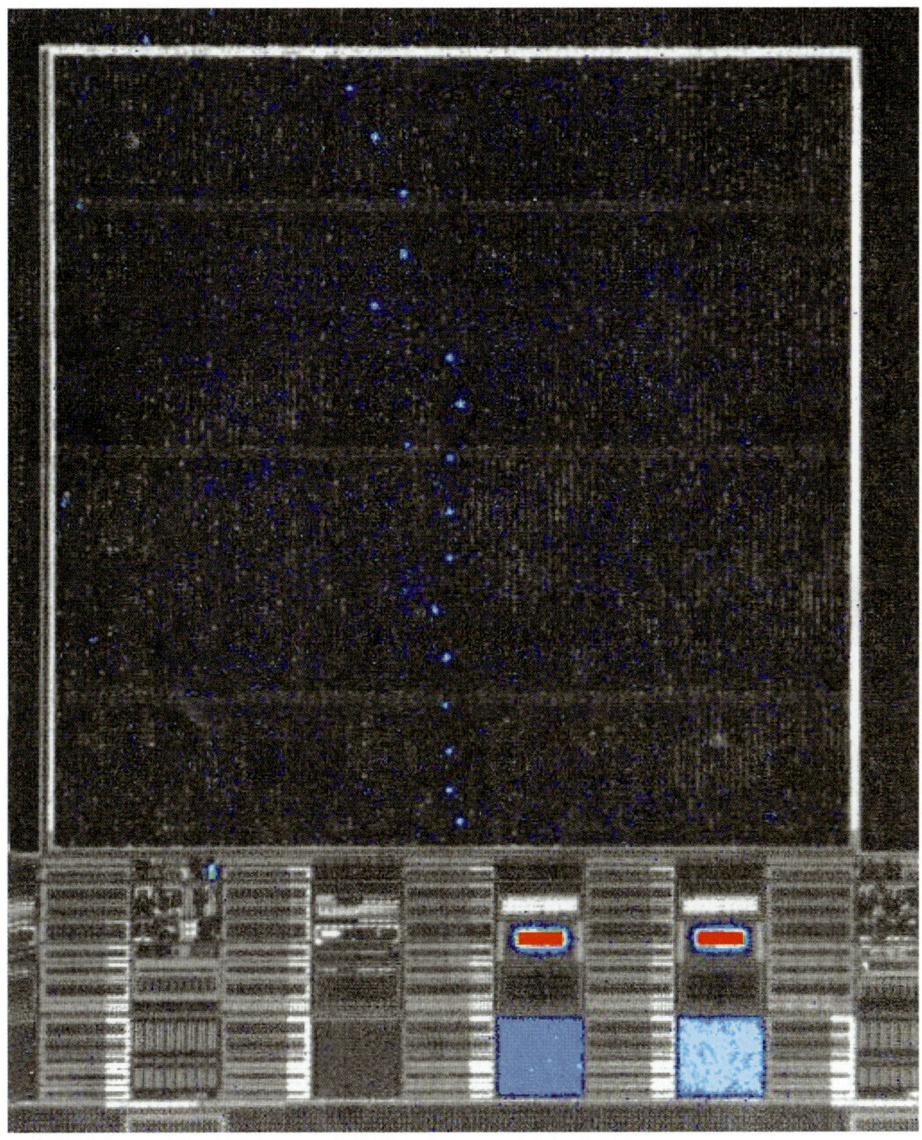

Figure 10.46: *Photo-emission image of operating ARM microcontroller core in 90 nm CMOS, taken through the backside of the silicon. Field of view is about 300x300 microns. The 16 strong emission points in the core are part of the clock tree. (Source: NXP Semiconductors)*

The key performance parameter for PEMs is the overall sensitivity of the system, which is a function of the wavelength, but also depends on the optical system. Currents well below a micro-ampere can be visualised at spatial resolutions less than $1\,\mu$m. In this way, latch-up, gate oxide defects, saturated MOS transistors, degraded (avalanching) junctions and unwanted forward-biased junctions can be detected.

The emitted photons can have both visible and near infra-red (IR) wavelengths. With the large numbers of metal layers, frontside analysis faces severe limitations. Because silicon is transparent to near infrared light, IR PEM can also be used for die backside analysis. With highly sensitive cameras (LN2 cooled InGaAs camera, 900-1600 nm wavelength), this greatly reduces image capture time and prevents optical obstruction by multi-level metal layers and flip-chip packaging.

10.6.4 More recent debug and failure analysis techniques

The continuously growing complexity and density of integrated circuits, both in terms of number of transistors and timing requirements have increased the variety of failure mechanisms. These failures can be originated by manufacturing defects or by design related failure mechanisms. Section 10.3.2 discussed DfM basics to improve yield and to reduce the number of manufacturing defects. Still many ICs are not first-time-right products and require a lot of support to reduce the debug, failure analysis and diagnosis time, in order to reduce time to market. *Design for debug* is a step in that direction and helps to improve observability, both at system level as well as at the physical level.

Developments in failure analysis (FA) techniques have enabled the way to access critical nodes from the backside of the wafer or chip, because of the above described inability to observe the transistors through analysis techniques from the frontside (figure 10.47).

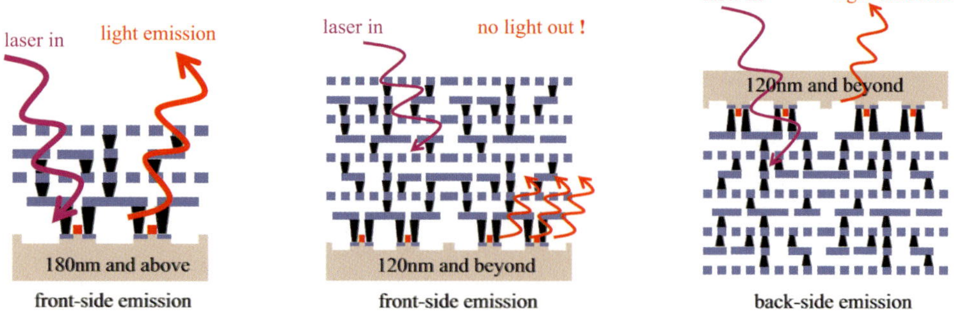

Figure 10.47: *Frontside FA techniques could be used in conventional technologies, but beyond 120 nm, backside techniques are required*

One technique, which is based on photon-emission microscopy but with improved sensitivity for backside usage is the time-resolved PEM technique. The others are based on stimulating the circuit with either laser beam or electron beam. The following subsections present a flavour of state-of the art failure analysis techniques. For a more detailed summary on electron- and laser-beam failure analysis techniques the reader is referred to references [10] and [11]. The chapter ends with a short discussion on techniques that can be applied already during the design phase to support the debug and failure analysis phase.

Time Resolved Photo Emission Microscopy (TR-PEM)

The basics of PEM are already discussed in the previous section, for the detection of breakdown, hot carriers, latch-up, gate-oxide defects, degraded junctions and even saturated transistors.

Picosecond imaging circuit analysis (PICA) is a form of *time-resolved photon-emission microscopy* developed by IBM. During switching of MOS transistors, light pulses are generated due to hot-carrier injection, figure 10.48.

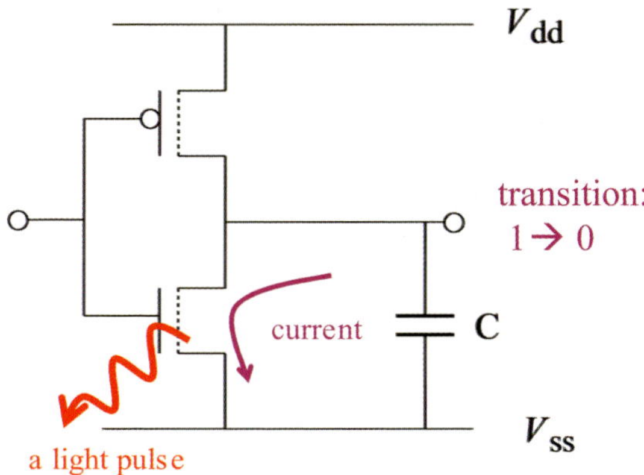

Figure 10.48: *The emission of a photon during switching transition of a logic gate*

When the substrate is thinned, a very sensitive, high-resolution, time-resolved camera can capture the integrated IR component of these light-pulses through the backside of the IC. The pulses are a measure for the switching moment and can be used to measure the timing behaviour of

any node in a digital circuit. The PICA technique images the switching
activity of several circuits in a relatively large field. Because not every
switch generates a photon, it takes many hours to a day to aggregate
sufficient photons for the creation of the switching-activity image in the
scanned field.

The technology is further developed by Credence Systems in their
Emiscope tool. It is based on single-point detection, rather than on field
imaging. It uses an avalanche photo diode to capture single photons from
a single switching node over a certain collection time and transports it
via an optical fiber to the time resolving equipment. It can create a
measured waveform in a few minutes to an hour, much less than was
needed by the original PICA system. The number of photons generated
by a switch is much less than one and is dependent on the voltage swing.
In a 90 nm technology the number of photons per detected switch is
in the order of 10^{-5}. For smaller technologies, with reduced supply
voltages, the aggregation of sufficient photons will become more time
consuming. Hot electrons generate one to two orders of magnitude more
photons than hot holes, which makes this technique better suited for the
timing analysis of nMOS transistor switching.

Individual light pulses of transistors can be visualised by a histogram
of detected photons versus time (figure 10.49). The signal is the result
of the integration of photons accumulated by the microscope objective.

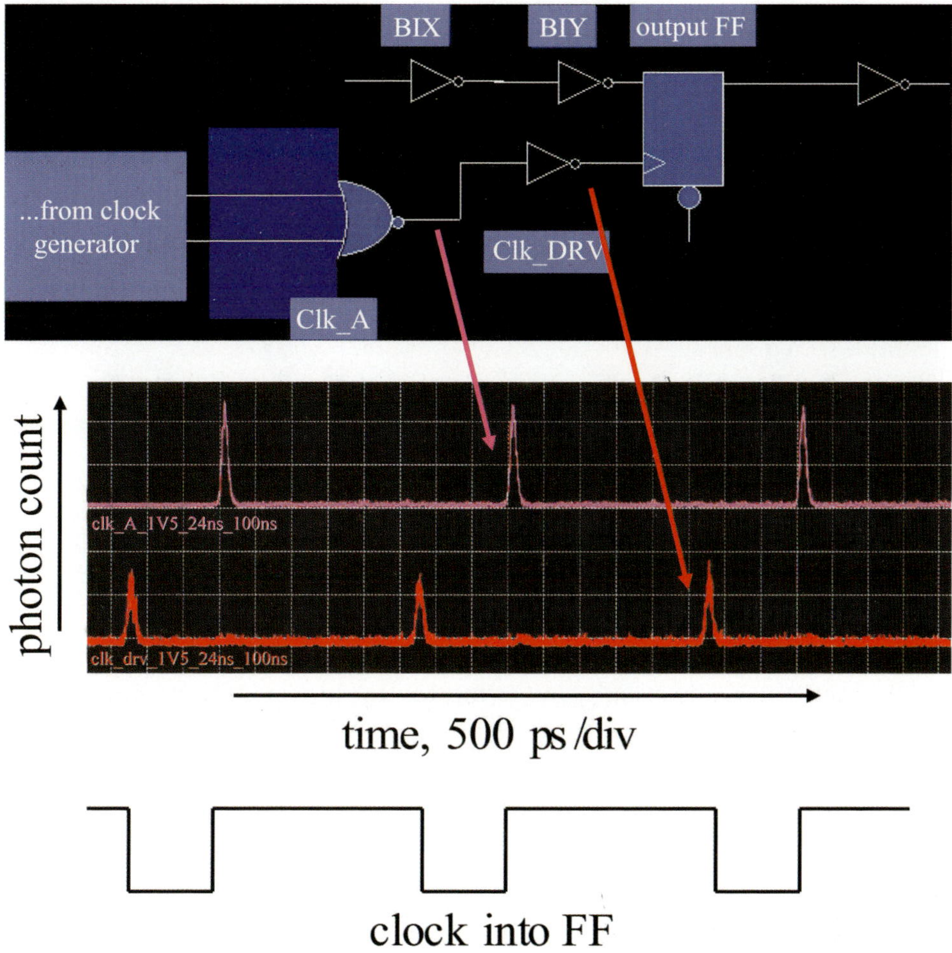

Figure 10.49: *A typical Time-Resolved PEM measurement result*

However, the silicon substrate has a relatively high refraction index ($n = 3.5$), which means that most rays are strongly refracted at the silicon-to-air interface and rays with angles more than seven degrees stay completely within the substrate. The result is a loss of 90% of the captured light. The solution to this problem is to bring a **silicon** *solid immersion lens (SIL)* in contact with the silicon substrate (figure 10.50).

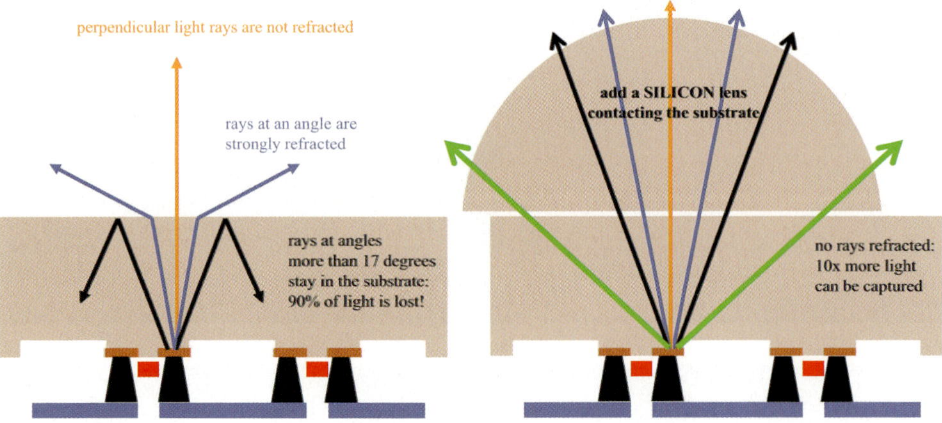

Figure 10.50: *A 90 nm CMOS IC viewed through the (backside) silicon with a conventional microscope objective (a) and with a solid immersion lense (SIL) (b). The size of the image is about 80x50 microns.*

In this way the rays are no longer refracted and the system can capture about ten times more light and, due to the higher NA, its improved resolution enables visualisation of even the smallest 90 nm CMOS devices (Figure 10.51.b). The SIL is also used to improve resolution of other laser-based FA techniques.

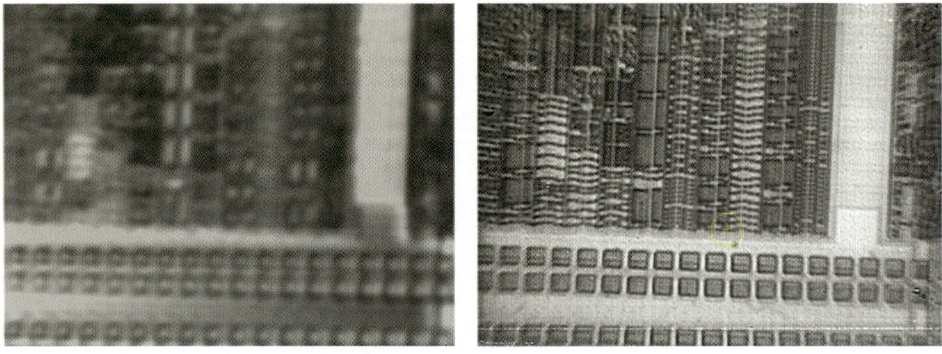

Figure 10.51: *a) Image captured with a normal lense b) same image captured by a solid-immersion lens (SIL)*

Scanning optical beam (SOM) techniques (or laser signal injection microscopy LSIM)

Basically we can divide scanning optical beam FA techniques into two categories. The first SOM category does not use a tester to generate test stimuli, but only applies a constant voltage source across the supply terminals to sense current changes, or a constant current source to sense voltage changes, both as a result of internal circuit stimulation by thermal or charge induction.

The second SOM category requires an IC tester to create the required input stimuli and generate the optimum operating conditions to enable detection of even the smallest change in electrical performance. These conditions are usually such, that the operating point is set at the edge of the Shmoo plot (figure 10.52).

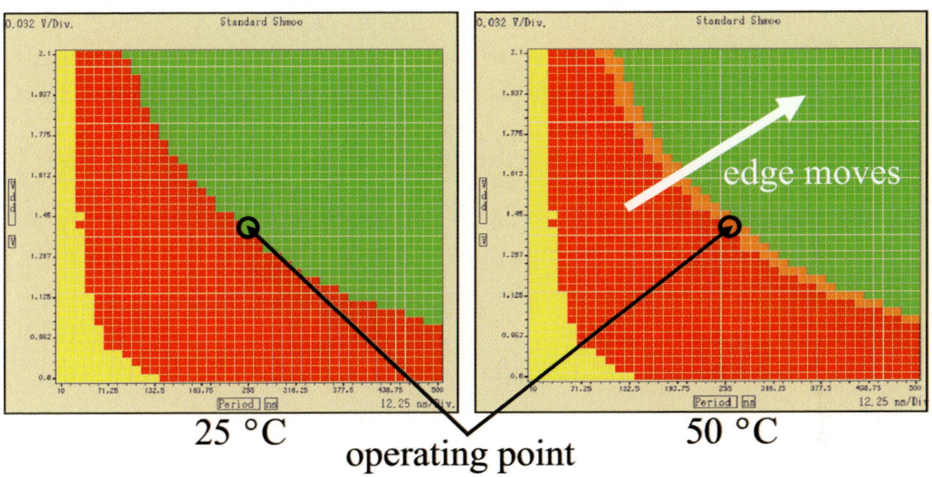

Figure 10.52: *Preferable operating point for failure analysis test*

In this point the chip is marginally operating and the circuit that causes the failure will be very sensitive to any injection of heat and/or charge into its critical node. Next, the chip is scanned with a laser beam, which induces thermal heating of the material (silicon, metal, etc.) in the laser spot or injects charge into a device. Once the spot has arrived at the critical node, it will introduce a change of its electrical behaviour, either in amplitude, or in timing. The example Shmoo plot refers to a chip that exhibits too much delay in one of its critical paths. When the laser

strikes this critical path, it will increase the delay because of the local
thermal heating that it causes in the spot of the beam. In this way it
allows the detection of critical paths in a logic design.

Many FA optical techniques can be used both from the top and backside of the chip, depending on what layer needs to be analysed. If the
failure happens to be in one of the in-between metal layers, it becomes
very difficult to create access to that layer. Optical backside analysis exploits the relative transparency of silicon to (near) infrared light.
Backside analysis requires thinning and polishing of the substrate since
the transmission of IR light through silicon decreases exponentially with
its thickness (figure 10.53). Particularly heavily doped substrates, which
are much less transparent, often need to be mechanically ground down
until a thickness in the order of $50\,\mu$m is reached and then further polished to achieve an adequate optical backside surface quality for proper
light injection and propagation during laser-beam stimulation. Many
CMOS circuits, today, employ lightly doped substrates (see chapter 3)
and do not require thinning at all. The package has a large influence
on the ease of use of backside failure analysis. Needless to say, that
backside analysis is easier for flip-chip packaged devices.

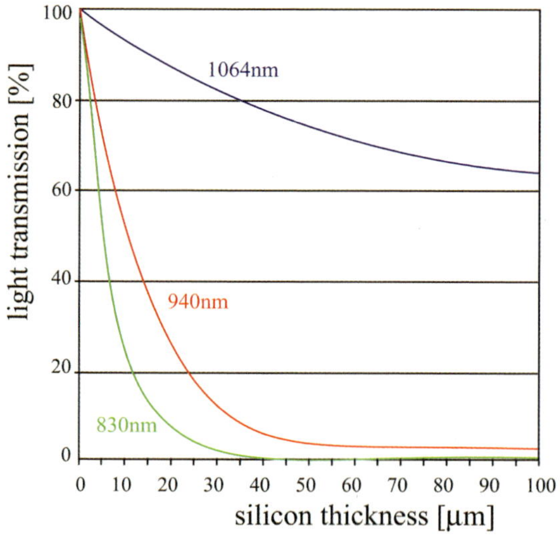

Figure 10.53: *Behaviour of light transmission vs silicon thickness at
different wavelengths of the light*

Many of the laser-beam stimulation techniques are based on the same principle of scanning a laser beam across the chip and measure potential changes in its electrical or logic behaviour.

The applied laser wavelength depends on its usage. Systems meant for thermal stimulation use a laser wavelength of 1340 nm and are most commonly used for detection of resistive changes in the backend of the process, e.g., the metal layers and vias, or for the local heating of transistors to measure changes in timing behaviour. A laser wavelength of 1064 nm is used for photonic stimulation to create electron-hole pairs in the p-n junction, which may also change local timing behaviour.

There exist many alternative laser-based FA techniques. Based on the above discussion we can categorise them as shown in table 10.2.

Table 10.2: Different categories of SOM failure analysis techniques

	Thermal stimulation	Photonic stimulation
Wave length of applied laser	1340 nm	1064 nm
Application of constant-voltage source or constant-current source to sense current or voltage changes, resp.	OBIRCH TIVA, XIVA SEI (Seebeck effect)	OBIC LIVA
Application of IC tester to observe test results	SDL RIL	LADA

In most failure analysis techniques, the response is visualised by an image of (part of) the ciruit or of the chip. It uses a *confocal laser-scanning microscope (CLSM)*, in which the laser beam is guided through a combination of mirrors across a chip. It creates a first image, which is based on the intensity of the reflected light. Next, an identical scan is performed, but now the current (or voltage, dependent on the specific FA technique), is measured with respect to the position of the chip. These current (or voltage) levels can be transferred into different grey levels, or colors.

When we define all current changes above $100\,\mu\text{A}$ to become red, then the current measurement image shows these red spots only on the positions where the response to a laser strike results in a larger current. This image is then used as an overlay on top of the other, so that the defect location can be easily matched with the position in the chip.

Optical-beam induced resistive change (OBIRCH), thermally-induced voltage alteration (TIVA) and externally-induced voltage alteration (XIVA)

These analysis techniques are very similar and based on the generation of changes in IC power demands because of changes in the resistance of a short (*OBIRCH* and *TIVA*) when illuminated by a scanning laser beam which induces heat into the spot area of the beam. TIVA systems use a constant current source at the supply terminals of the chip and measure the voltage changes. OBIRCH systems apply a constant voltage to the supply terminals and measure the current changes. Both constant-voltage and constant-current systems often offer about the same sensitivity. It depends on the impedance of the sample, which of the systems is the best one to use.

XIVA systems claim the same high sensitivity as TIVA by enabling constant-current sensing by supplying a constant-voltage to the terminals of the device [12]. Actually, XIVA can be applied to detect all defect mechanisms: junction defects, opens, short, defective vias, etc., because it may use thermal or photonic stimulation, depending on the applied wavelength of the laser. Application of the technique is more difficult than TIVA and OBIRCH.

Seebeck effect imaging (SEI)

The Seebeck effect imaging is based on the fact that electrical potential gradients are generated in conductors because of the creation of thermal gradients. In correct functioning ICs, the potential gradient in a metal line, produced by local laser-beam heating is compensated by a current in a transistor driving that line. However, if a void in the metal line, or an open via is isolating this line from a driver transistor, the potential of the open line will change because of the Seebeck effect, causing a change in the gate voltage of the connected transistors. These transistors will change their conducting states and cause a change in power demands of the chip. The SEI technique uses a constant current source applied to the supply terminals of the IC.

The laser may generate thermal gradients in a conductor of several tens of degrees Celcius, causing voltage changes in the order of a few tens of micro volts. The technique can also work from both the frontside and backside. For SEI no external bias is needed to observe a signal. SEI is today not very often used. One reason might be that the resulting voltage changes are very difficult to interpret.

Soft defect localisation (SDL) and resistive-interconnect localisation (RIL)

This technique can be applied to both the front and backside of the chip to detect soft defects (e.g., gate-oxide leakage, resistive vias, process spread, marginal timing). It is based on the fact that when a $1.3\,\mu m$ laser is targeted at a soft defect, the defect changes its electrical behaviour during testing. As discussed previously, the IC is operated at the edge of the Shmoo plot, such that the device only functions marginally. When the laser reaches a sensitive position in the chip, it might change the pass/fail status, e.g., by changing the resistance of a defect or by changing the conductive state of a transistor to change timing behaviour of critical signals.

Light-induced voltage alteration (LIVA), optical-beam induced current (OBIC) and laser-assisted device alteration (LADA)

LIVA analysis can be performed either from the frontside using a visible laser, or from the backside using a photon beam generated by an infrared laser. Next to junction defects, this technique can also be used for a localisation of opens in interconnections, contacts and vias, by creating an image of the entire chip. Such an image is the result of the monitoring of the voltage fluctuations of a constant-current power supply during a laser-beam scan across the chip, using a scanning optical microscope (SOM), which allows zooming-in to the physical defect location. Both LIVA and OBIC analysis are based on photon-induced electron-hole pairs, which are representative for the material to which the laser spot is focused. Recombination of the electron-hole pairs cause a change in the power demands of the chip, leading to voltage fluctuations that are dependent on the amount of generated electron-hole pairs. When, for example, the spot is focused onto a transistor junction which is connected to an open conductor, it will result in a change in conductive state of the transistor(s) connected to that junction and causes a change in the supply voltage. A photo-multiplier detector is used for the visible-light laser, while a germanium-diode detector is used for the infrared laser(s). The LIVA system produces an image of the voltage fluctuations in relation with the position on the chip. When a laser spot arrives at a junction, the junction will always cause a supply voltage fluctuation. However, it is difficult to interpret the measurement data, because one has to discriminate between the fluctuation caused by a correct (connected) junction and that of a defective junction.

The resolution of the OBIC technique is relatively low and can be used for analysing junction sizes of $0.18\,\mu\mathrm{m}$ technologies and larger. For smaller technologies its usage is limited to large transistor junctions and wells. For nanometer ICs this technique will lose more of its popularity.

Scanning electron-beam microscopy (SEM) techniques

Most, if not all, scanning electron beam techniques are used during FA after the location of the failing mechanism or defect has already been determined by one of the previously discussed FA techniques, due to the about two orders of magnitude better resolution. It also has a large depth-of-focus, which allows that during imaging a large part of the circuit is within focus. The basic idea behind the use of electron-beam microscopy for failure analysis is analogous to optical (laser) beam techniques, but with the difference that an electron beam is used as stimulus instead of an optical beam. SEM operation is based on the capture of backscattered and secondary electrons produced by the sample. *Backscattered electrons* are electrons from the original beam that are attracted by the gravitational force of the positive nucleus, such that it will partly circle around the nucleus and then leave it with a different angle, without loosing speed. *Secondary electrons* are electrons that are physically hit by the strike of an electron of the original beam, due to the fact the beam electron repels the device electron. If this repulsion is large enough, this secondary electron may be pushed out of the atom and may exit the device through the surface of the die, but much slower than the backscattered electrons. A positive charge on the detector must physically attract these secondary electrons to create an image.

The amount of secondary emission depends on the voltage on the scanned node (metal line). Metal lines at V_{dd} level absorb most of the secondary electrons and provide dark fields in the voltage-contrast image. Active voltage-contrast SEM techniques require the chip to operate to create the image. In passive voltage-contrast SEM techniques, the contrast is created by the beam. When the beam strikes an open metal, it will charge the metal line to a different potential then when it strikes a normal connected metal line.

Analogous to SOM FA techniques, with SEM techniques the sample also needs to be prepared (grinded) to create access for the beam to the point or layer of interest. In many cases also FIB tools are used for the deposition of additional test points (probe pads) on the metal connections to the suspicious node. In literature, a couple of scanning electron-beam FA techniques are described, particularly *(low-energy) charge-induced voltage alteration ((LE)CIVA)* and *electron-beam induced current (EBIC)*. However, these techniques are generally not very extensively used and a detailed discussion on their application is therefore beyond the scope of this book.

10.6.5 Observing the failure

After the test has identified a failing chip, the debug and failure analysis techniques support the identification of the failure location. If the failure is caused by a manufacturing defect, such as a short circuit or an open circuit, it is necessary to make a detailed materials analysis of the defect to understand its cause.

The next example shows the failure analysis process, starting with the Shmoo plot from a failing device all the way to a TEM cross-section of the physical defect. The example is related to a device, which has passed all tests, except for a delay-fault test (see section 10.2.1). Shmoo plots (figure 10.54) of the digital tests confirmed that a certain critical path suffered from an additional propagation delay of around 14 ns (compare Shmoo plots (a) and (b)) at a supply voltage of 1.8 V. The Shmoo plots show the operating area (green) as a function of the frequency (horizontal) and supply voltage (vertical). The first plot shows the Shmoo plot of the delay-fault test of a correct operating reference device, while the second one represents the Shmoo plot of the same test for the failing device.

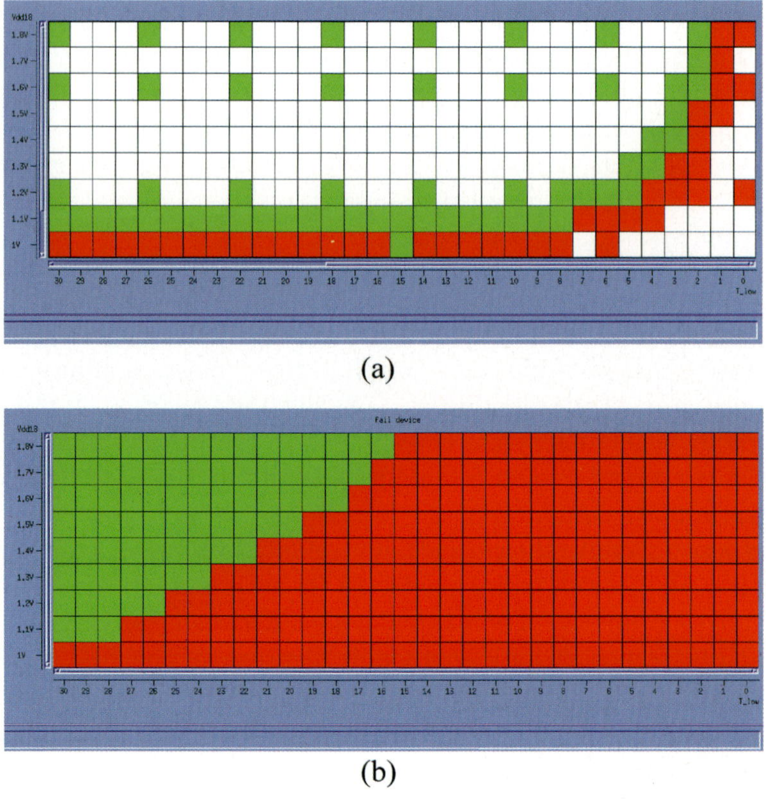

Figure 10.54: *Shmoo plots of a correct device (a) and of a failing device (b)*

By using a software tool that combines logged test-data with an image of the IC, the particular failing net could be localized. The result is depicted in figure 10.55, with the failing net highlighted.

Figure 10.55: *Image of the die with the failing net highlighted*

Next, backside SDL is used to visualise the response to the laser scan. Figure 10.56(left) shows a zoom-in backside image with GDS-II layout overlay and the failing net. No spots were visisble at the driver side of the net, but all three gates on the receiver side of the net showed a response to the laser scan. This suggests that there exists a relatively high resistance somewhere between the driver and the receiving gates.

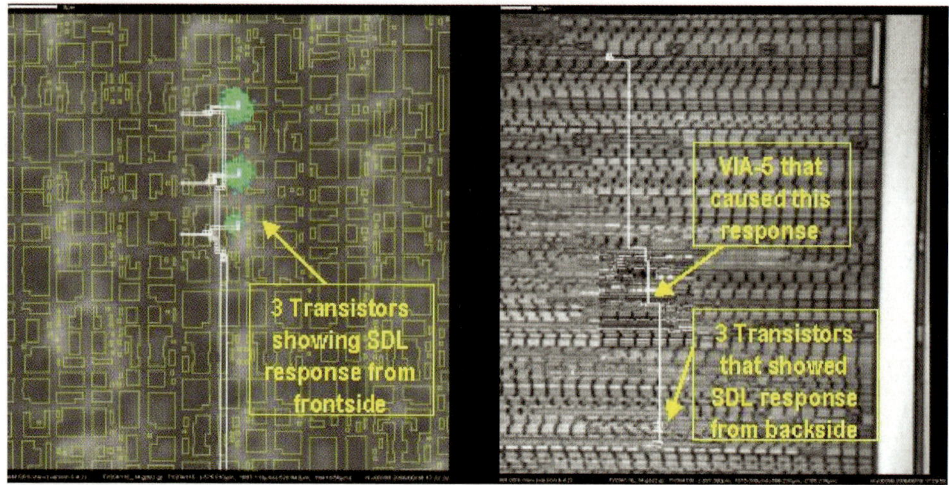

Figure 10.56: *Zoom-in of a GDS-II layout backside image with an overlay of the SDL laser response (left) and and an overlay of failing track on a frontside image of the IC (right)*

Because backside SDL did not show further responses to the laser scan, the device was opened from the top and a frontside laser scan was performed. This yielded a clear response from only one VIA-5 between the fifth and sixth metal layer. Because this via was in the top of the metal stack, it was not directly visisble with the backside SDL analysis. Figure 10.56(right) shows an overlay of the failing net on a frontside image of the IC. Because the response is not clearly visible in the figure, a more detailed view is presented in figure 10.57.a.

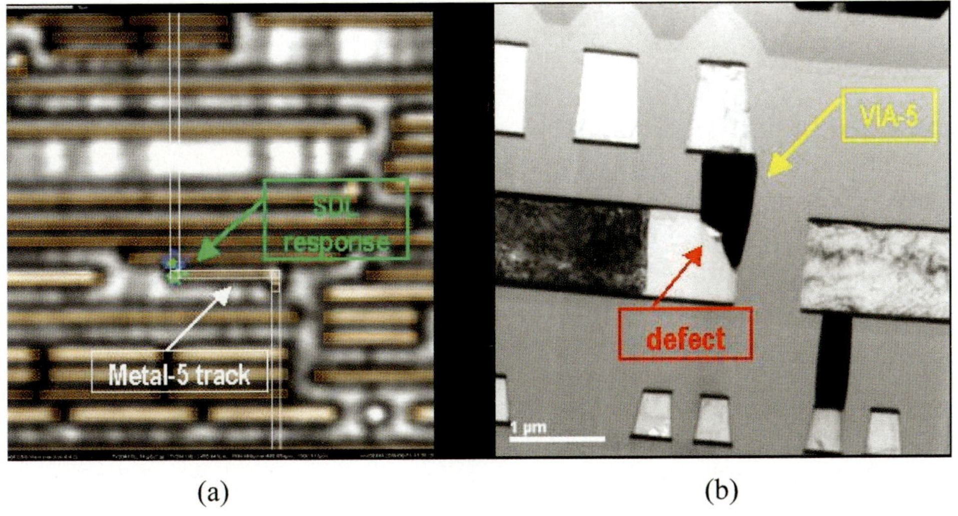

(a)   (b)

Figure 10.57: *(a) Zoom in of the SDL response of the failing VIA-5 in the overlay of the GDS-2 and (b) a TEM image of the defect*

The GDS-2 overlay helped to locate the coordinates of the defective VIA. Once the location is known, the failure analysis is continued to find the real cause of the failure. A thin cross section slice was cut from the chip by a focussed ion beam (FIB). Figure 10.57.b shows the cross section as viewed in a transmission electron microscope (TEM). It clearly shows the defect in the VIA-5. The VIA-etching process step did not stop at the top of the Titanium Nitride capping layer. Finally, the enlarged image of the defective via showed that there was only partly contact between the VIA and the METAL. This caused the increased resistance in the delay path, which was the cause of the original delay-fault test.

This example shows that the complete failure analysis process, from a failing test, down to the defect, is a costly, time-consuming and non-trivial effort, which may dramatically increase time-to-market of the final product. Therefore, a well-equipped failure analysis lab, combined with highly-skilled FA engineers is an essential part of the process of bringing a new design to volume production.

10.6.6   Circuit editing techniques

Once the diagnosis has been made, the chip can be repaired directly by making and breaking techniques. The ability to physically edit an IC (*circuit editing*) may reduce the number of respins and helps in reducing

679

time to market.

Traditionally laser beams were used for both cutting lines and laser-induced liquid-phase metal deposition to create interconnections on top of the scratch protection. A disadvantage of laser systems is that their resolution is limited, which has made them much less popular for circuit editing techniques in deep-submicron and nanometer ICs.

Focused Ion Beam (FIB) systems show better resolution and, for cutting conductors, spatial resolutions of less than 10 nm have already been demonstrated. Operation of a FIB system is similar to that of a SEM. Instead of an electron beam, a FIB system uses a focused beam of gallium ions, which is scanned across the chip, to image the sample in a vacuum chamber. At the location where the beam strikes the chip, ions and secondary electrons are emitted. The secondary electrons are captured and their intensity is used to create an image of the surface of the chip. An important advantage of a FIB system is that it can be used to remove material from the surface of a chip *(milling)* or to deposit dielectric or metal layers *(deposition)*. FIB is often used to physically edit a circuit on a chip. It can cut unwanted connections and deposit metal to change or add connections on top of the passivation layer or to create additional probe pads. Because holes can be made with high accuracy, even connections between different metal layers can be made, providing the capability to rewire ICs directly on the chip. These "design modifications" may fix design errors or implement spec changes.

Additionally, it enables the connection of an internal circuit node to a FIB-deposited metal area on top of the scratch protection. This will increase its load capacitance and can be used to correct timing violations. Figure 10.58 shows a schematic diagram of a basic FIB system [13].

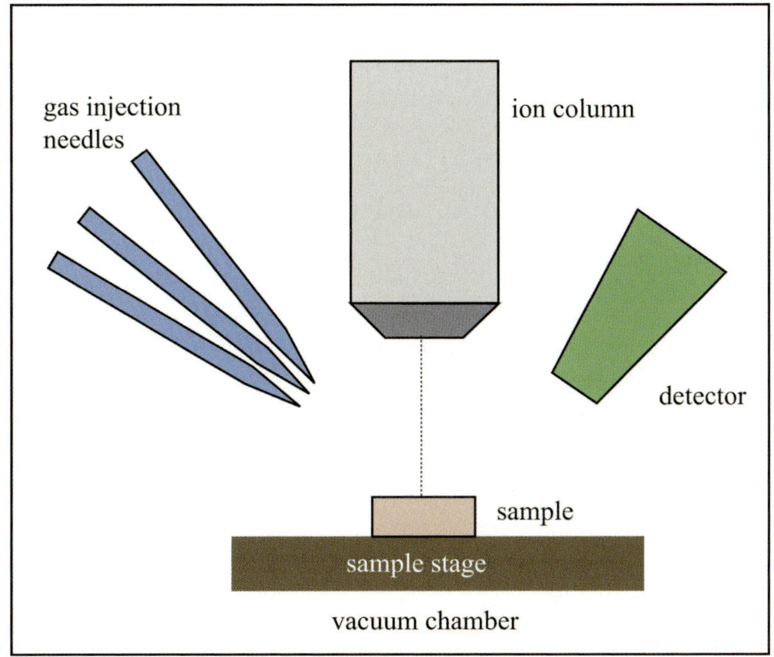

Figure 10.58: *Basic schematic diagram of a FIB system [13]*

It is equipped with a computer controlled gas injection system that can handle various different gases for the deposition of metal and dielectric material or for enhanced and selective etching.

A modern FIB system consists of complex and expensive equipment, which is capable of removing and depositing material (metal and dielectrics) and making smooth cross-sections for SEM or TEM analysis. It is sometimes combined with a SEM column for high-resolution imaging. To allow faster material removal at lower beam intensities, advanced FIB systems use gas assisted etching. With this technique, holes can be etched down to the first metal layer. In this respect, holes with aspect ratios of up to 30 with a minimum feature size below 25 nm can be created [14].

The deposition of the conductive material on top of the scratch protection is easy but time consuming. But the turnaround time of the modified chips is only in the order of several hours. The combination of a new mask and fabrication respin is very expensive, takes several months and introduces additional risks, since it is only based on simulations. A FIB circuit change allows the customer to perform all system-level tests and assures the next respin will include all the necessary changes.

Since FIB technology also allows small holes to be accurately cut through the wafer, it may be a valuable tool in inspecting flip-chip packaging.

10.6.7　Design for Debug and Design for Failure Analysis

The enormous density of VLSI chips with nanometer feature sizes and small operating voltages have reduced the margins in the electrical operation of the digital circuits. Voltage drop, temperature variations, cross-talk, supply and substrate noise, all depend on unforeseen operational conditions and, consequently, circuits may fail although they passed structural test procedures. There are even examples of ICs that fail during test, but pass the application. Particularly for nanometer CMOS ICs, the large number of metal layers with the increasing metal densities (metal fill (tiles), to support the CMP planarisation process), prevents physical probing of the signals for debug purposes. To enhance observation of important design and technology parameters, such as supply noise, capacitances, temperature, threshold voltage, etc., monitors can be embedded within the functional cores. Dedicated monitor circuits have been proposed in literature [15,16,17]. Another article [18] describes a fully integrated signal integrity self-test (SIST) system that combines a variety of on-chip signal integrity and performance sensors with a very simple digital readout mechanism. This technique supports debug by enabling additional observability by monitoring important electrical signals, which represent certain operating conditions of (different parts) of the IC. This type of *design for debug* techniques may also complement the physical failure analysis tools. Design for debug also includes support for fault observation at higher levels of design, e.g., software fault isolation. It will enable all flip-flops to be monitored and controlled at full functional speed. This requires additional on-chip hardware, which supports the debug software running separately from the digital tester.

Design for failure analysis include design strategies to facilitate I_{ddq} testing, design-in of additional test points for probing (E-beam, physical or other probing techniques), as well as the addition of markers to support on-chip navigation during the use of FIB or optical microscopy equipment.

10.7 Conclusions

The general requirement of a high fault coverage during the test of an IC is being challenged by an ever-increasing design complexity. Advanced test methods have been developed to maintain high fault coverages, both during IC testing and board testing. Additional hardware is included in the design to support these methods. This also reduces test time, which can be a relatively large contribution to the ultimate price of an IC.

The purpose of IC testing is not only to separate the good from the bad dies, but the test results can also be used for feedback on the operating margins that the chip has with respect to its specification. Testing is also closely related to yield. Therefore a single yield model served to present a flavour of the most important aspects that determine the yield.

Although packaging is not really a typical CMOS issue, the overview here shows the importance of choosing the right package. The temperature of the packaged die and the self-inductance of the package pins are important parameters which may dominate the performance and the operation integrity of a design.

The increased device complexity, combined with more levels of metal, reduces the fault observability. A design must therefore be supported by a design-for-debug approach to support both a rapid identification of the failure and the failure mechanism.

When the cause of a failure can not be traced with the combination of test/debug software and on-chip test/debug hardware, various techniques can be used to further analyse the failure. Because of the increasing complexity of integrated circuits, e.g., smaller feature sizes, increasing number of devices and metal layers and higher densities, semiconductor companies have installed very advanced failure analysis (FA) labs with complex FA tools, which offer sufficient observability both from the frontside and the backside of the die with very high resolution.

Finally, circuit editing techniques, performed with a focused ion beam tool, support the debug and failure analysis. It allows to physically remove and/or add dielectric and metal species on top of the scratch protection to enable circuit changes before mask changes are made and a respin is started.

10.8 References

[1] S. Wang, et al.,
'A Scalable Scan-Path Test Point Insertion Technique to Enhance Delay Fault Coverage for Standard Scan Designs',
ITC 2003, pp. 574-583

[2] L. Peters,
'DFM: Worlds Collide, Then Cooperate',
Semiconductor International, June 2005, www.semiconductor.net

[3] P. Rabkin,
'DFM for Advanced Technology Nodes: Fabless View',
Future Fab International, Issue 20, January 2006

[4] T. Lizotte,
'Laser Dicing of Chip Scale and Silicon Wafer Scale Packages',
2003 IEEE/CPMT/SEMl Int'l Electronics Manufacturing Technology Symposium, 2003

[5] JEDEC, Joint Electron Devices Engineering Councils,
www.jedec.org

[6] G.Q.Zhang, et al.,
'Mechanics of Microelectronics',
Springer, 2006, www.springer.com

[7] W. Mann,
'Leading Edge of Wafer-Level Testing',
South-West Test Workshop, 2004 ITC

[8] R.Tummala,
'Moore's Law Meets its Match',
IEEE Spectrum, June 2006, pp. 38-43

[9] K. de Kort,
'Techniques for Characterization and Failure Analysis of Integrated Circuits',
Analysis of Microelectronic Materials and Devices, John Wiley and Sons Ltd, 1991

[10] E.I.Cole,
'Beam-based defect localization methods',

Microelectronics Failure Analysis Desk Reference Fifth Edition, 2005, pp. 406-416

[11] F. Beaudoin, et al.,
'Principles of Thermal Laser Stimulation Techniques',
Microelectronics Failure Analysis Desk Reference Fifth Edition, 2005, pp. 417-425

[12] R. Aaron Falk,
'Advanced LIVA/TIVA Techniques',
Proceedings of the 27th International Symposium for Testing and Failure Analysis, 2001, pp. 59-65

[13] 'Introduction to Focused Ion Beams; Instrumentation, Theory, Techniques and Practice,
edited by Lucille A. Giannuzzi and Fred A. Stevie,
2005 Springer Science + Business Media, Inc., Boston

[14] 'Focused Ion Beam System',
Vienna University of Technology - Institute for Solid State Electronics
http://www.fke.tuwien.ac.at/silizium/alois/FIB_processing.htm

[15] M. Nourani, et al.,
'Detecting Signal-Overshoots for Reliability Analysis in High-Speed System-on-Chips',
IEEE Trans. on Reliability, vol. 51, no. 4, Dec. 2002, pp. 494-504

[16] E. Alon, et al.,
'Circuits and Techniques for High-Resolution Measurement of On-Chip Power Supply Noise',
IEEE J. Solid-State Circuits, vol. 40, no. 4, April 2005, pp. 820-828

[17] D. Schinkel, et al.,
'A 1-V 15W High-Precision Temperature Switch',
Proc. ESSCIRC, Sept. 2001, pp. 77-80

[18] V. Petrescu, et al.,
'A Signal-Integrity Self-Test Concept for Debugging Nanometer CMOS ICs',
ISSCC Digest of Technical Papers, February 2006, pp. 544-545

10.9　Exercises

1. Why is Design for Testability an increasingly important design requirement?

2. A given CMOS manufacturing process has the following parameters at a certain point in time: $Y_s = 0.75\,\mathrm{cm}^{-2}$, $D_0 = 1\,\mathrm{cm}^{-2}$.

   a) Express your opinion about this process.

   b) Calculate the expected yield for a chip with a die area of $80\,\mathrm{mm}^2$.

   c) Three months later, a $80\,\mathrm{mm}^2$ chip has a yield of $60.3\,\%$ and a $120\,\mathrm{mm}^2$ chip can be produced with a yield of $49.4\,\%$. Calculate M and D_0, assuming that the class of the clean room has not changed.

3. An IC dissipates $300\,\mathrm{mW}$ while its junction temperature is $32\,^\circ\mathrm{C}$. If the thermal resistance of its package is $30\,^\circ\mathrm{C/W}$, then what is the IC's ambient temperature?

4. What are the major differences between through-hole and SMD area array packages? Specify their respective advantages and disadvantages.

5. Explain the differences between test, debug and failure analysis of first silicon.

6. Why is Design for Debug a must for current and future ICs?

7. What is the drive for reducing the self-inductance (L) of the package pins?

8. What kind of tests are required to determine quality and reliability of packaged dies?

9. Explain what is meant by observability and discuss its trend with respect to future process generations.

10. In current and future processes, the transistors and lower metal layers are shielded by the upper ones. What could be done during the design phase to support failure analysis in this respect?

11. Explain how FIB supports failure analysis.

Chapter 11

Effects of scaling on MOS IC design and consequences for the roadmap

11.1   Introduction

The continuous scaling of CMOS devices according to Moore's law, has brought the design complexity in terms of number of transistors and performance requirements to such a high level, that design styles and methods need to be changed in order to manage this complexity and to enable full exploitation of the potentials of advanced and future CMOS technologies. A prediction of these potentials is presented in the *International Technology Roadmap for Semiconductors (ITRS)* [1], created by the *Semiconductor Industrial Association (SIA)*.

This chapter discusses the consequences of the scaling process for deep-submicron IC design, with the focus on future trends of power, speed, reliability and signal integrity. Nanometer CMOS design requires more focus on the physical design and on the consequences of further scaling. This will certainly have an impact on the semiconductors technology roadmap. In the race towards a multi-giga-transistor heterogeneous *System On a Chip (SOC)*, see figure 11.1, design methods and tools not only have to be changed to make the design manageable (system design aspects) but also to make a functional design (physical design aspects). Note the difference with figure 7.5, which only shows the system design aspects.

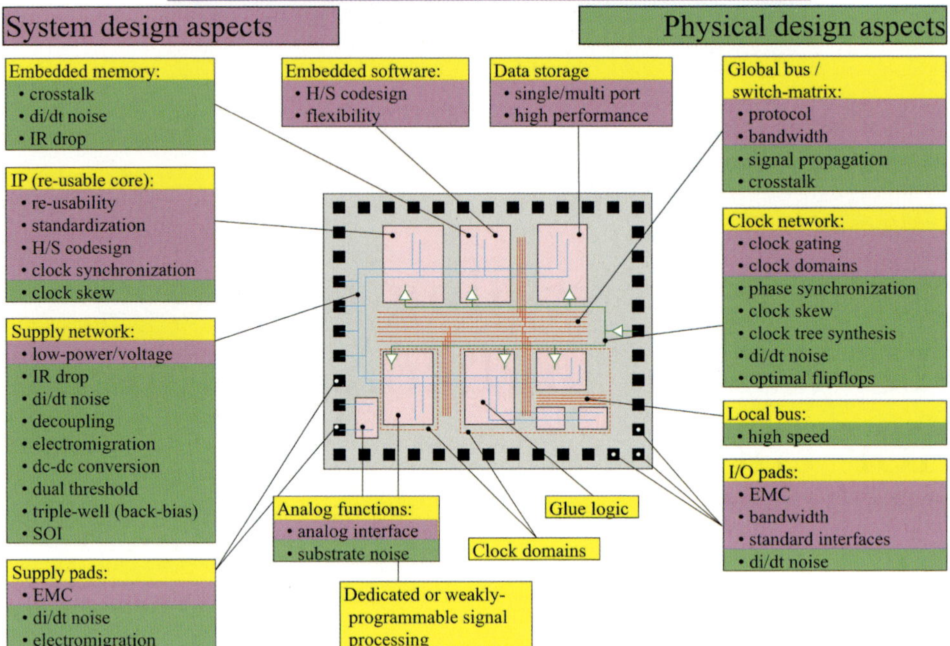

Figure 11.1: *Important aspects of a (heterogeneous) System on a Chip*

The complexity of such an SoC can only be managed by applying:

- a platform with integrated hardware/software architecture and application development tools

- system level synthesis to improve design efficiency

- design reuse

- increased design resources per chip.

The first three items deal with system level design aspects, see chapter 7. The increased design resources, however, are not only required to manage the SoC design complexity, they are also needed to cope with the increasing number of physical design aspects as depicted in the figure.

Previously, only analogue circuits were susceptible to these physical effects. In future process generations, these effects will dominate the SoC's performance and signal integrity, while some of these effects are already threatening the performance of today's complex VLSI chips.

Future VLSI design therefore requires a more analogue approach. Design is no longer about switches and ones and zeros only, but also about resistors, capacitors, inductors, noise, interference and radiation.

Basically, a VLSI chip is just a bunch of transistors that perform a certain function by the way that they are interconnected. The next sections focus on the influence of scaling on the basic elements, the transistor and the interconnections, and the consequences for the overall performance, reliability and signal integrity of deep-submicron IC designs. This increasing design complexity has caused a paradigm shift in the costs components of an IC: from fabrication costs towards design costs. Section 11.5 will therefore discuss trends in design complexity and costs in some more detail.

11.2   Transistor scaling effects

Over many technology generations, both the feature sizes and the supply voltage have scaled with the same factor. This type of scaling is called *constant-voltage scaling* and involves technology nodes from $0.8\,\mu$m (5 V) through $0.5\,\mu$m and $0.35\,\mu$m (both 3.3 V), $0.25\,\mu$m (2.5 V), 180 nm (1.8 V) to 120 nm (1.2 V). When scaling transistor sizes and bias voltages by a factor of s ($s \approx 0.7$), the transistor current scales by the same factor. To maintain performance, the threshold voltage is also required to scale with s. The threshold-dependent leakage current is an important factor that limits the pace of scaling. This subthreshold leakage current can be estimated by the following relation, which means that this current increases by about a factor of 18 for every 100 mV (depending on the subthreshold slope) decrease in V_{T}:

$$I_{\mathrm{subthreshold}}(scaled) = 18^{10(1-s)V_{\mathrm{T}}} \cdot I_{\mathrm{subthreshold}}$$

For large SoCs, this background leakage current will be larger than a gate oxide short current, for example, which will dramatically limit the potentials of I_{ddq} testing. The eventual reduction of the threshold voltage requires alternative techniques, such as the dual-V_{T} concept [3] and the triple-well concept [4], to limit the subthreshold power consumption during standby and test modes. Both concepts are discussed in chapter 8.

Another result of transistor scaling is the increased channel dope, caused by V_{T} correction and by the implants (drain extensions and halos) needed to suppress short-channel effects (chapter 2). As a consequence, the thinner depletion layers cause higher parasitic junction capacitances.

In combination with gate oxide thickness (t_{ox}) scaling, the higher channel dope mirrors depletion charge into the gate (gate depletion), which reduces the effective current control by the gate. Below a t_{ox} of about 2 nm [5], quantum-mechanical tunnelling of charge through the gate oxide may occur, resulting in additional standby currents and possibly a reliability problem. Finally, scaling of the channel length will increase the mismatch between 'equal' transistors, as a result of increased spread of the number of dopants in the transistor channel. A minimum transistor in a 0.25 μm process contains about 1100 dopant atoms. In the 65 nm node, the granularity on molecular level is almost reached. In this technology, a transistor with a W/L ratio of 100 nm/60 nm contains only between 60 to 80 doping atoms in its depletion region. While the threshold voltage is proportional to this number, the spread is proportional to the square root of it: $V_T \propto 80$ and $\sigma V_T \propto \sqrt{80}$, which is about equal to 11% of V_T. Section 9.4.3 includes a table which presents the variation in clock arrival times in a clocktree, as a result of this spread in V_T for different technology nodes. This V_T spread is additional to the process spread in V_T of about 60 to 90 mV. Further transistor scaling aspects and alternative device architectures to improve transistor performance are extensively discussed in section 3.9.4.

11.3 Interconnection scaling effects

Scaling of widths and spacings has caused the metal interconnections to start dominating the IC's performance, reliability and signal integrity. The output load of a logic gate is equal to the total of the fan-in capacitances of its connecting gates and the total wire load of the interconnections. Table 11.1 shows the increase in the average ratio between wire load and fan-in, for average standard cell blocks, caused by scaling. These numbers represent average values; for each individual chip, this ratio may be different from the table.

Table 11.1: Increasing interconnect dominance

Technology	Ratio: wire load/fan-in
350 nm	30/70
250 nm	33/67
180 nm	36/64
130 nm	45/55
90 nm	54/46
65 nm	66/34
45 nm	75/25
32 nm	81/19

The increasing resistance values of the on-chip interconnections lead to larger voltage drops and the increasing mutual capacitance values cause more cross-talk, while the combination leads to larger signal propagation delays. At 1 GHz, the required signal rise and fall times should be less than 50 ps to perform some computational tasks within the available 1 ns time frame. Even on-chip wires then cause interference with other modules. For such signal edges, line lengths of 3 mm and above become critical and require transmission line modelling. Figure 11.2 shows the *propagation delay* of an embedded metal track (metal track embedded between two minimum-spaced neighbours) in different technologies.

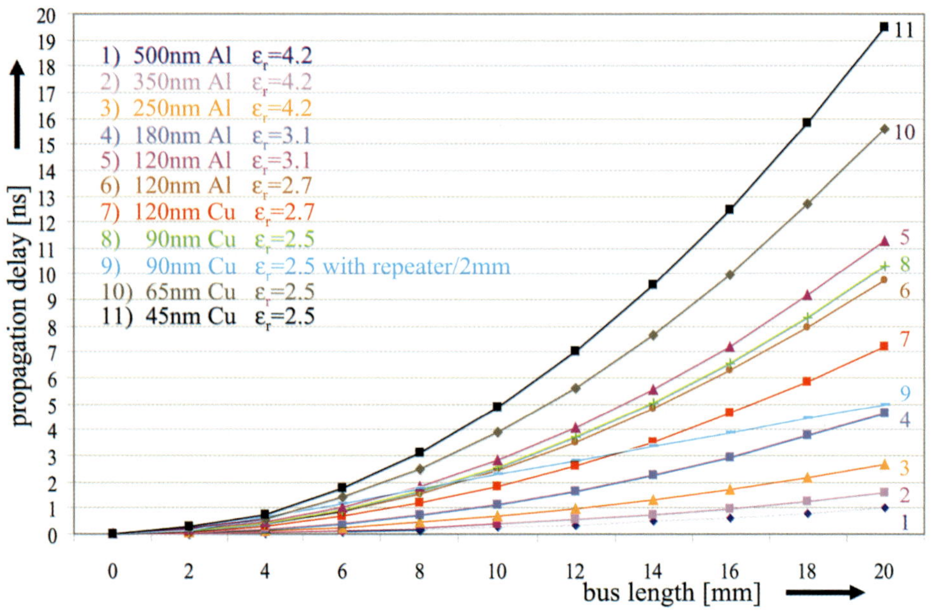

Figure 11.2: *Propagation delay of an embedded track in different technologies*

There are several approaches to reduce the negative effects of scaled interconnections. One is to reduce the capacitance, which is expressed as $C = \epsilon_0\epsilon_\mathrm{r}A/t_\mathrm{dielectric}$. Current values for the *interlevel metal (ILD)* dielectric coefficient ϵ_r are between 2.5 and 3.5. In the ITRS roadmap, edition 2006, values around $\epsilon_\mathrm{r} \approx 2$ are expected in a decade from now (with no known solutions yet), leading to a further capacitance reduction of a factor of 1.5. The second approach is to reduce the resistance. The sheet resistance of conventional aluminium alloys is around $3\,\mu\Omega$cm, while that of copper is about $1.8\,\mu\Omega$cm. However, the potentials of the reduced copper resistance cannot fully be exploited. Because copper diffuses through oxides, it cannot be deposited and etched like aluminium. By applying a damascene back-end flow, copper can be completely encapsulated within a barrier material, as shown in figure 3.28 (chapter 3). The effective sheet resistance of copper wiring depends on the barrier material and, for a metal layer used for global wiring, is expected to increase from about $2.5\,\mu\Omega$cm today to about $3.5\,\mu\Omega$cm by 2015 (see figure 3.51). This value is even more than that of the original aluminium metal wiring of the 180 nm CMOS nodes and above.

Figure 11.2 also shows the individual influence of copper and low-ϵ dielectrics. The signal propagation delay over a metal wire is proportional to the square of its length. The use of repeaters, however, reduces the propagation delay to a linear dependency on length. Particularly for longer wires, this may reduce the propagation delay by more than a factor of two (compare curves 8 and 9) The increasing clock skew and propagation delay for global signal wires are in direct contrast to the reducing clock period. Therefore there will be an increased drive to limit the size of the standard cells blocks (between one to several square millimeters), which will also limit local interconnect lengths and clock skew. Designs will therefore become *globally asynchronous* and *locally* (within blocks) *synchronous (GALS)*. To further relief the propagation delay problems, pipelines could be built into the global interconnects, but the bus latency will then become an important design parameter.

Figure 11.3 shows an example cross-section of a 65 nm CMOS circuit. The figure clearly demonstrates the increased dominance of the interconnect in current and future nanometer CMOS processes.

Figure 11.3: *Example cross-section of a 65 nm CMOS circuit*

As an example, in nanometer CMOS SRAMs not only the increase of the mutual capacitance between two minimum spaced wires is important, but also the increase of the mutual capacitance between two minimum spaced contacts or two minimum spaced vias. Figure 11.4 shows a 3-D cross section of an SRAM array. It requires an accurate 3-D extraction

694

tool to enable proper SRAM timing simulation.

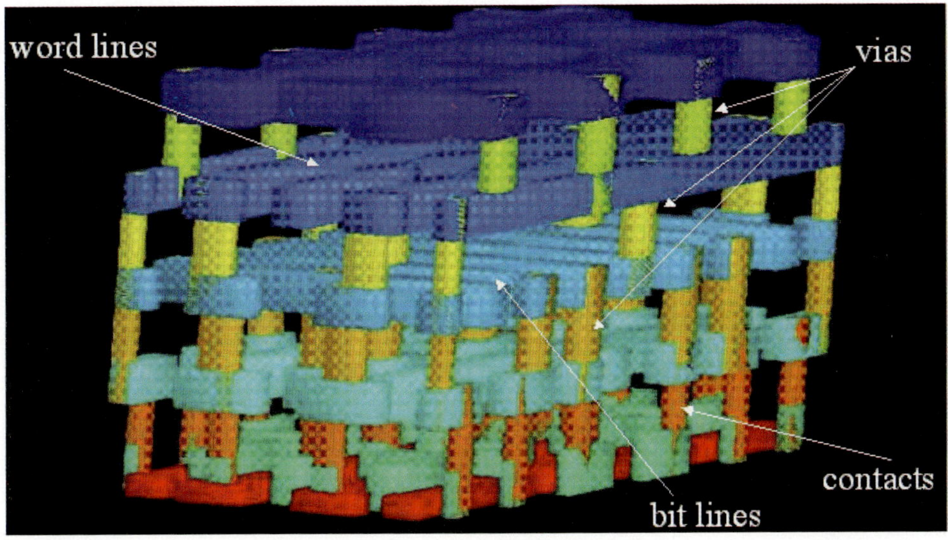

Figure 11.4: *3-D cross section of a nanometer CMOS SRAM array*

11.4 Scaling consequences for overall chip performance and robustness

For many technology generations in the past the supply voltage has been constant and equal to 5V. The scaling process over that period of time was called *constant-voltage scaling*. Figure 11.5 shows the evolution of the voltage scaling over the last couple of decades.

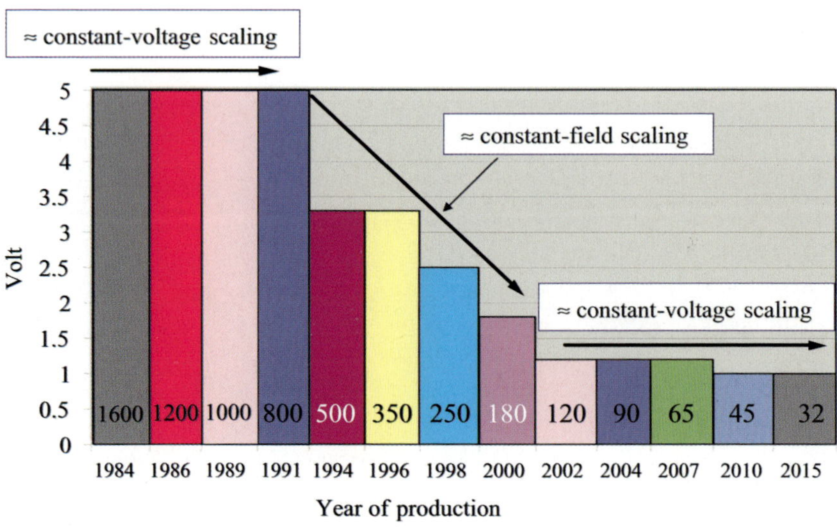

Figure 11.5: *The voltage scaling evolution for low-standby power (LSTP) CMOS processes [1][7]*

During the nineties, the advances in CMOS technology were not just related to scaling of the devices and the minimum features sizes, but also of the supply voltages. This is called *constant-field scaling*. Because of the shrinking voltages, the threshold voltages also reduced with about the same factor. At a certain moment in time, the subthreshold and gate-oxide leakage currents reached levels that were unacceptable for many mobile applications. Particularly these applications forced the semiconductor manufacturers to differentiate between general-purpose and low-leakage CMOS processes in the same technology node. In the *general-purpose CMOS processes (GP processes)* both the supply and threshold voltages are scaled further, supporting either high-performance or low-operating-power designs. These processes are meant for applications in which the switching power (active power) is the main contributor to the total power consumption and where the logic is continuously active and hardly in standby. From the 120 nm node, onwards, the supply voltage in these processes scaled roughly by about 0.1 V per technology node. In the *low-leakage CMOS processes*, also known as *low-standby power processes (LSTP)*, the supply and threshold voltage were no longer scaled and remained close to the nominal supply voltage of 1.2 V of the 120nm node (see figure 11.5), which means that this is also a period of constant-voltage scaling. In this "second" constant-voltage scaling era, the transistor current has a different relation with the supply voltage and therefore the transistor and chip show quite some different relation with the scaling parameter s, compared to the "first" constant-voltage scaling era.

It is obvious that these different scaling scenarios have a different impact on the basic transistor parameters and on the performance and robustness of CMOS ICs. Table 11.2 shows how the transistor performance, reliability and signal integrity parameters depend on the scaling factor s ($s \approx 0.7$) and the impact of the different scaling scenarios, when we continue the scaling process assuming that no additional design and technology measures/changes have been taken into account.

Table 11.2: Different scaling scenarios

The left margin groups the rows as: **basic parameters** (voltages … metal resist.), **performance** (gate delay … gate leak. curr/mm²), **robustness** (electromigr. … noise margin).

	relation (∝ means is proportional with)	Scaling factor ($s \approx 0.7$) p≠s≠q	p=q=1 before yr 1990	p=q=s	p=q=1 after yr 2000
voltages	V	p	1	s	1
	V_t	q	1	s	1
feature sizes	$W,\ L,\ t_{is},\ \text{dist}\quad t_{ox}(EOT)$	s q	s 1	s	s 1
devices per unit area	$\propto 1/A$	$1/s^2$	$1/s^2$	$1/s^2$	$1/s^2$
transistor bias current	$i = \beta/2(V_{gs}-V_t)^2$	p^2/q	1	s	$\approx\sqrt{s}$
capacitance	$C=\varepsilon_0\varepsilon_r A/t_{is}$	s	s	s	s
metal resist. (top metals)	$R=\rho\,\ell/(t_m W)\ (t_m\approx const)$	1	1	1	1
gate delay τ	CV/I (with $f \propto 1/\tau$)	sq/p	s	s	$\sqrt{s}$
average curr/unit area	$I = CVf/A$	p^2/qs^2	$1/s^2$	$1/s$	$1/(s\sqrt{s})$
pwr dissipation/gate D	CV^2f (with $f \propto 1/\tau$)	p^3/q	1	s^2	$\sqrt{s}$
pwr-delay product τD	$\propto CV^2$	p^2s	s	s^3	s
pwr density P	CV^2f/A	p^3/qs^2	$1/(s\sqrt{s})$	1	$1/(s\sqrt{s})$
subthr. leak. curr/mm²	expon. with V_t	$s^{-1}.18^{10(1-q)Vt}$	$1/s$	$s^{-1}.18^{10(1-s)Vt}$	$1/s$
gate leak. curr/mm²	expon. with V and t_{ox}	$10^{5(1-q)t_{ox}+^2\log p}$	1	$10^{5(1-s)t_{ox}+^2\log s}$	1
electromigr. (curr.dens.)	$I = CVf/A$	p^2/qs^2	$1/s^2$	$1/s$	$1/(s\sqrt{s})$
latch-up (for Vdd >>1V)	$\propto V/dist$	1 to p/s	1 to $1/s$	≈ 1	1 to $1/s$
ESD susceptibility	$\propto 1/t_{ox}$	$\approx 1/q$	≈ 1	$\approx 1/s$	≈ 1
hot-carrier lifetime	$f(V,V/dist.,material)$	$s^C.e^{B/V(1/p-1)}$	s^C	$s^C.e^{B/V(1/s-1)}$	s^C
NBTI V_t-shift	$\Delta Vt \propto f(V,L,t_{ox})$	$(p/s)^m$	$(1/s)^m$	1	$(1/s)^m$
variability (matching)	$\sigma_{\Delta Vt} \propto t_{ox}/\sqrt{(WxL)}$	q/s	$1/s$	1	$1/s$
cross-talk/unit length	$\propto 1/dist.$	$1/s$	$1/s$	$1/s$	$1/s$
induct. noise/unit area	$(di/dt)/A$	p^3/q^2s^3	$1/s^3$	$1/s^2$	$1/s^2$
voltage drop/unit length	$I.\sigma_{metal}/W_{metal}$	p^2/qs^3	$1/s^3$	$1/s^2$	$1/(s^2\sqrt{s})$
SER per Mb or Mflip-flop	$A_{diff} \cdot e^{(-Qcrit/<Qcoll>)}$	≈ 1	≈ 1	≈ 1	≈ 1
noise margin	V_{dd} and V_t	p and q	1	s	1

constant-voltage scaling (before yr 1990)
constant-field scaling
constant-voltage scaling (after yr 2000)

Let us assume that the transistor sizes are scaled with a factor of s, the voltages with a factor of p and both the threshold voltage and gate-oxide thickness with a factor of q. The first scaling column ($p \neq s \neq q$) shows how a parameter scales when the voltages scale with a different factor than the sizes. In the first constant-voltage scaling column ($p = q = 1$), only the sizes scale, while the voltages are kept constant. In the constant-field scaling column ($p = q = s$), both the sizes and voltages scale with the same factor, keeping the field in the channel constant. Finally, in the second constant-voltage scaling column (after yr 2000), the physical effects that cause mobility reduction (chapter 2), such as velocity saturation, are included in the relations. These are therefore different from the first constant-voltage scaling column (before yr 1990).

The table (column $p \neq s \neq q$) shows that signal integrity is more affected by scaling of the sizes, while the performance is more affected by voltage scaling. For several parameters the relation with the scaling factors s and p is not completely clear. The hot-carrier lifetime is described by the well-accepted empirical expression from Takeda, as discussed in chapter 9. For NBTI there is not such a well-accepted model. In the table the V_T-shift due to NBTI is taken as a parameter. Also here, lifetime issues are involved, but due to the complex behaviour, it is not included in the table.

When we combine this table with the voltage scaling evolution as shown in figur 11.5 we are able to visualise the trends in performance, variability and signal integrity and reliability. The following figures show these trends, assuming that we continue to use bulk-CMOS wafers and that both the transistor architecture and the supply voltage do not change dramatically. Figure 11.6 shows the improvements in various performance parameters over the last couple of decades and their expected improvements. The figure also assumes that the first year for volume production for the 32 nm and 22 nm nodes are delayed with respect to the two-year cycle with which technology nodes were introduced before.

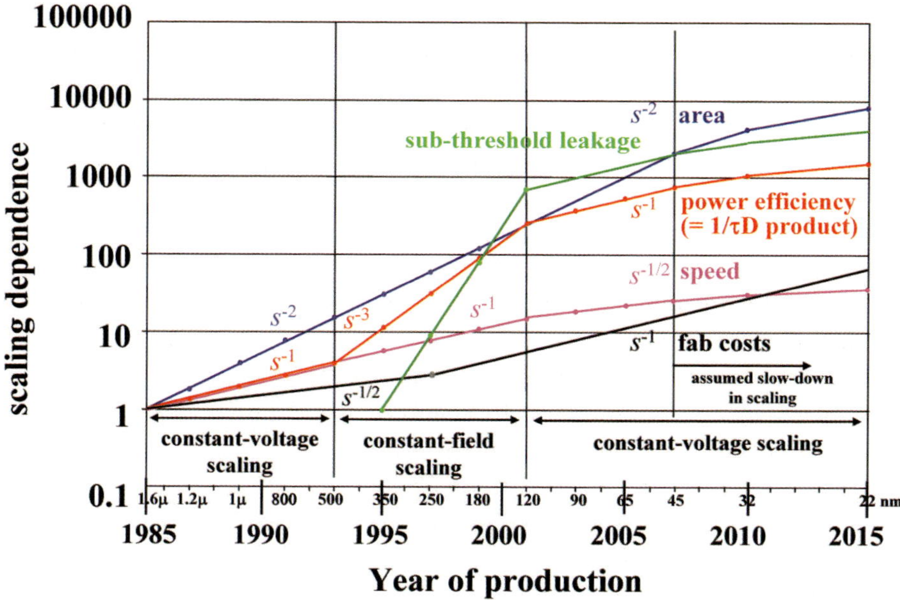

Figure 11.6: *Experienced and expected trends in various performance parameters in relation with the fabrication cost*

The figure shows that the constant-field scaling era has been particularly beneficial for power-efficiency improvement, due to the combined scaling of the sizes and the supply voltage. This allowed new generations of electronic devices to exhibit about two-and-a-half times more functionality for the same power needs, compared to their previous generation. In that same period of time, the subthreshold leakage power has increased by more than three orders of magnitude, which was a major drive to limit further supply and threshold-voltage scaling. The diagram also clearly shows that below 100 nm the improvements both in power efficiency and speed are only limited, and that it will remain so in the future, while the fab cost are still expected to increase at least with the same trend.

The above discussed trends have severe impact on the design of integrated circuits. For high-performance microprocessors, this has led to moving away from higher frequency architectures towards multi-core architectures. For vendors of general VLSI and ASIC ICs, not only the performance and functionality of an IC are differentiators, but certainly also the power consumption, both in active and standby modes, because it has severe consequences for the size and cost of the package for most of the products, as well as for the battery lifetime of mobile products. Power management at all hierarchy levels of design has therefore become a necessity for a successful introduction of a product into the market.

Figure 11.7 shows the variability and signal integrity trends, starting with the 500 nm CMOS technology node as a reference. Before that node the impact of most of these parameters on the behaviour of digital circuits was hardly visible.

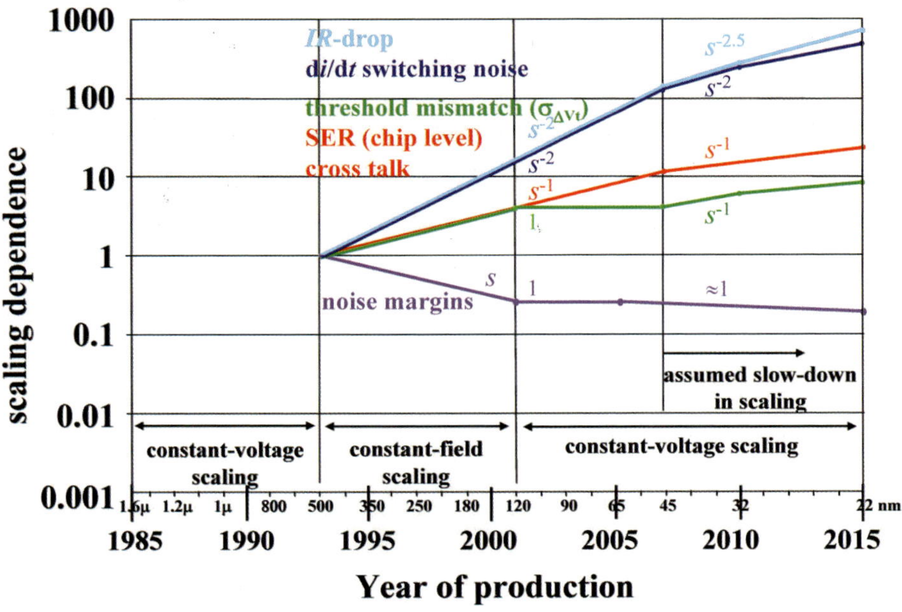

Figure 11.7: *Experienced and expected trends in variability and signal integrity parameters*

The diagram shows that the noise, IR-drop, cross-talk and variability parameters are all increasing, while the noise margins (V_{dd} and V_T) reduce or remain almost constant. This requires the design styles and design flows to continuously adapt to the increasing impact of these parameters. More decoupling capacitance, larger wire spacing in buses, limited di/dt in clock, bus and I/O drivers and *variability-aware design* are examples of how to deal with these effects. But the trend shows that what seems to be a sufficient solution in one technology node, will certainly be not sufficient in the next. Design solutions and tool development must therefore be targeted to bend the positive slopes of the parameters into the direction of the noise-margins slope.

The relation between the reliability parameters and the scaling factors, as presented in the table, is shown in the diagram in figure 11.8. This diagram is only meant to show the trends rather than representing an accurate estimation. Some lines therefore show a "question-mark" relation with the scaling factor.

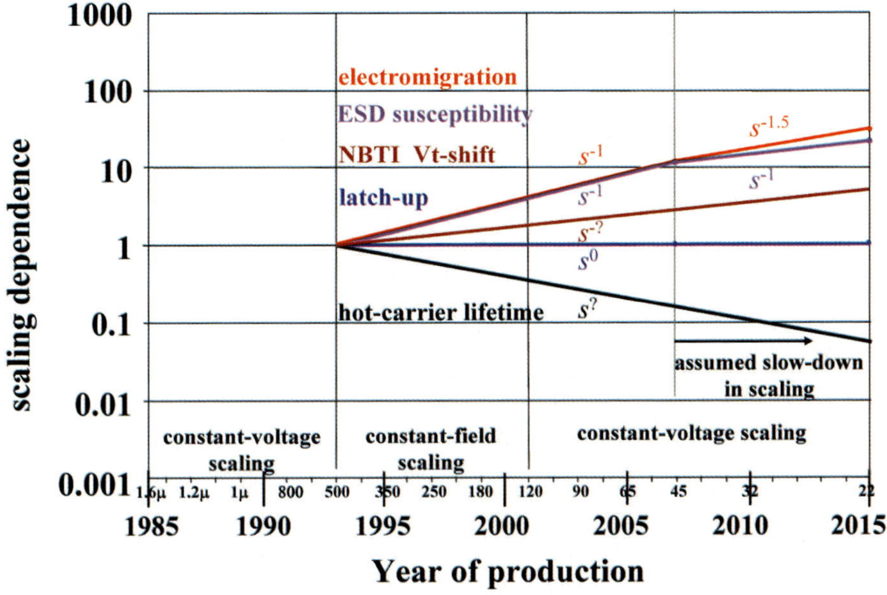

Figure 11.8: *Experienced and expected trends in reliability parameters*

It shows that almost all parameters are getting worse, again if we assume that both the transistor architecture as well as the supply voltage would not dramatically change. Generally, reliability parameters are very difficult to model and predict, because they depend on more factors that just the scaling of the sizes and/or voltages. Huge R&D efforts are required to understand all physical mechanisms that underlay these reliability parameters in order to build an accurate model that can be used to predict the behaviour of individual MOS transistors as well as of the ICs built from them.

Creative solutions, both in technology and design, are needed to keep the IC's robustness at a sufficiently high level in order to extend Moore's law for yet another decade. However, this will lead to a major increase of the complexity and total development and production costs of an IC.

11.5 Potential limitations of the pace of scaling

Moore's law (a quadrupling of IC complexity every three years) has proven its validity from the invention of the chip until now. It is sometimes called a self-fulfilling prophecy and is viewed as a measure for future trends and sets the pace of innovation. Almost according to this law, the Semiconductor Industrial Association has set up its roadmap for the next couple of years. Table 11.3 shows several important parameters of this roadmap [1].

The previously-discussed scaling trends show that there are potentially several key factors that may limit the pace of scaling. The complexity of MOS ICs increases exponentially with time, as can be seen from the table. However, the complexity of the design and test tasks is accelerated and forms a potential barrier to obtaining full exploitation of the available manufacture potentials. The overall success of the semiconductor industry will be increasingly dominated by how the complex design, engineering and test challenges will be addressed [1]:

- system complexity

    - huge amount of transistors on one single chip (10 million to > 1 billion)

    - convergence of consumer, computing and communications domains, which accelerates the introduction of new features on a single chip. This makes a design more heterogeneous, with a large variety of domain-specific, general-purpose IP and memory cores

    - validation of the total system through extensive (hybrid) simulation and emulation

    - different performance demands on a single chip, e.g., high performance and low power, which require multi-threshold voltages, multi-gate oxides, multi-clock and multi-voltage domains

- silicon complexity

    - huge amount of transistors on one single chip (10 million to > 1 billion)

    - increasing manifestation of deep-submicron/nanometer physical effects like cross-talk, voltage drop, supply noise, electromigration, variability, stress, leakage, etc.

- performance increase is no longer an implicit benefit of further scaling (beyond 90 nm CMOS) and we are approaching the frequency barrier. This results in a complex and very time consuming timing closure process
  - imperfect lithography
  - changing process defect mechanisms
- design-flow complexity
  - development of an application-domain specific design platform
  - increasingly complex design flow to cover all previous additional design tasks (deep-submicron/nanometer physical effects, power switches, multi voltage/frequency domains, adaptive voltage/frequency scaling, etc)
  - validation, verification and timing closure are increasingly complex
  - 3-D extraction tools required for memory and analog circuit design
  - design closure, which is the process of a (slow) convergence to a fully functional design that meets all constraints
  - test development and test coverage, also dealing with multi-clock and multi-voltage domains
- fabrication complexity
  - increasingly complex and expensive lithography
  - mask cost explosion
  - wafer fab cost explosion
- package complexity
  - increasing number of power and I/O pads/balls
  - MCM, SiP and SoP solutions
  - limited thermal conductivity improvement
- debug and failure analysis complexity
  - increasing variety of failure causes: defects, stress, proximity effects, process spread, noise, temperature

703

- less noise margins

- less first-time-right products

- metals shield access from top side

- complex backside stimulation and analysis tools

Figure 11.9 shows a summary of increasing design tasks. It also shows the exponential increase in average ASIC design costs because of the rapidly increasing design complexity.

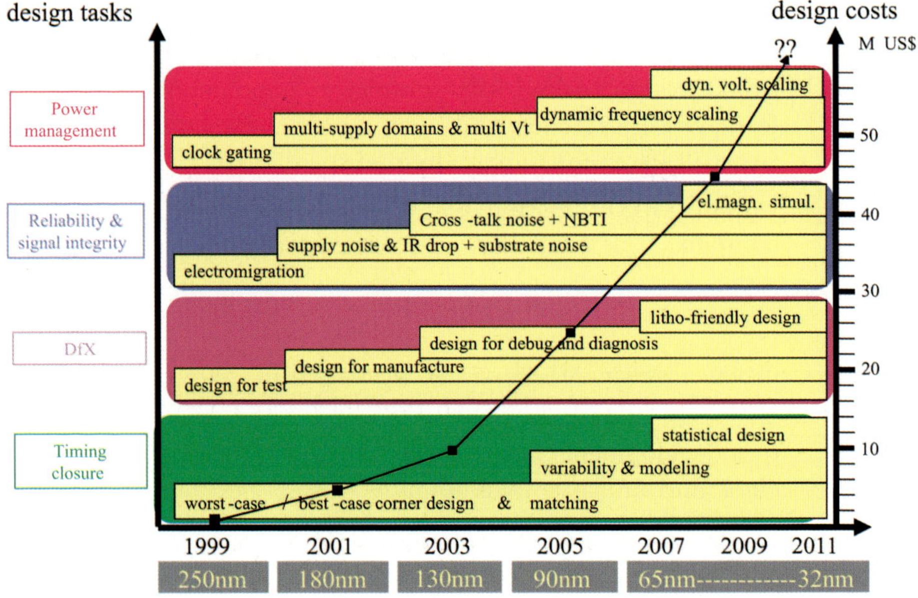

Figure 11.9: *Increasing number of design tasks and growing design costs for an average ASIC design*

The design will therefore have an increasing impact on the total price setting of an integrated circuit. These costs reflects the average ASIC category [1]. The average design costs (including the complete system, the architecture and software development) for a 45 nm ASIC may rise to about 50M US$. Assuming that the chip is meant for a consumer application, and that the profit per device is expected to be in the order of 1 US$, then a simple calculation learns, that at least a total volume of 50 million devices is required to reach break even with respect to the development costs. There are not so many applications that generate

market volumes for a single supplier in this order of magnitude. For the
32 nm and 22 nm nodes, the total development costs will only increase
further.

Figure 11.10 shows that the total design costs are increasing much
faster than the other cost contributors.

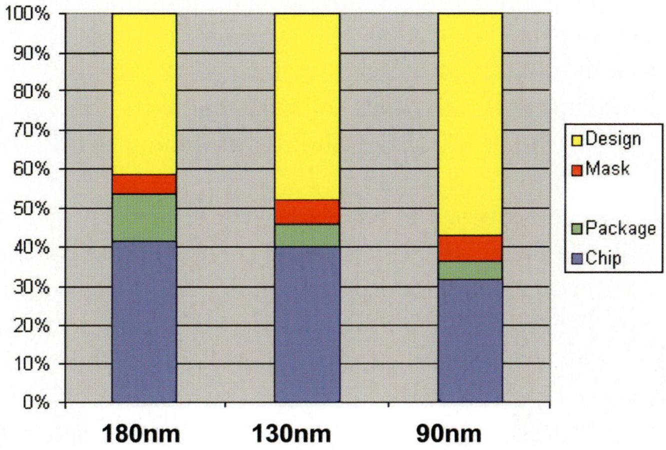

Figure 11.10: *Various contributions to the overall chip costs for different
technology nodes (Source: Leon Stok (IBM) ISPD2003)*

The total costs of a $1\,cm^2$ chip, designed to run at the maximum frequency achievable with a standard-cell design flow (say 700 MHz for
a 45 nm SoC), including the costs of software (for both platform and
application), are so high, that the required volume to recover the development and fabrication costs, about equals the number of transistors on
that chip. This means that soon, for several application areas, the move
to the next technology node may no longer be economically attractive.
When the fabrication costs have only become a minor part of the total
costs of an IC, scaling to the next technology node will hardly reduce
the price. Therefore, it may be possible that the 32 nm node, plus or minus one generation, may be the last economically viable technology for
most applications. For some cheap consumer products the 45 nm node
may be the final one, while this may be the 22 nm or 15 nm node for
high-performance processors, because both the price and profit per chip
in this category is at least an order of magnitude larger, which allows to
recover the huge development and production costs.

The ability to completely verify, test, debug and diagnose future complex designs will reduce dramatically. It is therefore likely that current design styles with fixed and dedicated logic will be replaced by design styles that allow flexibility and configurability. This flexibility can be enhanced by software solutions (programmability) as well as hardware solutions (*reconfigurable computing* such as embedded *FPGA* and/or sea-of-gates architectures). Remaining bugs can then be bypassed by changing the program or by remapping the function, respectively.

Another potential key factor in lowering the pace of process innovation, which is already discussed, is formed by the economics of the production facilities. From 1966 to 2007, the costs of a wafer factory increased by a factor of 500, from about \$10 million to \$3 billion respectively [6].

If this trend continues, the costs of a wafer factory will reach about \$10 billion by the year 2017. These investments can only be raised by a few individual large semiconductor companies and several semiconductor alliances. For 45 nm CMOS and beyond we will see an increasing number of semiconductor companies that outsource chip fabrication and will become fabless: "only the elite few will be able to afford it" [6]. It has already been discussed before, that, when the design costs become significantly larger than the fabrication costs, the drive for scaling an application to the next technology node will reduce. Moreover, since electrons run close to their saturation speed in the 65 nm technology node and beyond, only limited circuit performance improvements can be expected from this scaling. This has changed the focus from GHz to multi-core designs.

The third key factor that may limit the pace of scaling is represented by the increased manifestation of physical and electrical effects in deep-submicron technologies. Larger current slew rates (di/dt) and mutual signal track capacitances will bring the circuit noise to unacceptable levels. In addition to this, the noise margins of future processes will further decrease due to the continuous drive for further reduction of the supply and threshold voltages (figure 11.7). Every new technology requires additional design and/or technology measures to reduce the noise and increase the gap between the noise and the noise margin. However after scaling to the next technology, the problem is the same again and new measures are required. Relatively large additional chip areas must therefore be devoted to on-chip measures like decoupling capacitances and to more widely-spaced buses and other global signal interconnections etc. These deep-submicron effects, which are extensively discussed in chapter 9, reduce the chance of fully exploiting the potentials of the new process generations. Design for test, design for robustness, design for debug, litho-friendly design, design for manufacturability, etc., all add up to an increased design complexity and chip area. The level to which these additional measures will limit the efficient use of chip area cannot be predicted because it also depends on the creative design alternatives that will be developed in the near future.

Table 11.3: Important IC characteristics and their change according to the ITRS roadmap [1]

Year of First IC Shipment DRAM ½ pitch	...2007..... 65nm	...2009.... 50nm	...2011..... 40nm	...2013..... 32nm	...2015..... 25nm	...2017..... 20nm	...2019..... 16nm
Power: Single-Chip Package (Watts/mm^2)							
Low-cost	n.a.	n.a.	n.a.	n.a.	n.a.	n.a.	n.a.
Hand-held and memory (Watts)	3	3	3	3	3	3	3
Cost / Performance	0.74	0.83	0.85	0.98	0.98	1.08	1.08
High-Performance	0.61	0.64	0.64	0.64	0.64	0.64	0.64
Harsh	0.18	0.20	0.22	0.25	0.27	0.28	0.29
Chip Size (mm^2)							
Low-cost	100	100	100	100	100	100	100
Hand-held	100	100	100	100	100	100	100
Cost / Performance	140	140	140	140	140	140	140
High-Performance	660	730	750	750	750	750	750
Harsh	100	100	100	100	100	100	100
Core Voltage (Volts)							
Low-cost	0.9	0.7	0.6	0.5	0.4	0.4	0.4
Hand-held and memory	0.7	0.6	0.5	0.5	0.4	0.4	0.4
Cost / Performance	0.9	0.8	0.6	0.6	0.5	0.5	0.5
High-Performance	0.9	0.8	0.6	0.6	0.5	0.5	0.5
Harsh	1.2	1.2	1.0	0.9	0.9	0.8	0.8
Performance: On-Chip (MHz)							
Low-cost	735	885	1076	1243	1370	1510	1665
Hand-held	4676	5660	6846	7903	8718	9612	10597
Cost / Performance	6824	9827	14151	18679	22602	27345	33092
High-Performance	6824	9827	14151	20378	29344	41910	60350
Harsh	106	128	155	188	227	275	333
Performance: Chip-to-Board for Peripheral Buses (MHz)							
Low-cost	100	100	125	125	125	150	150
Hand-held	75	100	100	100	125	125	150
Cost / Performance	733	800	800	1000	1200	1200	1200
High-Performance	4880	7629	11900	18600	29100	41900	60300
Harsh	106	115	125	125	150	150	150
Memory (D/SRAM)	667	800	800	1000	1200	1200	1200
Logic (High-volume: Microprocessor) Cost-Performance product generation (Mtransistors/cm^2)							
SRAM transistor density	827	1348	2187	3532	5687	9130	14625
Logic transistor density	154	245	389	617	980	1555	2469
Package Pincount							
Low-cost	148-606	160-668	180-738	198-812	218-896	240-988	266-1088
High-Performance	4000	4620	5094	5616	6191	6826	7525

In this ITRS roadmap, the following definitions are used for the different IC-categories:

Low-cost	< consumer products, microcontrollers, disk drives, displays
Hand-held	< battery-powered products, mobile products, hand-held cellular and other hand-helds
Cost / Performance	< notebooks, desktop personal computers, telecommunications
High-Performance	< high-end workstations, servers, avionics, supercomputers, most demanding requirements
Harsh	< under the hood and other hostile environments

11.6 Conclusions

Conventionally, the drive for a continuous scaling of integrated circuits has been the shrinkage of the circuits and of the systems built from them, plus the increased performance that accompanied the advent of every new generation. However, scaling not only influences the system sizes and performance positively, it also has major negative effects on the reliability and signal integrity of deep-submicron ICs.

These effects have increased to such an extent that digital ICs can no longer be regarded as circuits that propagate ones and zeros in a certain order to perform certain functionality. The design of digital circuits increasingly requires an analogue approach to maintain reliability and signal integrity at a sufficiently high level. The manifestation of the responsible physical effects, which increase with scaling of the feature sizes, will be a further challenge, if not a threat to the reliability and signal integrity of future VLSI designs. An understanding of the effects of scaling is essential for the efficient exploitation of the full potential of modern nanometer IC manufacture processes.

These effects place high demands on the design and test strategies used for modern ICs and systems. Additional measures in the design are needed to maintain testability, observability, reliability and signal integrity at a sufficiently high level. In combination with power management solutions and DfX requirements, these measures all require additional chip area, which limit an efficient exploitation of the potentials of the new process generations. Moreover, they also contribute to the exploding cost of IC design.

Within a decade, we will face the fact that a move to the next process generation will no longer be commercially attractive for various categories of products. For cheap high-volume consumer products, however, this point in time will already be reached within a couple of years.

11.7 References

[1] Semiconductors Industrial Associations,
ITRS roadmap, yearly update, http://www.itrs.net

[2] M.Vertregt, et al.,
'Scalable high-speed analog circuit design',
2001 AACD, Kluwer Academic Publishers, pp 3-21, 2002.

[3] M. Izumikawa, et al.,
'A 0.25 μm 0.9 V 100 MHz DSP core',
IEEE-JSSC, Jan. 1997, pp 52-61.

[4] T. Kuroda, et al.,
'A 0.9 V, 150 MHz, 10 mW, 4 mm^2, 2-D DCT Core Processor with variable Threshold voltage Scheme',
IEEE-JSSC, Nov. 1996, pp 1770-1779.

[5] S.H. Lo, et al.,
'Quantum-Mechanical modelling of Electron tunnelling Current from the Inversion Layer of Ultra-Thin-Oxide in MOSFET's',
IEEE Electron Device Letters, Vol. 18, No 5, 1997, pp 209-211

[6] M. LaPedus,
'Costs cast ICs into Darwinian struggle',
EE Times, March 30, 2007

[7] J. Schoelkopf,
'ATRS: an alternative roadmap for semiconductors, technology evolution and impacts on system architecture',
12th IEEE International Symposium on Asynchronous Circuits and Systems, March 2006, Grenoble, France

11.8 Exercises

1. Explain the differences between the constant-voltage scaling and the constant-field scaling process. How did they influence the main driving force behind the scaling process?

2. Why was copper not used in the early MOS processes? What is the result of using copper instead of aluminium for the interconnection patterns of an IC?

3. An IC with channel lengths of 65 nm is manufactured in a 45 nm CMOS process and used in a particular application. Suppose this IC is scaled by a factor of 0.7 and manufactured in the same process. What would happen to the following parameters when this IC is used in the same application:

   a) the transistor gain factors β_n and β_p

   b) the threshold voltages V_{T_n} and V_{T_p}

   c) the chip's power dissipation

   d) the chip's power density

   e) the noise on the chip's supply and ground lines.

4. Suppose that the additionally required decoupling capacitance on a chip results in an area penalty of 20 percent. How could the capacitance density (i.e., capacitance value per unit area) be increased by technology means?

5. What would be the four biggest threats for the pace of scaling? Motivate your answer.

Index

bubbles, 117
buffer circuits, 209
Built-in Self Test, 604
built-in self-test and repair, 605
bulk silicon, 95
buried-channel CCD, 262
buried-oxide layer, 101
burn-in test, 316
burst mode, 300
bus latency, 693

cache register, 340
CAD tools, 367
CAM, 290, 294
capacitances, 42
capacitor-under-bitline, 317
carrier mobility reduction, 59
Cascode Voltage Swing Logic, 234
CBRAM, 349
CCD, 261, 262
CCD cell, 263
CCD operating frequency, 265
CCD shift register, 262
CCO, 524
CD, 120, 124
CDU, 124
cell abutment, 481
cell-based IC design, 369
channel
    ~ conductance, 31
    ~ dope, 5
    ~ hot electron injection, 336
    ~ hot-electron injection, 339
    ~ length, 120
    ~ length modulation, 64
    ~ stopper, 153, 164
    ~ stopper implant, 35
channel-free gate array, 416
channel-less gate array, 416
characterisation tests, 591

charge
    ~ bucket, 262
    ~ characteristic, 207
    ~ distribution, 12
    ~ sharing, 224, 232
    ~ transfer, 264
charge-coupled device, 261, 262
charge-pump, 194
charged-device model, 576
CHEI, 336
Chemical Mechanical Polishing,
    147
Chemical Vapour Deposition, 138
chip, vi
    ~ select, 297
chip-scale
    ~ package, 631, 639
    ~ packaging, 624
choice of logic implementation,
    235
circuit
    ~ density, 159
    ~ simulation, 48, 393
circuit editing techniques, 679
circuit-analysis program, 196
class-one clean room, 611
clean room, 611
    ~ conventional standard, 611
clock
    ~ activity, 495
    ~ generation, 523
    ~ jitter, 520, 527
    ~ signals, 225
    ~ skew, 227, 231, 236, 518
    ~ tree synthesis, 519
clock-phase synchronisation, 525
clocked CMOS circuits, 225
clocking strategies, 236
CMOS, 200

database set-up, 579
DDR, 322
decision tree, 386
decoupling capacitor, 537
delay fault, 597, 602
delay-locked loop, 527
ΔI_{ddq} test, 602
depletion
    $\sim$ layer, 12
    $\sim$ layer thickness, 66
    $\sim$ process, 12
    $\sim$ transistor, 33
depletion$\sim$ load, 194
deposition, 137, 680
depth of focus, 109
depth-of-focus, 113
design
    $\sim$ documentation, 579
    $\sim$ efficiency, 688
    $\sim$ for anything, 622
    $\sim$ for debug, 664, 682
    $\sim$ for failure analysis, 682
    $\sim$ for manufacturabilty, 175
    $\sim$ for testability, 608
    $\sim$ productivity, 350
    $\sim$ resources, 688
    $\sim$ rules, 240
    $\sim$ style, 706
    $\sim$ verification, 392
Design for Lithography, 118
design$\sim$ hierarchy, 579
design$\sim$ organisation, 579
design-rule-check program, 406
designing a CMOS inverter, 207
destructive read-out, 311, 347
DfL, 118
DfM, 175, 392, 620
DfM-rules, 620
DfT, 608

DfX, 622
DIBL, 77
dicing, 628
dielectric relaxation time, 38
Differential Split Level Logic, 235
diffusion, 142
    $\sim$ coefficient, 143
digital
    $\sim$ CMOS circuits, 218
    $\sim$ ICs, 368
    $\sim$ potentiometer, 394
direct slice writing, 368
direct writing techniques, 127
discharge characteristic, 207
dishing, 149
dislocations, 97
disturbances in the production
        environment, 610
DLL, 527
DMOS transistor, 270
DOF, 109, 113
DOMINO-CMOS, 229
donor, 9
dope profile, 144
Double Data Rate, 322
Double Pass-Transistor Logic, 478
Double Patterning Technology, 121
double-diffused MOS transistor,
        270
double-flavoured polysilicon, 137,
        158, 203
double-gate transistor, 173
DPL, 478
DPT, 121
drain, 4
    $\sim$ extension, 154, 165
    $\sim$ series resistance, 566
Drain-Induced Barrier Lowering
        effect, 77

energy

    $\sim$ band, 6

    $\sim$ band diagram, 15

    $\sim$ band theory, 5

    $\sim$ gap, 6

energy-delay product, 461

enhancement transistor, 33

epi layer, 95

epitaxial

    $\sim$ film, 137

    $\sim$ wafer, 95, 572

EPLD, 371

EPROM, 335

equivalence checking, 393

erosion, 150

error-correction code, 326

ESD, 100, 573, 645

eSRAM, 309

etching, 131

EUV, 124

exclusive OR, 223

EXOR gate, 223, 494, 499

Extended Data Out, 321

    $\sim$ DRAM, 322

externally-induced voltage alteration, 672

extreme data rate RDRAM, 326

Extreme-UV lithography, 124

FA, 654

fab-lite, 178, 443

fabless, 178, 443

failure analysis, 654

Fast Page Mode, 321

    $\sim$ DRAM, 321

fat zero, 264

FD-SOI, 102

Fermi level, 9

ferroelectric RAM, 346

FIB, 680

field oxide isolation, 417

Field Programmable Device, 420

field-effect principle, 1

field-programmable device, 371

FIFO, 291, 293

fill factor, 268

filler cells, 436

FinFET, 174

firm cores, 370

first time right silicon, 511

first-silicon debug, 654

flash memory, 339

flat-band

    $\sim$ condition, 16

    $\sim$ voltage, 16

flip-chip bonding, 631

flip-flop, 225, 226

floating gate, 336

Focused Ion Beam, 680

formal verification, 393

forward-bias effect, 30

FOUP mini environment, 613

four-transistor SRAM cell, 301

Fowler-Nordheim tunnelling, 337

FPGA, 420, 706

FPM, 321

FPM DRAM, 321

FRAM, 346

full adder, 197, 383

full-CMOS SRAM cell, 300

full-custom IC, 369

full-featured EEPROM, 337

fully-regular library, 119

fully-silicided, 137

functional level, 379

FUSI, 137

    $\sim$ gate, 168

gain factor, 58

GALS, 533, 693

gate, 4
  ~ array, 415
  ~ delay, 597
  ~ depletion, 137, 168, 690
  ~ forest, 416
  ~ inversion, 168
  ~ oxidation, 154
  ~ oxide, 135
  ~ oxide tunnelling, 690
gate-drain overlap capacitance, 155
gate-induced drain leakage, 82
gate-isolation technique, 417
gate-last CMOS process, 169
gate-oxide
  ~ leakage, 79
  ~ leakage current, 457
  ~ thickness, 136
gate-source overlap capacitance, 155
gated clock, 497, 521
GDSII, 441
general-purpose
  ~ CMOS processes, 696
  ~ process, 137
geometric layout
  ~ description language, 441
  ~ representation, 441
GIDL, 82
GLDL, 441
glitches, 494
global variations, 551
globally asynchronous and locally synchronous, 693
globally synchronous, locally asynchronous, 533
glue logic, 376
golden device, 602
GP process, 137, 696
graded-drain transistor, 564

Gray code counter, 488
ground bounce, 536

halo, 68
handcrafted layout, 406
handshake circuits, 493
hard cores, 370
hardware
  ~ accelerator, 392
  ~ description language, 369, 383
hardware/software codesign, 389
HCE, 563
HDD, 166
HDGA, 416
HDL, 369, 383
HDP, 133
hemispherical grain, 314
hetero-epitaxy, 138
heterogeneous system, 376
  ~ on a chip, 523, 687
hierarchical
  ~ design approach, 438
  ~ layout, 438
high-density gate array, 416
  ~ layout, 228
High-Density Plasma, 133
high-energy cosmic particles, 543
high-voltage CMOS, 273
Highly-Doped Drain, 166
hillocks, 560
hold-time violation, 519
hole mobility, 202
holes, 7
homo-epitaxy, 138
homogeneous sytem, 377
hot carrier, 82
hot electron, 336
hot-carrier effect, 83, 165, 563
hot-electron effect, 339

LSTP process, 137
LUT, 420, 422

machine model, 576
macro, 414
    ~ cell, 369
magnetic tunnel junction, 347
Magneto-resistive RAM, 347
majority charge carrier, 11
Manhattan skyline effect, 439
mapping, 391
mask, 94
    ~ ROM, 329
mask-less lithography, 128
mask-programmable
    ~ ROM, 329, 412
    ~ gate arrays, 415
masks, 241
master cell, 416
matching, 554
    ~ coeffient, 555
    ~ of transistors, 690
maximum storage time, 264
MCM, 341, 642
meet-in-the-middle strategy, 439
mega cell, 369
memory
    ~ address, 292
    ~ array, 290
    ~ bank, 324
    ~ banks, 297
    ~ cell, 290
    ~ controller, 323
    ~ matrix, 290
    ~ word, 292
merged memory logic, 352
metal gate, 169
METAL mask, 155
Metal-Oxide-Semiconductor (MOS)
        capacitor, 11

micro defects, 96
microcode instruction, 412
microcontrol unit, 379
microprocessor core, 391
military specifications, 4
milling, 680
minority carrier, 18
mismatch, 551
MISR, 605, 649
ML2, 128
MLC, 342
MLL, 128
MLR, 127
MML, 352
mobility, 32, 58
module generator, 438
molybdenum, 4
    ~ gate, 155
more than Moore, 644
MOS, 1
    ~ capacitance, 38, 41
    ~ formulae, 23
    ~ transistor leakage mecha-
        nisms, 74
    ~ transistor weak inversion
        operating region, 75
MOS transistor, 5
MPW, 126
MRAM, 347
MTCMOS, 457
MTJ, 347
multi-chip module, 642
multi-layer reticle, 127
Multi-Level Cell, 342
multi-level flash memory, 342
multi-port memory, 327
multi-project wafers, 126
Multiple Input Signature Regis-
        ter, 649

oxide-nitride-oxide, 313

p-channel MOS transistor, 32
p-type substrate, 4
package-on-a-package, 644
packaging, 623
page, 321, 339
PAL, 412
parallel
    $\sim$ connection of transistors,
      197, 220
    $\sim$ multiplier, 381
parallelism, 465
parametric
    $\sim$ fault, 600
    $\sim$ yield loss, 614
parasitic
    $\sim$ MOS transistor, 34
    $\sim$ capacitances, 100
    $\sim$ thyristor, 277
partial product, 381
pass transistor, 222
pass-gate logic, 477
pass-transistor logic, 223, 477
passivation layer, 155
path delay, 597
PCM, 348, 619
PD-SOI, 102
PECVD, 139
PEM, 662
penetration depth, 143
periodic system of elements, 8
Perovskite crystals, 347
Phase-Change Memory, 348
phase-locked loop, 523
Phase-Shift Mask, 113
photolithography, 105
photon emission microscopy, 662
photoresist layer, 129
physical design aspects, 687

PICA, 665
picosecond imaging circuit anal-
    ysis, 665
pinch-off
    $\sim$ point, 22
    $\sim$ region, 65
pinhole, 651
pipelining, 465
PLA, 410, 412
place and route, 438
placement and routing, 414, 438
planar
    $\sim$ DRAM cell, 312
    $\sim$ IC technology, 35
    $\sim$ silicon technology, 4
planarisation, 146
plasma, 139
    $\sim$ etching, 132
platform ASIC, 436
PLD, 371, 420
PLL, 523
pMOS transistor, 200
    $\sim$ gain factor, 208
    $\sim$ threshold voltage, 200
pocket implants, 68
point defects, 96
Poisson's law, 12
poly fuse, 328
POLY mask, 154
polycide process, 166
polycrystalline silicon, 94
    $\sim$ layer, 4
polygon pusher, 406
polymide layer, 328
polysilicon, 94
    $\sim$ gate, 154
    $\sim$ interconnect, 154
PoP, 644
positive photoresist, 129

SOI, 99
SOI-CMOS, 99
solid immersion lens, 667
SOM, 654, 669
SONOS, 344
SoP, 644
SOS-CMOS process, 100
source, 4
$\sim$ series resistance, 566
source-synchronous timing, 523
source/drain capacitance, 100
spacer, 565
$\sim$ lithography, 122
specification, 579, 653
speed and area, 235
Spin-On-Glass, 146
spurious transitions, 494
sputter etching, 132
SRAF, 117
SRAM, 291, 294
$\sim$ memory cell, 300
SRB, 169
SRPL, 479
SSO, 542
SSTA, 393
STA, 393, 558
stacked capacitance cell, 313
stand-alone memory, 290
standard
$\sim$ IC, 402
$\sim$ cell, 413
$\sim$ commodities, 372
$\sim$ logic IC, 372
$\sim$ product, 372
standard-cell, 413
$\sim$ height, 477
$\sim$ layout, 413, 414
$\sim$ library, 413
standby

$\sim$ current, 76, 455
$\sim$ mode, 346, 453
static
$\sim$ CMOS circuits, 219
$\sim$ CMOS flip-flop, 226
$\sim$ CMOS inverter characteristic, 206
$\sim$ RAM, 291, 294
$\sim$ RAM cells, 300
$\sim$ column access, 319
$\sim$ memory, 291
$\sim$ noise margin, 303
$\sim$ power consumption, 451
$\sim$ timing analysis, 393
$\sim$ IR-drop, 536
Static Timing Analysis, 558
statistical static timing analysis,
393
Statistical Timing Analysis, 558
STC, 313
steady-state current, 600
step coverage, 141
step-and-repeat operation, 108
STI, 135, 160
$\sim$ stress, 550, 555
stick diagram, 245
storage gate, 262, 263
strain-relaxed buffer, 169
strained silicon, 169
strong inversion, 16
structural
$\sim$ fault, 600
$\sim$ test, 597
structured ASIC, 436
stuck-at fault, 600
subresolution assist feature, 117
substrate, 95
$\sim$ bounce, 539
$\sim$ dope, 5

VPT, 316
VRAM, 293, 324

wafer, 95
    $\sim$ diameter, 610
    $\sim$ map, 618
    $\sim$ probing, 610
wafer-level packaging, 624
Wallace tree multiplier, 382, 495
waveform measurements, 610
wearout, 646
well biasing, 456
well-bias, 455
well-proximity, 555
    $\sim$ effect, 550
wet-etching method, 132
wire
    $\sim$ bonding, 629
    $\sim$ self-heating, 562
    $\sim$ spreading, 621
WLP, 624
word line, 296
work function, 16
worst-case
    $\sim$ corner, 305
    $\sim$ delay path, 374, 388
write enable, 297

x-decoder, 296
X-ray lithography, 125
XDR RDRAM, 326
XIVA, 672
XRL, 125

y-decoder, 296
yellow room, 611
yield, 610
yield control, 619
yield degradation
    $\sim$ UV light, 611
    $\sim$ chemical impurities, 614

    $\sim$ dust particles, 611
    $\sim$ electrostatic charge, 614
    $\sim$ humidity, 610
    $\sim$ temperature fluctuations, 610
    $\sim$ vibrations, 611

Z-RAM, 326
Zero Capacitor DRAM, 326
zero-temperature-coefficient, 72
ZTC, 72